Harri Lönnberg
**Chemistry of Nucleic Acids**

## Also of interest

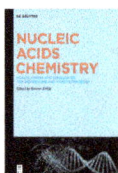

*Nucleic Acids Chemistry.*
*Modifications and Conjugates for Biomedicine and Nanotechnology*
Edited by: Ramon Eritja, 2021
ISBN 978-3-11-063579-9, e-ISBN (PDF) 978-3-11-063953-7

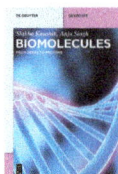

*Biomolecules.*
*From Genes to Proteins*
Shikha Kaushik and Anju Singh, 2023
ISBN 978-3-11-079375-8, e-ISBN (PDF) 978-3-11-079376-5

*Pharmaceutical Chemistry.*
*Drug Design and Action*
Joaquín M. Campos Rosa, 2023
ISBN 978-3-11-131654-3, e-ISBN (PDF) 978-3-11-131690-1

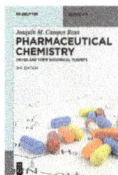

*Pharmaceutical Chemistry.*
*Drugs and Their Biological Targets*
Joaquín M. Campos Rosa, 2024
ISBN 978-3-11-131655-0; e-ISBN (PDF) 978-3-11-131688-8

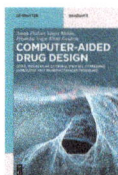

*Computer-Aided Drug Design.*
*QSAR, Molecular Docking, Virtual Screening, Homology and Pharmacophore Modeling*
Aman Thakur, Vineet Mehta, Priyanka Nagu and Kiran Goutam, 2024
ISBN 978-3-11-143474-2, e-ISBN (PDF) 978-3-11-143485-8

Harri Lönnberg

# Chemistry of Nucleic Acids

2nd, Revised and Extended Edition

DE GRUYTER

**Author**
Prof. Dr. Harri Lönnberg
Department of Chemistry
University of Turku
Henrikinkatu 2
20500 Turku
Finland
harri.lonnberg@utu.fi

ISBN 978-3-11-132532-3
e-ISBN (PDF) 978-3-11-132563-7
e-ISBN (EPUB) 978-3-11-132592-7

**Library of Congress Control Number: 2024938003**

**Bibliographic information published by the Deutsche Nationalbibliothek**
The Deutsche Nationalbibliothek lists this publication in the Deutsche Nationalbibliografie;
detailed bibliographic data are available on the Internet at http://dnb.dnb.de.

© 2024 Walter de Gruyter GmbH, Berlin/Boston
Cover image: CROCOTHERY/iStock/Getty Images Plus
Typesetting: Integra Software Services Pvt. Ltd.

www.degruyter.com

# Preface to the second edition

Interest in chemistry of nucleic acids has remained high since publishing of the first edition of this book in summer 2000. The success of RNA vaccines in combat against Sars-CoV2 virus infection received worldwide attention and urged studies toward application of in vitro transcribed messenger RNA as a drug. Site-directed RNA editing, targeting of RNA with small molecules and cleavage of messenger RNA with catalytic DNA-derived oligonucleotides are other emerging fields, as well as novel techniques for intracellular imaging and high-throughput sequencing. Progress in more established fields, including development of antivirals, antisense oligonucleotides, splice-switching oligonucleotides, RNA interference and aptamers (target recognizing oligonucleotides), has continued steadily. This second edition briefly surveys the progress in both emerging and established fields, but the main focus of the book is still in synthetic and mechanistic nucleic acid chemistry. Novel applications often depend on novel structures and novel conjugates obtainable by novel chemistry. I hope that graduate and undergraduate students of organic chemistry interested in working with nucleic acids in academia or industry will find this book useful. It may also be of interest for biochemists or cell biologists who like to broaden their understanding of the basic chemistry of molecules with which they work. The book tends to cover the literature published before March 2024.

Harri Lönnberg
April 2024

https://doi.org/10.1515/9783111325637-202

# Contents

# 1 Nucleosides: structure, nomenclature and solution equilibria

## 1.1 Structure and nomenclature

Nucleosides, the monomeric constituents of nucleic acids, are *N*-glycosylated derivatives of two different categories of heteroaromatic nitrogen bases, namely monocyclic pyrimidines and bicyclic purines. The pyrimidine bases are cytosine, uracil and thymine, and the purine bases are adenine and guanine. Uracil occurs only in RNA and thymine in DNA, while the other bases are common for both the types of nucleic acids. The structures and enumeration of these canonical nucleic acid bases and nucleosides are depicted in Figure 1.1. As indicated, the glycosyl group is attached to N1 of pyrimidine bases and N9 of purine bases. The enumeration of the glycosyl moiety starts from the anomeric carbon, that is, from the carbon atom bound to the nucleobase and not from the ring oxygen. In ribonucleosides (constituents of RNA), the glycosyl moiety is β-D-ribofuranosyl group and in 2′-deoxyribonucleosides (constituents of DNA) 2-deoxy-β-D-*erythro*-pentofuranosyl group. The latter group is often called 2-deoxy-β-D-ribofuranosyl group, but this name is not consistent with the nomenclature of carbohydrates [1]. The prefix "ribo" refers to a sugar having three stereogenic centers in addition to the anomeric (C1′) carbon. 2-Deoxypentoses contain only two nonanomeric stereogenic centers, and hence the correct prefixes are *erythro* and *threo*. The names of ribonucleosides are derived from the names of their base moieties: adenosine, guanosine, cytidine and uridine. The names of 2′-deoxyribonucleosides have, in turn, been formed from the names of the respective ribonucleosides by a prefix 2′-deoxy, with the exception of thymine derivative. This 2′-deoxyribonucleoside is, for historical reasons, called just thymidine.

The names of substituted or modified nucleosides are derived from the names of the parent nucleosides, as exemplified by a few illustrative examples in Figure 1.2. Substituents on the sugar and base moiety of nucleosides are indicated in the beginning of the name in alphabetical order. A missing ring nitrogen is indicated by a prefix "deaza" and an extra ring nitrogen by a prefix "aza." One should, however, note that this kind of nomenclature is applicable only as long as the sugar moiety is a five-membered D-sugar having a *ribo* (for ribonucleosides) or *erythro* configuration (for 2′-deoxyribonucleosides) and the anomeric configuration is β. If the configuration of the sugar moiety, enantiomeric form or ring size is changed, the name of the nucleoside is formed by adding the name of the sugar moiety as a substituent to the name of the base moiety. This is also the case when the base moiety is heavily modified. The compound is then named as a glycosylated heterocyclic compound. Sometimes abbreviations, such as *ara*-, *lyxo*- or *xylo*-adenosine, are used. This means that the sugar moiety is still a β-D-glycofuranosyl group, but the configuration is not any more *ribo*. In case the sugar ring has been opened by cleaving a C–C bond, the site of the missing

https://doi.org/10.1515/9783111325637-001

Nucleic acid bases

Purines

Pyrimidines

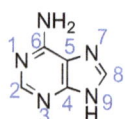
Adenine
(Ade)

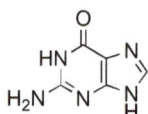
Guanine
(Gua)

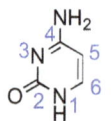
Cytosine
(Cyt)

Uracil
(Ura)

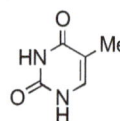
Thymine
(Thy)

Ribonucleosides (RNA)

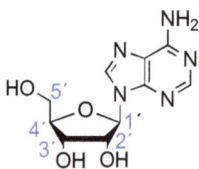
Adenosine
(A)

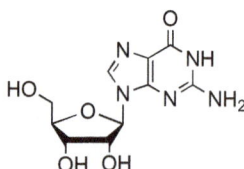
Guanosine
(G)

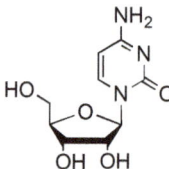
Cytidine
(C)

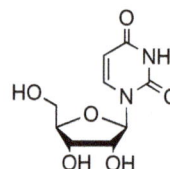
Uridine
(U)

2′-Deoxyribonucleosides (DNA)

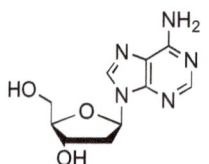
2′-Deoxyadenosine
(dA)

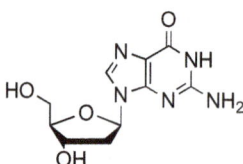
2′-Deoxyguanosine
(dG)

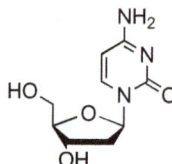
2′-Deoxycytidine
(dC)

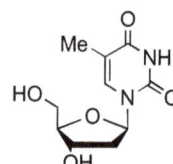
Thymidine
(dT)

**Figure 1.1:** The structure and enumeration of nucleic acid bases, ribonucleosides and 2′-deoxyribonucleosides.

bond is indicated by a prefix *seco*, for example, 2′,3′-*seco*-adenosine. In addition, some noncanonical nucleobases and nucleosides have trivial names that are commonly used. The 2-deamino analogs of guanine and guanosine are called hypoxanthine and inosine, respectively, and the 2-oxo derivatives of these are known as xanthine and xanthosine.

## 1.1.1 Rare nucleosides

In addition to the five canonical nucleobases (Ade, Gua, Cyt, Ura and Thy), nucleic acids contain numerous modified bases, which are introduced by various enzymatic reactions

**8-Methyladenosine**  $N^4,N^4$-**Dimethylcytidine**  **5′-O-Metyladenosine**  **2′,3′-O-Isopropylidene-adenosine**

**8-Aza-2′-deoxy-adenosine**  **7-Deaza-2′-deoxy-adenosine**  **2-Deoxy-2-thio-thymidine**  **2′,3′-Didehydro-2′,3′-dideoxyadenosine**

**1-(β-D-Arabinofuranosyl)-cytosine**  **1-(β-L-Ribofuranosyl)-cytosine**  **1-(β-D-Ribopyranosyl)-cytosine**  **1-(α-D-Ribofuranosyl)-cytosine**

**ara-Adenosine**  **lyxo-Adenosine**  **xylo-Adenosine**  **2′,3′-seco-Adenosine**

**Hypoxanthine**  **Inosine**  **Xanthine**  **Xanthosine**

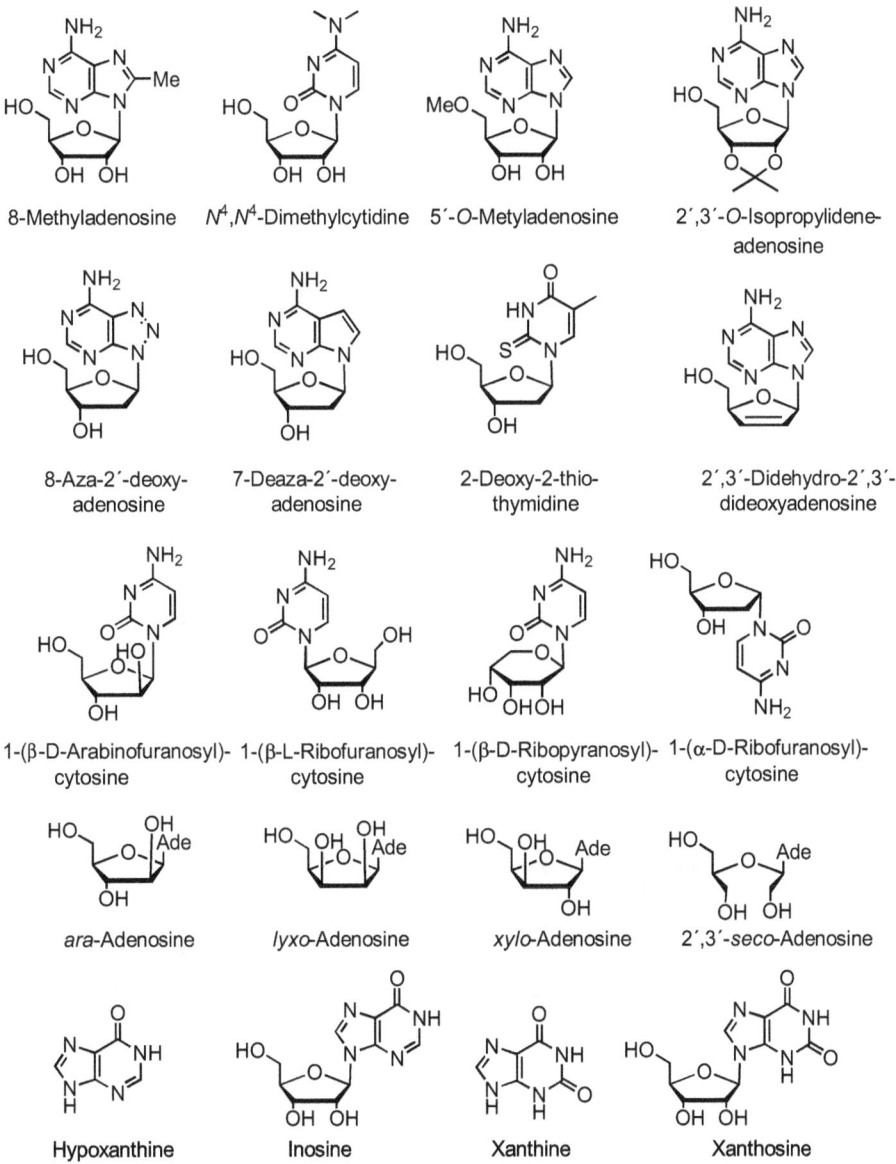

**Figure 1.2:** Examples of the names of substituted or modified nucleosides.

during or after replication (DNA) or transcription (RNA). In DNA, the number of noncanonical bases is still rather limited. The most frequently occurring minor bases are 5- and $N^4$-methylcytosines and $N^6$-methyladenine, the content of the most abundant 5-methylcytosine being around 4.5% [2]. The other rare bases are 5-hydroxymethylcytosine, 5-hydroxymethyluracil, 5-formylcytosine and 5-carboxycytosine. The 5-hydroxymethylcytosine and -uracil additionally occur in glycosylated form, as indicated in Figure 1.3. All these modified bases play a role in control of gene expression, though the exact mechanisms are still largely unknown [2]. 5-Methyl-, 5-hydroxymethyl-, 5-formyl- and 5-carboxy-cytosines have received special attention as epigenetic bases, that is, as modifications playing role in the formation of heritable changes in gene function [3].

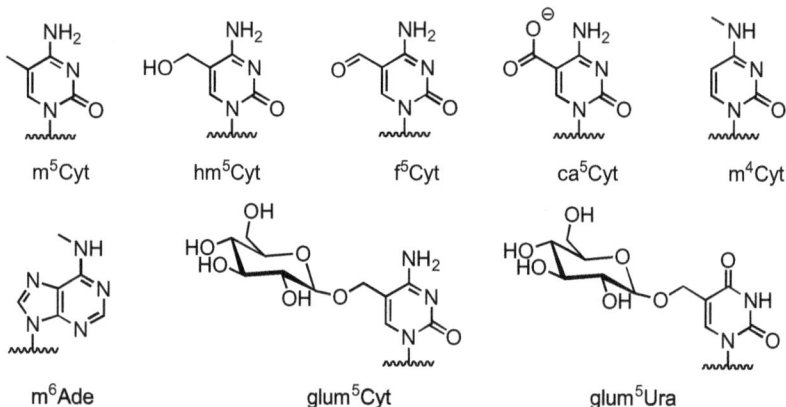

**Figure 1.3:** Noncanonical nucleobases in DNA: 5-methylcytosine (m⁵Cyt), 5-hydroxymethylcytosine (hm⁵Cyt), 5-formylcytosine (f⁵Cyt), 5-carboxycytosine (ca⁵Cyt), $N^4$-methylcytosine (m⁴Cyt), $N^6$-methyladenine (m⁶Ade), 5-(β-D-glucopyranosyl)oxymethylcytosine (glum⁵Cyt) and 5-(β-D-glucopyranosyl)oxymethyluracil (glum⁵Ura).

The noncanonical nucleosides are much more common in RNA than in DNA. So far more than 100 modifications have been identified [2]. Each of the canonical ribonucleosides occurs in various modified forms. The modifications range from simple methylation, acetylation and hydroxylation to methylthio, thio and seleno substitutions. In addition, glycosylations and alkylations with complex alkyl groups are common. The nucleobase itself may also be extensively modified. Examples of such hypermodified base moieties are given in Figure 1.4. Rather frequent occurrence of inosine is also worth noting.

The content of modified nucleosides is highest in transfer RNA, more than 10% of nucleosides being noncanonical. The most frequently modified region is the anticodon loop that recognizes messenger RNA (mRNA) or its immediate vicinity. In addition, ribosomal RNA and mRNA are rich in modifications. Undoubtedly, the occurrence of numerous noncanonical bases originates from adaptation of RNA to its many different biological functions discussed in subsequent chapters.

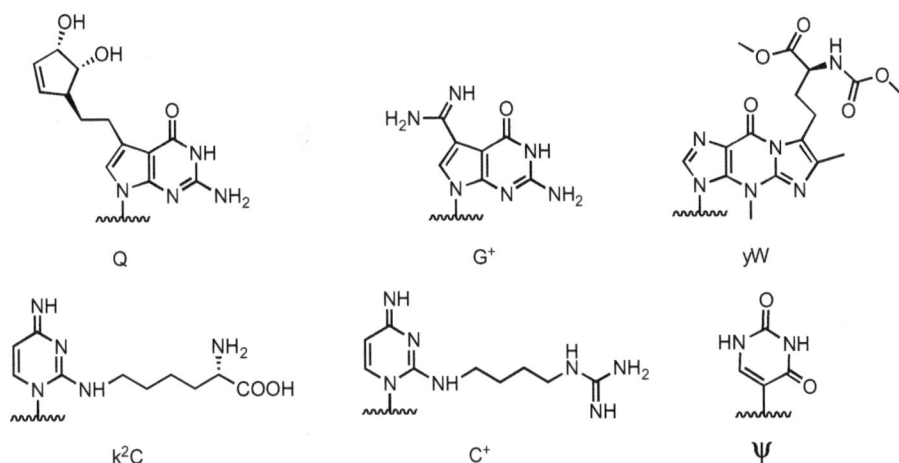

**Figure 1.4:** Some hypermodified bases in RNA: queuosine (Q), archaeosine (G⁺), wybutosine (yW), lysidine (k²C), agmatidine (C⁺) and pseudouridine (ψ).

## 1.1.2 Cyclonucleosides

Cyclonucleosides constitute a special group of nucleosides having an additional covalent linkage between the sugar and base moiety. They usually are of synthetic origin. Only one cyclonucleoside, 3,5′-anhydro-xanthosine, has been isolated from a natural source, that is, from an *Eryus* sp. marine sponge [4]. In addition, 5′,8-cyclo-2′-deoxyadenosine and 5′,8-cyclo-2′-deoxyguanosine have been identified among radical-induced modifications in mammalian DNA [5]. Cyclonucleosides are more rigid than nucleosides and, hence, useful model compounds in design of novel bioactive nucleoside analogs [6]. With purine cyclonucleosides, N3 of the purine base may be directly bound to C5′ displacing the hydroxyl group, as in the naturally occurring 3,5′-anhydro-xanthosine. Alternatively, an additional bridge between C8 and C2′, C3′ or C5′ may be mediated by oxygen, sulfur or nitrogen (Figure 1.5) [7]. When the bridged sugar carbon atom is 2′ or 3′, the sugar configuration must naturally be *arabino* or *xylo*, respectively. With pyrimidine cyclonucleosides, C6 or C2 is bridged to one of the sugar hydroxyls. The names of cyclonucleosides are derived from the name of the parent nucleoside from which the cyclonucleoside may in theory be obtained by removal of water. The bridged atoms and prefix "anhydro" is added in the beginning of the name.

3,5′-Anhydro-
guanosine

8,2′-Anhydro-8-hydroxy-9-
(β-D-arabinofuranosyl)-
adenine

8,3′-Anhydro-8-hydroxy-9-
(β-D-xylofuranosyl)-
adenine

8,5′-Anhydro-8-hydroxy-
adenosine

8,2′-Anhydro-8-mercapto-9-
(β-D-arabinofuranosyl)-
adenine

8,2′-Anhydro-8-amino-9-
(β-D-arabinofuranosyl)-
adenine

6,5′-Anhydrothymidine

6,5′-Anhydro-6-mercapto-
thymidine

2,2′-Anhydro-2-deoxy-2-mercapto-
1-(β-D-arabinofuranosyl)uracil

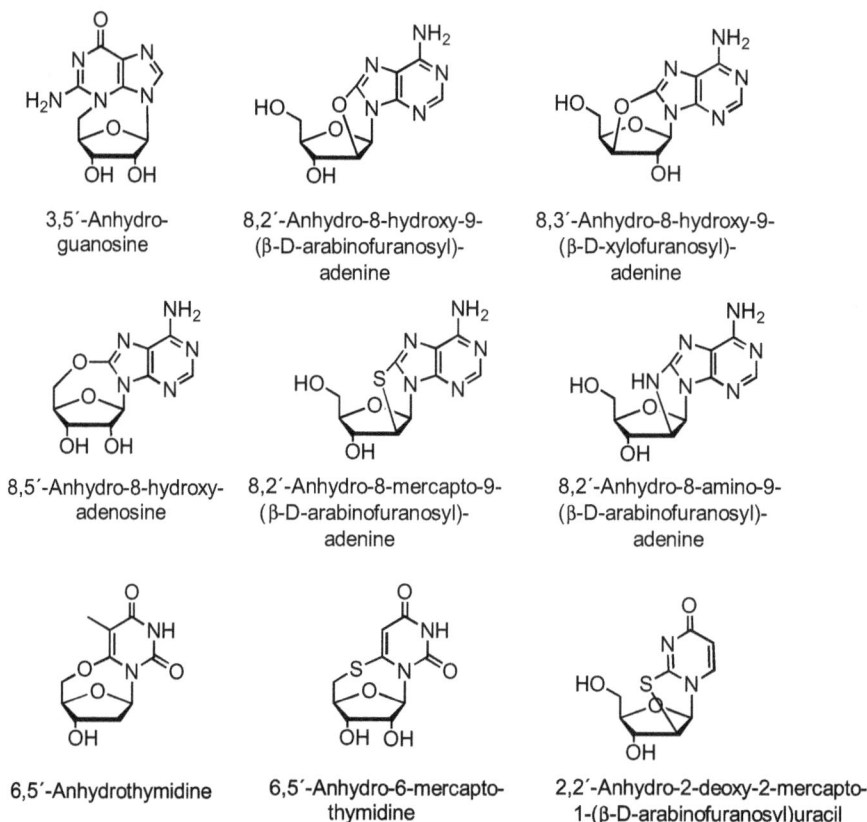

**Figure 1.5:** Examples of the names of cyclonucleosides.

### 1.1.3 C-Nucleosides and carbocyclic nucleosides

C-Nucleosides and carbocyclic nucleosides are hydrolytically stable analogs of nucleosides. In C-nucleosides, the N-glycosyl bond of nucleosides is replaced with a C–C bond. Accordingly, they do not undergo enzymatic or acid-catalyzed hydrolysis as nucleosides do and hence, have received interest as bioactive compounds. In carbocyclic nucleosides, the sugar ring oxygen is replaced with carbon. The carbocation formed upon departure of the nucleobase is not anymore stabilized by resonance with the neighboring ring oxygen, but the CN bond must be cleaved by a more difficult $S_N2$ displacement. C-Nucleosides can be named as C-glycosylated heterocycles, but more often trivial names are used, in particular, when the compound is naturally occurring. Examples of the structures of natural C-nucleosides are given in Figure 1.6. Several C-nucleosides have antibiotic or antitumor properties [8], and a few carbocyclic nucleosides exhibiting bioactivity have been isolated from natural sources (e.g., aristeromycin and neplanocin in Figure 1.6). Some synthetic carbocyclic nucleosides are used as anti-virals (e.g., abacavir and entecavir in Figure 1.6) [9]. As with *C*-nucleosides, trivial names are extensively used.

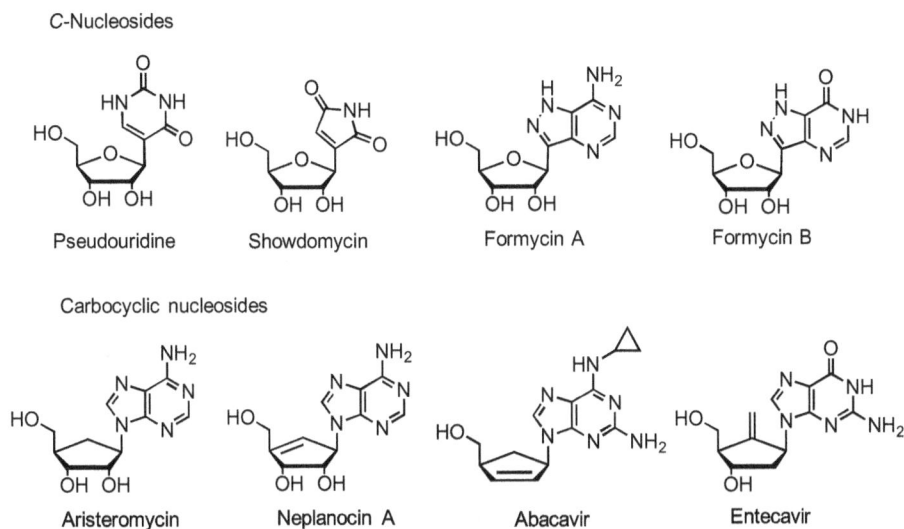

*C*-Nucleosides

| Pseudouridine | Showdomycin | Formycin A | Formycin B |

Carbocyclic nucleosides

| Aristeromycin | Neplanocin A | Abacavir | Entecavir |

**Figure 1.6:** Examples of *C*-nucleosides and carbocyclic nucleosides.

## 1.2 Tautomeric and protolytic equilibria

Nucleosides may in principle occur in various tautomeric forms as depicted in Figure 1.7. Since nucleic acids recognize each other by the formation of hydrogen-bonded base pairs between a purine and pyrimidine base, this kind of prototropic tautomerism self-evidently is a subject of crucial importance for high fidelity of the transfer of information. Unlike heteroaromatic compounds, in general, nucleosides occur as a single tautomer in aqueous solution. Amino and keto tautomers overwhelmingly predominate over imino and enol tautomers, respectively. With adenosine, for example, less than 0.01% is in the imino form where one of the $N^6$ protons is transferred to N1 [10]. Guanosine occurs as a keto/amino tautomer, the rapidly exchangeable proton being bound to N1 [11, 12]. The zwitterionic N7H-tautomer, for example, is five orders of magnitude less stable than the predominant N1H-form [13]. Cytidine likewise occurs in keto/amino form [12, 14–16] and uridine [12, 17, 18] and thymidine [14] in diketo form. With uridine, the content of enol tautomers is less than 0.03% [19].

Nucleobases, except uracil and thymine, are subject to annular tautomerism in addition to amino-imino and keto-enol tautomerism. In other words, the labile hydrogen atom at N9 of purines and N1 of pyrimidines may in principle be transferred to

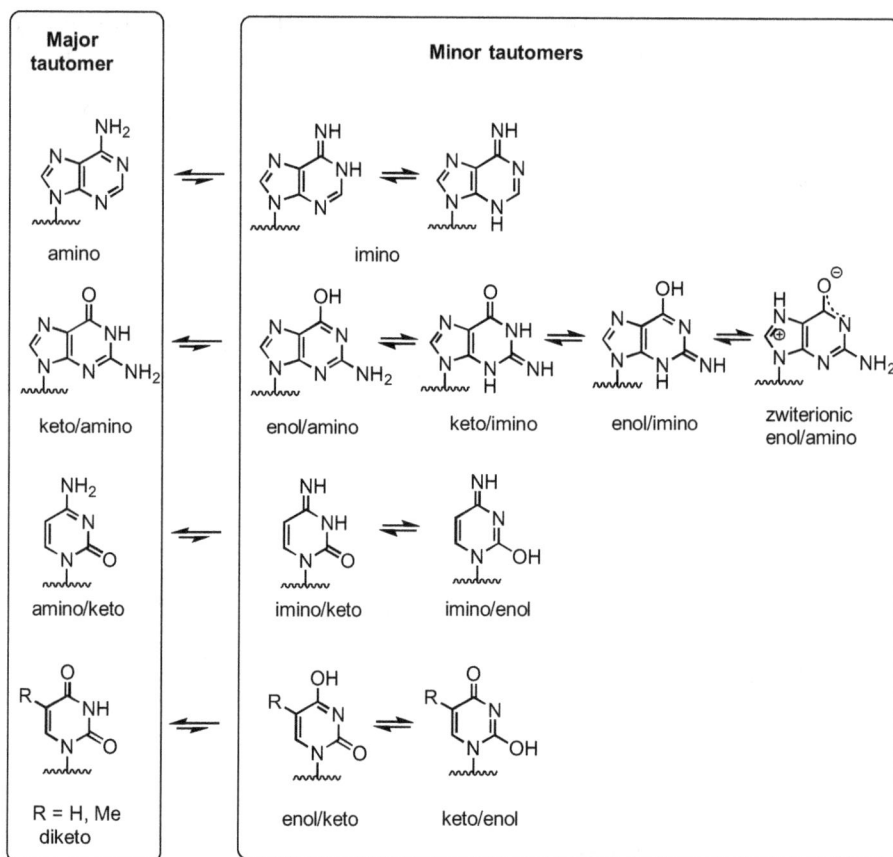

**Figure 1.7:** Tautomeric forms of nucleosides.

another ring nitrogen atom. In fact, several annular tautomers have been shown to exist in gas phase and in nonpolar media. The N1H,N7H-tautomer of guanine has been shown to be in gas phase even more stable than the canonical N1H,N9H-tautomer [20]. In aqueous environment, the canonical N9H-purine and N1H-pyrimidine tautomers usually predominate [21], although the situation is not quite clear with guanine. The estimations for the content of N3H,N7H-tautomer of guanine range from 10% [22] to more than 50% [20]. With adenine, the content of the minor N7H-tautomer is in aqueous solution 22% [23], whereas only 0.25% of cytosine is present as the N3H-tautomer [24].

All nucleosides are neutral molecules in the physiological pH range. In acidic solutions, consecutive protonations to mono- and dications take place. Table 1.1 records the $pK_a$ values for the canonical ribonucleosides. These values are called macroscopic $pK_a$ values since they simply refer to equilibrium between two differently charged species, for example, to conversion of neutral species to monocation, without taking into account that two different sites may compete for the proton. The macroscopic acidity constant, $K_a^m$, is related by eq. (1.1) to the microscopic acidity constants $K^1$ and $K^2$ that refer to protonation of a single site:

$$K_a^m = K^1 K^2 / (K^1 + K^2) \tag{1.1}$$

**Table 1.1:** Macroscopic p$K_a$ values for protonation and deprotonation of the base moiety of ribonucleosides at 25 °C.

| Nucleoside | p$K_a$ (NucH$_2{}^{2+}$) | p$K_a$ (NucH$^+$) | p$K_a$ (Nuc) |
|---|---|---|---|
| Adenosine | −1.4[a] | 3.61 ± 0.03[d] | – |
| Guanosine | −2.4[b] | 2.33 ± 0.02[e] | 9.03 ± 0.02[h] |
| Cytidine | −6.4[c] | 4.22 ± 0.02[f] | – |
| Uridine | | −2.4[g] | 9.18 ± 0.02[h] |

[a]In HClO$_4$ [25], [b]in HClO$_4$ [26], [c]in H$_2$SO$_4$ for cytosine [27], [d]Ref. [28], [e]Ref. [29], [f]Ref. [30], [g]in H$_2$SO$_4$ for uracil [27], [h]Ref. [31].
Abbreviation: Nuc stands for the neutral form of nucleosides. The values refer to ionic strength 0.1 M, except the negative ones that refer to concentrated mineral acids of various concentrations.

Let us consider protonation of adenosine as an example. The first macroscopic p$K_a$ value of adenosine is 3.61 ± 0.03 [28]. The first protonation mainly takes place at N1 [32], followed by protonation of N7 [25]. The basicity difference between N1 and N7 is, however, only 1.48 log units [33]. Accordingly, the concentration ratio of N1H$^+$ and N7H$^+$ monocations is 30, and hence 3.3% of the first protonation takes place at N7. The basicity of N3 is still lower, the difference between p$K_a$ values of N1H$^+$ and N3H$^+$ cations being 2.1 log units. In other words, around 0.7% of the first protonation takes place at N3. Protonation of the N1H$^+$ monocation then takes place at N7, the p$K_a$ value for the dication being −1.4 [25]. This means that the concentration of the mono- and dication is equal when the acidity function of the solution is −1.4. According to $H_o$ scale of Paul and Long [34], such acidity is achieved by molar concentrations of strong mineral acids, e.g., 4.5 M HNO$_3$ or 3.5 M HClO$_4$.

With guanosine, the N1 atom is protonated already under neutral conditions, the p$K_a$ value for N1H deprotonation being 9.03 ± 0.02 [31]. The monocation is obtained through protonation of N7 [35]. The p$K_a$ value of the monocation is 2.33 ± 0.02 [29]. The basicity difference between the N1 and N7 sites is, hence, much larger than with adenosine, and virtually no competition between these sites for proton exists. The p$K_a$ of the dication, having all the ring-nitrogen atoms (N1, N3, N7) protonated, is as low as −2.4 [26], being equivalent to the $H_o$ value of 5.4 M HClO$_4$.

Among pyrimidine nucleosides, cytidine is protonated at N3 [18, 36] with p$K_a$ = 4.22 ± 0.02 [30]. The N3 site of uridine and thymidine is protonated already in neutral solution, the p$K_a$ value for the deprotonation of N3H being 9.18 ± 0.02 [37] and 9.69 ±

0.03 [38], respectively. The protonation of the $O^4$ oxygen of pyrimidine nucleosides takes place only in molar concentrations of mineral acids. The $pK_a$ values reported for the free bases, cytosine and uracil are −6.4 and −2.4 [27], respectively. The dication of uracil is obtained by protonation of $O^2$ with $pK_a = -7.3$ [39].

Interestingly, the primary amino groups of nucleosides are not protonation sites. The lone electron pair of the amino group is involved in amidine resonance with the neighboring ring-nitrogen, and hence the electron density at the amino group is markedly lowered (Figure 1.8). The length of the C–N bond is reduced to 0.134 nm from 1.469 nm of a normal aliphatic C–N bond. The amino groups are actually rather easy to deprotonate even in aqueous solution. The $pK_a$ values are 17, 15 and 15.5 for adenosine, guanosine and cytidine, respectively [40]. This means, according to $H$-basicity function [41], that adenosine and guanosine become 50% deprotonated in 10 M and 3.5 M aq KOH, respectively.

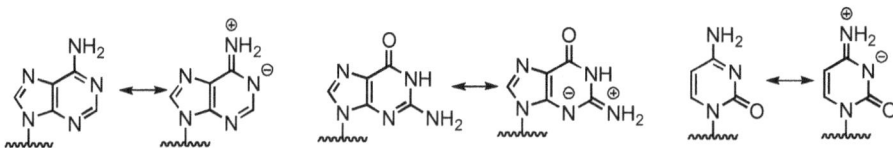

**Figure 1.8:** Amidine resonance in adenosine, guanosine and cytidine.

The base moiety $pK_a$ values of 2′-deoxyribonucleosides are virtually identical with those of ribonucleosides. Free nucleic acid bases are, in turn, from 0.2 to 0.5 pH units more basic than the respective base moiety of nucleosides because the electronegative glycofuranosyl group decreases the electron density with nucleosides. In addition, free nucleobases undergo deprotonation in pH range 9–13: N9H of adenine and guanine with $pK_a$ of 9.87 [42] and 12.4 [43], respectively, and N1H of cytosine with $pK_a$ of 12.1 [44]. With uracil and thymine this second deprotonation is difficult, the $pK_a$ value of uracil monoanion being 13.5 [45].

The protolyic equilibria of nucleosides are not, however, limited to the base moiety. The 2′-OH of ribonucleosides is sufficiently acidic to undergo deprotonation in normal pH range. The $pK_a$ values reported for all the canonical ribonucleasides are rather similar, viz. $12.15 \pm 0.04$, $12.54 \pm 0.02$, $12.66 \pm 0.01$ and $12.46 \pm 0.02$ for Ado, Guo, Ctd and Urd, respectively [46]. Electron withdrawal by the base moiety, ring-oxygen and 3′-OH weakens the $O^2$′-H bond, and the resulting 2′-oxyanion is additionally stabilized by H-bonding with the neighboring 3′-OH group. Evidently the influences of the base moiety and the ring-oxygen are most important for the acidification of 2′-OH since omission of adenine base and replacement of the ring-oxygen with carbon increased the $pK_a$ of adenosine by 0.97 and 0.90 pH units, respectively. Omission of 3′-OH and epimerization of the 2′-OH giving *ara* configuration resulted, in turn, only 0.49 and 0.16 units increase in the $pK_a$.

## 1.3 Conformational equilibria

Conformation of nucleosides is usually defined in terms of three distinct conformational equilibria: (i) sugar ring puckering, (ii) rotation around the C1′–N bond, and (iii) rotation around the C4′–C5′ bond (Figure 1.9). These three equilibria are discussed below in this order.

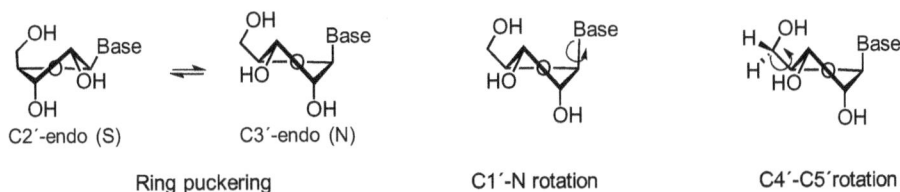

C2′-endo (S)     C3′-endo (N)

Ring puckering          C1′-N rotation          C4′-C5′rotation

**Figure 1.9:** Conformational equilibria of nucleosides.

### 1.3.1 Puckering of the sugar ring

The X-ray structures of nucleosides fall in two conformational families. The sugar ring adopts a twist-type puckering, where either C2′ or C3′ (as in Figure 1.9) deviates from the C1′–O$^{4'}$–C4′ plane toward the base moiety [47]. The members of the former and latter family are called C2′-endo and C3′-endo conformers or S- and N-conformers, respectively. In fact, these two families of conformers also prevail in solution. Accordingly, the ring puckering in solution is conventionally described by a two-state model, that is, by parameters that define the extent of ring puckering of the S- and N-conformers and additionally the equilibrium constant for the mutual interconversion of these two [48]. The parameters used to define the conformation of an individual conformer are pseudorotational phase angle, $P$, and maximal puckering amplitude, $\Phi_m$. Phase angle, $P$, indicates the location of conformer on the so-called pseudorotation cycle, depicted in Figure 1.10. The C3′-endo conformer, $^3_2$T, has been selected as the starting point on the cycle. In other words, its phase angle $P = 0$. Then all possible conformations are derived from this by stepwise movements of one carbon atom. The $^3_2$T-conformer is first converted to an envelope conformer, $^3$E, by moving C2′ upward into the C1′–O$^{4'}$–C4′ plane. When C4′ is then moved downward from resulting C2′–C1′–O$^{4'}$–C4′ plane, $^3_4$T is obtained. Transfer of C3′ into the plane C2′–C1–O$^{4'}$ gives $_4$E, etc. Altogether, the whole pseudorotation cycle consists of 10 twist and 10 envelope conformers alternating with each other. Accordingly, the distance between two neighboring conformers on the pseudorotation cycle is 18°. Nucleosides usually fall either in the region 0°–36° or 144°–180°. The former are called N-type and the latter S-type conformations.

The other pseudorotational parameter, the maximal puckering amplitude, $\Phi_m$, describes how distorted the twist conformation is. It is the maximal value for an endocy-

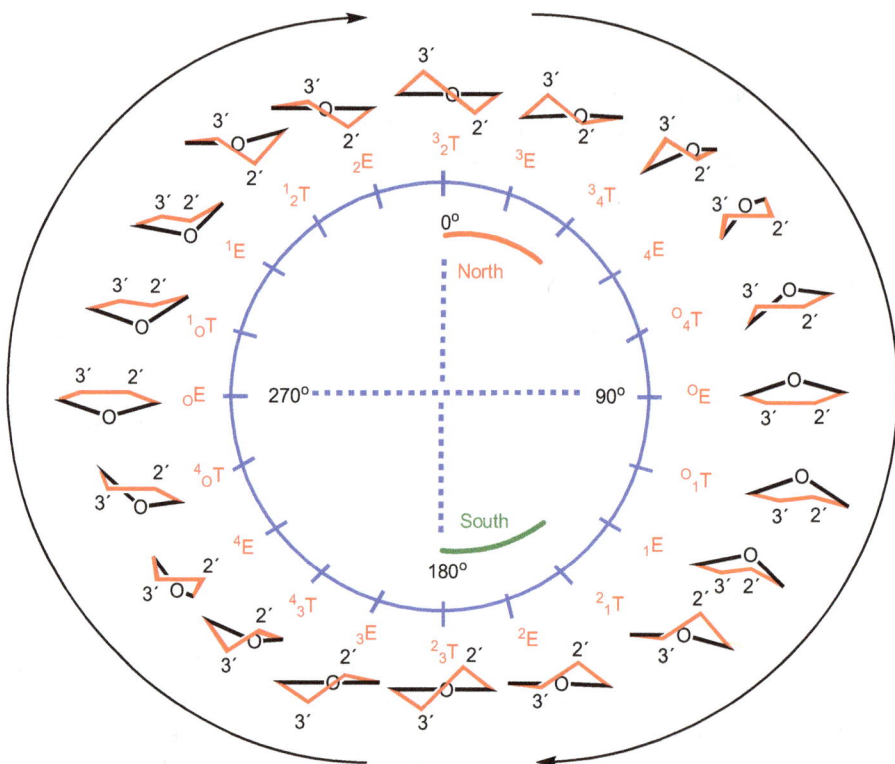

**Figure 1.10:** Pseudorotation cycle of nucleosides.

clic torsion angle. When $P = 0$, $\Phi_m$ is equal to the endocyclic torsion angle, $\Phi_0$, of the C2′–C3′ bond, that is, the angle between C1′–C2′ and C3′–C4′ bonds when viewed along the C2′–C3′ bond. In general terms, $\Phi_i$ depends on $\Phi_m$ and $P$ by eq. (1.2). For the definition of the other endocyclic torsion angles, see Figure 1.11:

$$\Phi_i = \Phi_m \cos P \quad (i = 0, 1, 2, 3, 4)$$

(1.2)

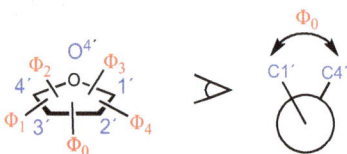

**Figure 1.11:** Endocyclic torsion angles as defined by Haasnoot et al. [49]. A Newman projection for the endocyclic torsion angle, $\Phi_0$, referring to the C2′–C3′ bond, is given as an example.

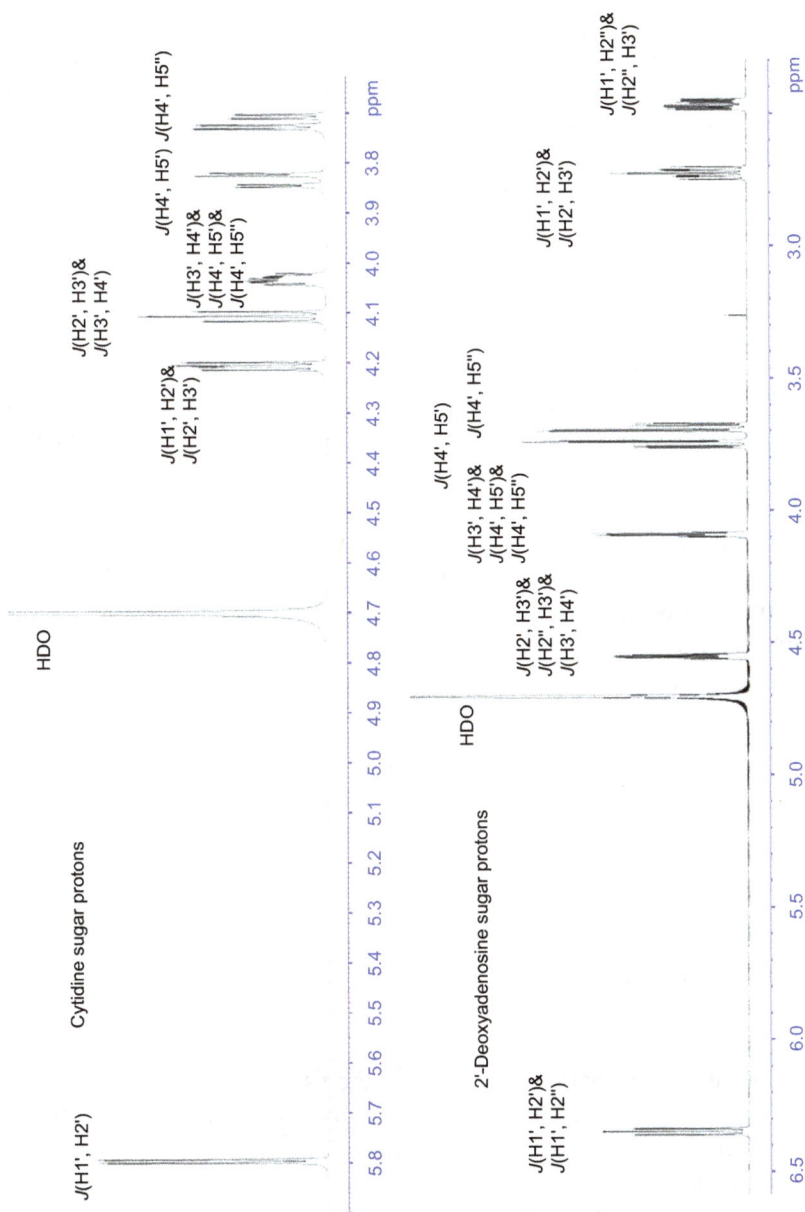

**Figure 1.12:** $^1$H NMR spectrum of cytidine and 2′-deoxyadenosine in D$_2$O. The vicinal H,H-coupling constants referring to each proton signal are indicated.

Pseudorotational parameters for both the N- and S-conformers and the equilibrium constant for their interconversion may be calculated on the basis of the vicinal H,H-coupling constants obtained from the splitting patterns of the $^1$H NMR signals of sugar protons. The sugar moiety signals of cytidine and 2′-deoxyadenosine are given in Figure 1.12 as typical examples of ribo- and deoxyribo-nucleosides.

Generalized Karplus equation (1.3) [50] may be applied to each of the vicinal $J_{HH}$ coupling constants:

$$J_{HH} = P_1\cos^2\varphi_{ij} + P_2\cos\varphi_{ij} + P_3 + \Sigma\Delta\chi_i\left\{P_4 + P_5\cos^2\left(\xi_i\varphi_{ij} + P_6|\Delta\chi_i|\right)\right\} \tag{1.3}$$

The first three terms refer to the dependence of the vicinal couplings ($J_{HH}$) within a given H-C-C-H fragment on the exocyclic torsion angle $\varphi_{ij}$. The remaining term accounts for the dependence of $J_{HH}$ on all electronegative nonhydrogen substituents, typically oxygen, on both carbon atoms of this particular fragment. That is why the summation ($\Sigma$). $\Delta\chi_i$ is the difference in Huggin's electronegativity between the non-hydrogen substituent and hydrogen. $\xi_i$ refers to orientation of substituents; $\xi_i = +1$ when the substituent is "a neighbor" of the hydrogens involved in the coupling, $\xi_i = -1$ when the substituent is "on the opposite side" of coupled hydrogen. For instance, $\xi_i = +1$ for $S^1$ and $S^3$ and $\xi_i = -1$ for $S^2$ and $S^4$ in Figure 1.13. In case α-substituents $S^i$ ($i = 1–4$) bear electronegative β-substituents, their influence on the term $\Delta\chi_i$ is taken into account by eq. (1.4):

$$\Delta\chi_i = \Delta\chi_i(\alpha) - P_7\Sigma\,\Delta\chi_j(\beta) \tag{1.4}$$

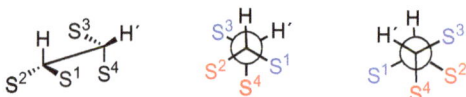

**Figure 1.13:** Definition of the sign, $\xi_i$, representing the orientation of an electronegative substituent in eq. (1.3).

The values determined for parameters $P_1$–$P_7$ on the basis of crystallographic data are listed in Table 1.2. As indicated, different sets of parameters are used for CH–CH and CH$_2$–CH fragments. When the exocyclic torsion angles $\varphi_{ij}$ have been obtained by eq. (1.3), these can be used to calculate the phase angle, $P$, and the maximal puckering amplitude, $\Phi_m$, by eqs. (1.5)–(1.7) for ribonucleosides and by eqs. (1.8)–(1.12) for 2′-deoxyribonucleosides [49]. In eqs. (1.8)–(1.12), 2′ refers to the C2′H on the β-face, that is, on the same side of the furanoid ring as the base moiety (β-face), whereas 2″ refers to the hydrogen on opposite the α-face.

**Table 1.2:** Parameters $P_1$–$P_7$ of generalized Karplus equation (eqs. (1.3) and (1.4)) [50].

| Fragment | $P_1$ | $P_2$ | $P_3$ | $P_4$ | $P_5$ | $P_6$ | $P_7$ |
|---|---|---|---|---|---|---|---|
| $CH_2$–CH | 13.22 | −0.99 | 0 | 0.87 | −2.46 | 19.9 | 0 |
| CH–CH | 13.24 | −0.91 | 0 | 0.53 | −2.41 | 15.5 | 0.19 |

Ribonucleosides:

$$\varphi_{1'2'} = 123.3° + 1.102\Phi_m\cos(P - 144°) \tag{1.5}$$

$$\varphi_{2'3'} = 0.2° + 1.090\Phi_m\cos P \tag{1.6}$$

$$\varphi_{3'4'} = -124.9° + 1.095\Phi_m\cos(P + 144°) \tag{1.7}$$

Deoxyribonucleosides

$$\varphi_{1'2'} = 121.4° + 1.03\Phi_m\cos(P - 144°) \tag{1.8}$$

$$\varphi_{1'2''} = 0.9° + 1.02\Phi_m\cos(P - 144°) \tag{1.9}$$

$$\varphi_{2'3'} = 2.4° + 1.06\Phi_m\cos P \tag{1.10}$$

$$\varphi_{2''3'} = 122.9° + 1.06\Phi_m\cos P \tag{1.11}$$

$$\varphi_{3'4'} = -124.0° + 1.09\Phi_m\cos(P + 144°) \tag{1.12}$$

Equations (1.5)–(1.7) or (1.8)–(1.12) enable calculation of $P$ and $\Phi_m$, but one has to bear in mind that these values refer to an equilibrium of an $N$-type and $S$-type conformer. To have a full description of the conformational equilibrium, $P$ and $\Phi_m$ should be bisected to contributions of these conformers ($P^N$, $P^S$, $\Phi_m{}^N$, $\Phi_m{}^S$) and additionally the equilibrium constant for their interconversion should be defined. In other words, five unknowns should be determined on the basis of eqs. (1.5)–(1.7) or (1.8)–(1.12). This is possible with deoxyribonucleosides since as many equations describing the dependence of $\varphi_{ij}$ on $P$ and $\Phi_m$ are available. With ribonucleosides, only three such equations exist, and hence, the number of parameters must be decreased by constraining two of the parameters to fixed values.

For many purposes, information on the equilibrium between N- and S-conformers is useful, although the conformational details of these conformers remain unknown. Instead of a complete pseudorotational analysis, the percentage of N or S form at the equilibrium is often estimated by approximations (1.13)–(1.15). In eq. (1.15), $J^N = 8.8$ Hz and 8.4 Hz for ribo- and deoxyribonucleosides, respectively, and $J^S = 1.1$ Hz:

$$S\% = 10 \times J_{H1',H2'}/\text{Hz} \tag{1.13}$$

$$N\% = 100 \times J_{H3',H4'}/\left(J_{H1',H2'} + J_{H3',H4'}\right) \tag{1.14}$$

$$S\% = 100 \times \left(J^N - J_{H3',H4'}\right)/\left(J^N - J^S\right) \tag{1.15}$$

### 1.3.2 Rotation around the C1′–N bond

Rotation around the C1′–N bond is defined by torsion group $\chi$ that in the case of purine nucleosides refers to sequence $O^{4'}$–C1′–N9–C4 and in the case of pyrimidine nucleosides to sequence $O^{4'}$–C1′–N1–C2. In other words, C4 in purines and C2 in pyrimidines play a special role. Their location with respect to the sugar moiety by definition determines the value of $\chi$. The conformation around the C1′-N bond is called *syn* when $\chi = 0° \pm 90°$ and *anti* when $\chi = 180° \pm 90°$. Crystallographic data of nucleosides show that most often the base moiety, that is, C4 of purine or C2 of pyrimidine, is oriented away from the sugar ring roughly along the $O^{4'}$-C1′ bond (A in Figure 1.14) [47]. In other words, most nucleosides prefer *anti*-conformation in solid state. In the alternative *syn*-conformation, the C4 (purine)/C2 (pyrimidine) atoms overlap the sugar ring (B in Figure 1.14). Sometimes the base is oriented away from the sugar ring along the C1′–C2′ bond. Such a rotamer is called a *high anti*-conformer (C in Figure 1.14).

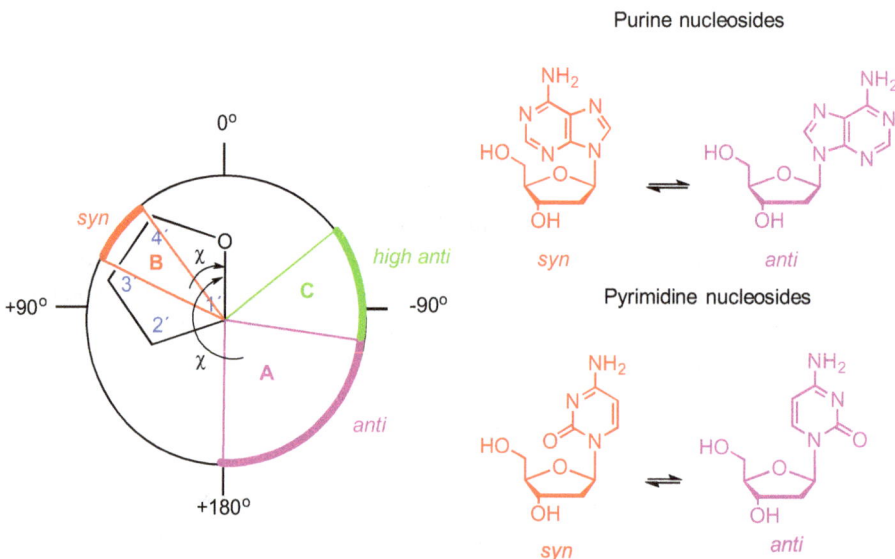

**Figure 1.14:** *Syn–anti*-confirmations of nucleosides.

The rotation around the *N*-glycosidic bond is in solution phase free; hence, indirect methods have to be used for determination of conformation around the *N*-glycosidic bond. Several different NMR techniques have been applied. Among these, the approach based on $^{13}$C,$^1$H coupling constants between the anomeric proton (H1′) and the carbon atoms next to the glycosylated nitrogen atom, together with parametrization of the appropriate Karplus equation with conformationally constrained cyclonucleosides [51], appears to be the most quantitative and generally applicable method. According to NMR studies, pyrimidine nucleosides favor the *anti*-conformation, usually

60–80% of conformers falling in this category [51]. Purine nucleosides also usually occupy *anti*-conformation [52], but the *anti*-form is not as dominant as with pyrimidine nucleosides. *Syn*- and *high anti*-conformers are encountered, and bulky substituents at C8 of purines and C6 of pyrimidines shift the equilibrium toward the *syn*-form.

### 1.3.3 Rotation around the C4′–C5′ bond

Rotation around the C4′–C5′ bond is usually described in terms of a three-state model depicted in Figure 1.15. Equilibrium between the three conformers is determined on the basis of vicinal H,H-coupling constants between the H5′ and H4′ protons [53]. With the most stable rotamer, 5′-OH overlaps the sugar ring, roughly bisecting the angle $O^{4'}$–C4′–C3′ (A in Figure 1.15). In the two remaining rotamers, the 5′-OH is oriented away from the sugar ring either along the C3′–C4′ bond (B in Figure 1.15) or along the $O^{4'}$-C4′ bond (C in Figure 1.15). Of these, the former conformation is favored. Various systems are used for naming of the conformers. Conformer A is called synclinical plus (*+sc*), gauche, gauche (*g*, *g*) or gauche plus (*g⁺*) conformation. Conformer B is either anitperiplanar (*app*), gauche, trans (*g,t*) or trans (*t*) and conformer C synclinical minus (*-sc*), trans, gauche (*t,g*) or gauche minus (*g⁻*).

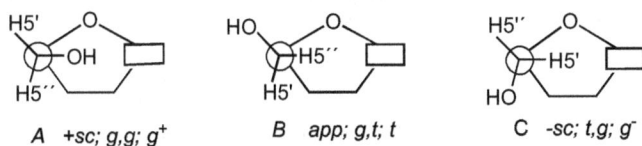

**Figure 1.15:** Preferred conformers for rotation around the C4′–C5′ bond. H5′ is the *pro-S* hydrogen and H5″ the *pro-R* hydrogen at C5′.

## 1.4 Metal ion complexes of nucleosides

Nucleosides form complexes with metal ions. The potential binding sites are N1 and N7 of purine nucleosides and N3 of pyrimidine nucleosides. N1 of guanosine and N3 of uridine and thymidine are protonated at physiological pH and metal ions, hence, have to compete for these sites with proton, which makes binding pH-dependent. N3 of purine nucleosides is not available for metal ions, owing to the steric hindrance that the N9-bound sugar moiety results in. Among metal ions, $Pd^{2+}$, $Hg^{2+}$, $Ag^+$ and $Cu^{2+}$ exhibit highest affinity to nucleosides, followed by 3d transition metal ions [54]. Interaction with alkali, alkaline earth metal and lanthanide ions is weak. Table 1.3 records selected examples of logarithmic formation constants, $K_M$, for the 1:1 complexes of nucleosides:

$$K_M = [M^{z+} L] / \{[M^{z+}][L]\}$$ 

(1.16)

**Table 1.3:** Selected examples of logarithmic formation constants, $\log(K_M/M^{-1})$, for 1:1 metal ion complexes of ribonucleosides.

| Nucleoside | Binding site | Metal ion | $T$ (°C) | $I$/M | $\log (K_M/M^{-1})$ | Reference |
|---|---|---|---|---|---|---|
| Ado | $N1$ | dienPd$^{2+}$ | 34.0 | 0.5 (KNO$_3$) | 4.5 | [55] |
| Ado | $N7$ | dienPd$^{2+}$ | 34.0 | 0.5 (KNO$_3$) | 3.9 | [55] |
| Ado | $N1$ | MeHg$^+$ | | | 3 | [56] |
| Ado | $N1$ | Ag$^+$ | 25.0 | | 2.02 | [72] |
| Ado | $N1$ and $N7$ | Zn$^{2+}$ | 25.0 | 1.0 (NaClO$_4$) | 0.2 | [73] |
| Ado | $N1$ and $N7$ | Cu$^{2+}$ | 25.0 | 1.0 (NaClO$_4$) | 0.96 | [73] |
| Ado | $N1$ and $N7$ | Ni$^{2+}$ | 25.0 | 1.0 (NaClO$_4$) | 0.4 | [73] |
| Guo | $N7$ | MeHg$^+$ | | | 4.5 | [56] |
| Guo | $N7$ | Zn$^{2+}$ | 25.0 | 1.0 (NaClO$_4$) | 0.8 | [29] |
| Guo | $N7$ | Cu$^{2+}$ | 25.0 | 1.0 (NaClO$_4$) | 1.9 | [29] |
| Guo | $N7$ | Ni$^{2+}$ | 25.0 | 1.0 (NaClO$_4$) | 1.4 | [29] |
| Guo N1-anion | $N1$ | MeHg$^+$ | | | 8.1 | [56] |
| Guo N1-anion | $N1$ | Cu$^{2+}$ | 25.0 | 0.1 (KNO$_3$) | 5.3 | [74] |
| Ctd | $N3$ | dienPd$^{2+}$ | 34.0 | 0.5 (KNO$_3$) | 5.4 | [75] |
| Ctd | $N3$ | MeHg$^+$ | | | 4.6 | [56] |
| Ctd | $N3$ | Zn$^{2+}$ | 21.0 | 1.0 (NaClO$_4$) | 0.56 | [13] |
| Ctd | $N3$ | Cu$^{2+}$ | 21.0 | 1.0 (NaClO$_4$) | 2.04 | [13] |
| Ctd | $N3$ | Ni$^{2+}$ | 21.0 | 1.0 (NaClO$_4$) | 0.95 | [13] |
| Urd N3-anion | $N3$ | dienPd$^{2+}$ | 34.0 | 0.5 (KNO$_3$) | 8.60 | [75] |
| Urd N3-anion | $N3$ | MeHg$^+$ | | | 9.0 | [56] |
| Urd N3-anion | $N3$ | Zn$^{2+}$ | 35.0 | 0.1 (KNO$_3$) | 3.57 | [76] |
| Urd N3-anion | $N3$ | Cu$^{2+}$ | 35.0 | 0.1 (KNO$_3$) | 5.90 | [76] |
| Urd N3-anion | $N3$ | Ni$^{2+}$ | 35.0 | 0.1 (KNO$_3$) | 3.57 | [76] |
| Thd N3-anion | $N3$ | dienPd$^{2+}$ | 34.0 | 0.5 (KNO$_3$) | 8.67 | [75] |
| Thd N3-anion | $N3$ | Cu$^{2+}$ | 20.0 | 1.0 (NaNO$_3$) | 4.7 | [77] |

The preferred binding site in adenosine has not been definitely established. The NMR studies on the formation of (dien)Pd$^{2+}$ complex show that the affinity to $N1$ site compared to $N7$ site is fourfold [55]. In this particular case, the situation is clear since the ligand exchange reactions of Pd$^{2+}$ are sufficiently slow in NMR time scale to allow the determination of the microscopic formation constants for N1 and N7 binding. Unfortunately, this is not possible with kinetically more labile metal ions, but indirect evidence has to be utilized to determine the binding site. MeHg$^+$ ion most likely binds to N1 [56,57] and displaces N$^6$-proton under basic conditions [58]. With 3d transition metal ions, the N1 vs. N7 competition has been evaluated by linear stability-basicity correlations for pyridine and imidazole-type nitrogen atoms [59]. According to such an approach, N7 binding is favored, but the affinity is less than double compared to N1-binding.

As mentioned above, the binding mode of guanosine is pH-dependent. In slightly acidic solutions, N1 remains protonated and the metal ion is coordinated to N7 [60] with $K_M$ values somewhat higher than those of the corresponding adenosine complexes. The affinity of metal ions to deprotonated N1 atom is, however, much higher than to neutral N7 site. With MeHg$^+$, the affinity difference is 4,000-fold. Hence, N1 coordination gradually takes over on approaching the $pK_a$ value of N1H. The cross-over pH, where both binding modes are as favorable, is 5.6. With 3d transition metal ions, the cross-over pH is higher, around 7.

Binding of a metal ion to one ring-nitrogen lowers the electron density and, hence, the basicity of the other ring-nitrogen atoms. The effect of a divalent metal ion is, however, smaller than that of a univalent proton, and it depends on the identity of the metal ion. N7 protonation of 9-methyladenine, for example, reduces the $pK_a$ value of N1H$^+$ by 3.57 log units, while the acidifying effect of N7-coordinated Pt$^{2+}$ is only 2.14 units [61]. With 9-methylguanine, the N1H $pK_a$ value is reduced to 2.35 units by N7-protonation and 1.22 units by Pt$^{2+}$ coordination. It is worth noting that the very slow ligand exchange kinetics of Pt$^{2+}$ allows accurate determination of the acidifying effects. The effect of 3d transition metal ions is greater than that of Pt$^{2+}$, but still smaller than the effect of protonation [62].

The coordination chemistry of pyrimidine nucleosides is rather straightforward. N3 serves as the binding site with both cytidine and uridine [63]. The complexes of cytidine, in particular, those with soft (dien)Pd$^{2+}$ and MeHg$^+$ ions, are from one to two orders of magnitude more stable than those of neutral ionic forms of purine nucleosides. N3 of uridine (and thymidine) is protonated under neutral conditions and metal ions have to compete for this site with proton, analogously to N1-binding of guanosine. The affinity to N3-anion of uridine is somewhat higher than to N1-anion of guanosine.

Interaction of nucleosides with bidentate complexes of Pt$^{2+}$ has been a subject of extensive interest since the discovery of *cis*-diamminedichloro complex of Pt$^{2+}$ as an anticarcinogenic compound [64], the biological target of which most likely is DNA. A special feature of Pt$^{2+}$ is slow ligand exchange. While the Pd$^{2+}$ complexes of nucleosides reach equilibrium in seconds, the equilibration of corresponding Pt$^{2+}$ complexes takes hours, in some cases, even days. Although the Pt$^{2+}$ complexes are more stable than the closely related Pd$^{2+}$ complexes [65], their formation is a kinetically controlled process rather than thermodynamically controlled process. Upon treatment of double-stranded DNA with the *cis*-diamminedichloro complex of Pt$^{2+}$, the preferred binding site is N7 of 2′-deoxyguanosine, followed by N7 of 2′-deoxyadenosine [66]. As mentioned above, (dien)Pt$^{2+}$ at N7 of 9-methylguanine increases the acidity of N1H by 1.22 log units. Accordingly, it also retards binding of metal ions to N1-anion.

Soft metal ions seem to be able to stabilize minor tautomers of nucleosides [67]. N1-bound (dien)Pt$^{2+}$ of 9-methyladenine, for example, migrates under alkaline conditions to N$^6$ displacing a proton. Upon acidification, the proton is attached to N1, resulting in the formation of $N^6$-metallated imino tautomer [68] (Figure 1.16). The $pK_a$ value of N1H of this species is 7.65, i.e., 4 unit higher than the $pK_a$ of N1 of the amino tautomer. Binding of (dien)Pt$^{2+}$ to 1-methylcytosine likewise gives an $N^4$-metallated imino tautomer with p$K_a$ of N3H = 7.5 [69] (Figure 1.16). Replacement of N3H of uridine or thymidine with Pt(NH$_3$)$_2$Cl$_2$ increases, in turn, the basicity of carbonyl oxygen to such an extent that enolization O$^2$ or O$^4$ takes place. According to theoretical calculations, the enolized oxygen is O$^4$ [70]. Interestingly, C5-mercurated 1,3-dimethyl uracil forms in aqueous acid a mixed-nucleobase complex with 9-methyladenine, where the C5-bound Hg$^{2+}$ is additionally coordinated to $N^6$ of the imino tautomer of 9-methyl adenine [71].

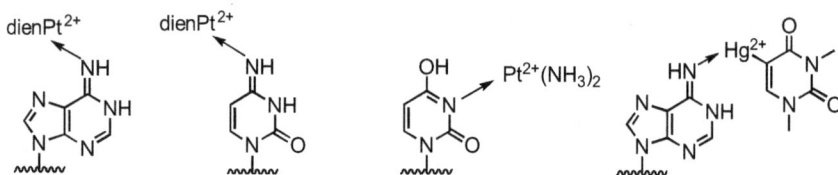

**Figure 1.16:** Examples of metal ion-stabilized rare tautomers of nucleosides.

## 1.5 Optical properties of nucleosides

### 1.5.1 UV absorption

Nucleic acid bases are intensively UV-absorbing, which greatly facilitates detection and quantification of nucleosides. Quantification of concentration and duplex stability of oligonucleotides, for example, is usually based on UV absorption of nucleobases. Since the sugar and phosphate groups are not UV absorbing at wavelengths higher than 230 nm, the UV spectra of nucleobases, nucleosides and nucleotides are virtually identical. Table 1.4 records the absorptivity at 260 nm and pH 7.0 and the wavelengths of absorption maxima at pH 1, 7 and 11 together with the respective absorptivity.

**Table 1.4:** Absorptivity of nucleosides at 260 nm and pH 7 ($\epsilon_{260}$) and wavelengths of absorption maxima ($\lambda_{max}$) at pH 1, 7 and 11 together with the respective absorptivity.

| | $\epsilon_{260}$ at pH 7[a,b] | $\lambda_{max}$ ($\epsilon_{max}$) at pH 7[a,b,c] | $\lambda_{max}$ ($\epsilon_{max}$) at pH 1[b,c,d] | $\lambda_{max}$ ($\epsilon_{max}$) at pH 11[b,c,d] |
|---|---|---|---|---|
| A | 15.02 | 259 (15.04) | 257 (15.14) | 259 (15.49) |
| C | 7.07 | 271 (8.74) | 279 (12.88) | 272 (8.91) |
| G | 12.08 | 252 (14.09) | 257 (12.30) | 257 (11.22) |
| U | 9.66 | 262 (9.78) | 262 (10.00) | 261 (7.24) |
| T | 8.56 | 267 (9.49) | 267 (10.00) | |

[a]Values refer to nucleoside 5′-phosphates [78]. [b]$\epsilon$ values given as M$^{-1}$ cm$^{-1}$. [c]Wavelength given as nm. [d][79].
The values refer to dilute solutions where intermolecular association can be neglected.

## 1.5.2 Circular dichroism (CD)

Nucleosides are chiral molecules, owing to the presence of stereogenic centers in the sugar moiety. Accordingly, nucleosides rotate plane polarized light and exhibit circular dichroism (CD). In other words, the absorptivity for right- and left-handed circularly polarized UV radiation is different [80]. The difference between the absorptivity of the left and right circularly polarized light, $\varepsilon_L - \varepsilon_R$, is usually measured as a function of the wavelength in the region 200–320 nm, where the sugar moiety does not absorb. In this region, nucleosides usually show a positive Cotton effect. This means that $\varepsilon_L - \varepsilon_R$ is negative at wavelengths shorter than the absorption maximum, $\lambda_m$, and positive at longer wavelengths. The CD is believed to mainly refer to mutual orientation of the sugar and base moiety. The positive Cotton effect most likely refers to *anti*-conformation. Consistent with this view, the Cotton effect is with pyrimidine nucleosides, known to strongly favor *anti*-conformation, greater than with purine nucleosides. Double-helical DNA, having nucleosides locked to *anti*-form, exhibit a very strong positive Cotton effect. The CD of monomeric nucleoside units evidently contributes to this, in addition to asymmetric helical overall structure.

## 1.5.3 Fluorescent nucleosides

Canonical nucleosides are not fluorescent, but some of their base-modified analogs are. Of particular interest are those analogs that closely resemble the canonical ones. Such analogs can be incorporated into oligonucleotides, or enzymatically even in nucleic acids, in place of canonical nucleosides. Since the emission efficiency of fluorescent nucleoside analogs usually is rather sensitive to environment, interaction of the labeled oligonucleotide probe with other nucleic acids, proteins or small molecules may be examined. Figure 1.17 shows several fluorescent nucleosides used for this purpose. Among them 2-aminopurine (2-AP) and 6-methyl-3,7-dihydro-2*H*-pyrrolo[2,3-d] pyrimidin-2-one ("pyrroloC," pC) nucleosides are the most extensively used analogs. 2-AP nucleosides are able to take the role of adenine nucleosides in double-stranded DNA or RNA. The difference between the emission and excitation wavelength (Stokes shift) is reasonably large (370 vs 310 nm) and the quantum yield, 0.68 in water, is high. [81]. In addition, the intensity of the fluorescence emission is sensitive to the polar nature of the microenvironment [82]. The Stokes shift of pyrroloC nucleosides is even longer than that of 2-AP nucleosides, 110 nm (460 vs 350 nm), but the quantum yield is lower, 0.2 [83]. H-bonding with guanine is efficient and sensitivity to microenvironment is high. The photophysical properties of its phenyl counterpart, PhpC, are even better, but the larger size may cause problems. This may also be the case with the recently introduced 5-(benzo[*b*]thiophen-2-yl)uridine that has emission at the wavelength of visible light [84]. Thienoguanosine is a recent emissive analog of guanosine that exhibits high quantum yield, long fluorescence lifetime and good sensitivity to

local structural changes [85]. Figure 1.17 additionally shows a couple of other nucleoside analogs that have been successfully used in monitoring nucleic acid–protein
and nucleic acid–small molecule interactions [86].

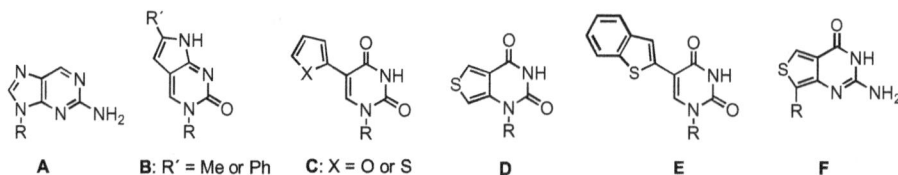

**A**    **B: R′ = Me or Ph**    **C: X = O or S**    **D**    **E**    **F**

**Figure 1.17:** Fluorescent isomorphic analogs of canonical nucleosides: (A) 2-aminopurine nucleoside
(2-AP), (B) 6-methyl-(pC) and 6-phenyl-3,7-dihydro-2H-pyrrolo-[2,3-d]pyrimidin-2-one (PhpC) nucleosides,
(C) 5-(furane-2-yl)-and 5-(thiophene-2-yl)-uracil nucleosides, (D) thieno[3,4-d]pyrimidine-2,4(1H,3H)-dione
nucleoside, (E) 5-(benzo[b]thiophen-2-yl)uridine and (F) thienoguanosine.

## 1.6 Base stacking

The solution equilibria and optical properties of nucleosides are influenced by their
tendency to associate in aqueous solution [87]. This tendency is more marked with
purine nucleosides than with pyrimidine nucleosides. The reason is not hydrogen
bonding, as one might expect. Unlike in case of H-bonding, the association results in
upfield shift of $^1$H and $^{13}$C resonances [88] and replacement of nitrogen-bound hydrogen with alkyl groups enhances association [89]. For instance, the equilibrium constants for dimerization of adenosine and $N^6,N^6$-dimethyladenosine are 4.5 M$^{-1}$ and
22.2 M$^{-1}$, respectively. Addition of an organic solvent into an aqueous solution of nucleoside weakens the association, also in contrast to the behavior of H-bonding [90].

$^1$H NMR studies [88, 89] have shown that the bases stack vertically. They largely
overlap, but not completely. Although base-stacking is a dynamic process, some orientations of the bases with respect to each other still are favored. These are depicted in
Figure 1.18. When the pyrimidine rings as well as the imidazole rings partly overlap,
the stacking geometry is "head to head." If the pyrimidine ring of nucleoside 1 overlaps with the imidazole ring of nucleoside 2, the geometry is "head to tail." In both
cases, the substituent at N9 may be situated at the same edge of the stack or at opposite edges. The former mode of stacking is called "face to back" and the latter "face to
face." Among the eight alternative geometries, the head-to-head orientations are believed to be favored.

At least pentameric adducts are formed at high concentrations [91]. A so-called
isodesmic model is obeyed, better with purines than pyrimidines [87]. This means
that the equilibrium constants for the successive steps are approximately equal and
the steps are isoenthalpic. The process is very fast, $k \approx 2 \times 10^9$ M$^{-1}$ s$^{-1}$ [92]. Salt effects

Head to head                          Head to tail

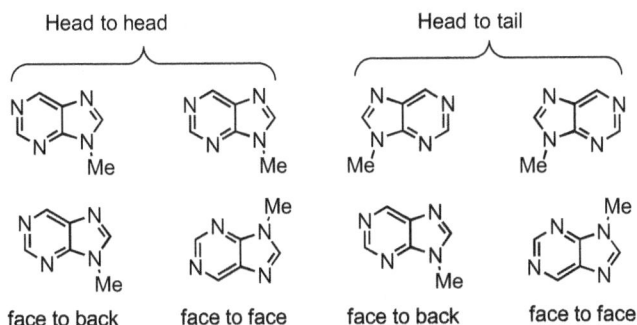

face to back        face to face        face to back        face to face

**Figure 1.18:** Mutual orientation of two 9-methylpurines upon stacking.

are of minor importance [93]. As solution equilibria in general, stacking is weakened at elevated temperatures.

The thermodynamics of base-stacking differ totally from that of classical hydrophobic bonding. Stacking is an enthalpy-driven process. $\Delta H°$ varies from 0 to −40 kJ mol$^{-1}$ and the entropy change is also negative, falling in the range from 0 to −100 J K$^{-1}$ mol$^{-1}$. In other words, stacking suffers from an entropy penalty. For comparison, hydrophobic bonding is an entropy-driven process, the enthalpy change being usually close to 0 [94].

The key question is: why do nucleosides stack in water. A classical qualitative explanation is based on dipole-induced dipole-dipole interactions [95]. The bonds in nucleic acid bases may be regarded as microscopic dipoles, owing to the difference on electronegativity of the atoms involved. Upon stacking, these microscopic dipoles mutually enforce the formation of a complementary attractive electron distribution to each other. The strength of stacking, hence, depends on the polarizing power of the base, which is largely equivalent to the number of C–N and C=O bonds, and on the polarizability of the π-electron cloud, that is, the ability to adjust to requirements of the polarizing bonds of the partner base. Both alkyl and heteroatom substituents increase the polarizability and, hence, stacking. In addition to permanent dipoles, the charges within each base are subject to continuous fluctuations, inducing complementary dipoles in the partner. This gives rise to attraction by London dispersion forces. Unfortunately, no mathematical description that would enable prediction of stacking strength without exploitation of experimental data is available. In particular, the crucial role of water in stacking has not been thoroughly explained.

## Further reading

Altona C, Sundaralingam M. Conformational analyisis of the sugar ring in nucleosides and nucleotides. A new description using the concept of pseudorotation. J Am Chem Soc 1972, 94, 8205–8212.

Carell T, Brandmayr C, Hienzsch A, Müller M, Pearson D, Reiter V, Thoma I, Thumbs P, Warner M. Structure and function of noncanonical nucleobases. Angew Chem Int Ed 2012, 51, 7110–7131.

Kypr J, Kejnovska I, Bednarova K, Vorlickova M. Comprehensive Chiroptical Spectroscopy, Vol. 2: Applications in Sterochemical Analysis of Synthetic Compounds, Natural Products and Biomolecules, 1st ed. Berova N, Polavarapu PL, Nakanishi K, Woody RW, eds. Wiley, 2012, 573–584.

Lippert B, Gupta D. Promotion of rare nucleobase tautomers by metal binding. Dalton Trans 2009, 0, 4619–4634.

Mieczkowski A, Agrofoglio LA. Potential and perspectives of cyclonucleosides. Curr Med Chem 2010, 17, 1527–1549.

Secrist JA III. Nucleoside and nucleotide nomenclature. Curr Protoc Nucleic Acid Chem 2001, May; Appendix 1: Appendix 1D. doi: 10.1002/0471142700.nca01ds00.

Sigel H. Acid-base properties of purine residues and the effect of metal ions: Quantification of rare nucleobase tautomers. Pure Appl Chem 2004, 76, 1869–1886.

Srivatsan SG, Sawant AA. Fluorescent ribonucleoside analogues as probes for investigating RNA structure and function. Pure Appl. Chem 2011, 83, 213–232.

Stambasky J, Hocek M, Kocovsky P. C-Nucleosides: Synthetic strategies and biological applications. Chem Rev 2009, 109, 6729–6764.

## References

[1]  McNaught AD. Nomenclature of carbohydrates. Pure Appl Chem 1996, 68, 1919–2008.

[2]  Carell T, Brandmayr C, Hienzsch A, Müller M, Pearson D, Reiter V, Thoma I, Thumbs P, Warner M. Structure and function of noncanonical nucleobases. Angew Chem Int Ed 2012, 51, 7110–7131.

[3]  Schön A, Kaminska E, Schelter F, Ponkkonen E, Korytiaková E, Schiffers S, Carell T. Analysis of an active deformylation mechanism of 5-formyl-deoxycytidine (fdC) in stem cells. Angew Chem Int Ed 2020, 59, 5591–5594.

[4]  Capon RJ, Trotter NSJ. N3,5′-Cycloxanthosine, the first natural occurrence of a cyclonucleoside. Nat Prod 2005, 68, 1689–1691.

[5]  Chatgilialoglu C, Ferreri C, Terzidis MA. Purine 5′,8-cyclonucleoside lesions: Chemistry and biology. Chem Soc Rev 2011, 40, 1368–1382.

[6]  Mieczkowski A, Agrofoglio LA. Potential and perspectives of cyclonucleosides. Curr Med Chem 2010, 17, 1527–1549.

[7]  Mieczkowski A, Roy V, Agrofoglio LA. Preparation of Cyclonucleosides. Chem Rev 2010, 110, 1828–1856.

[8]  Stambasky J, Hocek M, Kocovsky P. C-Nucleosides: Synthetic strategies and biological applications. Chem Rev 2009, 109, 6729–6764.

[9]  Wang J, Rawal RK, Chu CK. Recent advances in carbocyclic nucleosides: Synthesis and biological activity. In Zhang L-H, Xi Z, Chattopadhyaya J, eds. Medicinal Chemistry of Nucleic Acids. Hoboken: John Wiley & Sons, 2011, 1–100.

[10]  Wolfenden R. Tautomeric equilibria in inosine and adenosine. J Mol Biol 1969, 40, 307–310.

[11]  Miles HT, Howard FB, Frazier J. Tautomerism and protonation of guanosine. Science 1963, 142, 1458–1463.

[12]  Lord RC, Thomas GJ, Jr. Raman spectral studies of nucleic acids and related molecules – I Ribonucleic acid derivatives. Spectrochim Acta A 1967, 23, 2551–2591.

[13]  Kim S-H, Martin RB. Binding sites and stabilities of transition metal ions with nucleosides and related ligands. Inorg Chim Acta 1984, 91, 19–24.

[14]  Miles HT. Infrared spectra and tautomeric structure of nucleosides and nucleotides in $D_2O$ solution. II. Biochim Biophys Acta 1958, 27, 46–52.

[15]  Miles HT. The tautomeric structure of deoxycytidine. J Am Chem Soc 1963, 85, 1007–1008.

[16]  Ulbricht TVL. The tautomeric structure or deoxycytidine. Tetrahedron Lett 1963, 0, 1027–1030.

[17]  Katritzky AR, Waring AJ. Tautomeric azines. Part I. The tautomerism of 1-methyluracil and 5-bromo -1-methyluracil. J Chem Soc 1962, 0, 1540–1544.

[18]  Roberts BW, Lambert JB, Roberts JD. Nitrogen-15 magnetic resonance spectroscopy. VI. Pyrimidine derivatives. J Am Chem Soc 1965, 87, 5439–5441.

[19]  Poulter CD, Frederick GD. Uracil and its 4-hydroxy-1(H) and 2-hydroxy-3(H) protomers. $pK_a$'s and equilibrium constants. Tetrahedron Lett 1975, 0, 2171–2174.

[20]  Hanus M, Ryjacek F, Kabelac M, Kubar T, Bogdan TV, Trygubenko SA, Hobza P. Correlated ab initio study of nucleic acid bases and their tautomers in the gas phase, in a microhydrated environment and in aqueous solution. Guanine: Surprising stabilization of rare tautomers in aqueous solution. J Am Chem Soc 2003, 125, 7678–7688.

[21]  Shukla MK, Leszczynski J. Tautomerism in nucleic acid bases and base pairs: A brief overview. WIREs Comput Mol Sci 2013, 3, 637–649.

[22]  Gorb L, Leszczynski J. Intramolecular proton transfer in mono- and dihydrated tautomers of guanine: An ab initio post hartree–fock study. J Am Chem Soc 1998, 120, 5024–5032.

[23]  Dreyfus M, Bensaude O, Dodin G, Dubois JE. Tautomerism of purines. I. N(7)H .dha. N(9)H equilibrium in adenine. J Am Chem Soc 1975, 97, 2369–2376.

[24]  Dreyfus M, Bensaude O, Dodin G, Dubois JE. Tautomerism in cytosine and 3-methylcytosine. A thermodynamic and kinetic study. J Am Chem Soc 1976, 98, 6338–6349.

[25]  Benoit RL, Frechette M. Protonation de l'adenine, de la purine et de I'adenosine en milieu acide fort. Can J Chem 1984, 62, 995–1000.

[26]  Zoltewicz JA, Clark DF, Sharpless TW, Grahe G. Kinetics and mechanism of the acid-catalyzed hydrolysis of some purine nucleosides. J. Am. Chem. Soc 1970, 92, 1741–1750.

[27]  Benoit RL, Frechette M. [1]H and [13]C nuclear magnetic resonance and ultraviolet studies of the protonation of cytosine, uracil, thymine, and related compounds. Can J Chem 1986, 64, 2348–2352.

[28]  Tribolet R, Sigel H. Self-association and protonation of adenosine 5'-monophosphate in comparison with its 2'- and 3'-analogues and tubercidin 5'-monophosphate (7-deaza-AMP). Eur J Biochem 1987, 163, 353–363.

[29]  Lönnberg H, Vihanto P. Complexing of inosine and guanosine with divalent metal ions in aqueous solution. Inorg ChimActa 1981, 56, 157–161.

[30]  Voet D, Gratzer WB, Cox RA, Doty P. Absorption spectra of nucleotides, polynucleotides, and nucleic acids in far ultraviolet. Biopolymers 1963, 1, 193–208.

[31]  Ogasawara N, Inoue Y. Titration properties of homodinucleoside monophosphates. Determination of overlapping ionization constants and intramolecular stacking equilibrium quotients of ApA, CpC, GpG, and UpU. J Am Chem Soc 1976, 98, 7048–7053.

[32]  Gonnella NC, Nakanishi H, Holtwick JB, Horowitz JB, Kanamori K, Leonard NJ, Roberts JD. Studies of tautomers and protonation of adenine and its derivatives by nitrogen-15 nuclear magnetic resonance spectroscopy. J Am Chem Soc 1983, 105, 2050–2055.

[33]  Kapinos LE, Operschall BP, Larsen E, Sigel H. Understanding the acid-base properties of adenosine: The intrinsic basicities of N1, N3 and N7. Chem Eur J 2011, 17, 8156–8164.

[34]  Paul MA, Long FA. $H_0$ and related indicator acidity function. Chem Rev 1957, 57, 1–45.

[35]  Marskowski V, Sullivan GR, Roberts JD. Nitrogen-15 nuclear magnetic resonance spectroscopy of some nucleosides and nucleotides. J Am Chem Soc 1977, 99, 714–718.

[36] Becker ED, Miles HT, Bradley RB. Nuclear magnetic resonance studies of methyl derivatives of cytosine. J Am Chem Soc 1965, 87, 5575–5582.

[37] Knobloch B, Da Costa CP, Linert W, Sigel H. Stability constants of metal ion complexes formed with N3-deprotonated uridine in aqueous solution. Inorg Chem Commun 2003, 6, 90–93.

[38] Moreno-Luque CF, Freisinger E, Costisella B, Griesser R, Ochocki J, Lippert B, Sigel HJ. Comparison of the acid–base properties of 5- and 6-uracilmethylphosphonate (5Umpa2– and 6Umpa2–) and some related compounds. J Chem Soc Perkin Trans 2001, 2, 0, 2005–2011.

[39] Frederick GD, Poulter D. Extension of the HA acidity function into oleum mixtures. J Am Chem Soc 1975, 97, 1797–1801.

[40] Taylor SE, Buncel E, Norris AR. Metal ion-biomolecule interactions. II. Methylmercuration of the deprotonated amino groups in adenine, guanine, and cytosine derivatives, and its relationship to amino group acidity. J Inorg Biochem 1981, 15, 131–141.

[41] Yagil G. Effect of ionic hydration in equilibria and rates in concentrated electrolyte solutions. III. The H⁻ scale in concentrated hydroxide solutions. J Phys Chem 1967, 71, 1034–1044.

[42] Izatt RD, Christensen JJ, Rytting JH. Sites and thermodynamic quantities associated with proton and metal ion interaction with ribonucleic acid, deoxyribonucleic acid, and their constituent bases, nucleosides, and nucleotides. Chem Rev 1971, 71, 439–481.

[43] Christensen JJ, Rytting JH, Izatt RM. Thermodynamic pK, ΔH°, ΔS°, and ΔCp° values for proton dissociation from several purines and their nucleosides in aqueous solution. Biochemistry 1970, 9, 4907–4913.

[44] Gningue D, Aaron -J-J. Fluorimetric determination of dissociation constants and pH-controlled fluorescence analysis of purines and pyrimidines. Talanta 1985, 32, 183–187.

[45] Ganguly S, Kundu KK. Protonation/deprotonation energetics of uracil, thymine, and cytosine in water from e.m.f./spectrophotometric measurements. Can J Chem 1994, 72, 1120–1126.

[46] Velikyan I, Acharya S, Trifonova A, Földesi A, Chattopadhyaya J. The pKa's of 2'-hydroxyl group in nucleosides and nucleotides. J Am Chem Soc 2001, 123, 2893–2894.

[47] Sundaralingam M. Stereochemistry of nucleic acids and their constituents. IV. Allowed and preferred conformations of nucleosides, nucleoside mono-, di-, tri-, tetraphosphates, nucleic acids and polynucleotides. Biopolymers 1969, 7, 821–860.

[48] Altona C, Sundaralingam M. Conformational analyisis of the sugar ring in nucleosides and nucleotides. A new description using the concept of pseudorotation. J Am Chem Soc 1972, 94, 8205–8212.

[49] Haasnoot CAG, de Leeuw FAAM, de Leeuw HPM, Altona C. The relationship between proton-proton NMR coupling-constants and substituent electronegativities. 2. Conformational-analysis of the sugar ring in sucleosides and nucleotides in solution using generalized Karplus equation. Org Magn Reson 1981, 15, 43–52.

[50] Haasnoot CAG, de Leeuw FAAM, Altona C. The relationship between proton-proton NMR coupling-constants and substituent electronegativities. 1. An empirical generalization of the Karplus equation. Tetrahedron 1980, 36, 2783–2792.

[51] Davis DB, Rajani P, Sadikot H. Determination of glycosidic bond conformations of pyrimidine nucleosides and nucleotides using vicinal carbon–proton coupling constants. J Chem Soc Perkin Trans 1985, 2, 0, 279–285.

[52] Stolarski R, Hagberg C-E, Shugar D. Studies on the dynamic syn-anti equilibrium in purine nucleosides and nucleotides with the aid of H-1 and C-13 NMR-spectroscopy. Eur. J. Biochem 1984, 138, 187–192.

[53] Haasnoot CAG, de Leeuw FAAM, de Leeuw HPM, Altona C. Interpretation of vicinal proton-proton coupling-constants by a generalized Karplus relation – Conformational analysis of the exocyclic C4'-C5' bond in nucleosides and nucleotides. Rec. Trav. Chim. Pays Pas 1979, 98, 576–577.

[54] Lönnberg H. Proton and metal ion interactions with nucleic acid bases, nucleosides and nucleoside monophosphates. In Burger K, ed. Biocoordination Chemistry: Coordination Equilibria in Biologically Active Systems. Ellis Horwood, 1990, 284–346.

[55] Kim S-H, Martin RB. Stabilities and 1H NMR studies of (diethylenetriamine)Pd(II) and (1,1,4,7,7-pentamethyldien)Pd(II) with nucleosides and related ligands. Inorg Chim Acta 1984, 91, 11–18.

[56] Simpson RB. Association constants of methylmercuric and mercuric ions with nucleosides. J Am Chem Soc 1964, 86, 2059–2065.

[57] Mansy S, Peticolas WL, Tobias RS. Raman spectra of methyl derivatives of 5′-adenosine monophosphate, tubercidin, inosine, uridine and cytidine. Perturbation of nucleoside vibrations by electrophilic attack at different sites. Spectrochim Acta A 1979, 35, 315–329.

[58] Hoo D-I., McConnell B. Nucleic acid-methylmercury (II) interactions and the proton NMR lifetimes of nucleobase exchangeable protons. J Am Chem Soc 1979, 101, 7470–7477.

[59] Kinjo Y, Tribolet R, Corfu NA, Sigel H. Stability and structure of xanthosine metal ion complexes in aqueous solution, together with intramolecular adenosine metal ion equilibria. Inorg Chem 1989, 28, 1480–1489.

[60] Buchanan GW, Stothers JB. Diamagnetic metal ion – Nucleoside interactions in solution as studied by 15N nuclear magnetic resonance. Can J Chem 1982, 60, 787–791.

[61] Griesser R, Kampf G, Kapinos LE, Komeda S, Lippert B, Reedijk J, Sigel H. Intrinsic acid–base properties of purine derivatives in aqueous solution and comparison of the acidifying effects of platinum(II) coordinated to N1 or N7: Acidifying effects are reciprocal and the proton "Outruns" Divalent Metal Ions. Inorg Chem 2003, 42, 32–41.

[62] Sigel H. Acid-base properties of purine residues and the effect of metal ions: Quantification of rare nucleobase tautomers. Pure Appl Chem 2004, 76, 1869–1886.

[63] Kotowycz G. A Carbon-13 nuclear magnetic resonance study of binding of copper (II) to pyrimidine nucleotides and nucleosides. Can J Chem 1974, 52, 924–929.

[64] Rosenberg B, Van Camp L, Trosko JE, Mansour VH. Platinum compounds: A new class of potent antitumour agents. Nature 1969, 222, 385–386.

[65] Lim MC, Martin RB. The nature of cis amine Pd(II) and antitumor cis amine Pt(II) complexes in aqueous solutions. J Inorg Nucl Chem 1976, 38, 1911–1914.

[66] Fichtinger-Schepman AMJ, van der Veer JL, den Hartog JHJ, Lohman PM, Reedijk J. Adducts of the antitumor drug cis-diamminedichloroplatinum(II) with DNA: Formation, identification, and quantitation. Biochemistry 1985, 24, 707–712.

[67] Lippert B, Gupta D. Promotion of rare nucleobase tautomers by metal binding. Dalton Trans 2009, 0, 4619–4634.

[68] Arpalahti J, Klika K. Platination of the exocyclic amino group of the adenine nucleobase by Pt$^{II}$ migration. Eur J Inorg Chem 1999, 0, 1199–1201.

[69] Sanz Miguel PJ, Lax P. Lippert B. (Dien) M$^{II}$ (M = Pd, Pt) and (NH$_3$)$_3$Pt$^{II}$ complexes of 1-methylcytosine: Linkage and rotational isomerism, metal-promoted deamination, and pathways to dinuclear species. J. Inorg. Biochem 2006, 100, 980–991.

[70] von der Wijst T, Fonseca Guerra C, Swart M, Bickelhaupt FM, Lippert B. Rare tautomers of 1-methyluracil and 1-methylthymine: Tuning relative stabilities through coordination to Pt$^{II}$ complexes. Chem Eur J 2009, 15, 209–218.

[71] Zamora F, Kunsman M, Sabat M, Lippert B. Metal-stabilized rare tautomers of nucleobases. 6. Imino tautomer of adenine in a mixed-nucleobase complex of mercury (II). Inorg Chem 1997, 36, 1583–1587.

[72] Phillips R, George P. Metal-ATP binding. I. Thermodynamic data for adenosine-silver binding. Biochim Biophys Acta 1968, 162, 73–78.

[73] Lönnberg H, Arpalahti J. Stability constants for some transition metal complexes of adenosine and 9-(β-D-ribofuranosyl)purine in aqueous solution. Inorg Chim Acta 1981, 55, 39–42.

[74] Ghose R, Dey K. Ternary copper(II) complexes of some purine derivatives using nitrilotriacetic acid as a primary ligand. Acta Chim Acad Sci Hung 1981, 108, 9–12.

[75] Scheller KH, Scheller-Krattiger V, Martin RB. Equilibriums in solutions of nucleosides, 5'-nucleotides, and diethylenetriaminepalladium(2+). J Am Chem Soc 1981, 103, 6833–6839.

[76]   Rabindra Reddy P, Malleswara Rao VB. Metal complexes of uridine in solution. J Chem Soc Dalton Trans 1986, 0, 2331–2334.

[77]   Fiskin AM, Beer M. Determination of base sequence in nucleic acids with the electron microscope. IV. Nucleoside complexes with certain metal ions. Biochemistry 1965, 4, 1289–1294.

[78]   Cavaluzzi MJ, Borer PN. Revised UV extinction coefficients for nucleoside-5'-monophosphates and unpaired DNA and RNA. Nucleic Acids Res. 2004, 32, e13.

[79]   Beaven GH, Holiday ER, Johnson EA. Optical properties of nucleic acids and their components. In Chargaff E, Davidson JN, ed. The Nucleic Acids. vol. 1, Academic Press, 1954, 493–553.

[80]   Kypr J, Kejnovska I, Bednarova K, Vorlickova M. Comprehensive Chiroptical Spectroscopy, Vol. 2: Applications in Sterochemical Analysis of Synthetic Compounds, Natural Products and Biomolecules, 1st ed. Berova N, Polavarapu PL, Nakanishi K, Woody RW, eds. Wiley, 2012, 573–584.

[81]   Ward DC, Reich E, Stryer L. Fluorescence studies of nucleotides and polynucleotides. I. Formycin, 2-aminopurine riboside, 2,6-diaminopurine riboside, and their derivatives. J Biol Chem 1969, 244, 1228–1237.

[82]   Evans K, Xu D, Kim Y, Nordlund TM. 2-Aminopurine optical spectra: Solvent, pentose ring, and DNA helix melting dependence. J Fluoresc 1992, 2, 209–216.

[83]   Berry DA, Jung K-Y, Wise DS, Sercel AD, Pearson WH, Mackie H, Randolph JB, Somers RL. Pyrrolo-dC and pyrrolo-C: Fluorescent analogs of cytidine and 2'-deoxycytidine for the study of oligonucleotides. Tetrahedron Lett 2004, 45, 2457–2461.

[84]   Li J, Fang X, Ming X. Visibly emitting thiazolyl-uridine analogues as promising fluorescent probes. J Org Chem 2020, 85, 4602–4610.

[85]   Kuchlyan J, Martinez-Fernandez L, Mori M, Gavvala K, Ciaco S, Boudier C, Richert L, Didier P, Tor Y, Improta R, Mély Y. What makes thienoguanosine an outstanding fluorescent DNA Probe? J Am Chem Soc 2020, 142, 16999–17014.

[86]   Srivatsan SG, Sawant AA. Fluorescent ribonucleoside analogues as probes for investigating RNA structure and function. Pure Appl. Chem 2011, 83, 213–232.

[87]   Ts'o POP, Melvin IS, Olson AC. Interaction and association of bases and nucleosides in aqueous solutions. J Am Chem Soc 1963, 85, 1289–1296.

[88]   Cheng DM, Kan LS, Ts'o POP, Giessner-Prettre C, Pullman B. Proton and carbon-13 nuclear magnetic resonance studies on purine. J Am Chem Soc 1980, 102, 525–534.

[89]   Broom AD, Schweizer MO, Ts'o POP. Interaction and association of bases and nucleosides in aqueous solutions. V. Studies of the association of purine nucleosides by vapor pressure osmometry and by proton magnetic resonance. J Am Chem Soc 1967, 89, 3612–3622.

[90]   Chan SI, Schweizer MP, Ts'o POP, Helmkamp GK. Interaction and association of bases and nucleosides in aqueous solutions. III. A nuclear magnetic resonance study of the self-association of purine and 6-methylpurine. J Am Chem Soc 1964, 86, 4182–4188.

[91]   Ts'o POP, Chan SI. Interaction and association of bases and nucleosides in aqueous solutions. II. Association of 6-methylpurine and 5-bromouridine and treatment of multiple equilibria. J Am Chem Soc 1964, 86, 4176–4181.

[92]   Heyn MP, Nicola CU, Shwarz G. Kinetics of the base-stacking reaction of N6-dimethyladenosine. An ultrasonic absorption and dispersion study. J Phys Chem 1977, 81, 1611–1617.

[93]   Marenchic MG. Sturtevant JM. Calorimetric investigation of the association of various purine bases in aqueous media. J Phys Chem 1973, 77, 544–548.

[94]   Nemethy G, Scheraga HA. The structure of water and hydrophobic bonding in proteins. III. The thermodynamic properties of hydrophobic bonds in proteins. J Phys Chem 1962, 66, 1773–1789.

[95]   Lawaczeck R, Wagner KG. Stacking specificity and polarization. Comparative synopsis of affinity data. Biopolymers 1984, 13, 2003–2014.

# 2 Nucleosides: synthesis, transformation reactions and hydrolytic stability

## 2.1 Synthesis of nucleosides and their congeners

### 2.1.1 Synthesis of the N-glycosidic bond

The conventional synthesis of nucleosides is based on nucleophilic attack of the base moiety on the anomeric carbon of the glycosyl moiety with concomitant displacement of acylated anomeric hydroxyl function. The reaction is an $S_N1$ rather than $S_N2$-type substitution, and hence, the attack from both the α- and β-faces is possible, giving an anomeric mixture. With ribo sugars, acylated 2′-OH on the α-face, however, converts the reaction stereoselective. The carbonyl oxygen of the acylated 2′-OH steers the entering nucleobase to β-position. The so-called Vorbrüggen method, that is, a Lewis acid-catalyzed reaction of a silylated nucleobase with a fully esterified ribofuranose in a polar aprotic solvent is usually applied [1]. The course of reaction when promoted by trimethylsilyl triflate is outlined in Figure 2.1. Silylation of the 1-*O*-acetyl group of 1-*O*-acetyl-2,3,5-tri-*O*-benzoyl-β-D-ribofuranose (A in Figure 2.1) results in departure of the silylated acetyloxy group with concomitant formation of a cyclic oxocarbenium ion. The latter is immediately stabilized by an attack of the 2′-*O*-carbonyl oxygen on the anomeric carbon from the α-face. The dioxolane cation formed steers the silylated base to attack on the anomeric carbon from the β-face, affording the desired stereochemistry. The remaining silyl and acyl protections are then easily removed by methanolic sodium methoxide. One should, however, note that the formation of this thermodynamically controlled product may be accompanied by several kinetically controlled products obtained by the attack of some other heteroatom of the attacking base [2]. Upon elongated treatment, these are converted to the thermodynamic product. Besides peracylated glycofuranoses, glycosyl trifluoroacetimidate (B in Figure 2.1) has been used as a glycosyl donor [3].

An older but still viable method is a reaction of peracylated glycosyl bromide or chloride with a heavy metal salt of the nucleobase having the primary amino groups protected. The reaction is carried out in an inert solvent, typically xylene, at a refluxing temperature. When Ag(I) salt is used, the reaction is known as Fischer-Helferich reaction [4], on using Hg(II) salt as Davoll–Lowy reaction [5]. With purines, the *N*9-regioselectivity is good and β-stereoselectivity moderate. Pyrimidines give first *O*-glycosides, which are rearranged to *N*-glycosides, predominantly to *N*1-glycosides. Instability of glycosyl halides and low solubility of the metal salts of nucleobases diminish the applicability of these approaches.

Weakly nucleophilic bases, such as 2,6-dichloropurine, have been glycosylated by a simple solvent-free fusion with moderate success. Ribofuranose peracetate, an appropriately protected base and *p*-toluenesulfonic acid are melted to a homogeneous

https://doi.org/10.1515/9783111325637-002

mixture at elevated temperature (up to 150 °C) [6]. The C1'-acetoxy group departs as acetic acid that is continuously removed under reduced pressure, and the protected base attacks on the developing oxocarbenium ion. The harsh conditions, however, severely limit the general applicability of this method.

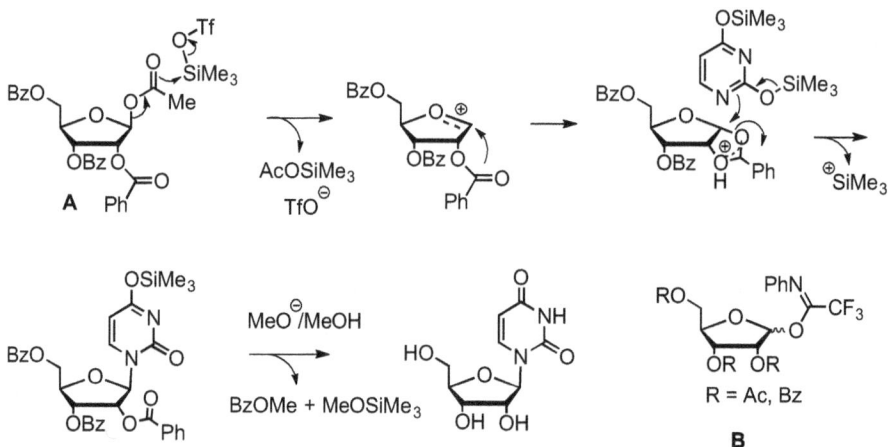

**Figure 2.1:** Mechanism of Vorbrüggen ribonucleoside synthesis using 1-O-acetyl-2,3,5-tri-O-benzoyl-β-D-ribofuranose (A) as a starting material. Alternatively, 2,3,5-tri-O-benzoylribofuranose trifluoroacetimidate (B) may be used as the glycosyl donor.

Recently a direct one-pot synthesis of ribonucleosides by a modified Mitsunobu reaction has been developed [7]. Treatment of 5'-O-protected D-ribose in MeCN with tributylphosphine and 1,1'-(azodicarbonyl)dipiperidine results in the formation of a 1,2-anhydrosugar intermediate, and subsequent addition of the nucleobase then gives the desired nucleoside with 100% stereoselectivity (Figure 2.2). Overall yields up to 70% have been reported.

2'-Deoxyribonucleosides and their congeners that miss a participating substituent at C2' are inevitably obtained as an anomeric mixture. The silylated base approach described in Figure 2.1 may naturally be applied to the preparation of 2'-deoxynucleosides, as well, but the steroselectivity is lost. Proper choice of sugar protecting groups and solvent may increase the proportion of the desired β-anomer in the product mixture. Conventional approach involves a nucleophilic attack of the sodium salt of nucleobase on a toluoyl protected α-glycosyl chloride in MeCN (Figure 2.3A) [8]. Up to 80% yields of the desired β-anomers have been obtained. As with ribonucleosides, Mitsunobu reaction has been introduced as an alternative for the preparation of 2'-deoxyribonucleosides [9].

Another approach for the preparation of nucleosides is an enzymatic, or in some cases chemical, transglycosylation. A desired sugar modification is often most conveniently introduced into thymidine or uridine since base moiety protection is avoided.

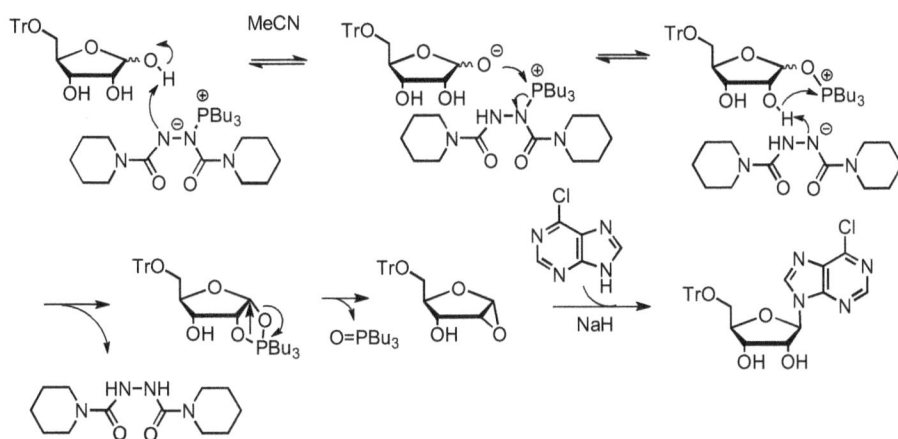

**Figure 2.2:** Mechanism of the synthesis of ribonucleosides by Mitsunobu reaction.

The modified sugar moiety is then transferred to another nucleobase by enzyme-catalyzed transglycosylation depicted in Figure 2.3. [10]. Enzymes used for this purpose are nucleoside phosphorylases (NP) or *N*-deoxyribosyltransferases, among which the *Escherichia coli* uridine (UP; EC 2.4.2.3), thymidine (TP; EC 2.4.2.4) and purine nucleoside (PNP; EC 2.4.2.1) phosphorylases are most frequently used. Pyrimidine nucleosides are commonly used as sugar donors and purine bases as acceptors [10, 11]. Recently, 7-methyl-2′-deoxyguanosine hydroiodide has been successfully used as a glycosyl donor for the synthesis of numerous base-modified 2′-deoxyribonucleosides [12].

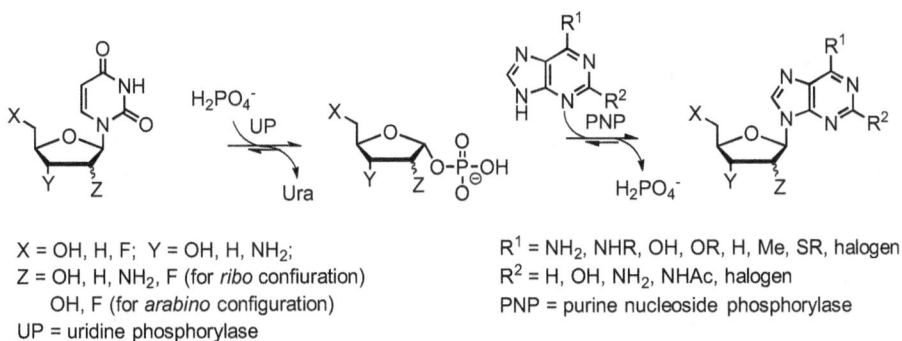

X = OH, H, F; Y = OH, H, NH$_2$;
Z = OH, H, NH$_2$, F (for *ribo* confiuration)
    OH, F (for *arabino* configuration)
UP = uridine phosphorylase

R$^1$ = NH$_2$, NHR, OH, OR, H, Me, SR, halogen
R$^2$ = H, OH, NH$_2$, NHAc, halogen
PNP = purine nucleoside phosphorylase

**Figure 2.3:** Enzymatic transglycosylation using sugar-modified uridines as starting materials [11].

Chemical transglycosylation takes place under the same conditions as the Vorbrüggen reaction, that is, with silylated bases in polar aprotic solvent using trimethylsilyl triflate or SnCl$_4$ as a catalyst [2]. A pyrimidine nucleoside usually serves as the glycosyl donor and a silylated purine base as the acceptor, although reaction from purine nucleoside to pyrimidine base still is feasible [13]. Chemical transglycosylation can also be used for the transfer of the base moiety. The donor is a purine nucleoside and the acceptor a peracylated sugar or its acyclic congener [2]. The regio- and stereoselectivity of chemical transglycosylation reactions vary from modest to moderate.

### 2.1.2 Synthesis of cyclonucleosides

Cyclonucleosides are obtained either by a nucleophilic attack of a sugar hydroxyl function on a base moiety carbon atom or by an attack of a base moiety heteroatom on a carbon atom of the sugar moiety [14]. In case $O^{5'}$ serves as a nucleophile or C5' as an electrophile, cyclic 2',3'-$O$-isopropylidine protection enhances the reaction. Figure 2.4 shows examples of synthetic reactions of purine cyclonucleosides. Either N3 displaces the 5'-tosyloxy group (Figure 2.4A) [15, 16] or 8-bromopurine nucleoside is in situ converted to an 8-oxo [17, 18], 8-amino [19] or 8-thio [20, 21] analog that then undergoes cyclization by the displacement of 2'-sulfonyloxy group (Figure 2.4B, C and D, respectively). 5'-Oxo-bridged cyclonucleoside is obtained by the attack of 5'-oxyanion on C8 of 8-bromopurine (Figure 2.4E) [14, 22]. Though all the examples in Figure 2.4 are depicted for adenosine, most of the reactions can also been carried out with guanosine [16, 18, 21].

Nucleophilic displacement of a mesylate or tosylate group on the sugar moiety by the $O^2$-atom of a pyrimidine nucleoside also gives a cyclonucleoside. Uridine forms readily $O^2$,2'- and $O^2$,5'-cyclonucleosides and thymidine $O^2$,3'- and $O^2$,5'-cyclonucleosides (see Figure 2.5). These are useful synthetic intermediates since the anhydro bridge may be opened by an attack of numerous nucleophiles from the α-face [23]. The cyclization is alternatively achieved by an intramolecular Mitsunobu reaction consisting of transient activation with diethyl azodicarboxylate and triphenylphosphine, Ph$_3$P.

When a 2',3'-$O$-isopropylidene-protected pyrimidine nucleoside is treated with excess of $N$-bromosuccinimide in DMF, a cyclic 5,5-dibromo-5,6-dihydro-6,5'-$O$-anhydro intermediate is obtained, which by MeONa treatment in MeOH yields the 5-bromo-6,5'-$O$-anhydro nucleoside (Figure 2.5A) [24]. 6,2'-$O$-Anhydrouridine has been obtained from 1-(β-D-arabinofuranosyl)-5-bromouracil by reluxing in methanolic NaOMe (Figure 2.5B) [25]. Cyclonucleoside formation is easier when the attacking nucleophile is sulfur or nitrogen. Accordingly, 2',3'-$O$-isopropylidine-5'-deoxy-5'-acetylthio- and 5'-deoxy-5'-amino-5-bromouridine are converted to 6,5'-$S$- and 6,5'-$N$-anhydrouridine by refluxing in methanolic sodium methoxide or pyridine, respectively (Figure 2.5C and D) [26, 27]. The attack of 5'-hydroxyl group, in turn, gives the 5'$O$-C6 cyclo nucleoside only as an unstable intermediate in low yield.

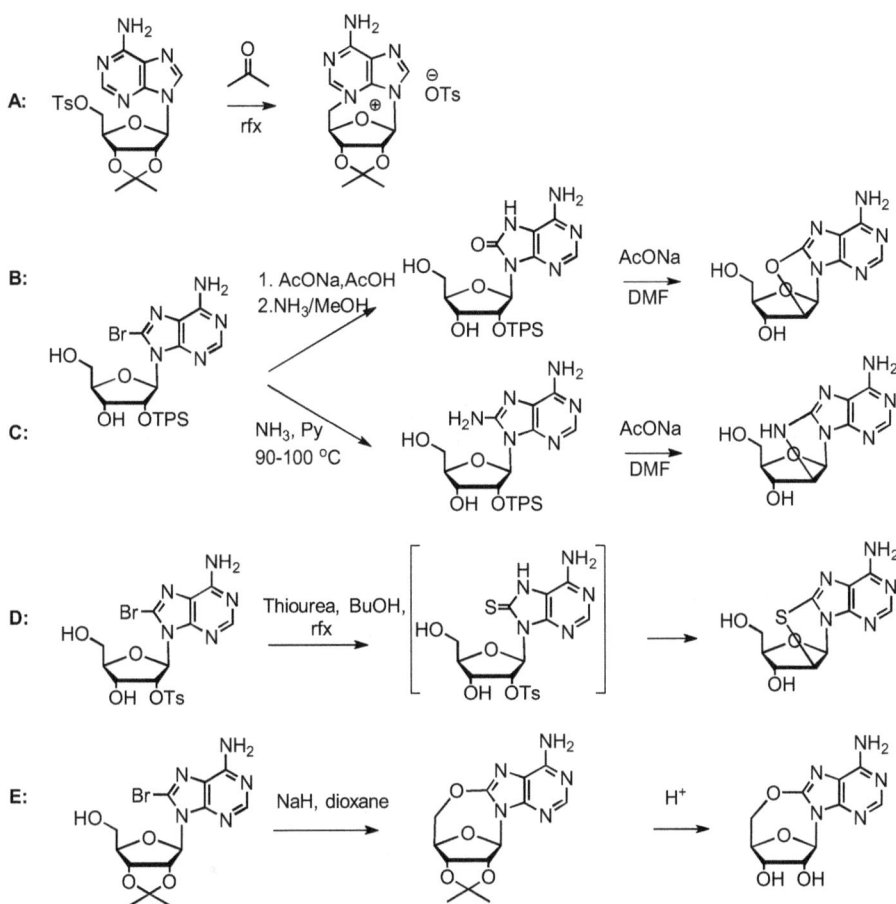

**Figure 2.4:** Illustrative examples of synthetic reactions of purine cyclonucleosides.

## 2.1.3 Synthesis of C-nucleosides

C-Nucleosides have been a subject of continuous interest of medicinal chemists, owing to their obvious potential in chemotherapy. The classical approach is to attach a reactive functional group to the anomeric carbon of a prefabricated sugar moiety and then build the base moiety by stepwise derivatization of this functionality [28]. Synthesis of pyrimidine and pyrazole C-nucleosides has been outlined as an illustrative example in Figure 2.6 [29]. The first step, conversion of glycofuranose peracetete to glycosyl nitrile with the aid of trimethylsilylcyanide, is one of the most frequently utilized initial reactions of C-nucleoside synthesis. Other popular alternatives include Wittig reaction of reducing sugars, that is, sugars that contain an unsubstituted hydroxyl function at the anomeric carbon, and attachment of an alkene or alkyne group

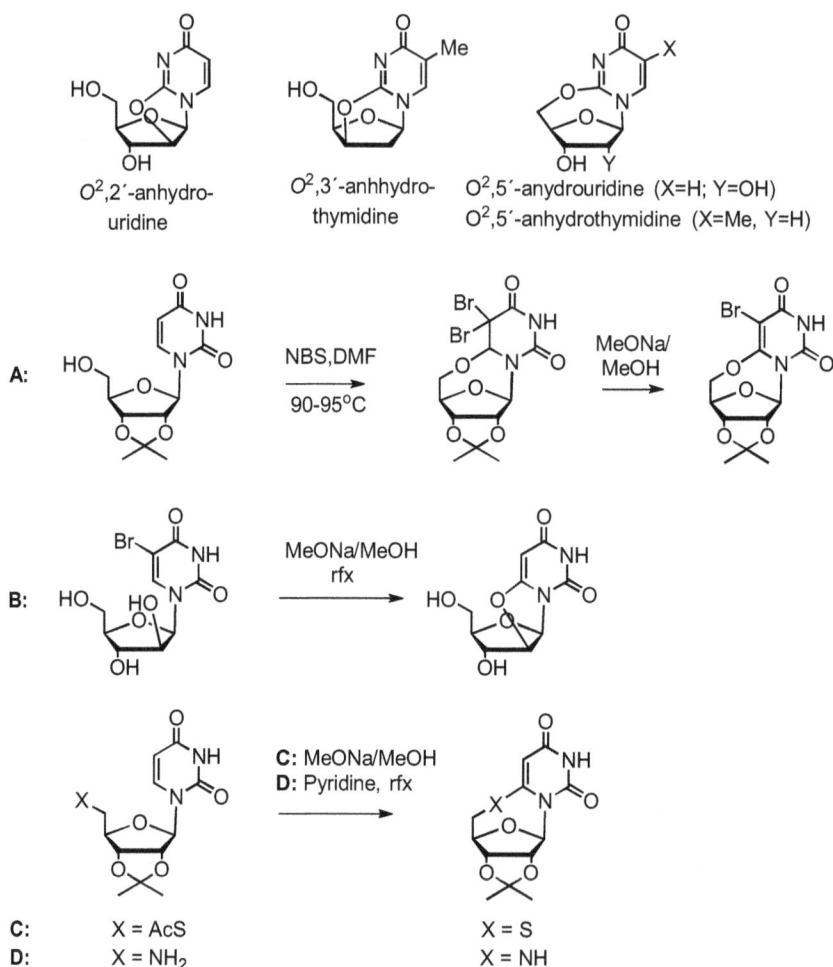

O$^2$,2′-anhydro-
uridine

O$^2$,3′-anhhydro-
thymidine

O$^2$,5′-anydrouridine (X=H; Y=OH)
O$^2$,5′-anhydrothymidine (X=Me, Y=H)

A:

NBS,DMF
⎯⎯⎯⎯→
90-95°C

MeONa/
MeOH
⎯⎯⎯→

B:

MeONa/MeOH
rfx
⎯⎯⎯⎯→

C: MeONa/MeOH
D: Pyridine, rfx
⎯⎯⎯⎯→

C:    X = AcS          X = S
D:    X = NH$_2$          X = NH

**Figure 2.5:** Illustrative examples of synthetic reactions of pyrimidine cyclonucleosides.

to the anomeric position by a metal ion assisted coupling reaction, followed by a dipo-lar cycloaddition [28].

Build-up of the sugar moiety on a prefabricated aglycon offers a complementary ap-proach for the preparation of C-nucleosides. Synthesis of 2,6,7-trichloro-3-(β-D-ribofurano-syl)imidazo[1,2-a]pyridine is one of the early applications of this kind of an approach (Figure 2.7) [30]. C-Formyl-substituted aglycon is subjected to Wittig reaction with (R)-[2-(2,2-dimethyl-1,3-dioxolan-4-yl)ethyl]triphenylphosphonium iodide. Hydrolysis of the diox-olane ring, iodocyclization of the resulting alkenol, elimination of HI and hydroxylation then gives the desired product.

The most straightforward approach for the preparation of C-nucleosides, how-ever, is direct coupling of prefabricated glycon and aglycon moieties [28]. Several al-

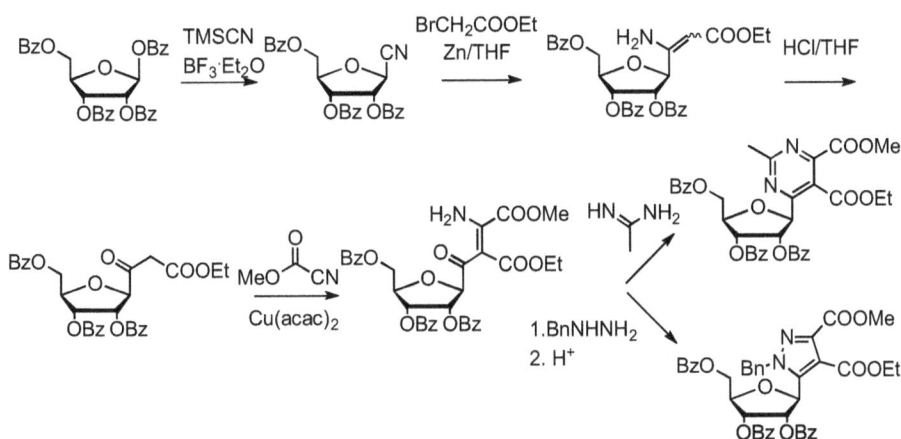

**Figure 2.6:** Synthesis of pyrimidine and pyrazole *C*-nucleosides by stepwise build-up of the base moiety on a prefabricated sugar moiety [29].

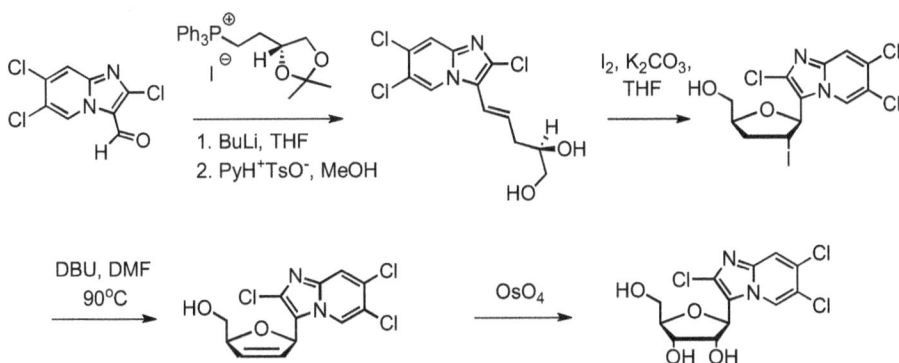

**Figure 2.7:** An example of stepwise build-up of the sugar moiety on a prefabricated base moiety [30].

ternative coupling methods can be used, among which Pd-catalyzed Heck coupling of an aglycon halide to a 1,2-glycal is a common approach (Figure 2.8A) [31]. The stereochemistry largely depends on whether O3′ bears a protecting group. Large protecting groups force the aglycon to approach from the β-face, resulting in good stereoselectivity, especially in the presence of AsPh₃ substituent. Unsubstituted 3′-OH, in turn, steers the aglycon to α-position. The regioselectivity is excellent. The other frequently used approaches include Lewis acid-catalyzed electrophilic aromatic substitution of the aglycon with glycon (Figure 2.8B) [32], and nucleophilic attack of aglycon on a furanosyl halide (Figure 2.8C) [33], 1,2-anhydrofuranose (Figure 2.8D) [34] or furanolactone (Figure 2.8E) [35]. The advantage of electrophilic aromatic substitution of aglycon (Reaction B) is its simplicity, but the regio- and stereoselectivity are usually poor and difficult to predict. The situation is similar with coupling of organometallic aglycon and

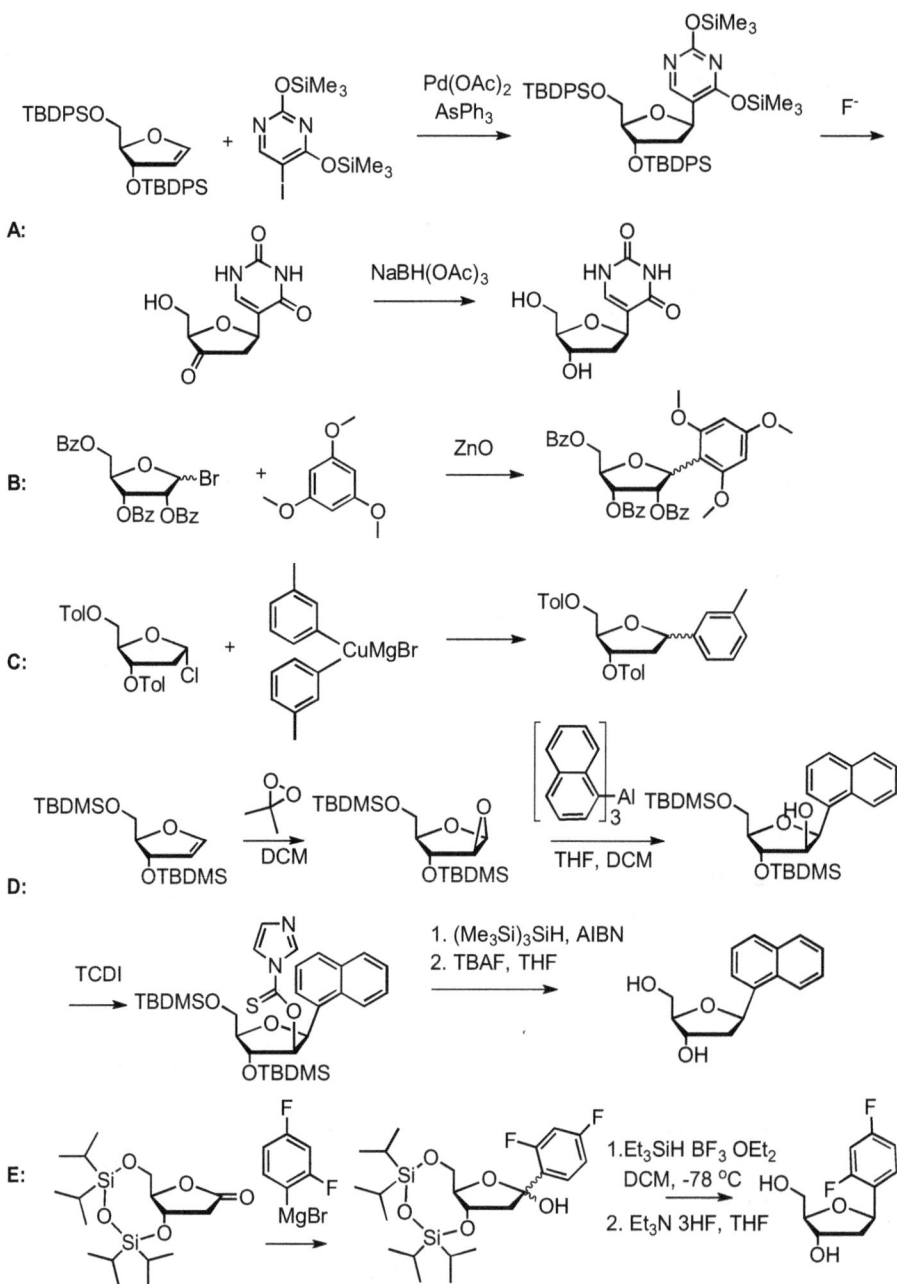

**Figure 2.8:** Alternative approaches for the synthesis of C-nucleosides from prefabricated base and sugar moieties.

glycosyl halide (Reaction C). The reaction is simple but yields an anomeric mixture where the α-anomer even predominates. To enrich the mixture with respect to the β-anomer, an acid-catalyzed anomerization has to be carried out. Best overall efficiency has been obtained on using an arene magnesium-cuprate as a catalyst. Interestingly, coupling with 1,2-anhydroarabinofuranose has been shown to exhibit excellent β-selectivity (reaction D). Attack of a metal salt of aglycon on a furanolactone (reaction E) is, apart Heck coupling, the most frequently applied approach for the synthesis of C-nucleosides. Various methods have been used for the reduction of the hemiketal intermediate, giving the β-anomer a good yield [28].

## 2.1.4 Synthesis of carbocyclic nucleosides

As discussed above, the formation of the N-glycosidic bond of nucleosides is an $S_N$-1-type nucleophilic substitution since electron donation from the neighboring ring-oxygen stabilizes the carbocation center developed at C1′ upon departure of the leaving group. With carbocyclic nucleosides, this kind of carbocation stabilization is missing and substitution proceeds by an $S_N2$ rather than $S_N1$ mechanism, and hence a more powerful nucleophile is required in the synthesis of carbocyclic nucleosides than in the synthesis of normal nucleosides. Instead of using nucleobase salts or silylated nucleobases as nucleophiles [36, 37], Mitsunobu reaction is often applied (Figure 2.9A) [38]. Since amino groups usually make heteroaromatic bases sparingly soluble in organic solvents, their chloro-substituted analogs are generally used, and conversion to amino compounds takes place after the synthesis of the C–N bond.

Another extensively used approach is Pd-catalyzed coupling of nucleobase to a cyclopentene bearing an activating acyloxy, alkoxycarbonyloxy or N,N-ditosylamino group in allylic position (Figure 2.9B) [39]. Alternatively, a good leaving group in allylic position of an α-acivated cyclopentene is displaced (Figure 2.9C) [40]. The regio- and stereoselectivity depends on the nature of the catalyst and nucleophile employed as well as on the presence of steric obstacles on the cyclopentene ring. The third alternative is construction of the nucleobase in a stepwise manner on the amino function of an appropriate cyclopentylamine (Figure 2.9D) [41]. Recently, a stereoselective multistep synthesis for 1′-substituted carbocyclic nucleosides has been developed (Figure 2.9E) [42].

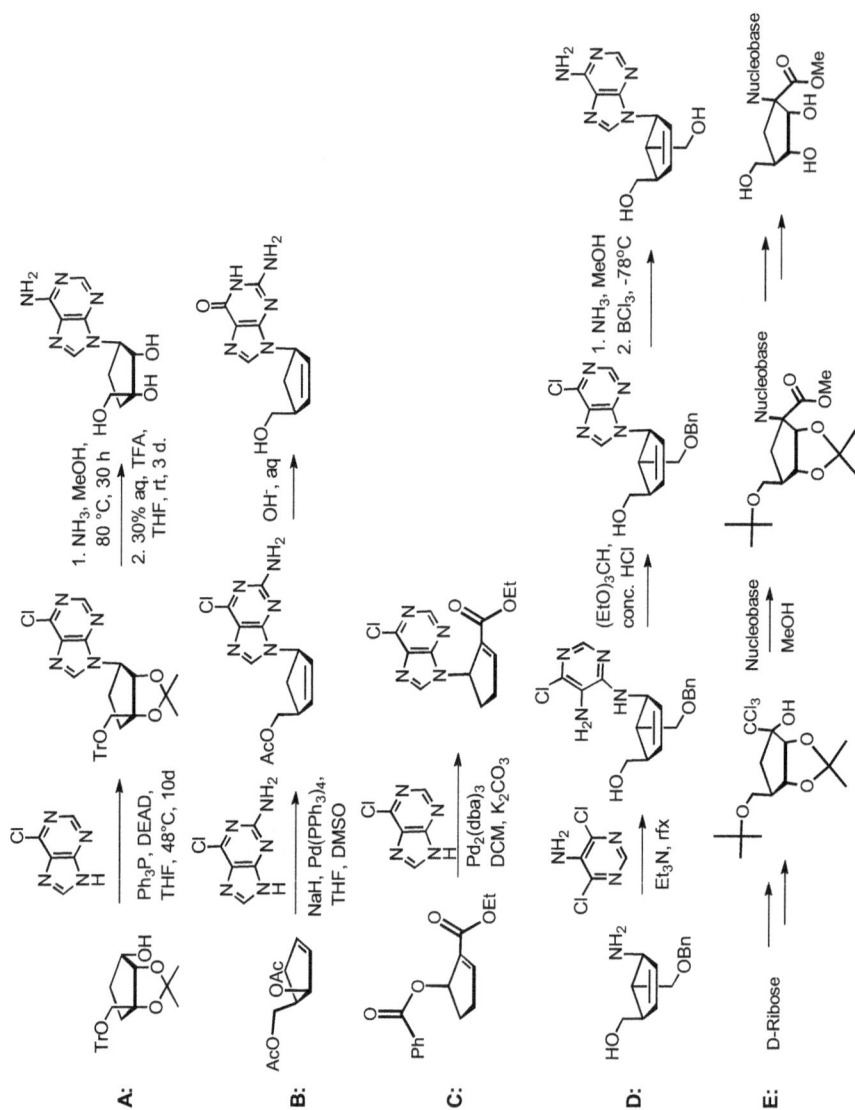

**Figure 2.9:** Alternative approaches for the synthesis of carbocyclic nucleosides.

## 2.2 Transformation reactions of nucleosides

### 2.2.1 Transformation of purine nucleosides

In some cases, it may be advantageous to prepare a structurally modified nucleoside by transformation from a commercially available precursor nucleoside, instead of synthesizing the *N*-glycosidic bond. Retention of the original regio- and stereoisomerism is a clear advantage of this approach. 6-Chloropurine nucleosides, for example, are extensively used precursors. The ribo- and 2′-deoxyribo-nucleoside of 6-chloropurine is obtained by the chlorination of 2′,3′,5′-tri-*O*-acetylinosine [43] or 3′,5′-di-*O*-trifluoroacetyl-2′-deoxyinosine [44], respectively, with thionyl chloride in DMF. Both nucleosides are also commercially available. The 6-chloro substituent is susceptible to nucleophilic displacement giving a variety of purine nucleosides as outlined in Figure 2.10 [45, 46].

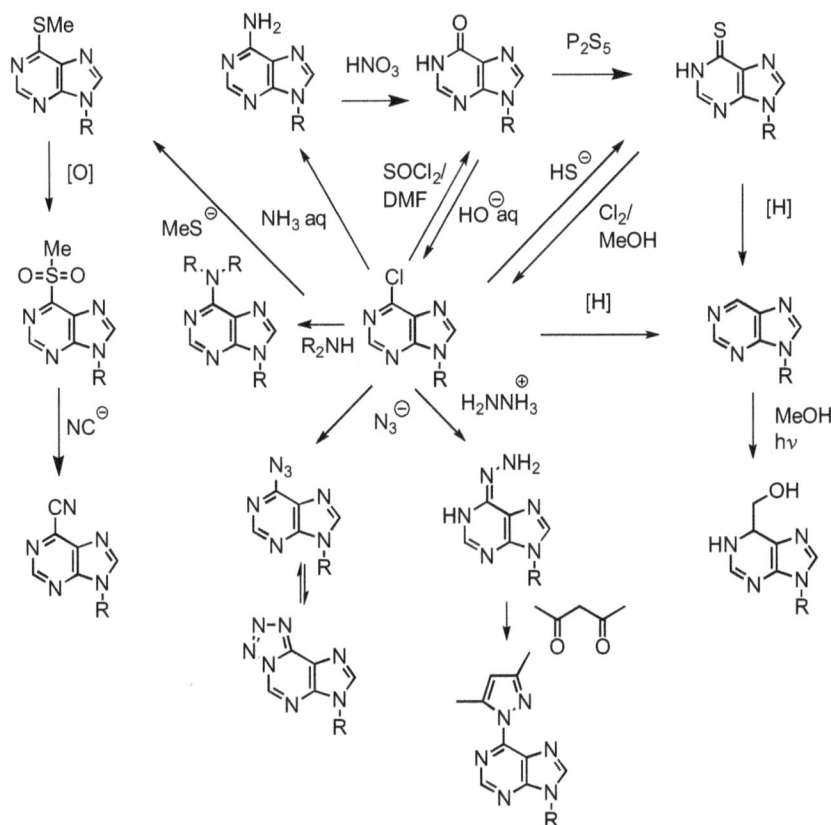

**Figure 2.10:** Transformation of 6-chloropurine nucleoside to other C6-substituted nucleosides.

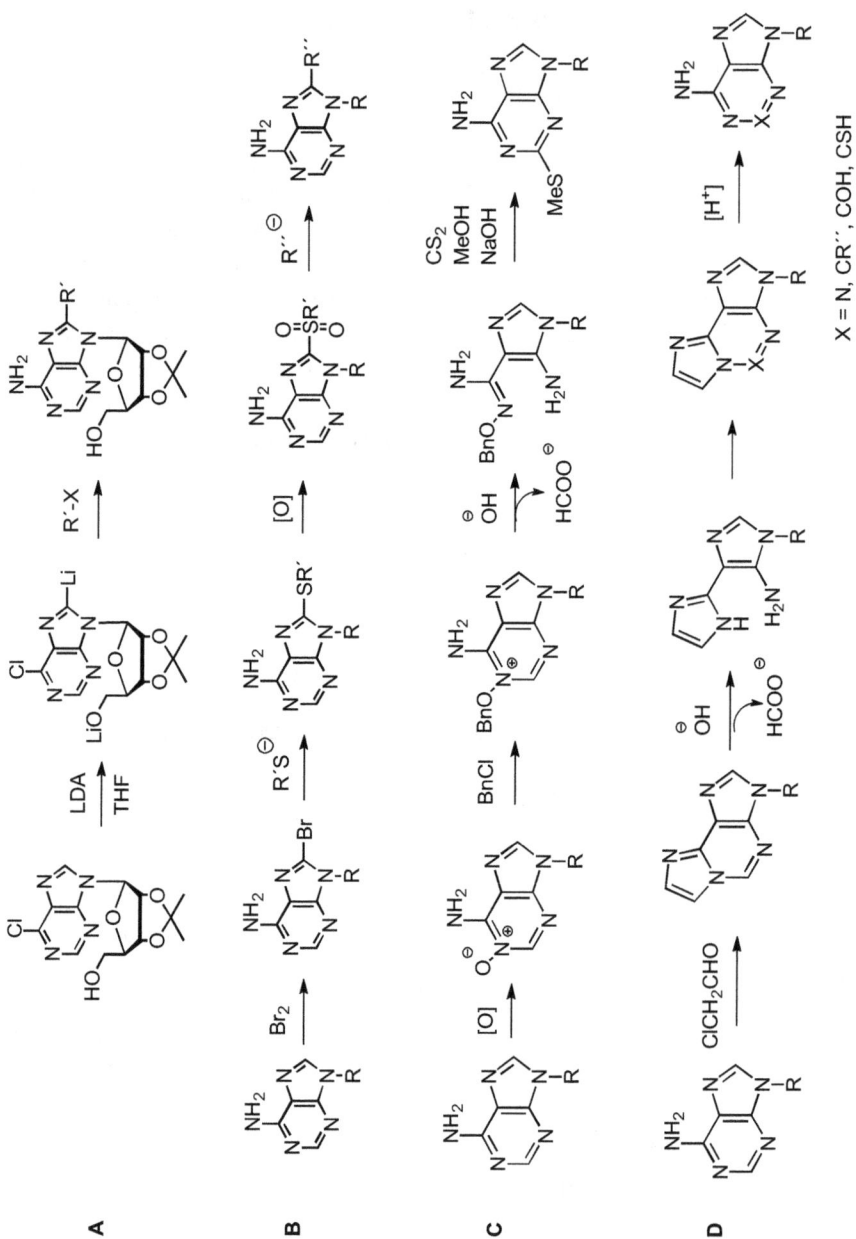

**Figure 2.11:** C8- and C2-substitution of purine nucleosides.

6-Chloropurine ribonucleoside additionally withstands alkylation of C8. 2′,3′-O-Isopro-prylidine-protected nucleoside undergoes lithiation of C8 when treated with lithium dii-sopropylamide in THF at low temperature ($T < -70$ °C) [47]. Subsequent treatment with electrophiles, including alkyl halides and aldehydes, then gives the alkylated product (Figure 2.11A). The 6-chloro substituent may finally be displaced as indicated above. An alternative approach to C8 substitution is initial bromination of C8 of the purine nucleo-side in aqueous solution, followed by nucleophilic displacement of the 8-bromo substit-uent, first by alkyl (or aryl) sulfide ion to 8-alkylthio nucleoside and then by oxidation to 8-alkylsulfonyl nucleoside and subsequent carbanion displacement to 8-alkyl nucleo-side (Figure 2.11B) [48].

Substitution of the C2 site of adenosine is less straightforward requiring interme-diary opening of the pyrimidine ring by cleavage of the N1–C2 bond. This can be achieved by the oxidation of adenosine to $N$1-oxide followed by alkylation to $N$1-benzyloxy derivative (Figure 2.11C) [49]. This greatly accelerates the attack of hydrox-ide ion on C2 of the electron deficient pyrimidine ring resulting in cleavage of the N1–C2 bond and departure of the N3-formyl group as a formate ion. Treatment with an electrophile, such as $CS_2$, at elevated temperature completes the reaction by reforma-tion of the pyrimidine ring and release of the auxiliary benzyloxy ion. An alternative method to induce the cleavage of N1–C2 bond is treatment with haloacetaldehyde that converts adenosine to $1,N^6$-ethenoadenosine (Figure 2.11D). The latter is hydrolyzed and deformylated under alkaline conditions to a 5-amino-4-(imidazole-2-yl)imidazole nucleoside [50]. Treatment by various electrophiles then results in recyclization to 2-substituted $1,N^6$-ethenoadenosine and the etheno bridge is finally removed by acid-catalyzed hydrolysis.

As discussed in Section 1.2, the primary amino groups of nucleosides are neither basic nor nucleophilic. The ring nitrogen atoms are alkylated more readily than the exocyclic amino groups. The primary $N^6H_2$ of adenosine undergoes alkylation, but the reaction takes place via Dimroth rearrangement. Adenosine is initially alkylated at N1. This makes the neighboring C2 susceptible to attack of hydroxide ion, which leads to cleavage of the N1–C2 bond, rotation around the C5–C6 bond and formation of a bond between C2 and the original $N^6$-atom (Figure 2.12). The alkylated nitrogen atom,

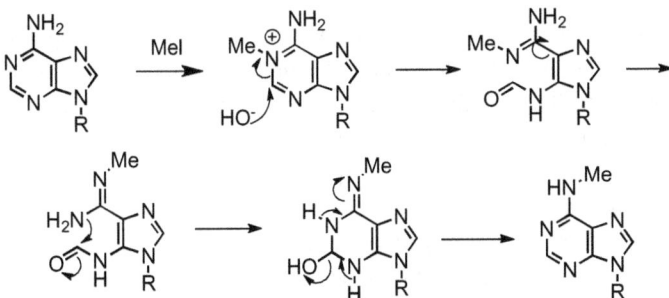

**Figure 2.12:** Alkylation of adenosine via Dimroth rearrangement.

hence, takes the exocyclic position. When $N^6$-alkylation under milder conditions is required, 4-nitrothiophenol can be used as a C2-attacking nucleophile that opens the pyrimidine ring in a reversible manner allowing Dimroth rearrangement [51]. With guanosine, the preferred alkylation site is N7.

Besides N1 methylation, N1,$N^6$-annulation enhances the opening of the pyrimidine ring. Treatment with an appropriate electron withdrawing annulating agent and subsequent ammonolytic deformylation of N3 yields substituted imidazole nucleosides, examples of which are given in Figure 2.13 [52]. Alternatively, pyrimidine ring of purine nucleosides may be opened by treating N1-alkylated inosine with alkalis. Ring-opened nucleosides may in some cases undergo re-closure by the formation of a seven-membered diazepine ring [53].

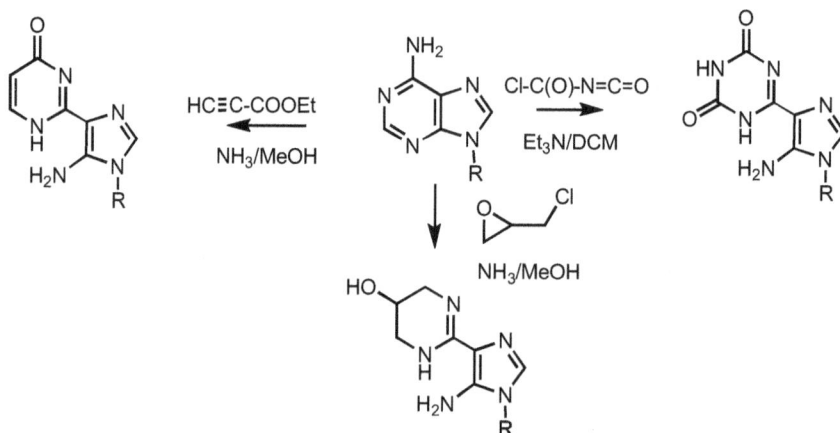

**Figure 2.13:** Imidazole nucleosides obtained by ammonolytic deformylation of pyrimidine ring-opened 1,$N^6$-annulated purine nucleosides.

### 2.2.2 Transformation of pyrimidine nucleosides

Reactions of pyrimidine nucleosides fall in three categories: nucleophilic, electrophilic and radical substitutions [54]. As regards nucleophilic substitutions, a special feature of pyrimidine nucleosides is susceptibility of the C6 atom to a nucleophilic attack. Bromination in hydroxylic solvents serves as an example (Figure 2.14A). Addition of a nucleophilic solvent molecule to the 5,6-double bond is followed by binding of electrophilic bromonium ion to C5. Departure of the solvent molecule then returns the aromatic nature of the pyrimidine base. Cytidine reacts less readily than uridine and is subject to concurrent deamination to uridine and formation of nonchromophoric products by opening of the pyrimidine ring. Attack of bisulfite ion on C6 of cytidine results in deamination to uridine [55] or transamination to an $N^4$-alkylated cytidine (Figure 2.14B) [56].

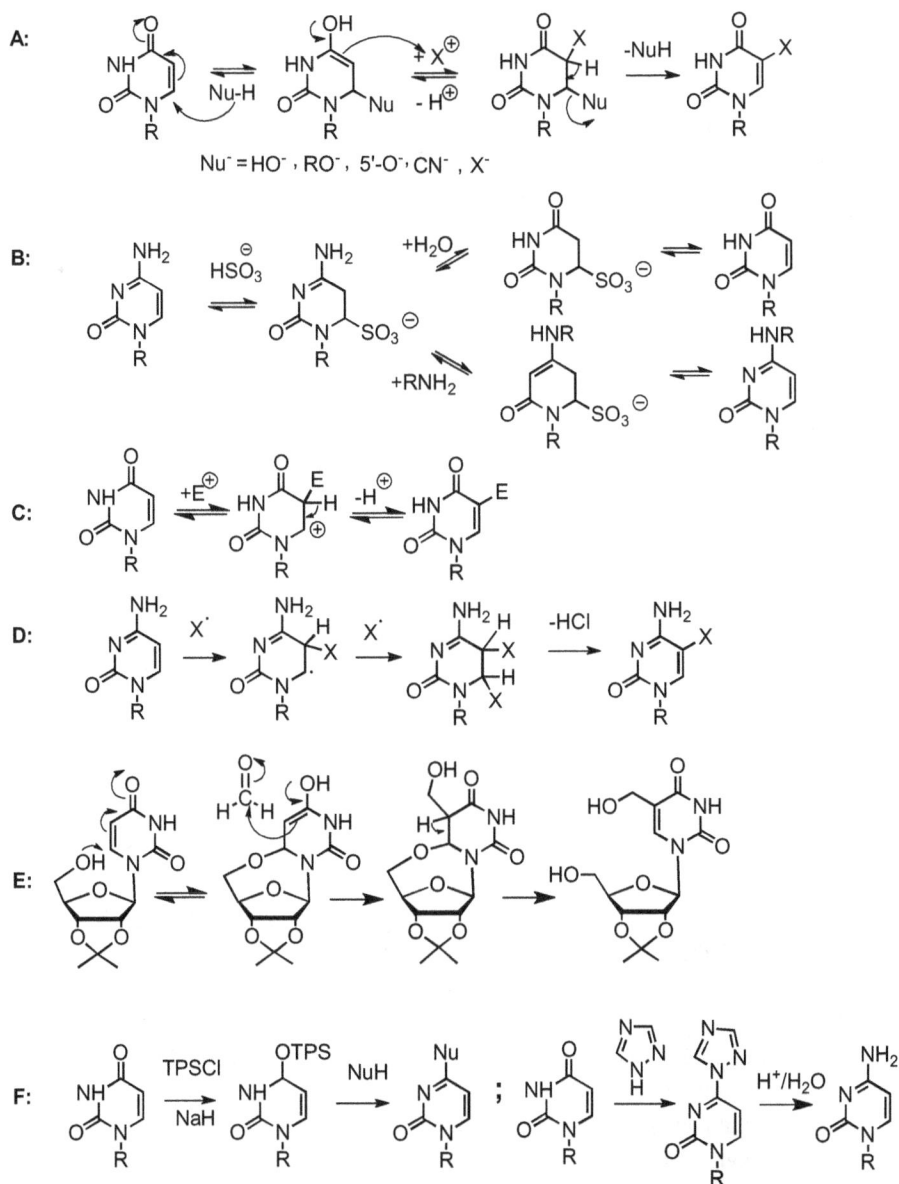

**Figure 2.14:** Transformation reactions of pyrimidine nucleosides.

Electrophilic C5-substitutions of pyrimidines proceed through the addition of electrophile to the 5,6-double bond and subsequent elimination of proton from C5 (Figure 2.14C). Halogenation most likely proceeds by an analogous electrophilic mechanism [54] and photosensitized halogenation by an analogous radical mechanism (Figure 2.14D). Interestingly, the 5'-hydroxy function participates in electrophilic C5-alkylation, especially when the sugar moiety is 2′,3′-*O*-isopropylidene-protected. Attack of the 5′-hydroxy group on C6, followed by electrophilic attack of formaldehyde on C5, yields 5-hydroxymethylpyrimidine (Figure 2.14E) [57]. Since 5-methyl, 5-hydroxymethyl, 5-formyl and 5-carboxy-2′-deoxycytidines have recently received increasing interest as epigenetic modifications, that is, as modifications related to heritable changes in gene function [58], their syntheses have also been in the focus of reinvestigations aimed at finding most convenient and cost-effective transformation reactions [59, 60].

4-Substituted pyrimidine nucleosides are conveniently obtained by converting sugar moiety-protected uridine or thymidine first to its 4-*O*-(2,4,6-triisopropylphenylsulfonyl) derivative and displacing then the sulfonyl group by an attack of oxygen, nitrogen or carbon nucleophiles on C4 (Figure 2.14F) [61]. 1,2,4-Triazole may be used for the same purpose [62]. The 4-(1,2,4-triazole-1-yl) intermediate can additionally be hydrolyzed to cytidine. This is a highly useful reaction since many transformations are easier to carry out with uridine than with cytidine.

### 2.2.3 Metal-ion-promoted transformations

A classical example of metal-ion-promoted reactions is mercury(II)chloride-catalyzed C5-alkylation of pyrimidine nucleosides. Both uridine and cytidine react with mercury(II) chloride giving C5-chloromercury-pyrimidines, which may be further converted to 5-alkylated nucleosides (Figure 2.15A) [63]. Uridine has been shown to undergo $Pd^{2+}$-catalyzed C5 alkenylation on using *tert*-butyl peracetate as an oxidant (Figure 2.15B) [64]. The latter reaction is an example of palladium-catalyzed cross-coupling reactions that have more recently gained increasing popularity in synthetic nucleoside chemistry [65]. 8-Halopurines and 5-halopyrimidines are conventionally used as starting materials to which alkynyl, aryl or alkenyl groups are coupled by Sonogashira [66], Suzuki [67] and Heck [68] reactions, respectively (Figure 2.15C). Sonogashira coupling with $Pd(PPh_3)/CuI$ catalyst in DMF allows alkynylation of unprotected nucleosides. 8-Bromo-2′-deoxyadenosine, 5-iodo-2′-deoxyuridine and 5-iodo-2′ deoxycytidine, for example, have been alkynylated with acidic, basic and hydrophobic terminal alkynes [69]. This reaction is very useful since the alkynyl derivatives can subsequently be reduced to their alkenyl and alkyl counterparts on Pd/C [70]. Suzuki coupling, in turn, offers a method for the arylation of unprotected nucleosides with arylboronic acids or their esters. Syntheses of 5-(1-pyrenyl)- [71] and 8-(1-pyrenyl)-2′-deoxyguanosine [72] by $Pd(dppf)Cl_2$ and $Pd(PPh_3)_2$ catalysis, respectively, in a mixture of THF, MeOH and water serve as illustrative examples. Several 8-aryladenosines have been obtained by $Pd(PPh_3)_2$ catalysis in aqueous DME [73]. Heck coupling has been less frequently used for the derivatization of unprotected nucleosides. The examples include $Pd(OAc)_2$- [74] and $K_2PdCl_4$-promoted [75] alkenylation of 5-iodo-2-deoxyuridine in water and $Pd(OAc)$ $2/P(o-tolyl)_3$-promoted alkenylation of 2-iodoadenosine in MeCN [76].

Application of cross-coupling reactions is not limited to halogen-substituted nucleosides but unmodified nucleosides may be derivatized with arylhalides [77]. Unprotected adenine nucleosides, for example, undergo Pd(OAc)$_2$/CuI-promoted C8-arylation in DMF at elevated temperature [78]. Direct arylation of guanosine is, in striking contrast, much more difficult and low yielding [79]. Sugar moiety-protected uridines are, in turn, subject to Pd-promoted oxidative cross-coupling of C5 with some alkenes, including maleimides [80], alkyl acrylates [81] or styrene in MeCN. Either a stoichiometric amount of Pd(OAc)$_2$ is required, or *tert*-butyl perbenzoate is used as reoxidant of palladium.

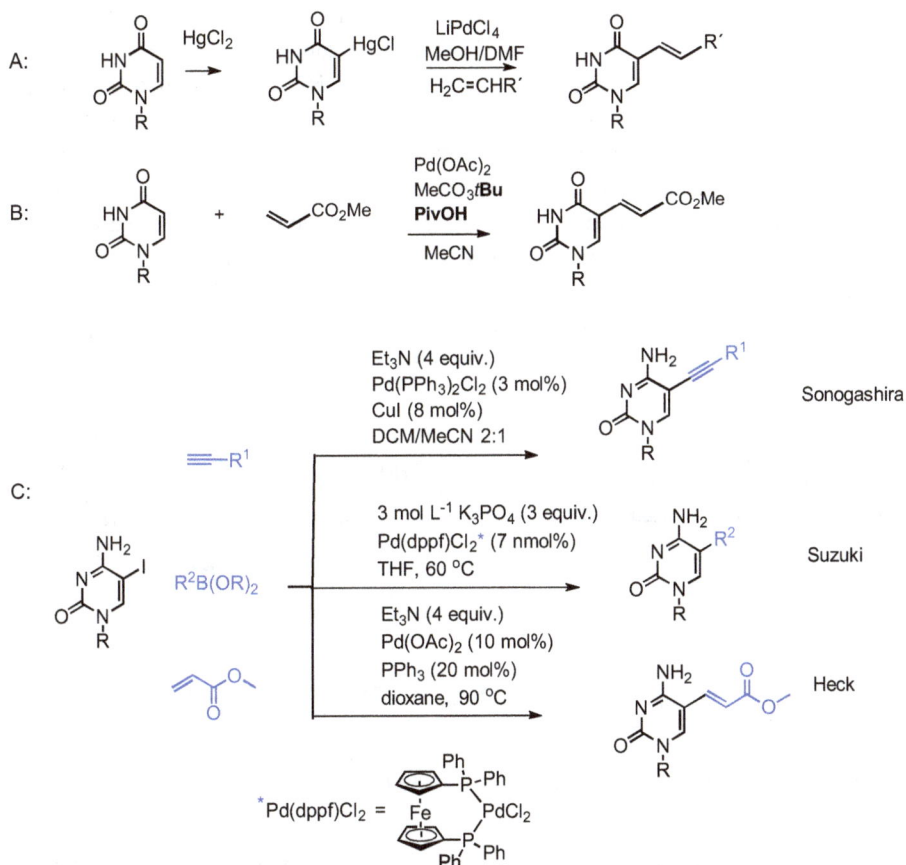

**Figure 2.15:** Examples of metal-ion-promoted transformations of unmodified and halogen-substituted nucleosides.

## 2.3 Hydrolytic stability of nucleosides

### 2.3.1 Acid-catalyzed hydrolysis

The $N$-glycosidic bond of nucleosides is hydrolyzed in acid and not under basic conditions. Purine nucleosides are much more susceptible to hydrolysis than their pyrimidine counterparts and 2′-deoxyribonucleosides more susceptible than their ribonucleoside counterparts. The reaction still is rather slow. 2′-Deoxyadenosine and 2′deoxyguanosine are hydrolyzed almost as rapidly, the half-life being approximately 1 h at pH 2 and 50 °C ($I = 0.1$ M). Their ribonucleoside counterparts are 500 times more stable. The reactive species is, depending on pH, a mono- or dication (Figure 2.16). With adenosine, the first protonation with $pK_a = 3.6$ takes place at N1 and the second with $pK_a = -1.4$ at N7 (cf. Section 1.2). With guanosine, the first protonation site is N7 and the second N3, the $pK_a$ values being 2.3 and −2.4, respectively. Protonation of the base moiety withdraws the electrons of the $N$-glycosidic bond toward N9, resulting in unimolecular fission of the bond. [82] A resonance-stabilized oxocarbenium ion is obtained, onto which water rapidly attacks. The low pre-equilibrium concentration of the dication is compensated by much faster bond cleavage and, hence, the dependence of log $k$ of the hydrolysis remains inversely proportional to pH on passing the $pK_a$ of the monocation [79]. Electronegative oxygen atom at C2′ reduces the electron density at the anomeric carbon and, hence, retards the development of carbocation center at this site.

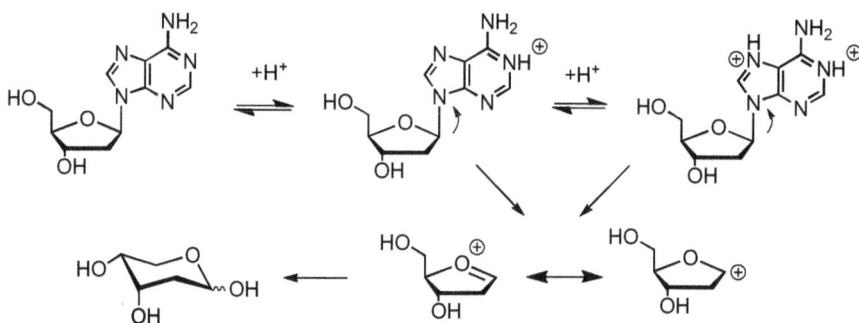

**Figure 2.16:** Mechanism of the acid-catalyzed hydrolysis of adenosine.

Polar nature of substituents on the base moiety does not markedly affect the rate of depurination. The effects on the pre-equilibrium protonation and rate-limiting heterolysis are opposite and they largely cancel each other [83]. Acylated 6-amino group of adenine nucleosides, however, makes an exception. This group that is frequently used as a base moiety protection group makes the $N$-glycosidic bond more prone to acid-catalyzed hydrolysis [84]. $N^6$-Benzoylation, for example, markedly retards protonation of the adenine base ($pK_a = 1.76$), but at the same time changes the preferred site of protonation from N1 to N7. Consequently, the hydrolysis is accelerated by one order of magnitude at pH > 2 in spite of reduced basicity. In more acidic solutions, where hydrolysis via an N1,N7-dication predominates, the acceleration gradually disappears. In other words, the N7-protonated adenine is a better leaving group than its N1-protonated counterpart.

Another interesting fact is unexpectedly large rate acceleration of 8-amino substitution. 8-Amino-, 8-methylamino- and 8-diethylamino-adenosines are hydrolyzed at pH 3, 35, 20 and $3.5 \times 10^4$ times as fast as adenosine [85]. With guanosine, the same rate accelerations at pH 4.6 are 25-, 20- and $2 \times 10^3$-fold, respectively [86]. Cationic N3-alkylated purine nucleosides are also hydrolyzed exceptionally rapidly under acidic conditions. 3-Methylinosine, for instance, is hydrolyzed $10^4$ times as fast as inosine [87]. The acceleration is far too great to be accounted by steric acceleration. Evidently N3-alkyl purines are electronically exceptionally good leaving groups.

Purine ribonucleosides missing the C6 substituent undergo acid-catalyzed depurination by the mechanism described above (Figure 2.16) only when pH is lower than 1. In more diluted acids, the reaction is initiated by the opening of the imidazole ring (Figure 2.17) [88]. It should be, however, noted that this concerns only ribonucleosides. With purine 2′-deoxyribonucleosides, depurination is so fast that no competition with imidazole-ring opening is observed.

As regards the reaction via imidazole ring opening (Figure 2.17), attack of water on C8 results in cleavage of the C8-N9 bond, giving an acyclic Schiff base. Since this step is reversible, isomerization of the starting material to pyranoid and α-furanoid nucleosides takes place. Attack of water on the anomeric carbon then gives the so-called carbinolamine intermediate of Schiff base hydrolysis. Formation of this species is acid-catalyzed since the imidazole ring is opened only by an attack of water on protonated purine base. At sufficiently low pH, breakdown of the carbinolamine that requires deprotonation of the C1′-OH, however, becomes rate-limiting. Accordingly, at low pH (pH < 1), depurination through an oxocarbenium ion intermediate (Figure 2.16) becomes faster than the reaction through a Schiff base intermediate (Figure 2.17).

Cytosine nucleosides undergo acid-catalyzed depyrimidination by pre-equilibrium protonation of N3 followed by rate-limiting formation of an oxocarbenium intermediate, that is, analogously to purine nucleosides [89]. Depyrimidination, however, is almost one order of magnitude slower than depurination. While 2′-deoxycytidine reacts only by this mechanism, cytidine undergoes hydrolytic deamination concurrent with depyrimidination. Deamination even predominants at low temperatures, owing to more negative entropy of activation [90]. 6-Methyl substituent accelerates depyrimidination by almost two orders of magnitude, while deamination is not accelerated [91].

The acid-catalyzed hydrolysis of uridine and thymidine is slow, slower than that of the corresponding cytosine nucleosides. In striking contrast to other nucleosides, uridine and thymidine undergo anomerization and isomerization to pyranoid nucleosides concurrent with hydrolysis [92]. The aromatic nature of 4-oxopyrimidines is weaker than the other nucleobases. As an indication of this, uridine equilibrates

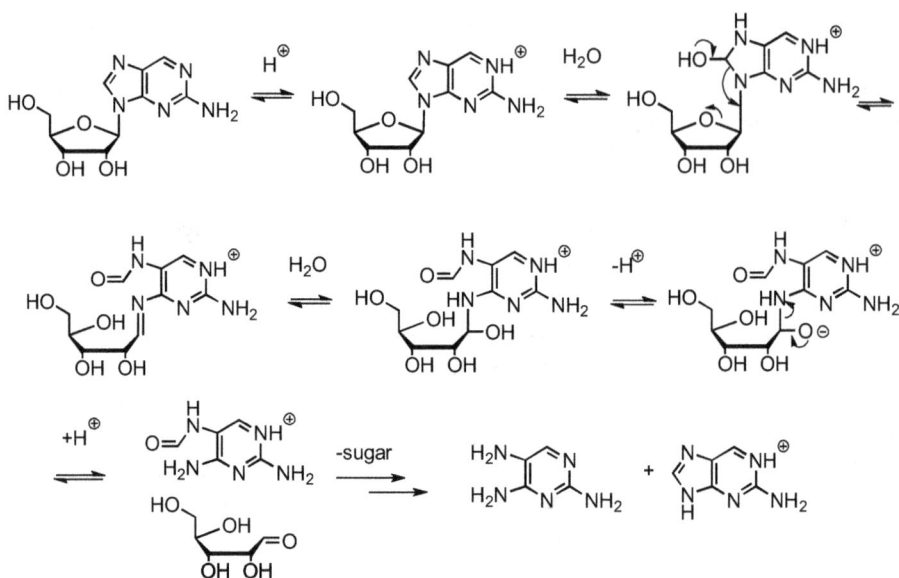

**Figure 2.17:** Mechanism for the acid-catalyzed hydrolysis of 6-unsubstituted purine ribonucleosides at low acidicty (pH > 2) [88].

under acidic conditions with its covalent hydrate obtained by attack of water on C6 and concomitant saturation of the 5,6-double bond [93]. Evidently, the aromatic character of the uracil (and thymine) base is so weak that the mechanism described for glycosylamines (Figure 2.18) is followed in striking contrast to other nucleosides.

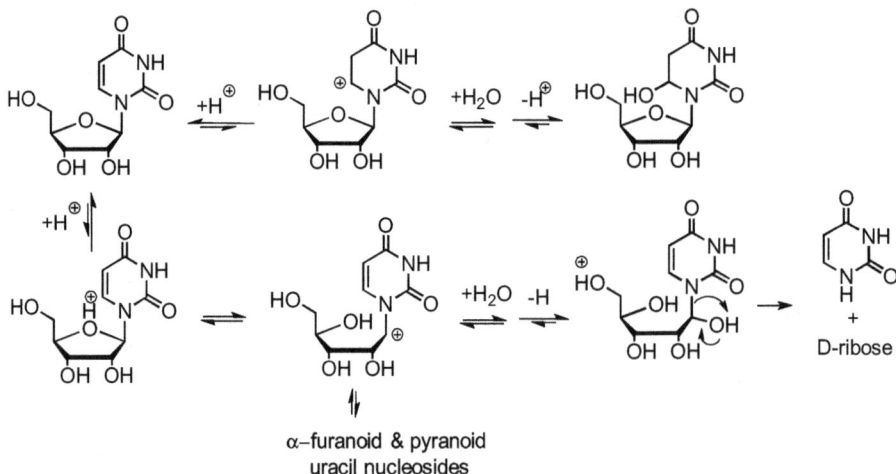

α–furanoid & pyranoid
uracil nucleosides

**Figure 2.18:** Mechanism for the acid-catalyzed hydrolysis of 4-oxopyrimidine nucleosides [93].

### 2.3.2 Base-catalyzed and nucleophilic reactions

Nucleosides are under basic conditions much more stable than in acid. The *N*-glycosidic bond is not hydrolyzed, but the base moiety is rather degraded under very basic conditions. Guanosine, uridine and thymidine are extremely stable, owing to deprotonation of the base moiety, which protects the base against nucleophiles. Adenosine undergoes imidazole ring-opening by an attack of hydroxide ion on C8 [94]. Concomitant conversion of nucleoside to glycosylamines results in isomerization of the sugar moiety followed by imidazole ring re-closure. Deformylation to 4,5,6-triaminopyrimidine riboside, however, competes with the recyclization (Figure 2.19). Some hydrolytic deamination to inosine may also occur as a minor side reaction.

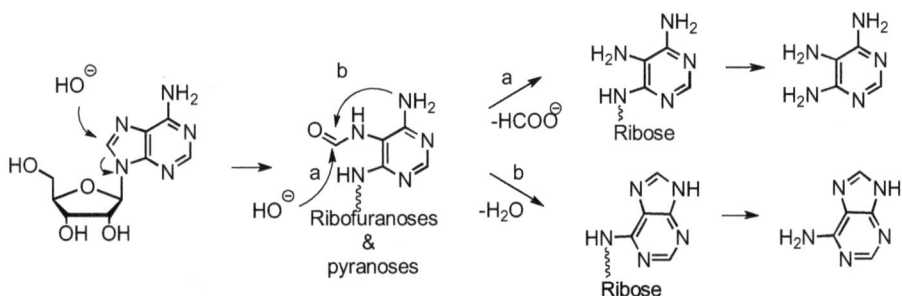

**Figure 2.19:** Degradation of adenosine in aqueous alkali [94].

Cytidine is partly deaminated to uridine and partly degraded to UV-inactive products under basic conditions. As discussed above, cytidine is deaminated even in acid, but the mechanisms of the acid- and base-catalyzed reactions are different. In acid, the reaction is initiated by an attack of water on C6 of N3-protonated base moiety, followed by the displacement of the 4-amino group by water (Figure 2.20) [91]. Under basic conditions, direct displacement of the 4-amino group by hydroxide ion takes place [95]. Deamination of 2′-deoxycytidine is of considerable interest since it is, besides depurination, a potential source of spontaneous mutagenesis.

The susceptibility of C6 of pyrimidine nucleosides to nucleophiles was utilized in the first-generation sequencing method, the Maxam-Gilbert method [96]. Treatment with hydazine cleaves cytosine form DNA as 3-aminopyrazole and thymine as 4-methyl-3-pyrazolone (Figure 2.21). In the presence of high concentration of NaCl, the cleavage of thymine is inhibited and the hydrazine reaction, hence, is cytosine-selective. For the purine selective cleavage, depurination with formic acid was applied and guanine selective cleavage was obtained by methylation of N7 of guanine with DMSO, followed by alkaline imidazole ring opening. Treatment with piperidine at elevated temperature then resulted in chain cleavage of DNA at the abasic sites.

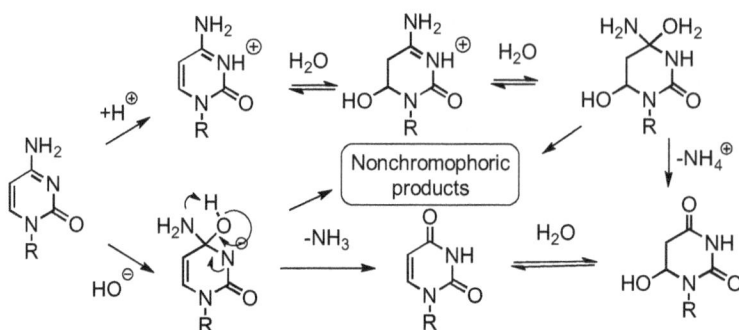

**Figure 2.20:** Deamination of cytidine under acidic and basic conditions [91, 95].

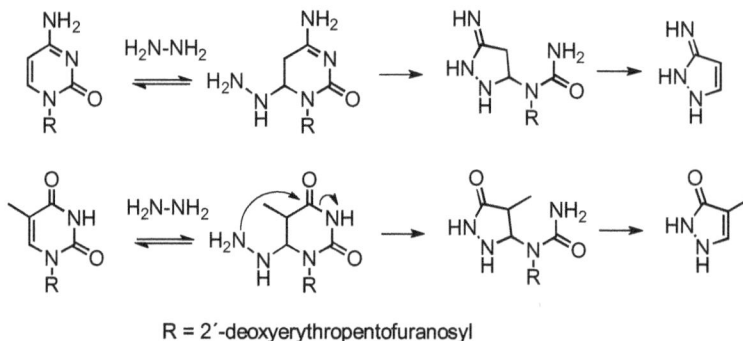

R = 2′-deoxyerythropentofuranosyl

**Figure 2.21:** Hydrazine-induced modification of cytosine and thymine base in Maxam-Gilbert sequencing.

## 2.3.3 The effect of metal ions on hydrolysis of nucleosides

Metal ions compete with protons for the ring-nitrogen atoms of nucleosides, as discussed in more detail in Section 1.4. Owing to this competition, metal ions retard acid-catalyzed hydrolysis of the N-glycosidic bond [97]. The rate-retardation is more marked with 2′-deoxyguanosine than with 2′-deoxyadenosine. With dGuo, protons and metal ions compete for the same site, N7, whereas in dAdo, protons preferably bind to N1 and metal ions to N7.

Studies on depurination of kinetically stable (dine)Pt(II) complexes help to understand the effect of metal ion binding in more detail. Binding to N7 retards the acid-catalyzed depurination of 2′-deoxyinosine (used as a model of dGuo) by two orders of magnitude at pH 1. At pH > 3, only a pH-independent spontaneous depurination of the N7-complex is observed [98]. Replacement of N1H of deoxyinosine with (dine)Pt(II) has virtually no effect on depurination at pH > 1, that is, under conditions where the acid-catalyzed reaction proceeds via N7 protonation. In other words, a marked, rate retardation is observed, when the metal ion and proton compete for the same site; otherwise the influence is modest. The metal complex of a purine base clearly is a worse leaving group than its protonated counterpart. N7-Bound metal ion, however, polarizes the *N*-glycosidic bond sufficiently to induce bond cleavage that, however, is two orders of magnitude slower than that induced by protonation.

## Further reading

Boryski J. Reactions of transglycosylation in the nucleoside chemistry. Curr Org Chem 2008, 12, 309–325.

Boutureira O, Matheu MI, Diaz Y, Castillon S. Advances in the enantioselective synthesis of carbocyclic nucleosides. Chem Soc Rev 2013, 42, 5056–5072.

Fox JJ. Pyrimidine nucleoside transformatons via anhydronucleosides. Pure Appl Chem 1969, 18, 223–255.

Liang Y, Wnuk SF. Modification of purine and pyrimidine nucleosides by direct C-H bond activation. Molecules 2015, 20, 4874–4901.

Mathe C, Perigaud C. Recent approaches in the synthesis of conformationally restricted nucleoside analogs. Eur J Org Chem 2008, 1489–1505.

Mieczkowski A, Roy V, Agrofoglio LA. Preparation of cyclonucleosides. Chem Rev 2010, 110, 1828–1856.

Shaughnessy KH. Palladium-catalyzed modification of unprotected nucleosides, nucleotides, and oligonucleotides. Molecules 2015, 20, 9419–9454.

Stambasky J, Hocek M, Kocovsky P. *C*-Nucleosides: Synthetic strategies and biological applications. Chem Rev 2009, 109, 6729–6764.

Vorbrüggen H, Höfle G. On the mechanism of nucleoside synthesis. Chem Ber 1981, 114, 1256–1268.

## References

[1]   Vorbrüggen H, Höfle G. On the mechanism of nucleoside synthesis. Chem Ber 1981, 114, 1256–1268.

[2]   Boryski J. Reactions of transglycosylation in the nucleoside chemistry. Curr Org Chem 2008, 12, 309–325.

[3]   Liao J, Sun J, Yu B. Effective synthesis of nucleosides with glycosyl trifluoroacetimidates as donors. Tetrahedron Lett 2008, 49, 5036–5038.

[4]   Fischer E, Helferich B. Synthetische Glucoside der Purine. Chem Ber 1914, 47, 210–235.

[5]   Davoll J, Lowy BA. A new synthesis of purine nucleosides. The synthesis of adenosine, guanosine and 2,6-diamino-9-β-D-ribofuranosylpurine. J Am Chem Soc 1951, 73, 1650–1655.

[6]   Diekmann E, Friedrich K, Fritz H-G. Didesoxy-ribonucleoside durch Schmelzkondensation. J Prakt Chem 1993, 335, 415–424.

[7]   Downey AM, Richter C, Pohl R, Roithova J, Hocek M. Synthesis of nucleosides through direct glycosylation of nucleobases with 5-*O*-monoprotected or 5-modified ribose: Improved protocol, scope, and mechanism. Chem Eur J 2017, 23, 3910–3917.

[8]   Kazimierczuk Z, Cottam HB, Revankar GR, Robins RK. Synthesis of 2′-deoxytubercidin, 2′-deoxyadenosine, and related 2′-deoxynucleosides via a novel direct stereospecific sodium salt glycosylation procedure. J Am Chem Soc 1984, 106, 6379–6382.

[9]   Seio K, Tokugawa M, Kaneko K, Shiozawa T, Masaki Y. A systematic study of the synthesis of 2′-deoxynucleosides by Mitsunobu reaction. Synlett 2017, 28, 2014–2017.

[10] Kamel S, Yehia H, Neubauer P, Wagner A. Enzymatic synthesis of nucleoside analogues by nucleoside phosphorylases. In Ferdandez-Lucas J, Camarasa Rius M-J, eds. Enzymatic and Chemical Synthesis of Nucleic Acid Derivatives. 1st ed. Wiley-VCH Verlag GmbH & Co. KGaA, 2019, 1–28.

[11] Mikhailopulo IA, Miroshnikov AI. Biologically important nucleosides: Modern trends in biotechnology and application. Mendeleev Commun 2011, 21, 57–68.

[12] Drenichev MS, Alexeev CS, Kurochkin NN, Mikhailov SN. Use of nucleoside phosphorylases for the preparation of purine and pyrimidine 2′-deoxynucleosides. Adv Synth Catal 2018, 360, 305–312.

[13] Suhadolnik RJ, Uematsu T. Synthesis of 3′-deoxyuridine via transglycosylation of uracil with 3′-deoxy-adenosine (cordycepin). Carbohydr Res 1978, 61, 545–548.

[14] Mieczkowski A, Roy V, Agrofoglio LA. Preparation of cyclonucleosides. Chem Rev 2010, 110, 1828–1856.

[15] Clark VM, Todd AR, Zussman J. Nucleotides. Part VIII. CycloNucleoside Salts. A novel rearrangement of some toluene-*p*-sulphonylnucleosides. J Chem Soc 1951, 0, 2952–2958.

[16] Matsuda A, Tezuka M, Niizuma K, Sugiyama E, Ueda T. Synthesis of carbon-bridged 8,5′-cyclopurine nucleosides: Nucleosides and nucleotides – XXIV. Tetrahedron 1978, 34, 2633–2637.

[17] Ikehara M, Kaneko M. Studies of nucleosides and nucleotides. XLII. Purine cyclonucleosides. (9). Synthesis of adenine cyclonucleosides having 8,2′-and 8,3′-O-anhydro linkages. Chem Pharm Bull 1970, 18, 2401–2406.

[18] Ikehara M, Maruyama T. Studies of nucleosides and nucleotides – LXV: Purine cyclonucleosides-26. A versatile method for the synthesis of purine O-cyclo-nucleosides. The first synthesis of 8,2′-anhydro-8-oxy 9-β-D-arabinofuranosylguanine. Tetrahedron 1975, 31, 1369–1372.

[19] Kaneko M, Ikehara M. Synthesis and properties of 8,2′-N-cycloadenosines. Tetrahedron Lett 1971, 12, 3113–3116.

[20] Ikehara M, Tada T. Studies of nucleosides and nucleotides. XXXII. Purine cyclonucleosides. 3. Synthesis of 2′-deoxy- and 3′-deoxyadenosine from adenosine. Chem Pharm Bull 1967, 15, 94–100.

[21] Ogilvie KK, Slotin L, Westmore JB, Lin D. Synthesis of 8,2′-thioanhydroguanosine. Can J Chem 1972, 50, 1100–1104.

[22] Divakar KJ, Reese CB. The cyclization of 8-carboxamido-2′-O-tosyladenosine. A new preparation of 9-β-D-arabinofuranosyladenine (ara-A). J Chem Soc Chem Commun 1980, 0, 1191–1193.

[23] Fox JJ. Pyrimidine nucleoside transformatons via anhydronucleosides. Pure Appl Chem 1969, 18, 223–255.

[24] Sako M, Saito T, Kameyama K, Hirota K, Maki Y. A facile synthesis of 5′-O,6-cyclo-5,5-dihalogeno-5,6-dihydropyrimidine nucleosides. Synthesis 1987, 0, 829–831.

[25] Otter BA, Falco EA, Fox JJ. Nucleosides. LII. Transformations of pyrimidine nucleosides in alkaline media. 1. Conversion of 5-haloarabinosyluracils to imidazoline nucleosides. J Org Chem 1968, 33, 3593–3600.

[26] Inoue H, Ueda T. Synthesis of 6, 5′-S- and 6, 5′-N-cyclouridines (nucleosides and nucleotides. XXII). Chem Pharm Bull 1978, 26, 2664–2667.

[27] Gissot A, Massip S, Barthélémy P. Intramolecular Michael additions in uridine derivatives: Isolation of the labile 5′O-C6 cyclonucleoside. ACS Omega 2020, 5, 24746–24753.

[28] Stambasky J, Hocek M, Kocovsky P. C-Nucleosides: Synthetic strategies and biological applications. Chem Rev 2009, 109, 6729–6764.

[29] Veronese AC, Morelli FC. A new and efficient route to the synthesis of pyrazole and pyrimidine C-nucleoside derivatives. Tetrahedron Lett 1998, 39, 3853–3856.

[30] Gudmundsson KS, Drach JC, Townsend LB. Synthesis of the first C3 ribosylated imidazo[1,2-a] pyridine C-nucleoside by enantioselective construction of the ribose moiety. J Org Chem 1998, 63, 984–989.

[31]  Zhang HC, Daves GD, Jr. Syntheses of 2′-deoxypseudouridine, 2′-deoxyformycin B, and 2′,3′-dideoxyformycin B by palladium-mediated glycal-aglycon coupling. J Org Chem 1992, 57, 4690–4696.

[32]  Kalvoda L, Farkas J, Sorm F. Synthesis of showdomycin. Tetrahedron Lett 1970, 26, 2297–2300.

[33]  Hainke S, Singh I, Hemmings J, Seitz O. Synthesis of C-aryl-nucleosides and O-aryl-glycosides via cuprate glycosylation. J Org Chem 2007, 72, 8811–8819.

[34]  Singh I, Seitz O. Diastereoselective synthesis of β-aryl-C-nucleosides from 1,2-anhydrosugars. Org Lett 2006, 8, 4319–4322.

[35]  Tawarada R, Seio K, Sekine M. Synthesis and properties of oligonucleotides with iodo-substituted aromatic aglycons: Investigation of possible halogen bonding base pairs. J Org Chem 2008, 73, 383–390.

[36]  Boutureira O, Matheu MI, Diaz Y, Castillon S. Advances in the enantioselective synthesis of carbocyclic nucleosides. Chem Soc Rev 2013, 42, 5056–5072.

[37]  Campian M, Putala M, Sebesta R. Bioactive carbocyclic nucleoside analogs. Syntheses and properties of Entecavir. Curr Org Chem 2014, 18, 2808–2832.

[38]  Lee JA, Moon HR, Kim HO, Kim KR, Lee KM, Kim BT, Hwang KJ, Chun MW, Jacobson KA, Jeong LS. Synthesis of novel apio carbocyclic nucleoside analogues as selective A3 adenosine receptor agonists. J Org Chem 2005, 70, 5006–5013.

[39]  Crimmins MT, King BW. An efficient asymmetric approach to carbocyclic nucleosides: Asymmetric synthesis of 1592U89, a potent inhibitor of HIV reverse transcriptase. J Org Chem 1996, 61, 4192–4193.

[40]  Kang B, Zhang Q-Y, Qu GR, Guo HM. The enantioselective synthesis of chiral carbocyclic nucleosides via palladium-catalyzed asymmetric allylic amination of alicyclic MBH adducts with purines. Adv Synth Catal 2020, 362, 1955–1960.

[41]  Katagiri N, Nomura M, Sato H, Kaneko C, Yusa K, Tsuruo T. Synthesis and anti-HIV activity of 9-[c-4,t-5-bis(hydroxymethyl)cyclopent-2-en-r-l-yl]-9H-adenine. J Med Chem 1992, 35, 1882–1868.

[42]  Sung K, Aswar VR, Song J, Jarhad DB, Jeong LS. Stereoselective approach for the synthesis of diverse 1′-modified carbanucleosides. Org Lett 2023, 25, 8377–8381.

[43]  Ikehara M, Uno H. Studies of nucleosides and nucleotides. XXVI. Further studies on the chlorination of inosine derivatives with dimethylformamide-thionyl chloride complex. Chem Pharm Bull 1965, 13, 221–223.

[44]  Robins MJ, Basom GL. Nucleic acid related compounds. 8. Direct conversion of 2′-deoxyinosine to 6-chloropurine 2′-deoxyriboside and selected 6-substituted deoxynucleosides and their evaluation as substrates of adenosine deaminase. Can J Chem 1973, 51, 3161–3169.

[45]  Mizuno Y. The Organic Chemistry of Nucleic Acids. Amsterdam: Elsevier, 1986.

[46]  Robins MJ. Chemical transformations of naturally occurring nucleosides to otherwise difficultly accessible structures. Annals New York Acad Sci 1975, 255, 104–120.

[47]  Tanaka H, Uchida Y, Shinozaki M, Hayakawa H, Matsuda A, Miyasaka T. A simplified synthesis of 8-substituted purine nucleosides via lithiation of 6-chloro-(2′,3′-O-isopropylidene-β-D-ribofuranosyl) purine. Chem Pharm Bull 1983, 31, 878–890.

[48]  Ikehara M, Tazawa I, Fukui T. Nucleosides and nucleotides. XXXIX. Synthesis of 8-substituted purine nucleotides by the direct replacement reactions. Chem Pharm Bull 1969, 17, 1019–1024.

[49]  Kikugawa K, Suehiro H, Yanase R, Aoki A. Platelet aggregation inhibitors. IX. Chemical transformation of adenosine into 2-thioadenosine derivatives. Chem Pharm Bull 1977, 25, 1959–1969.

[50]  Sattsangi PD, Barrio JR, Leonard NJ. 1,N6-Etheno-bridged adenines and adenosines. Alkyl substitution, fluorescence properties, and synthetic applications. J Am Chem Soc 1980, 102, 770–774.

[51]  Liu H, Zeng T, He C, Rawal VH, Zhou H, Dickinson BC. Development of mild chemical catalysis conditions for m1A to m6A rearrangement on RNA. ACS Chem Biol 2022, 17, 1334–1342.

[52]  Leskovskis K, Zaķis JM, Novosjolova I, Turks M. Applications of purine ring opening in the synthesis of imidazole, pyrimidine, and new purine derivatives. Eur J Org Chem 2021, 0, 5027–5052.

[53]  Daley S, Cordell GA. Homopurine alkaloids: A brief overview. Nat Prod Commun 2020, 15, 1–14.

[54]  Bradshaw TK, Hutchinson DW. 5-Substituted pyrimidine nucleosides and nucleotides. Chem Soc Rev 1977, 94, 978–982.

[55]  Shapiro R, Servis RE, Welcher M. Reactions of uracil and cytosine derivatives with sodium bisulfite. A specific deamination method. J Am Chem Soc 1970, 92, 422–424.

[56]  Shapiro R, Weisgras JM. Bisulfite-catalyzed transamination of cytosine and cytidine. Biochem Biophys Res Commun 1970, 40, 839–843.

[57]  Scheit KH. Die synthese der 5'-diphosphate von 5-methyl-uridin, 5-hydroxymethyl-uridin und 3.5-dimethyl-uridin. Ber Dtsch Chem Ges 1966, 99, 3884–3891.

[58]  Yuan BF. Assessment of DNA epigenetic modifications. Chem Res Toxicol 2020, 33, 695–708.

[59]  Monfret O, Liu D, Rollando P, Bourdreux Y, Urban D, Doisneau G, Guianvarch D. Preparation of 5-hydroxymethyl-pyrimidine based nucleosides: A reinvestigation. Eur J Org Chem 2023, 26, e202300298 (9 pages).

[60]  Tran A, Zheng S, White DS, Curry AM, Cen Y. Divergent synthesis of 5-substituted pyrimidine 2'-deoxynucleosides and their incorporation into oligodeoxynucleotides for the survey of uracil DNA glycosylases. Chem Sci 2020, 11, 11818–11826.

[61]  Bischofberger N. Synthesis of 4-C substituted pyrimidine nucleosides. Tetrahedron Lett 1987, 28, 2821–2824.

[62]  Reese CB, Ubasawa A. Reaction between 1-arenesulphonyl-3-nitro-1,2,4-triazoles and nucleoside base residues. Elucidation of the nature of side-reactions during oligonucleotide synthesis. Tetrahedron Lett 1980, 21, 2265–2268.

[63]  Bigge CF, Kalaritis P, Deck JR, Mertes MP. Palladium-catalyzed coupling reactions of uracil nucleosides and nucleotides. J Am Chem Soc 1980, 102, 2033–2038.

[64]  Zhao Q, Xie R, Zeng Y, Li W, Xiao G, Li Y, Chen G. Palladium-catalyzed C–H olefination of uridine, deoxyuridine, uridine monophosphate and uridine analogues. RSC Adv 2022, 12, 24930–24934.

[65]  Shaughnessy KH. Palladium-catalyzed modification of unprotected nucleosides, nucleotides, and oligonucleotides. Molecules 2015, 20, 9419–9454.

[66]  Sonogashira K. Development of Pd–Cu catalyzed cross-coupling of terminal acetylenes with sp2-carbon halides. J Organomet Chem 2002, 653, 46–49.

[67]  Miyaura N, Suzuki A. Palladium-catalyzed cross-coupling reactions of organoboron compounds. Chem Rev 1995, 95, 2457–2483.

[68]  Heck RF, Nolley JP. Palladium-catalyzed vinylic hydrogen substitution reactions with aryl, benzyl, and styryl halides. J Org Chem 1972, 37, 2320–2322.

[69]  Jäger S, Rasched G, Komreich-Leshem H, Engesser M, Thum O, Famulok M. A versatile toolbox for variable DNA functionalization at high density. J Am Chem Soc 2005, 127, 15071–15072.

[70]  Sagi G, Ötvös L, Ikeda S, Andrei G, Snoeck R, De Clercq E. Synthesis and antiviral activities of 8-alkynyl-, 8-alkenyl-, and 8-alkyl-2'-deoxyadenosine analogs. J Med Chem 1994, 37, 1307–1311.

[71]  Amann N, Pandurski E, Fiebig T, Wagenknecht HA. Electron injection into DNA: Synthesis and spectrscopic properties of pyrenyl-modified oligonucleotides. Chem Eur J 2002, 8, 4877–4883.

[72]  Mayer E, Valis L, Huber R, Amann N, Wagenknecht HA. Preparation of pyrene-modified purine and pyrimidine nucleosides via Suzuki-Miyaura cross-couplings and characterization of their fluorescent properties. Synthesis 2003, 15, 2335–2340.

[73]  Kohyama N, Katashima T, Yamamoto Y. Synthesis of novel 2-aryl AICAR derivatives. Synthesis 2004, 17, 2799–2804.

[74]  Herve G, Len C. First ligand-free, microwave-assisted, Heck cross-coupling reaction in pure water on a nucleoside. Application to the synthesis of antiviral BVDU. RSC Adv 2014, 4, 46926–46929.

[75]   Ding H, Greenberg MM. Hole migration is the major pathway involved in alkali-labile lesion formation in DNA by the direct effect of ionizing radiation. J Am Chem Soc 2007, 129, 772–773.

[76]   Zhao Y, Baranger AM. Design of an adenosine analogue that selectively improves the affinity of a mutant U1A protein for RNA. J Am Chem Soc 2003, 125, 2480–2488.

[77]   Liang Y, Wnuk SF. Modification of purine and pyrimidine nucleosides by direct C-H bond activation. Molecules 2015, 20, 4874–4901.

[78]   Cerna I, Pohl R, Hocek M. The first direct arylation of purine nucleosides. Chem Commun 2007, 0, 4729–4730.

[79]   Storr TE, Bauman CG, Thatcher RJ, De Ornellas S, Whitwood AC, Fairlamb IJS. Pd(0)/Cu(I)-Mediated direct arylation of 2′-deoxyadenosines: Mechanistic role of Cu(I) and reactivity comparisons with related purine nucleosides. J Org Chem 2009, 74, 5810–5821.

[80]   Itahara T. Oxidative coupling of uracil derivatives with maleimides by palladium acetate. Chem Lett 1986, 15, 239–242.

[81]   Hirota K, Isobe Y, Kitade Y, Maki Y. A simple synthesis of 5-(1-alkenyl)uracil derivatives by palladium-catalyzed oxidative coupling of uracils with olefins. Synthesis 1987, 0, 495–496.

[82]   Zoltewicz JA, Clark DF, Sharpless TW, Grahe G. Kinetics and mechanism of the acid-catalyzed hydrolysis of some purine nucleosides. J Am Chem Soc 1970, 92, 1741–1750.

[83]   Lönnberg H, Lehikoinen P. Mechanisms for the solvolytic decompositions of nucleoside analogues. X. Acidic hydrolysis of 6-substituted 9-(β-D-ribofuranosyl)purines. Nucleic Acids Res 1982, 10, 4339–4349.

[84]   Remaud G, Zhou -X-X, Chattopadhyaya J, Oivanen M, Lönnberg H. The effect of protecting groups of the nucleobase and the sugar moieties on the acidic hydrolysis of the glycosidic bond of 2′-deoxyadenosine: A kinetic and 15N NMR spectroscopic study. Tetrahedron 1987, 43, 4453–4461.

[85]   Hovinen J, Glemarec C, Sandström A, Sund C, Chattopadhyaya J. Spectroscopic, kinetic and semiempirical molecular orbital studies on 8-amino-, 8-methylamino- & 8-dimethylamino-adenosines. Tetrahedron 1991, 47, 4693–4708.

[86]   Jordan F, Niv H. Glycosyl conformational and inductive effects on the acid catalysed hydrolysis of purine nucleosides. Nucleic Acids Res 1977, 4, 697–709.

[87]   Itaya T, Matsumoto H. 3-Methylinosine. Chem Pharm Bull 1985, 33, 2213–2219.

[88]   Hovinen J, Shugar D, Lönnberg H. Acid-catalyzed hydrolysis of 2-amino-9-(β-D-ribofuranosyl)purine and its acyclo and 2-methyl analogues. Competition between depurination and opening of the imidazole ring. Nucleosides Nucleotides 1990, 9, 697–712.

[89]   Shapiro R, Danzig M. Acidic hydrolysis of deoxycytidine and deoxyuridine derivatives. General mechanism of deoxyribonucleoside hydrolysis. Biochemistry 1972, 11, 23–29.

[90]   Lönnberg H, Käppi R. Competition between the hydrolysis and deamination of cytidine and its 5-substituted derivatives in aqueous acid. Nucleic Acids Res 1985, 13, 2451–2456.

[91]   Kusmierek J, Käppi R, Neuvonen K, Shugar D, Lönnberg H. Kinetics and mechanisms of hydrolytic reactions of methylated cytidines under acidic and neutral conditions. Acta Chem Scand 1989, 43, 196–202.

[92]   Cadet J, Teoule R. Nucleic acid hydrolysis. I. Isomerization and anomerization of pyrimidic deoxyribonucleosides in an acidic medium. J Am Chem Soc 1974, 96, 6517–6519.

[93]   Prior JJ, Santi DV. On the mechanism of the acid-catalyzed hydrolysis of uridine to uracil. Evidence for 6-hydroxy-5,6-dihydrouridine intermediates. J Biol Chem 1984, 259, 2429–2434.

[94]   Lehikoinen P, Mattinen J, Lönnberg H. Reactions of adenine nucleosides with aqueous alkalies: Kinetics and mechanism. J Org Chem 1986, 51, 3819–3823.

[95]   Lönnberg H, Suokas P, Käppi R, Darzynkiewicz E. Reactions of pyrimidine nucleosides with aqueous alkalies: Kinetics and mechanisms. Acta Chem Scand 1986, B40, 798–805.

[96]   Maxam AM, Gilbert W. A new method for sequencing DNA. Proc Natl Acad Sci USA 1977, 74, 560–564.

[97]  Arpalahti J, Käppi R, Hovinen J, Lönnberg H, Chattopadhyaya J. The effect of metal ion complex formation on acidic depurination of 2′-deoxyadenosine and 2′-deoxyguanosine. Tetrahedron 1989, 45, 3945–3954.

[98]  Arpalahti J, Jokilammi A, Hakala H, Lönnberg H. Depurination of (dine)Pt(II) complexes of purine deoxyribonucleosides. Comparison with the effects of (dine)Pd(II) ion complexing. J Phys Org Chem 1991, 4, 301–309.

# 3 Nucleotides

## 3.1 Nomenclature

Nucleotides are phosphoric acid esters of nucleosides. Any of the sugar hydroxyl functions may be phosphorylated. The name of nucleotide is derived from the name of nucleoside by adding an ending "phosphate" preceded by a number indicating the site of phosphorylation, for example, adenosine 5′-phosphate. Often ending "monophosphate" is used instead of "phosphate," to underline the difference from "diphosphate" and "triphosphate" obtained by esterification with di- or tri-phosphoric acid. Nucleosides may also form cyclic diesters of phosphoric acid, either 2′,3′- or 3′,5′-cyclic phosphates. Other biologically interesting nucleotide structures include nucleotide sugars that serve as building blocks in biosynthesis of carbohydrates, α,ω-dinucleoside oligophosphates, above all 7-methylguanosine(5′)triphospho(5′)guanosine, the 5′-terminal cap structure of messenger RNA, and cyclic di-GMP, a second messenger in bacteria. Examples of nucleotide structures are shown in Figure 3.1.

Sometimes monophosphates of canonical nucleosides are called adenylic, guanylic, cytidylic and uridylic acids. 3′-Adenylic acid, for instance, is an alternative name for adenosine 3′-phosphate. These names are especially used when fragments of nucleic acids are named: adenylyl(3′,5′)uridine or adenylyl(3′,5′)guanylyl(3′,5′)cytidine.

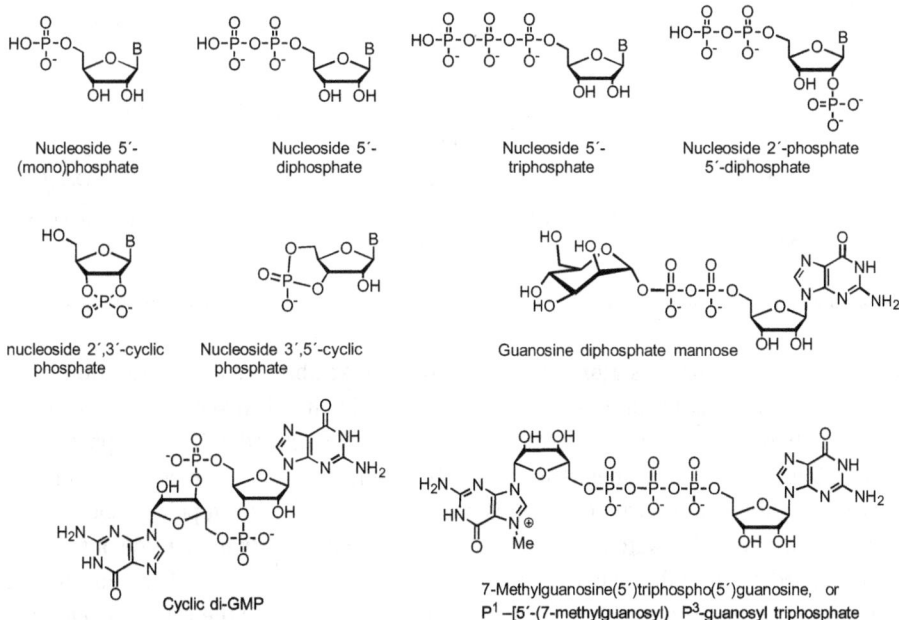

**Figure 3.1:** Illustrative examples of nucleotide structures.

https://doi.org/10.1515/9783111325637-003

The names of common nucleotides are often replaced by abbreviations, such as 3'-AMP, 5'-ATP and 2',3'-cAMP. pN stands for nucleoside 5'-phosphate and Np for nucleoside 3'-phosphate. Accordingly, NpN stands for a 3',5'-dinucleoside monophosphate.

Figure 3.2 shows examples of nucleotide analogs having one of the nonbridging oxygen atoms of nucleoside monophosphates replaced with nitrogen, sulfur, chlorine or carbon, and the compounds obtained are phosphoramidates, phosphorothioates, phosphorochloridates or alkylphosphonates, respectively. In case a bridging oxygen in a nucleoside di- or tri-phosphate is replaced with nitrogen, the compound is an imidophosphate and the site of substitution is indicated by defining the neighboring phosphorus atoms by letters α, β and γ (α refers to the phosphorus atom bonded to the 5'-oxygen). Similar nomenclature is used for sulfur and methylene substitutions.

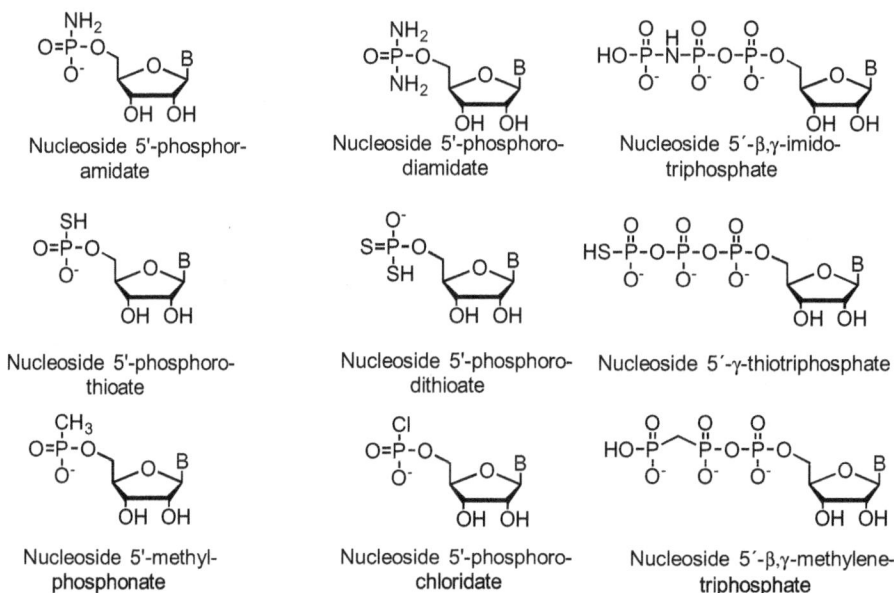

Nucleoside 5'-phosphor-amidate

Nucleoside 5'-phosphoro-diamidate

Nucleoside 5'-β,γ-imido-triphosphate

Nucleoside 5'-phosphoro-thioate

Nucleoside 5'-phosphoro-dithioate

Nucleoside 5'-γ-thiotriphosphate

Nucleoside 5'-methyl-phosphonate

Nucleoside 5'-phosphoro-chloridate

Nucleoside 5'-β,γ-methylene-triphosphate

**Figure 3.2:** Nitrogen-, sulfur-, chlorine- and carbon-substituted analogs of nucleoside phosphoesters.

Although the phosphorus atom in biomolecules invariably occurs at oxidation level V, P(III) compounds are at least as commonly used in chemical synthesis of nucleotides and oligonucleotides. P(III) esters of nucleosides are called nucleoside phosphites (Figure 3.3). The phosphite monoester, however, is only a minor tautomer, the major tautomer being the one where the proton is bound to the lone electron pair of phosphorus instead of oxygen. The tautomeric equilibrium is overwhelmingly on the side of this tautomer, called nucleoside hydrogen phosphonate or H-phosphonate. The equilibrium constant is of the order of $10^9$ [1]. Phosphite triesters that miss the rapidly exchangeable proton are naturally locked to the phosphite form. The situation is similar to phosphoramidites, the nitrogen analogs of phosphite esters. *O,N,N*-Trialkylphosphoramidites and $N,N,N',N'$-tetraalkylphosphordiamidites occur as single tautomers depicted in Figure 3.3 for methylated derivatives. Phosphoramidites are extensively used in synthesis of oligonucleotides or their congeners.

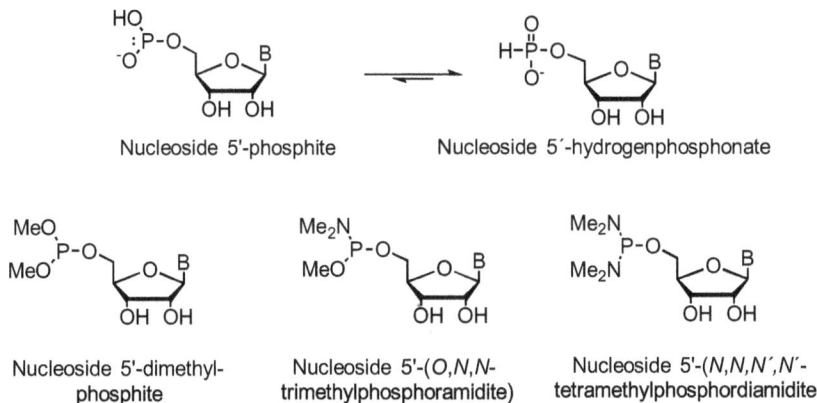

Nucleoside 5'-phosphite

Nucleoside 5′-hydrogenphosphonate

Nucleoside 5'-dimethyl-
phosphite

Nucleoside 5'-(O,N,N-
trimethylphosphoramidite)

Nucleoside 5'-(N,N,N',N'-
tetramethylphosphordiamidite

**Figure 3.3:** Tautomeric equilibrium of nucleoside phosphite esters and the single tautomers of fully alkylated phosphoramidite and phosphordiamidite.

## 3.2 Protolytic equilibria and metal ion complexing

The protolytic equilibria of nucleoside 5'-mono-, di- and tri-phosphates are shown in Figure 3.4. With monophosphates, deprotonation from neutral phosphate to monoanion takes please already under very acidic conditions ($pK_a < 1$) and deprotonation from monoanion to dianion with $pK_a = 6.0–6.2$ ($I = 0.1$ M, 25 °C) [2]. The terminal phosphate of 5'-di- and tri-phosphates behaves similarly, the $pK_a$ values being 6.2–6.4 and 6.3–6.5, respectively [2, 3]. The nonterminal phosphate group of 5'-diphosphates is deprotonated at pH < 1. With 5'-triphosphates, one of the nonterminal phosphates is expectedly deprotonated already at pH < 1 and the other around pH 2, evidently owing to simultaneous binding of proton to two phosphoryl oxyanions. The $pK_a$ value of 2',3'- and 3',5'-cyclic phosphates is less than 1.

   Alkaline earth metal ions and 3d transition metal ions are coordinated approximately as effectively to pyrimidine nucleoside 5'-monophosphates as to simple alkyl phosphates, suggesting that the base moiety does not participate [4]. The binding mode most likely is monodentate inner-sphere coordination to a nonbridging phosphoryl oxygen atom. As mentioned in Table 3.1, the 3d transition metal ion complexes are approximately one order of magnitude more stable than those of the alkaline earth metal ions. Soft metal ions bind only weakly.

   As regards purine nucleoside 5'-monophosphates, alkaline earth metal ions bind as tightly to purine and pyrimidine nucleotides, consistent with nonparticipation of the

$$\text{Nuc-O-}\overset{\overset{\text{O}}{\|}}{\underset{\underset{\text{OH}}{|}}{\text{P}}}\text{-OH} \quad\xrightarrow{pK_a < 1}\quad \text{Nuc-O-}\overset{\overset{\text{O}}{\|}}{\underset{\underset{\text{O}^-}{|}}{\text{P}}}\text{-OH} \quad\xrightarrow{pK_a = 6.0\text{-}6.2}\quad \text{Nuc-O-}\overset{\overset{\text{O}}{\|}}{\underset{\underset{\text{O}^-}{|}}{\text{P}}}\text{-O}^-$$

$$\text{Nuc-O-}\overset{\overset{\text{O}}{\|}}{\underset{\underset{\text{OH}}{|}}{\text{P}}}\text{-O-}\overset{\overset{\text{O}}{\|}}{\underset{\underset{\text{OH}}{|}}{\text{P}}}\text{-OH} \quad\xrightarrow{pK_a < 1}\quad \text{Nuc-O-}\overset{\overset{\text{O}}{\|}}{\underset{\underset{\text{O}^-}{|}}{\text{P}}}\text{-O-}\overset{\overset{\text{O}}{\|}}{\underset{\underset{\text{O}^-}{|}}{\text{P}}}\text{-OH} \quad\xrightarrow{pK_a = 6.2\text{-}6.4}\quad \text{Nuc-O-}\overset{\overset{\text{O}}{\|}}{\underset{\underset{\text{O}^-}{|}}{\text{P}}}\text{-O-}\overset{\overset{\text{O}}{\|}}{\underset{\underset{\text{O}^-}{|}}{\text{P}}}\text{-O}^-$$

$$\text{Nuc-O-}\overset{\text{O}}{\underset{\text{OH}}{\text{P}}}\text{-O-}\overset{\text{O}}{\underset{\text{OH}}{\text{P}}}\text{-O-}\overset{\text{O}}{\underset{\text{OH}}{\text{P}}}\text{-OH} \quad\xrightarrow{pK_a < 1}\quad \text{Nuc-O-P-O-P-O-P-OH} \;(\text{O}^-,\,\text{O}^-,\,\text{O}^-,\,H^+) \quad\xrightarrow{pK_a = 1.8\text{-}2.0}\quad \text{Nuc-O-P-O-P-O-P-OH}$$

$$\xrightarrow{pK_a = 6.3\text{-}6.5}\quad \text{Nuc-O-P-O-P-O-P-O}^-$$

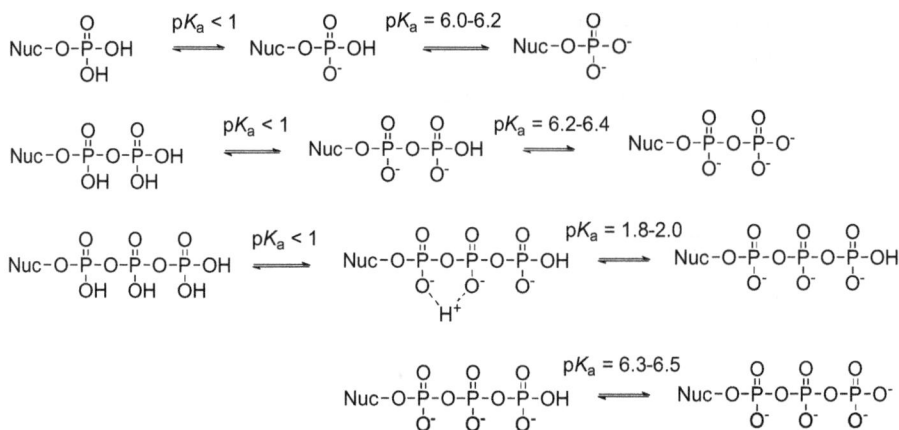

**Figure 3.4:** Protolytic equilibria of nucleoside 5′-mono-, di- and tri-phosphates at $I = 0.1$ M, 25 °C.

**Table 3.1:** Logarithmic stability constants ($\log K = [ML]/[M^{2+}][L^{2-}]$) for the metal ion complexes of nucleoside 5′-monophosphate dianions at $I = 0.1$ M and 25 °C [5].

|        | $Mg^{2+}$ | $Ca^{2+}$ | $Ni^{2+}$ | $Cu^{2+}$ | $Zn^{2+}$ |
|--------|-----------|-----------|-----------|-----------|-----------|
| 5′-CMP | 1.54      | 1.40      | 1.94      | 2.84      | 2.06      |
| 5′-UMP | 1.56      | 1.44      | 1.97      | 2.77      | 2.02      |
| 5′-AMP | 1.60      | 1.46      | 2.49      | 3.14      | 2.38      |

base moiety. In striking contrast, the 3d transition metal complexes of purine nucleoside monophosphates are more stable than their pyrimidine counterparts, as exemplified by 5′-AMP in Table 3.1 [5]. The increased stability evidently results from formation of a macrochelate by simultaneous binding of the transition metal ion to a nonbridging phosphoryl oxygen and N7 of the base moiety. It has been argued that the chelate formation is direct; interaction with both the phosphate and base moieties is inner-sphere coordination, not mediated by a water molecule [6]. As mentioned in Table 3.2, the macrochelate formation is more favored with 6-oxopurine nucleotides, 5′-IMP and 5′-GMP, than with 5′-AMP. Among various metal ions, this binding mode is particularly favored by $Ni^{2+}$.

Transition metal complexes of purine nucleoside 2′- and 3′-monophosphates also exhibit increased stability compared to their pyrimidine counterpart, but the stabilization is more modest than with 5′-monophosphates, decreasing in the order 5′-NMP > 2′-NMP > 3′-NMP [6]. The stabilization has been attributed to macrochelate bridging of a phosphoryl oxygen and N3. With 2′-NMPs, this is possible in the favored *anti*-conformation, whereas 3′-NMPs must adopt an unfavored *syn*-conformation to enable the macrochelate formation.

**Table 3.2:** Percentage of the metal ion complexes of purine nucleoside 5′-phosphates in a macrochelated form [4].

|          | Mn$^{2+}$   | Co$^{2+}$  | Ni$^{2+}$  | Cu$^{2+}$   | Zn$^{2+}$   | Cd$^{2+}$  |
|----------|-------------|------------|------------|-------------|-------------|------------|
| 5′-AMP   | 15 ± 14     | 49 ± 9     | 71 ± 4     | 46 ± 10     | 45 ± 13     | 43 ± 8     |
| 5′-IMP   | 29 ± 12     | 78 ± 4     | 89 ± 2     | 69 ± 6      | 62 ± 7      | 64 ± 5     |
| 5′-GMP   | 40 ± 10     | 83 ± 3     | 93 ± 1     | 81 ± 4      | 72 ± 5      | 70 ± 4     |

Metal ion complexes of nucleoside 5′-triphosphate tetraanions are up to three orders of magnitude more stable than complexes of the corresponding nucleoside 5′-monophosphate dianions (Table 3.3) [8]. 5′-ATP forms more stable complexes with 3d transition metal ions than 5′-CTP or 5′-UTP, most likely due to macrochelate formation between the terminal phosphate and N7 of the purine base. It has been argued that the direct inner sphere coordination (Figure 3.5A) is not as dominating as with 5′-monophosphates, but outer sphere binding to N7 (Figure 3.5B) also markedly contributes to stabilization. Table 3.4 records the percentage of the metal ion complexes of purine nucleoside 5′-triphosphates in a macrochelated form, and in case of 5′-ATP, the distribution between inner and outer spheres macrochelates.

**Table 3.3:** Logarithmic stability constants (log $K = [ML^{2-}]/[M^{2+}][L^{4-}]$) for the metal ion complexes of nucleoside 5′-triphosphate tetraanions at $I = 0.1$ M and 25 °C [2].

|          | Mg$^{2+}$ | Ca$^{2+}$ | Ni$^{2+}$ | Cu$^{2+}$ | Zn$^{2+}$ |
|----------|-----------|-----------|-----------|-----------|-----------|
| 5′-CTP   | 4.20      | 3.85      | 4.52      | 6.03      | 5.03      |
| 5′-UTP   | 4.27      | 3.94      | 4.47      | 5.87      | 5.01      |
| 5′-ATP   | 4.29      | 3.91      | 4.86      | 6.34      | 5.16      |

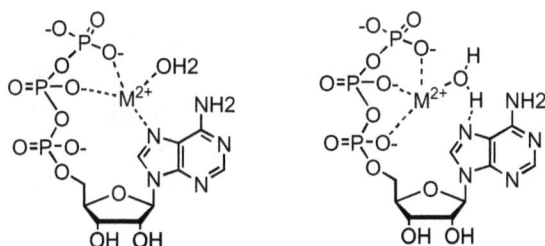

**Figure 3.5:** Inner- (A) and outer-sphere (B) coordination of 3d transition metal ions to 5′-ATP.

**Table 3.4:** Percentage of the metal ion complexes of 5′-ATP [2], 5′-ITP [7] and 5′-GTP [4] in a macrochelated form, and in case of 5′-ATP, the distribution between inner and outer spheres macrochelates.

|  | $Mn^{2+}$ | $Co^{2+}$ | $Ni^{2+}$ | $Cu^{2+}$ | $Zn^{2+}$ | $Cd^{2+}$ |
|---|---|---|---|---|---|---|
| 5′-ATP | 17 ± 10 | 38 ± 9 | 56 ± 4 | 67 ± 2 | 28 ± 7 | 46 ± 4 |
| 5′-ATP inner | 9 | 24 | 31 | 67 | 14 | 28 |
| 5′-ATP outer | 9 | 14 | 25 | 0 | 14 | 18 |
| 5′-ITP | 37 ± 15 | 41 ± 4 | 60 ± 4 | 55 ± 7 | 26 ± 9 | 55 ± 6 |
| 5′-GTP | 38 ± 14 | 52 ± 5 | 74 ± 3 | 67 ± 7 | 28 ± 10 | 51 ± 4 |

## 3.3 Synthesis of nucleotides and their congeners

### 3.3.1 Synthesis of nucleoside monophosphates

Unprotected nucleosides may be converted to their 5′-phosphates with $POCl_3$ in trimethyl phosphate (Yoshikawa reaction) [9] or in a mixture of acetonitrile and wet pyridine (Sowa–Ouchi reaction, Figure 3.6A) [10]. Instead of $POCl_3$, phosphoryltris (1,2,4-triazolide) that is obtained by treating $POCl_3$ with excess of 1,2,4-triazole may be used. This is less reactive than $POCl_3$; hence, formation of a 5′,5′-dinucleoside monophosphate as a side reaction is minimized. An alternative approach is phosphorylation of nucleoside first to a phosphotriester, because neutral triester is easier to purify than ionic monoester. Bis(2,2,2-trichloro-2,2-dimethylethyl) phosphorochloridate, for instance, is used as a phosphorylating agent that has good 5′-OH selectivity (Figure 3.6B) [11]. After purification, the phosphate group may be readily deprotected with cobalt(I) phthalocyanine anion, $[Co^{I}Pc]^-$ [12]. Unprotected pyrimidine nucleosides can be converted to 5′-phosphates with dibenzyl phosphate under Mitsunobu conditions (Figure 3.6C) [13]. On using purine nucleosides, the secondary hydroxyl function has to be protected and the reaction carried out in pyridine [14].

Instead of P(V) phosphorylating agents, P(III) compounds can be used. They esterify (phosphitylate) alcohols more rapidly and quantitatively than P(V) compounds. In other words, P(III) compounds are more susceptible to nucleophilic attack of 5′-OH. All other hydroxyl and amino functions of the starting nucleoside must be protected. The phosphite triesters obtained are then easily oxidized to phosphate triesters, usually with $I_2$ in aqueous MeCN in the presence of lutidine [15] or with *tert*-butylperoxide under anhydrous conditions [16]. Three illustrative examples on the utilization of P(III) chemistry are given in the following.

One commonly used phosphitylation agent is 2-cyanoethyl methyl *N,N*-diisopropylphosphoramidite (Figure 3.7A). To achieve coupling, this phosphoramidite is converted to a more reactive intermediate with the aid of an activator that serves both as an acid and a nucleophile. A conventional activator is tetrazole that displaces the diisopropyla-

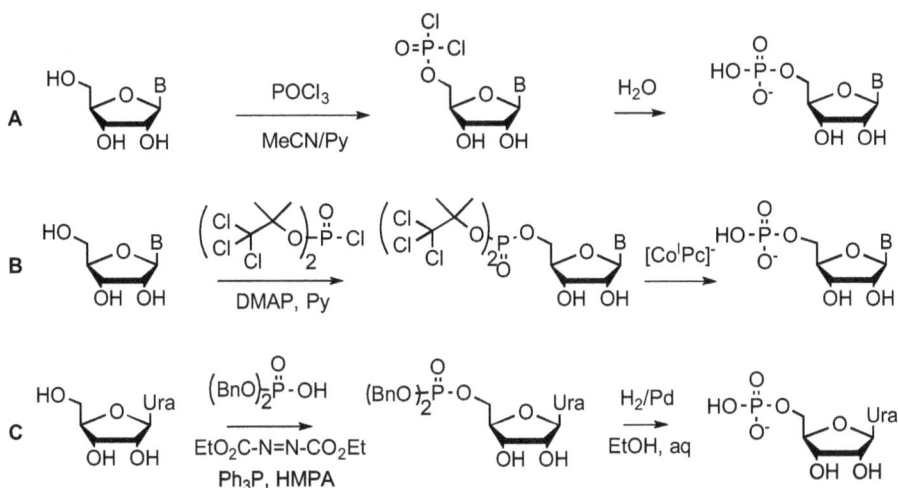

**Figure 3.6:** Conversion of unprotected nucleosides to their 5′-monophosphates: (A) phosphorylation via phosphorodichloridate intermediate, (B) phosphorylation via a phosphotriester intermediate and (C) phosphorylation by Mitsunobu reaction.

mino group by protonating the departing nitrogen atom and attacking on the phosphorus atom as the tetrazolide anion. The sugar hydroxyl function displaces tetrazole giving a phosphite triester that is oxidized in a separate step to a phosphate triester. Instead of tetrazole, acidic 1,3-azolium salts are used as activators [17]. Finally, a two-step deprotection is carried out: demethylation with a thionucleophile followed by base-catalyzed elimination of the 2-cyanoethyl group as acrylonitrile. The reason for utilization of a phosphoramidite bearing two different alkyl groups is extreme susceptibility of bis(2-cyanoethyl) phosphoramidite to nucleophiles. The compound is too unstable to be used without extreme caution. The presence of 2-cyanoethyl, however, is essential since it enables conversion of phosphodiester to monoester by simple ammonolytic elimination instead of more sluggish nucleophilic substitution. Bis[(4-methoxytrityl)benzyl)] N,N-diisopropylphosphoramidite is another viable phosphitylation agent [18]. The (4-methoxytrityl)benzyl groups are removed during P(III) to P(V) oxidation with $I_2$ in aqueous pyridine.

When dialkyl phosphorochloridite is used as a phosphitylation agent, no additional activator is needed, but only a base to neutralize the released HCl. Saligenylchlorophosphite is a chloridite that offers some advantages over its acyclic counterparts. Coupling takes place at low temperature in MeCN in the presence of a sterically hindered base (Figure 3.7B) [19]. The cyclic phosphite triester obtained is then oxidized to phosphate ester and the saligenyl group is removed by hydrolysis in aqueous acetonitrile containing $NEt_3$.

H-phosphonate chemistry offers still a very useful method for the conversion of appropriately protected nucleosides to nucleoside monophosphates. Treatment of nucleoside with commercially available diphenyl *H*-phosphonate in pyridine yields a phenyl H-phosphonate diester intermediate that by addition of aqueous triethylamine is hydrolyzed to nucleoside H-phosphonate diester (Figure 3.7C) [20]. This, when subjected to oxidation with aqueous iodine, gives the desired nucleoside phosphate.

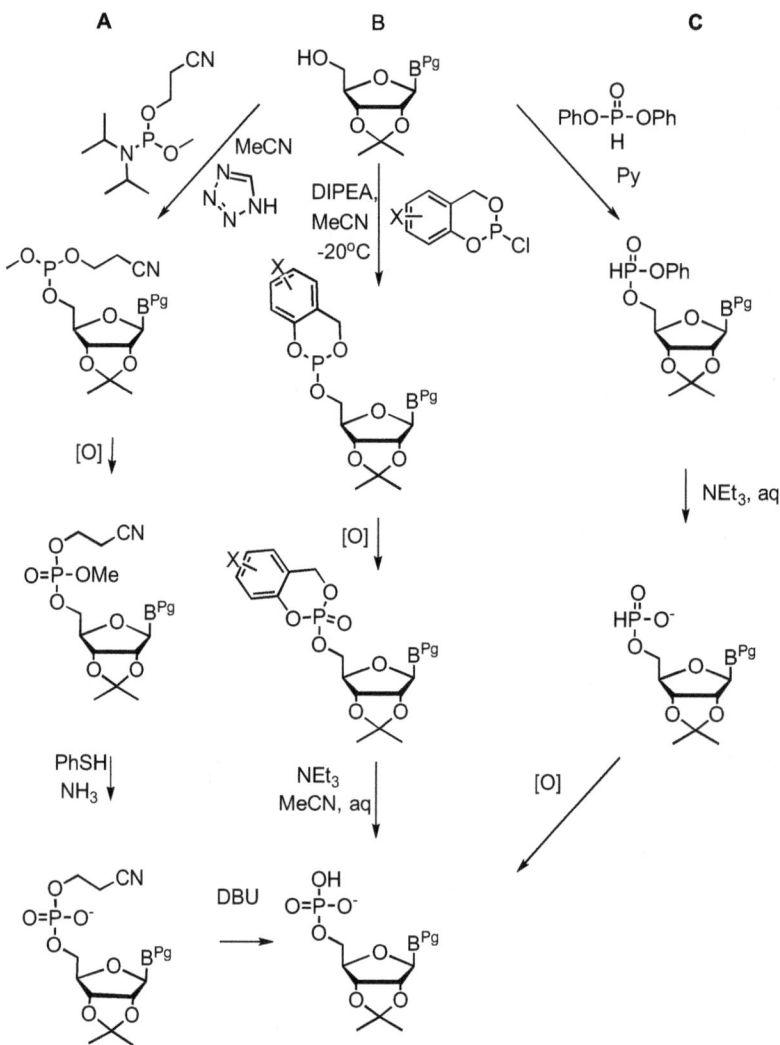

**Figure 3.7:** Conversion of fully protected nucleosides to nucleoside monophosphates by P(III) chemistry: (A) phosphoramidite chemistry, (B) saligenylchlorophosphite chemistry and (C) diphenyl H-phosphonate chemistry.

### 3.3.2 Synthesis of nucleoside phosphorothioates and phosphoramidates

As regards the congeners of nucleoside monophosphates, phosphorothioate monoesters are easily obtained by replacing $POCl_3$ with $PSCl_3$ or any other P(V) phosphorylating agent derived from $PSCl_3$ [21]. Alternatively, an appropriately protected nucleoside may be deprotonated with *tert*-butylmagnesium chloride and reacted with 2-chloro-2-thio-1,3,2-dioxaphospholane (Figure 3.8A) [22]. No base moiety protection is needed. This procedure has been recommended for acid-labile nucleosides, such as 2′,3′-dideoxy purine nucleosides.

Nucleoside phosphorodithioates are obtained by a multistep pathway based on H-phosphonate chemistry (Figure 3.8B) [23]. An appropriately protected nucleoside is first treated in DCM with tris(1,2,4-triazolo)phosphine that is prepared in situ from $PCl_3$ and 1,2,4-triazole. The resulting bis(1,2,4-triazolo)phosphoramidite is then subjected to hydrogen sulfenolysis with $H_2S$ [24] and the nucleoside H-phosphonodithioate obtained is coupled with 9-fluorenylmethanol under oxidative conditions. Elimination of the 9-fluorenylmethyl group by aqueous ammonia finally gives the desired phosphorodithioate.

Nucleoside phosphorodiamidates are, in turn, obtained by the treatment of nucleoside 5′-phosphorodichloridates with dilute aqueous ammonia [25]. Hydrolytic removal of one of the amino groups gives the 5′-phosphoramidate analog (Figure 3.8C).

### 3.3.3 Synthesis of nucleoside diphosphates

Both P(III) and P(V) chemistry may be utilized in the synthesis of nucleoside 5′-diphosphates. On applying P(V) chemistry, the main approach consists of activation of nucleoside 5′-monophosphate with carbonyldiimidazole, giving a phosphoroimidazolidate, and subsequent displacement of imidazole with tris(tetrabutylammonium) phosphate (Figure 3.9A) [26]. According to a more recent version, the phosphoroimidazolide is prepared by coupling imidazole to the triethylammonium salt of nucleoside monophosphate in DMF containing $PPh_3$ and 2,2′-dithiodipyridine as activators [27]. The imidazole ligand is then displaced with tris(tributylammonium) phosphate in the presence of $ZnCl_2$. Nucleoside 5′-diphosphates have also been prepared in moderate yields from 5′-O-tosylnucleosides using tris(tetrabutylammonium) pyrophosphate as a nucleophile [28].

Synthesis via a benzyl phosphoropiperidate intermediate offers an example of sequential utilization of P(III) and P(V) chemistry (Figure 3.9B) [29]. This intermediate is obtained by phosphitylation with benzyl *N,N*-diisopropylchlorophosphoramidite, followed by oxidative coupling of piperidine in a mixture of $CCl_4$ and $Et_3N$ in MeCN. Reductive removal of the benzyl group and treatment with a mixture of 4,5-dicyanoimidazole and bis(tetrabutylammonium) hydrogenphosphate in DMF then gives the 5′-diphosphate. A related strategy consists of phosphitylation of nucleoside monophosphate with bis(9-fluorenylmethyl) *N,N*-diisopropylphosphoramidite, oxidation of P(III) to P(V) and removal

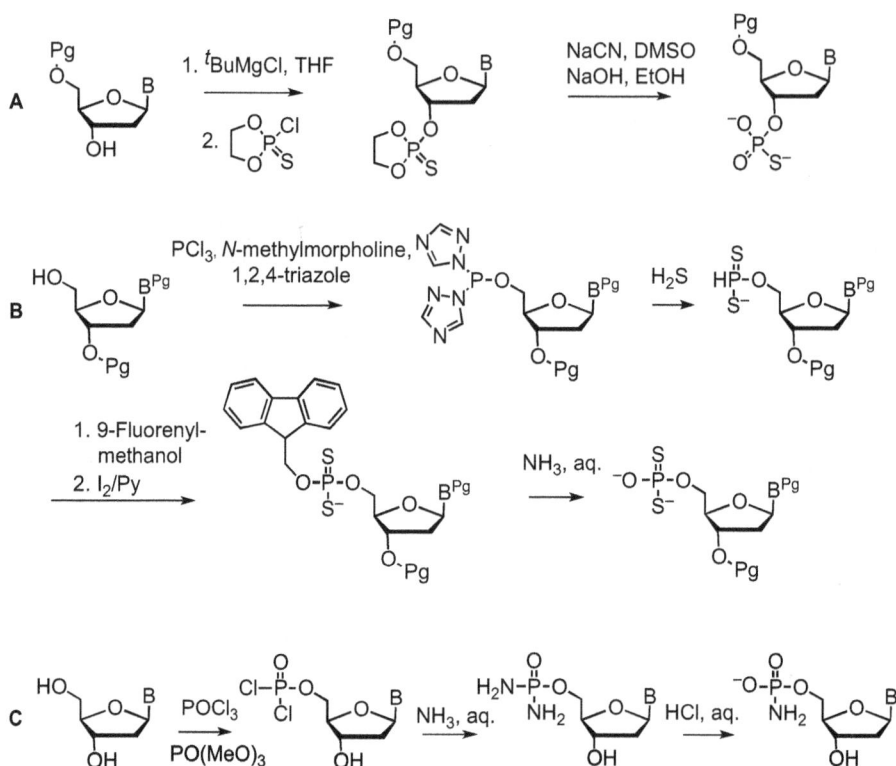

**Figure 3.8:** Synthesis of (A) nucleoside phosphoromonothioates, (B) nucleoside phosphorodithioates and (C) nucleoside phosphorodiamidates and phosphoromonoamidates.

of the 9-fluorenylmethyl groups by piperidine-catalyzed elimination [30]. It is worth noting that the sterically demanding bis(9-fluorenylmethyl) substitution makes the phosphitylation regioselective. The cyclic saligenyl phosphate, obtained as indicated in Figure 3.7B, may be converted to 5′-diphosphate by nucleophilic attack of bis(tetrabutylammonium) hydrogenphosphate (Figure 3.9C) [19].

Sugar nucleotides, such as guanosine 5′-diphospate mannose in Figure 3.1, are obtained in moderate yield by one-step synthesis from appropriate hexose and NDP using 2-chloro-1,3-dimethylimidazolinium chloride as the condensation reagent [31].

### 3.3.4 Synthesis of nucleoside triphosphates

Syntheses of nucleoside 5′-triphosphates closely resemble the preparation of 5′-diphosphates discussed above. On applying P(V) chemistry, unprotected nucleoside is first converted to 5′-phosphorodichloridate by the Yoshikawa reaction (cf. Figure 3.6A) and then reacted with pyrophosphate dianion to obtain a cyclic triphosphate. The latter is then

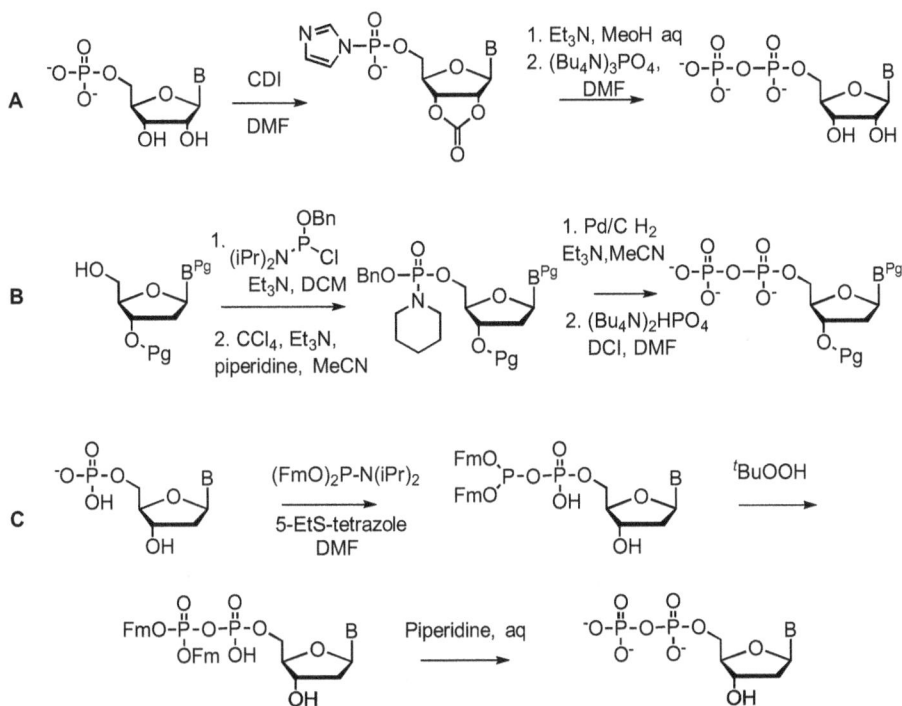

**Figure 3.9:** Alternative syntheses of nucleoside 5′-diphosphates: (A) P(V) chemistry, (B) sequential P(III) and P(V) chemistry and (C) P(III) chemistry.

hydrolyzed to a linear 5′-triphosphate under basic conditions (Figure 3.10A) [32]. Alternatively, nucleoside 5′-monophosphate is converted to phosphoromorpholidate by DCC activation and subsequently to 5′-triphosphate by prolonged treatment with pyrophosphate dianion in DMSO (Figure 3.10B) [33]. It is also possible to utilize a technique similar to that described in Figure 3.9A, that is, activate a nucleoside 5′-monophosphate with imidazole and displace the phosphorus-bound imidazole with a pyrophosphate dianion [34].

When P(III) chemistry is utilized, 5′-O is first phosphitylated by salicoyl phosphorochloridite in a mixture of pyridine, dioxane and DMF (Figure 3.10C) [35]. The attack of a pyrophosphate ion yields a cyclic product that upon $I_2$ oxidation is converted to an acyclic triphosphate. One can also apply repeatedly the methodology described in Figure 3.9C, namely phosphitylate nucleoside 5′-diphosphate with bis(9-fluorenylmethyl) N,N-diisopropylphosphoramidite [30]. Finally, saligenylphosphate, the intermediate (X = NO₂) depicted in Figure 3.7B, can be used as a starting material; ring opening with a pyrophosphate trianion yields 5′-triphosphate [36].

$P^1,P^3$-Dinucleosidyl-5′,5′-triphosphates are usually obtained by the condensation reaction between nucleoside 5′-phosphoroimidazolide and 5′-diphosphate in anhydrous DMF in the presence of $ZnCl_2$. The role of $ZnCl_2$ is to increase electrophilicity of the imidazolide [37]. Alternatively, 5′-phosphorothioate of $P^1$-nucleoside activated with 2,4-dinitrochlorobenzene can be reacted with the 5′-diphosphate of $P^3$-nucleoside [38].

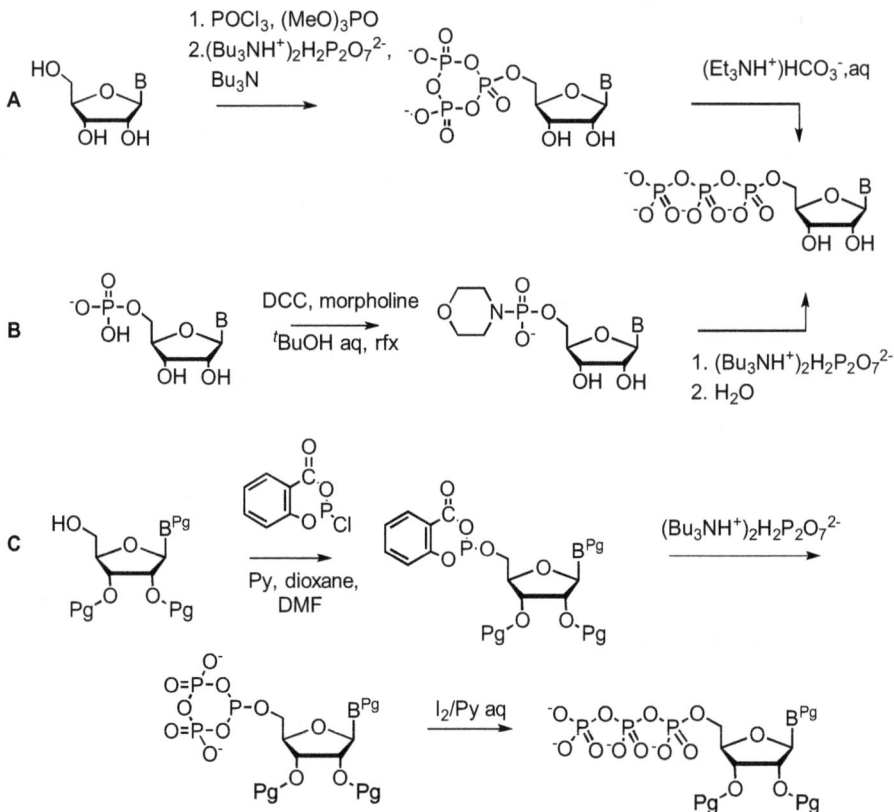

**Figure 3.10:** Alternative approaches for the synthesis of nucleoside 5′-triphosphates. An attack of pyrophosphate on nucleoside 5′-phosphorodichloridate (A), on nucleoside 5′-phosphoromorpholidate (B) or on a salicoyl-protected nucleoside 5′-phosphite (C).

### 3.3.5 Synthesis of nucleoside oligophosphates

Nucleoside 5′-tetraphosphates are obtained by phosphorylation of unprotected nucleosides with a cyclic dihydrogen tetrametaphosphate in DMF under nitrogen (Figure 3.11A) [39], or by phosphorylation of nucleoside 5′-monophosphates with tetrabutylammonium salt of cyclic trimetaphosphate via mesitylene chloride/N-methylimidazole activation in DMF (Figure 3.11B) [40]. Treatment under basic conditions then gives the linear 5′-tetraphosphate. Dinucleoside $P^1,P^4$-tetraphosphates can be prepared by opening the 5′-(cyclic metatriphosphate) of P¹-nucleoside by $Zn^{2+}$-promoted attack of the 5'-monophosphate of P⁴-nucleoside (Figure 3.11C) [41]. Another high yielding route is conversion of cyclic trimetaphosphate to DABCO-activated pyrophosphate followed by displacement of the DABCO groups in a stepwise manner with appropriate nucleoside monophosphates (Figure 3.11D) [42]. Related techniques have been applied for the preparation of terminally modified nucleoside 5'-oligophosphates [43].

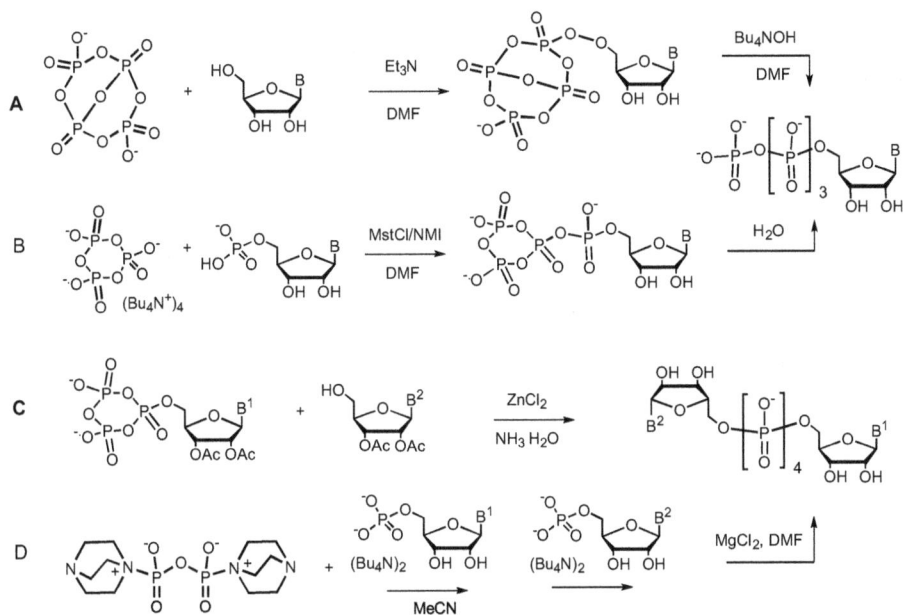

**Figure 3.11:** Alternative approaches for preparation of nucleoside 5'-tetraphosphates and dinucleoside P¹, P⁴-tetraphosphates.

## 3.4 Hydrolysis of nucleotides

### 3.4.1 Hydrolysis of nucleoside monophosphates

Nucleoside 5'-phosphates are hydrolyzed to nucleosides under mildly acidic conditions. The pH-rate profile is bell-shaped with the maximum around pH 5 at 90 °C [44]. The reaction, however, is slow; the half-life for dephosphorylation of 5'-UMP is 64 h at this temperature. The shape of the pH-rate profile strongly suggests that the reactive ionic form is the phosphate monoanion, as with simple alkyl phosphates [45]. In all likelihood, the dissociative mechanism suggested for the hydrolysis of simple alkyl phosphates (Figure 3.12) [46] is followed. The base moiety does not seem to participate. The reaction most likely takes place via a minor tautomer of the 5'-phosphate monoanion having the bridging oxygen atom protonated and all nonbridging oxygen atoms nonprotonated. The P–O⁵' bond is then cleaved without a nucleophilic participation. The released metaphosphate ion is, however, so unstable that it really can depart only when already preassociated with a water molecule. This means that the presence of a water molecule does not accelerate the cleavage as an attacking nucleophile, but is necessary to stabilize the metaphosphate ion by conversion to orthophosphate ion immediately after passing the transition state [47]. The reactive species is the monoanion, because fully deprotonated metaphosphate dianion is maximally resonance stabilized and the departing 5'-O still bears a proton that makes it a feasible leaving group. Phosphoromonothioates are hydrolyzed to nucleosides two orders of magnitude faster than their phosphate counterparts [48], evidently due to higher stability of thiometaphosphate ion [49].

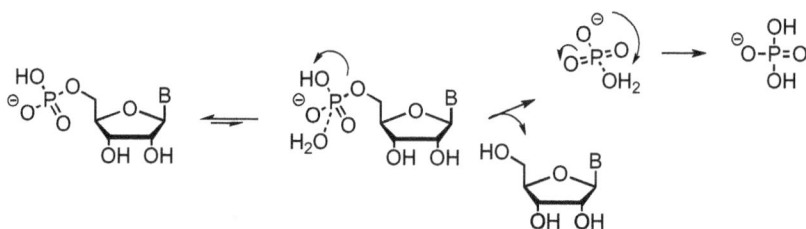

**Figure 3.12:** A preassociation mechanism for the hydrolysis of nucleoside 5-monophosphates.

With 2'- and 3'-monophosphates, the situation is more complicated. Hydrolytic dephosphorylation to nucleoside is 5 times as fast as with 5'-monophosphates [44, 50], but most likely the mechanism is still the same. Migration of the monophosphate group between 2'- and 3'-OH, however, severely competes with the dephosphorylation at pH ~ 2, and under more acidic conditions, migration becomes the predominant reaction. The migration takes place via a pentacoordinated oxyphosphorane intermediate, an unstable species introduced into the mechanistic descriptions of phosphoester reactions by Westheimer [51]. The geometry of oxyphosphorane is a trigonal bipyramid, having three equatorial and two apical ligands (Figure 3.13). Upon formation of the phosphorane intermediate, the entering nucleophile takes an apical position, and upon breakdown of the intermediate, the departing nucleophile may leave only through an apical position. When sufficiently long-lived, the phosphorane may undergo the so-called Berry pseudorotation [52]: one of the equatorial ligands remains equatorial, while the other two take an apical position. The apical ligands, in turn, are converted equatorial. In case two of the ligands are members of a five-membered ring, which is the case with the phosphorane intermediate obtained by the attack of 2'-OH or 3'-OH on the neighboring phosphate, one of these ligands must be apical and the other equatorial. Electronegative ligands favor apical position and electron-rich or bulky ligands equatorial position. These are the Westheimer rules in a nut shell.

**Figure 3.13:** Berry pseudorotation of a trigonal bipyramidal phosphorane intermediate.

As concerns the mutual isomerization of nucleoside 2'- and 3'-monophosphates, a phosphorane intermediate is obtained having either $O^{2'}$ or $O^{3'}$ in an apical position (Figure 3.14). These two phosphoranes are equilibrated via pseudorotation, which ultimately leads to equilibration between 2'- and 3'-monophosphates. The equilibrium constant [3'-NMP]/[2'-NMP] is approximately 2 [50]. It is also worth noting that the same phosphorane intermediate is obtained by hydrolysis of a 2',3'-cyclic phosphate, the equilibrium lying overwhelmingly on the side of acyclic monophosphates.

Since protonation of the phosphate group facilitates the nucleophilic attack on phosphorus and, hence, formation of the phosphorane intermediate, the migration is acid-catalyzed at pH < 2, showing both first- and second-order dependence of rate on hydronium ion concentration. As the acidity is increased, an attack on tri-protonated monocationic 5'-phosphate group gradually becomes faster than the attack on neutral diprotonated group [50]. Between pH 2 and 6, the interconversion of 2'- and 3'-phosphates is pH-independent, proceeding via the predominant monoanionic phosphate. The half-life in this pH region is 10 h at 90 °C ($I = 0.1$ M). As mentioned above, hydrolytic dephosphorylation competes with interconversion, the half-life for the monoanion being 12 h. At pH > 6, the reaction is continuously retarded due to increasing predominance of the unreactive phosphate dianion. With purine nucleoside monophosphates, depurination starts to compete with interconversion and dephosphorylation at pH < 3.

The effect of divalent metal ions on hydrolytic dephosphorylation of NMPs is very modest, usually slightly rate-retarding [53]. In striking contrast, lanthanide ions efficiently catalyze this reaction. The mechanism appears to be quite complex, since the reaction order in hydroxide ion concentration continuously increases on approaching the pH where lanthanide hydroxide starts to precipitate. Evidently, the catalytically active species is a polynuclear cluster containing several hydroxyl functions, as shown for the hydrolysis of 3',5'-cAMP [54]. Some amine complexes of $Co^{3+}$ are also rather efficient catalysts of dephosphorylation [55].

Hydrolysis of nucleoside 2',3'-cyclic phosphates is of special interest since the internucleosidic phosphodiester linkages of RNA are cleaved by transesterification to a 3'-terminal 2',3'-cyclic phosphate. The reaction is exothermic and catalyzed by both acids and bases. The acid-catalyzed hydrolysis is of second order in hydronium ion concentration at pH < 2.5 and the base-catalyzed reaction is of first order in hydroxide ion concentration at pH > 7.5, and the rate constants at 50 °C being $k_{H+} = 0.58$ M$^{-2}$ s$^{-1}$ and $k_{OH-} = 0.66$ M$^{-1}$ s$^{-1}$ [56]. Accordingly, the hydrolysis is approximately as fast at pH 2 and 10, the half-life being of

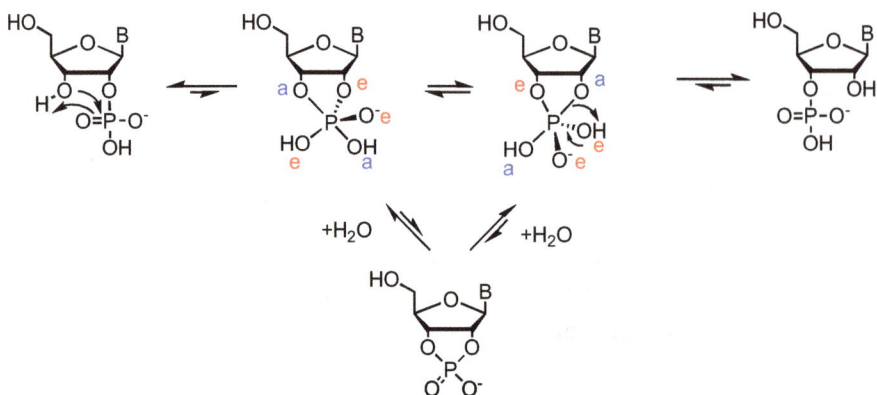

**Figure 3.14:** Interconversion of nucleoside 2′- and 3′-monophosphates and hydrolysis of nucleoside 2′,3′-cyclic phosphate via a pentaoxyphosphorane intermediate.

the order of 3 h at this temperature. On going from pH 2.5 to 5.0, the reaction order in hydronium ion concentration gradually changes from 2 to 0, remains 0 until pH 6.5 and then gradually changes to −1, in other words, to first-order dependence on hydroxide ion concentration. The pH-independent hydrolysis at pH 5.0–6.5 is very slow, and the half-life at 50 °C being almost 200 days [56]. In the pH range from 0 to 6.5, the reaction evidently proceeds by an attack of water, depending on pH, on diprotonated (cationic), monoprotonated (neutral) or anionic cyclic phosphate. At higher pH, the hydroxide ion attacks on the cyclic phosphate monoanion. The pentacoordinated species obtained most likely has a finite lifetime and undergoes breakdown by departure of $O^{2'}$ or $O^{3'}$, giving 40% 2′-NMP and 60% 3′-NMP [50]. The identity of the base moiety has only a minor effect on the kinetics or product distribution [57].

Hydrolysis of nucleoside 3′,5′-cyclic phosphates has, in turn, received interest, owing to the role of adenosine and guanosine 3′,5′-cyclic phosphates as secondary messengers. These six-membered cyclic phosphodiesters are considerably more stable than 2′,3′-cyclic phosphates. The half-life of hydrolysis to a mixture of 3′- and 5′-phosphates is around 1 h in 1 M HCl at 90 °C [58], while the half-life with 2′,3′-cyclic phosphates is of this order at pH 2.5 [50]. Under alkaline conditions, the reaction proceeds by complete inversion at phosphorus, giving a 4:1 mixture of 3′- and 5′-phosphates [59].

## 3.4.2 Hydrolysis of nucleoside triphosphates

Hydrolysis of nucleoside 5′-triphosphates to 5′-diphosphate and orthophosphate closely resembles the hydrolysis of 5′-NMPs. The reaction is dissociative: the cleavage of $P^{\gamma}$–$OP^{\beta}$ bond is far advanced in the transition state, while bond formation between $P^{\gamma}$ and the attacking water molecule is still at very early stage [60]. The half-life is 2.5 h at pH 7.2,

15 min in 0.1 M acid, but 25 h in 0.1 M alkali at 95 °C, consistent with the view that transfer of proton from $P^\gamma$–OH to the bridging oxygen of $P^\gamma$ plays a decisive role in the hydrolysis of NTPs (Figure 3.15A).

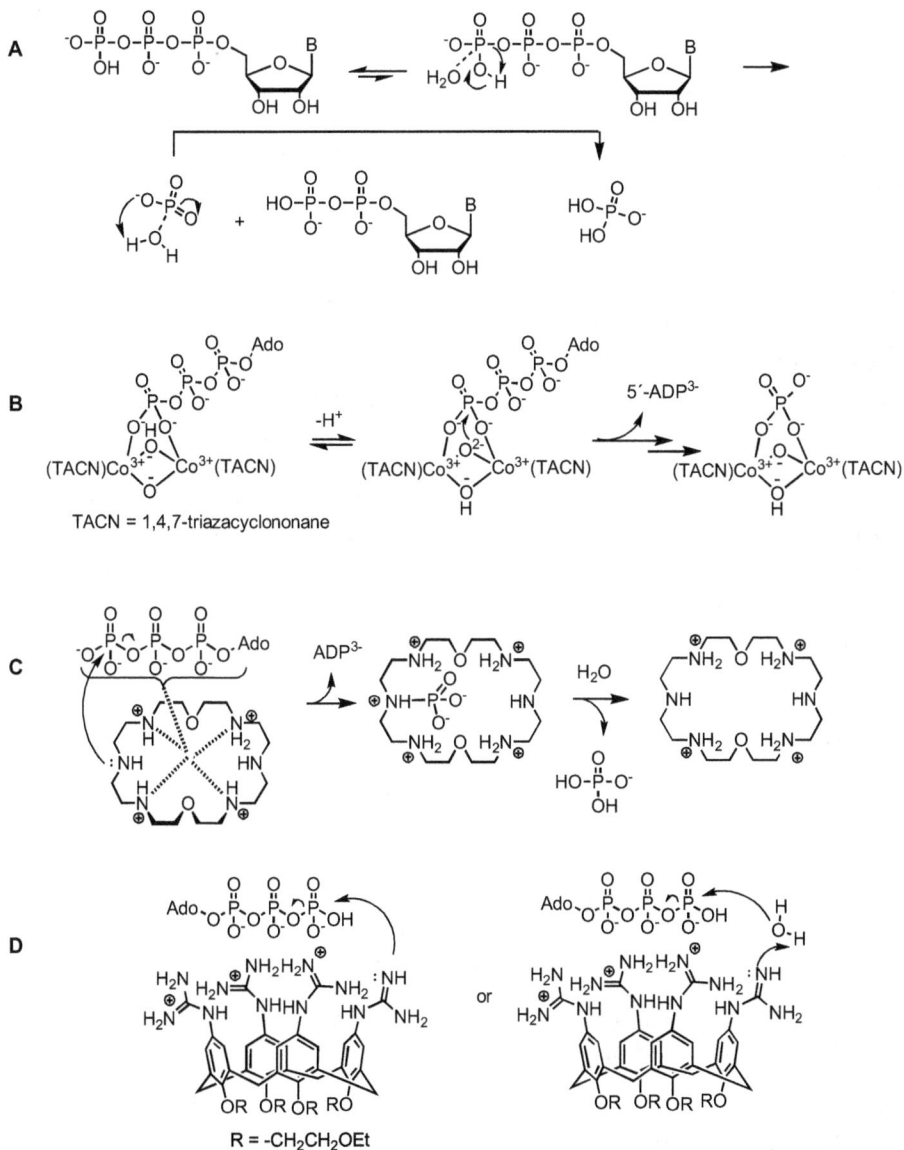

**Figure 3.15:** Mechanisms for the hydrolysis of adenosine 5′-triphosphate: (A) metal ion-independent hydrolysis [58], (B) (1,4,7-triazacyclonone)Co$^{3+}$-catalyzed hydrolysis [59], (C) hydrolysis catalyzed by 1,13-dioxa-4,7,10,16,19,22-hexaazacyclotetracosane tetracation [60, 61] and (D) hydrolysis by cone tetraguanidinocalix[4]arene [62].

$Mg^{2+}$, a common cofactor of enzymatic reactions of NTPs, has only a minor effect on the rate of nonenzymatic hydrolysis of nucleoside 5′-triphosphates. The $Mg^{2+}$ complex of ATP tetra-anion is hydrolyzed 3 times as fast as uncomplexed tetra-anion [61]. The situation, however, is different in the presence of (1,4,7-triazacyclonone) complex of $Co^{3+}$. This binds to the terminal phosphate forming a dinuclear $Co^{3+}$ complex. One of the bridging hydroxide ligands of the $Co^{3+}$ complex attacks $P^\gamma$ resulting in departure of 5′-NDP (Figure 3.15B) [61]. Interestingly, this associative cleavage reaction is subject to rather effective catalysis by $Mg^{2+}$ ions. Evidently, binding of $Mg^{2+}$ to α- and β-phosphates reduces the electron density of the pyrophosphate accelerating the departure of 5′-NDP by three orders of magnitude.

Cyclic polyamines offer a more straightforward means for hydrolysis of nucleoside 5′-triphosphates to 5′-diphosphates. Among several polyamines studied, 1,13-dioxa -4,7,10,16,19,22-hexaazacyclotetracosane, $[24]$-$N_6O_2$, has turned out to be most efficient. It forms a stable complex with ATP and accelerates the cleavage of the terminal phosphate by a factor of $10^3$ at pH 8.5, the hydrolysis rate remaining constant over a pH range from 2.5 to 8.5. The reaction proceeds by an attack of ring nitrogen N7 on the terminal phosphorus atom resulting in formation of a phosphoramidate intermediate that is finally hydrolyzed to orthophosphate and the original $[24]$-$N_6O_2$ (Figure 3.15C) [62]. The half-life at pH 7.6 and 70 °C is 30 min [63]. A comparable cleaving activity has been achieved by the upper rim cone tetraguanidinocalix[4]arene trication in 80% aqueous DMSO (Figure 3.15D) [64]. The protonated guanidine groups evidently anchor the triphosphate and the neutral guanidine group either attacks directly on the terminal phosphate or deprotonates a water molecule concerted with its attack on γ-phosphate.

## Further reading

Burgess K, Cook D Syntheses of nucleoside triphosphates. Chem Rev 2000, 100, 2047–2059.

Hollenstein M Nucleoside triphosphates – Building blocks for the modification of nucleic acids. Molecules 2012, 17, 13569–13591.

IUPAC_IUB Comission on Biochemical Nomenclature. Nomenclature of Phorphorus Containing Compounds of Biochemical Importance. Proc Natl Acad Sci USA 1977, 74, 2222-2230.

Kore AR, Srinivasan B Recent advances in the synthesis of nucleoside phosphates and triphosphates. Curr Org Synth 2013, 10, 903–934.

Mikkola S Nucleotide sugars in chemistry and biology. Molecules 2020, 25, 5755 (27 pages).

Roy B, Depaix A, Perigaud C, Peyrottes S Recent trends in nucleotide synthesis. Chem Rev 2016, 116, 7854–7897.

Shepard SM, Jessen HJ, Cummins CC Beyond triphosphates: Reagents and methods for chemical oligophosphorylation. J Am Chem Soc 2022, 144, 7517–7530.

Sigel H Interactions of metal ions with nucleotides and nucleic acids and their constituents. Chem Soc Rev 1993, 22, 255–267.

# References

[1]  Guthrie JP Tautomerization equilibria for phosphorous acid and its ethyl esters, free energies of formation of phosphorous and phosphonic acids and their ethyl esters, and pK, values for ionization of the B-H bond in phosphonic acid and phosphonic esters. Can J Chem 1979, 57, 236–239.

[2]  Smith RM, Martell AE, Chen Y Critical evaluation of stability constants for nucleotide complexes with protons and metal ions and the accompanying enthalpy changes. Pure Appl Chem 1991, 63, 1015–1080.

[3]  Sigel H, Tribolet R, Malini-Balakrishnan R, Martin RB Comparison of the stabilities of monomeric metal ion complexes formed with adenosine 5'-triphosphate (ATP) and pyrimidine-nucleoside 5'-triphosphate (CTP, UTP, TTP) and evaluation of the isomeric equilibria in the complexes of ATP and CTP. Inorg Chem 1987, 26, 2149–2157.

[4]  Sigel H Interactions of metal ions with nucleotides and nucleic acids and their constituents. Chem Soc Rev 1993, 22, 255–267.

[5]  Massoud SS, Sigel H Metal ion coordinating properties of pyrimidine-nucleoside 5'-monophosphates (CMP, UMP, TMP) and of simple phosphate monoesters, including D-ribose 5'-monophosphate. Establishment of relations between complex stability and phosphate basicity. Inorg Chem 1988, 27, 1447–1453.

[6]  Massoud SS, Sigel H Evaluation of the metal-ion-coordinating differences between the 2'-, 3'- and 5'-monophosphates of adenosine. Eur J Biochem 1989, 179, 451–458.

[7]  Kinjo Y, Corfu NA, Massoud SS, Sigel H Comparison of the extent of macrochelate formation in metal ion ($M^{2+}$) complexes of inosine 5'-monophosphate and inosine 5'-triphosphate. J Inorg Biochem 1991, 43, 463.

[8]  Scheller KH, Hofstetter F, Mitchell PR, Prijs P, Sigel H Macrochelate formation in monomeric metal ion complexes of nucleoside 5'-triphosphates and the promotion of stacking by metal ions. Comparison of the self-association of purine and pyrimidine 5'-triphosphates using proton nuclear magnetic resonance. J Am Chem Soc 1981, 103, 247–260.

[9]  Yoshikawa M, Kato T, Takenishi TA Novel method for phosphorylation of nucleosides to 5'-nucleotides. Tetrahedron Lett 1967, 8, 5065–5068.

[10]  Sowa T, Ouchi S The facile synthesis of 5'-nucleotides by the selective phosphorylation of a primary hydroxyl group of nucleosides with phosphoryl chloride. Bull Chem Soc Japan 1975, 48, 2084–2090.

[11]  Schneiderwind-Stöcklein RGK, Ugi I The 2-trichloromethyl-2-propyl group as a protecting group in oligonucleotide synthesis. Z Naturforsch 1984, 39b, 968–971.

[12]  Eckert H, Listl M, Ugi I The 2,2,2-trichloro-tert-butyloxycarbonyl group (TCBOC) – An acid- and base-resistant protecting group removable under mild conditions. Angew Chem Int Ed Engl 1978, 17, 361–362.

[13]  Kimura J, Fujisawa Y, Yoshizawa T, Fukuda K, Mitsunobu O Preparation of pyrimidine nucleoside 5'-phosphates and N3,5'-purine cyclonucleosides by selective activation of the 5'-hydroxy group. Bull Chem Soc Japan 1979, 52, 1191–1196.

[14]  Saady M, Lebeau L, Mioskowski C Synthesis of adenosine-5'-phosphates and 5'-alkylphosphonates via the Mitsunobu reaction. Tetrahedron Lett 1995, 36, 2239–2242.

[15]  Mateucci MD, Caruthers MH Synthesis of deoxyoligonucleotides on a polymer support. J Am Chem Soc 1981, 103, 3185–3191.

[16]  Hayakawa Y, Uchiyama M, Noyori R Nonaqueous oxidation of nucleoside phosphites to the phosphates. Tetrahedron Lett 1986, 27, 4191–4194

[17]  Tsukamoto M, Oyama K-I. Recent application of acidic 1,3-azolium salts as promoters in the solution-phase synthesis of nucleosides and nucleotides. Tetrahedron Lett 2018, 59, 2477–2484.

[18]  Romanucci V, Zarrelli A, Guaragna A, Di Marino C, Di Fabio G New phosphorylating reagents for deoxyribonucleosides and oligonucleotides. Tetrahedron Lett 2017, 58, 1227–1229.

[19] Wolf S, Zismann T, Lunau N, Warnecke S, Wendicke S, Meyer C A convenient synthesis of nucleoside diphosphate glycopyranoses and other polyphosphorylated bioconjugates. Eur J Cell Biol 2010, 89, 63–75.

[20] Jankowska J, Sobkowski M, Stawinski J, Kraszewski A. Studies on aryl H-phosphonates. I. An efficient method for the preparation of deoxyribo- and ribonucleoside 3′-H-phosphonate monoesters by transesterification of diphenyl H-phosphonate. Tetrahedron Lett 1994, 35, 3355–3358.

[21] Eckstein F Nucleoside phosphorothioates. J Am Chem Soc 1970, 92, 4718–4723.

[22] Szczepanik MB, Desaubry L, Johnson RA Synthesis of nucleoside 3′-thiophosphates in one step procedure. Nucleosides & Nucleotides 1999, 18, 951–953.

[23] Seeberger PH, Yau E, Caruthers MH 2′-Deoxynucleoside dithiophosphates: Synthesis and biological studies. J Am Chem Soc 1995, 117, 1472–1478.

[24] Brill WK-D, Yau EK, Caruthers MH Oxidative and nonoxidative formation of internucleotide linkages. Tetrahedron Lett 1989, 30, 6621–6624.

[25] Simoncsits A, Tomasz J Nucleoside 5′-phosphordiamidates, synthesis and some properties. Nucleic Acids Res 1975, 2, 1223–1224.

[26] Cramer F, Neunhoeffer H. Zür Chemie der Energiereichen Phosphate 15. Reaktionen von Adenosin-5′-Phosphorsaure-Imidazolid –Eine neue Synthese von Adenosindiphosphat und Föavin-Adenin-Dinucleotid. Chem Ber 1962, 95, 1664–1669.

[27] Kore AR, Parmar G Convenient synthesis of nucleoside 5′-diphosphates from the corresponding ribonucleoside-5′-phosphoroimidazole. Synth Commun 2006, 36, 3393–3399.

[28] Davisson VJ, Davis DR, Dixit VM, Poulter CD Synthesis of nucleoside 5′-diphosphates from 5′-O-tosyl nucleosides. J Org Chem 1987, 52, 1794–1801.

[29] Sun Q, Gong SS, Sun J, Liu S, Xiao Q, Pu SZ A P(V)-N activation strategy for the synthesis of nucleoside polyphosphates. J Org Chem 2013, 78, 8417–8426.

[30] Cremosnic GS, Hofer A, Jessen HJ Iterative synthesis of nucleoside oligophoshates with phosphoramidites. Angew Chem Int Ed 2014, 53, 286–289.

[31] Miyagawa A, Toyama S, Ohmura I, Miyazaki S, Kamiya T, Yamamura H One-step synthesis of sugar nucleotides. J Org Chem 2020, 85, 15645–15651.

[32] Ludwig J A new route to nucleoside 5′-triphosphates. Acta Biochim Biophys Acad Sci Hung 1981, 16, 131–133.

[33] Moffatt JG A general synthesis of nucleoside 5′-triphosphates. Can J Chem 1964, 42, 599–604.

[34] Hoard DE, Ott DG Conversion of mono- and oligodeoxyribonucleotides to 5′-triphosphates. J Am Chem Soc 1965, 87, 1785–1788.

[35] Ludwig J, Eckstein F Rapid and efficient synthesis of nucleoside 5′-O-(1-thiotriphosphates), 5′-triphosphates and 2′,3′-cyclophosphorothioates using 2-chloro-4H-1,3,2-benzodioxaphosphorin-4-one. J Org Chem 1989, 54, 631–635.

[36] Warnecke S, Meier C Synthesis of nucleoside di- and tri-phosphates and dinucleoside polyphosphates with cycloSal-nucleotides. J Org Chem 2009, 74, 3024–3030.

[37] Kadokura M, Wada T, Urashima C, Sekine M Efficient synthesis of γ-methyl-capped guanosine 5′-triphosphate as a 5′-terminal unique structure of U6 RNA via a new triphosphate bond formation involving activation of methyl phosphorimidazolidate using $ZnCl_2$ as a catalyst in DMF under unhydrous conditions. Tetrahedron Lett 1997, 38, 8359–8362.

[38] Hasegawa S, Inagaki M, Kato S, Li Z, Kimura Y, Abe H Synthesis of nucleoside oligophosphates by electrophilic activation of phosphorothioate. Org Biomol Chem 2023, 21, 3997–4001.

[39] Shepard SM, Windsor IW, Raines RT, Cummins CC Nucleoside tetra- and pentaphosphates prepared using a tetraphosphorylation reagent are potent inhibitors of ribonuclease A. J Am Chem Soc 2019, 141, 18400–18404.

[40] Mohamady S, Taylor SD Synthesis of nucleoside tetraphosphates and dinucleoside pentaphosphates via activation of cyclic trimetaphosphate. Org Lett 2013, 15, 2612–2615.

[41] Han Q, Gaffney BL, Jones RA One-flask synthesis of dinucleoside tetra- and pentaphosphates. Org Lett 2006, 8, 2075–2077.

[42] Shepard SM, Cummins CC N-donor base adducts of $P_2O_5$ as diphosphorylation reagents. Chemrxiv-2021-8v8nx.

[43] Shepard SM, Kim H, Bang QX, Alhokbany N, Cummins CC Synthesis of α,δ-disubstituted tetraphosphates and terminally-functionalized nucleoside pentaphosphates. J Am Chem Soc 2021, 143, 463–470.

[44] Oivanen M, Lönnberg H Kinetics of reactions of pyrimidine nucleoside 2′- and 3′-monophosphates under acidic and neutral conditions: Concurrent migration, dephosphorylation and deamination. Acta Chem Scand 1990, 44, 239–242.

[45] Cox JR, Ramsay OB Mechanisms of nucleophilic substitution in phosphate esters. Chem Rev 1964, 64, 317–352.

[46] Westheimer F Monomeric metaphosphates. Chem Rev 1981, 81, 313–326.

[47] Jencks WP A primer for the Bema Hapothle. An empirical approach to the characterization of changing transition-state structures. Chem Rev 1985, 85, 511–527.

[48] Ora M, Oivanen M, Lönnberg H Hydrolytic dethiophosphorylation and desulfurization of the monothioate analogues of uridine monophosphates under acidic conditions. J Chem Soc Perkin Trans 2 1996, 0, 771–774.

[49] Burgess J, Blundell N, Cullis PM, Hubbard CD, Misra R Evidence for free monomeric thiometaphosphate anion in aqueous solution. J Am Chem Soc 1988, 110, 7900–7901.

[50] Oivanen M, Lönnberg H Kinetics and mechanisms for reactions of adenosine 2′- and 3′-monophosphates in aqueous acid: Competition between phosphate migration, dephosphorylation and depurination. J Org Chem 1989, 54, 2556–2560.

[51] Westheimer FH Pseudo-rotation in the hydrolysis of phosphate esters. Acc Chem Res 1968, 1, 70–78.

[52] Berry RS Correlation of rates of intramolecular tunneling processes, with application to some group V compounds. J Chem Phys 1960, 32, 933–938.

[53] Kuusela S, Lönnberg H Metal ions that promote the hydrolysis of nucleoside phosphoesters do not enhance intramolecular phosphate migration. J Phys Org Chem 1993, 6, 347–356.

[54] Sumaoka J, Yashiro M, Komiyama M Remarkably fast hydrolysis of 3',5'-cyclic adenosine monophosphate by cerium(iii) hydroxide cluster. J Chem Soc, Chem Commun 1992, 0, 1707–1708.

[55] Chin J, Banaszczyk M Highly efficient hydrolytic cleavage of adenosine monophosphate resulting in a binuclear Co(III) Complex with a novel doubly bidentate μ4-phosphato bridge. J Am Chem Soc 1989; 111, 4103–4105.

[56] Eftink MR, Biltonen RL Energetics of ribonuclease A catalysis. 2. Nonenzymatic hydrolysis of cytidine cyclic 2',3-phosphate. Biochemistry 1983, 22, 5134–5140.

[57] Abrash HI, Cheung -C-CS, Davis JC The nonenzymic hydrolysis of nucleoside 2',3'-phosphates. Biochemistry 1967, 6, 1298–1303.

[58] Oivanen M, Rajamäki M, Varila J, Hovinen J, Mikhailov S, Lönnberg H Additional evidence for the exceptional mechanism of the acid-catalysed hydrolysis of 4-oxopyrimidine nucleosides: Hydrolysis of 1 -(I -alkoxyalkyl)uracils, seconucleosides, 3'-C-alkyl nucleosides and nucleoside 3',5'-Cyclic monophosphates. J Chem Soc Perkin Trans 2 1994, 0, 309–314.

[59] Mehdi S, Coderre JA, Gerlt JA Stereochemical course of the base-catalyzed hydrolysis of cyclic 2′-deoxyadenosine 3′,5′-[$^{17}O,^{18}O$]monophosphate. Tetrahedron 1983, 39, 3483–3492.

[60] Admiraal SJ, Herschlag D Mapping the transition state for ATP hydrolysis: Implications for enzymatic catalysis. Chem & Biol 1995, 2, 729–739.

[61] Williams NH Magnesium ion catalyzed ATP hydrolysis. J Am Chem Soc 2000, 122, 12023–12024.

[62]   Hosseini MV, Lehn J-M, Mertes MP Efficient molecular catalysis of ATP-hydrolysis by protonated macrocyclic polyamines. Helv Chim Acta 1983, 66, 2454–2466.

[63]   Hosseini MW, Lehn J-M, Maggiora L, Mertes KB, Mertes MP Supramolecular catalysis in the hydrolysis of ATP facilitated by macrocyclic polyamines: Mechanistic studies. J Am Chem Soc 1987, 109, 537–544.

[64]   Salvio R, Casnati A, Mandolini L, Sansone F, Ungaro R ATP cleavage by cone tetraguanidinocalix[4]arene. Org Biomol Chem 2012, 10, 8941–8943.

# 4 Nucleosides and nucleotides in chemotherapy

## 4.1 Antiviral nucleosides

The first antiviral nucleoside used as a drug was 5-iodo-2′-deoxyuridine (idoxuridine), accepted in 1963 for the treatment of herpes simplex virus keratitis [1]. This thymidine analog was, however, cardiotoxic and could be used only topically. A real boost for the discovery of antiviral nucleosides was appearance of a new epidemic disease, acquired immune deficiency syndrome (AIDS), in the USA in 1983 [2]. This disease was caused by a retrovirus, human immunodeficiency virus (HIV), against which development of vaccines proved to be difficult, owing to rapidly mutating protein coating of the virus. Even today, no vaccine is available.

The genome of retrovirus consists of two single-stranded RNAs. Once inside the host cell cytoplasm, the virus uses its own reverse transcriptase enzyme to transcribe its RNA genome to DNA. This new DNA is then incorporated into the host cell genome by an integrase enzyme. The host cell accepts the viral DNA as part of its own genome. It translates and transcribes the viral genes along with its own genes, producing the proteins required to assemble new copies of the virus. Obvious targets of antiviral development against retroviruses, hence, are the viral reverse transcriptase and integrase.

Nucleoside drugs used to combat against HIV are inhibitors of reverse transcriptase [1, 3]. The structures of nucleoside analogs approved for clinical use are depicted in Figure 4.1. They all contain a canonical or modestly modified base moiety and a modified 2′,3′-dideoxyfuranosyl group instead of the ribofuranosyl moiety. The first one, Zidovudine (3′-azido-3′-deoxythymidine, AZT), was licensed as early as 1987 and the rest in 1990s. Usually a combination of 2′,3′-dideoxyribonucleosides is used to increase the barrier for development of drug resistance: lamivudine and emtricitabine are the most frequently used components in the mixture. The nucleoside analog first undergoes phosphorylation to 5′-triphosphate by kinases of the host cell, and the triphosphate then inhibits the viral reverse transcriptase. The use of these nucleoside analogs has largely tamed the lethal HIV infection into a chronic condition that can be controlled by combination chemotherapy.

Nucleoside analogs also play a role in chemotherapy of hepatitis C virus (HCV), another single-stranded RNA virus. Two antiviral nucleosides, sofosbuvir and ribavirin, are in clinical use [3]. Their action is based on stepwise phosphorylation to triphosphates that then compete with natural nucleotides for hepatitis polymerases, resulting in chain termination after incorporation into the growing RNA chain. Ribavirin shows activity, besides HCV, against a number of RNA and DNA viruses, including influenza virus and respiratory syncytial virus. It is evidently able to interfere in the metabolism of viral RNA replication by several different mechanisms [4].

In many viruses, the genome is DNA, not RNA. Accordingly, the target for antiviral action is viral DNA polymerase. A nucleoside analog must first become converted

https://doi.org/10.1515/9783111325637-004

Reverse transcriptase inhibtors

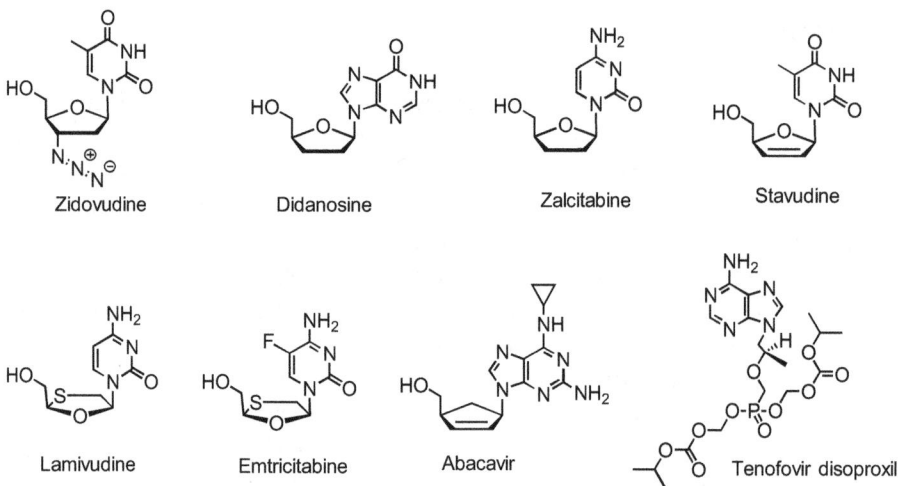

Zidovudine          Didanosine          Zalcitabine          Stavudine

Lamivudine          Emtricitabine          Abacavir          Tenofovir disoproxil

RNA polymerase inhibitors

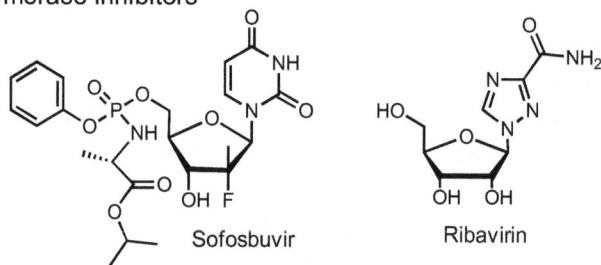

Sofosbuvir          Ribavirin

**Figure 4.1:** Antiviral nucleosides against RNA viruses.

to 5′-triphosphate to be able to serve as a competitive inhibitor of DNA polymerase, and hence, become incorporated into the viral DNA. The first phosphorylation that is rate-limiting may be catalyzed not only by kinases of the host cell but also by kinases of the virus. The viral kinase is often less specific than the kinase of the host cell, and hence, the antiviral nucleoside analog may become phosphorylated only in infected cells and consequently incorporated into the growing DNA chain only in infected cells inhibiting viral replication. Brivudin (BVDU; Figure 4.2), which shows high and selective antiviral activity against herpes simplex virus 1 and varicella zoster virus (VZV), is an example of such an antiviral nucleoside [5].

Acyclic analogs of 2′-deoxyguanosine constitute another group of Herpes simplex antivirals (Figure 4.2). The parent compound in this series is acyclovir [6]. The mechanism of action is similar to that of BVDU. Among the subsequently developed analogs,

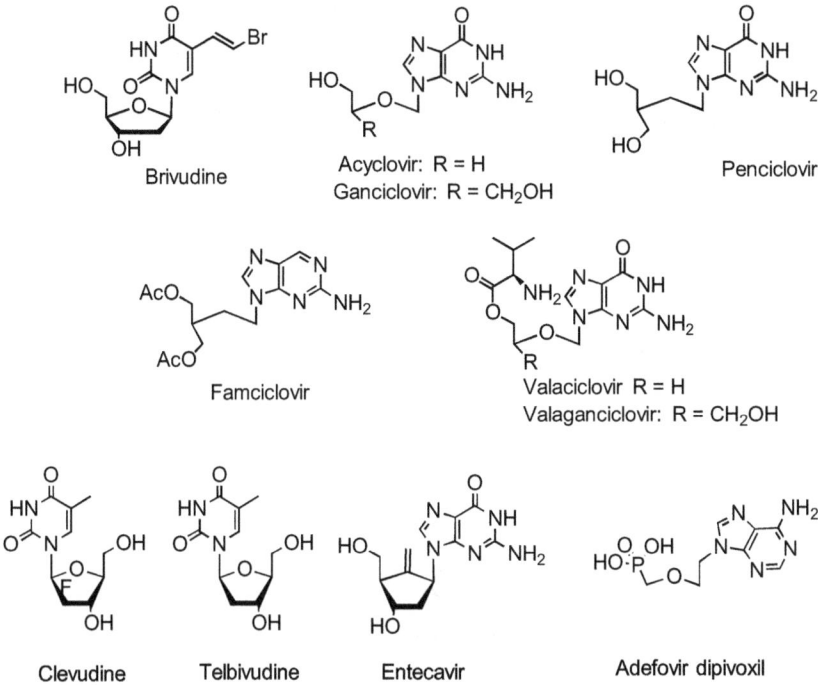

**Figure 4.2:** Antiviral nucleosides against DNA viruses.

ganciclovir is used for the treatment of human cytomegalovirus infections and penciclovir against VZV [1]. To improve water solubility, penciclovir has been converted to its diacetylated 6-deoxyanalog, famciclovir. The latter releases penciclovir in vivo by esterase-catalyzed deacetylation and xanthine oxidase-catalyzed oxidation of the 2-aminopurine moiety to guanine. Valaciclovir, that is, the L-valyl ester of acyclovir, is a prodrug absorbed readily in the intestines through a stereospecific transport system of amino acids. Inside the cell, the valyl moiety is rapidly removed by esterases. Valganciclovir, in turn, is a prodrug of ganciclovir [7].

Lamivudine, one of the reverse transcriptase inhibitors, is additionally licensed to treat hepatitis B virus (HBV) that contains a circular double-stranded DNA genome. In addition, four nucleoside antivirals have been accepted for the treatment of hepatitis B: clevudine, telbivudine, entecavir and adefovir dipivoxil (Figure 4.2). They all inhibit replication/transcription of DNA polymerase.

Severe acute respiratory syndrome coronavirus type 2 (SARS-CoV-2) caused a worldwide pandemic in 2019. The foremost means to fight against this transmissible and pathogenic disease is vaccination (cf. Section 12.5), but antivirals are also needed to save the lives of transmitted patients. SARS-CoV-2 is a single-stranded RNA virus. That is why several antivirals approved earlier for medication against other RNA viruses are now under active drug repurposing against SARS-CoV-2. So far, only one drug, remdesivir, has received permission for emergency use. Remdesivir is an inhibitor of RNA-dependent RNA polymerase. It lowers the respiratory tract infection of SARS-CoV-2, but the influence on mortality appears to be rather modest [8]. The other nucleoside analogs in clinical trials are AT-527 and molnupiravir (Figure 4.3).

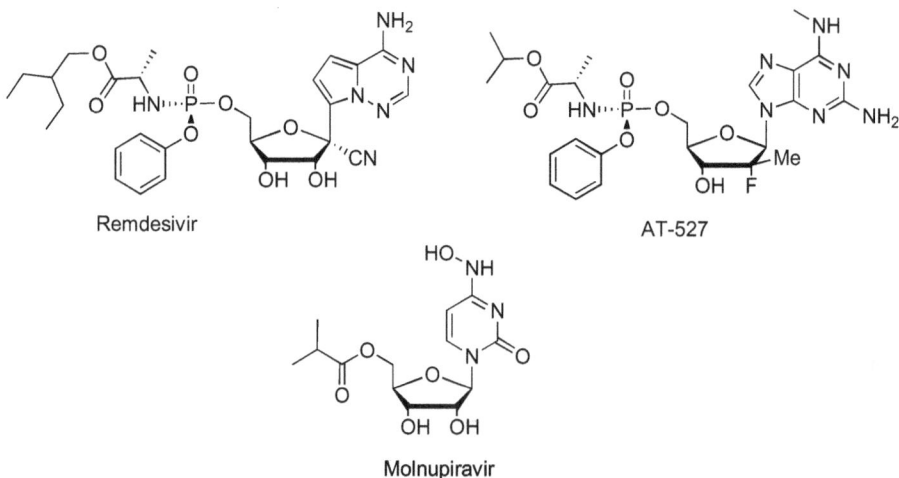

**Figure 4.3:** Nucleoside analogue antivirals inhibiting SARS-CoV-2.

## 4.2 Anticancer nucleosides

Analogs of nucleosides and nucleobases were among the first chemical agents used in cancer chemotherapy [9]. They usually serve as antimetabolites, that is, competitors for canonical nucleosides, resulting in cytotoxicity. Many of the approved anticancer nucleosides are aimed for the treatment of different types of leukemia, but some also against solid tumors [10]. Cellular uptake takes place by membrane-bound transporter proteins, which may allow cell-type selectivity [11, 12]. Inside the cell, the obvious mechanism of cytotoxicity is then inhibition of host-cell kinases and DNA polymerase. Among the approved nucleoside drugs, depicted in Figure 4.4, clofarabine, nelarabine, cytarabine, and cladribine are used for the treatment of leukemia. Gemcitabine is active against solid tumors, including breast cancer, ovarian cancer, nonsmall cell lung cancer, pancreatic cancer and bladder cancer [10]. Capetitabine, in turn, is used for the medication of gastric cancer, colorectal cancer and breast cancer, and floxuridine against colorectal cancer. Azacitidine and its 2′-deoxy analog decitabine are demethylating agents exhibiting antiproliferative activity against cancer cells. Deoxycoformycin is an inhibitor of adenosine deaminase. Additionally, some nucleobase analogs are used as cancer chemotherapeutics. These include thioguanine (2-amino-6-thiopurine;

leukemia), 6-mercaptopurine (leukemia), 5-fluorouracil (various solid tumors) and pe-
metrexed (analog of folic acid; nonsmall cell lung cancer).

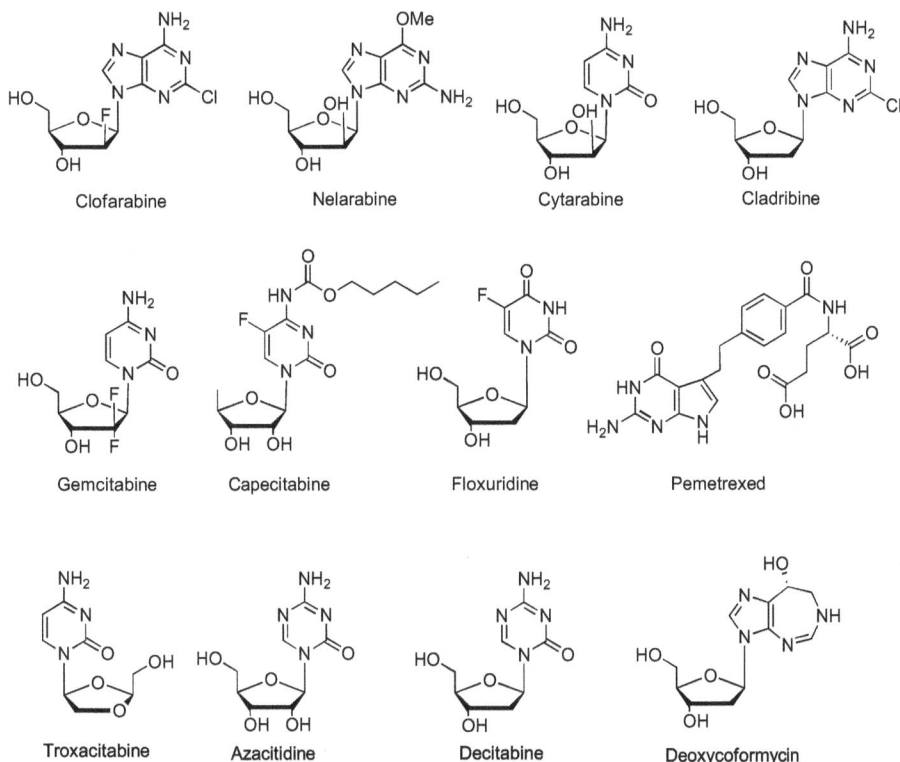

**Figure 4.4:** Nucleoside and nucleobase analogs approved as anticancer drugs.

## 4.3 Antibiotic nucleosides

The appearance of antibiotic-resistant strains of bacteria causes an increasing threat
for common health hastening the discovery of novel antibiotics [13]. Hundreds of thousands of people die annually because of drug-resistant bacterial infections, and this
number is steadily increasing. Repurposing of drugs that are accepted for medical use
offers a possible way to find novel antibiotics. Since nucleosides participate in many
different ways in the life cycle of bacteria interacting with numerous enzymes, their
analogs may have the potential to interfere with bacterial processes vital for cellular
functioning. Owing to multiple ways of interaction, they could well be less vulnerable
to the development of drug resistance [14]. So far, nucleoside and nucleobase analogs
have not been extensively studied as antibiotics, in striking contrast to their role as anticancer and antiviral agents [15]. Figure 4.5 shows some FDA-approved nucleoside ana-

log drugs that additionally have promising antibacterial properties. 5-Fluorocytosine is an antifungal drug, which is actually a prodrug of 5-fluorouracil [16]. It inhibits the growth of *Staphylococcus aureus* and *Staphylococcus epidermidis* [17], and it additionally reduces the biofilm formation of *Escherichia coli* K-12 [18]. 6-Mercaptopurine and its prodrug azathioprine are, in turn, inhibitors of *Mycobacterium avium* subspecies *paratuberculosis* [19]. Azidothymidine antiviral protects patients also from *Salmonella* bacteremia infections [20]. 2′-Deoxy-2′,2′-difluorocytidine, known as the cancer drug gemcitabine, is active against gram-positive bacteria [21].

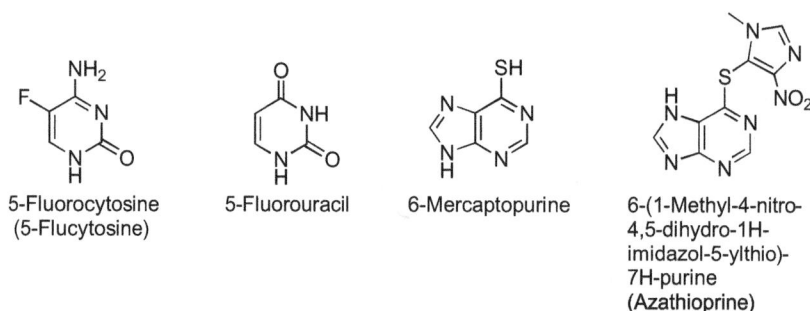

5-Fluorocytosine
(5-Flucytosine)

5-Fluorouracil

6-Mercaptopurine

6-(1-Methyl-4-nitro-
4,5-dihydro-1H-
imidazol-5-ylthio)-
7H-purine
(Azathioprine)

**Figure 4.5:** Nucleobase analogs showing potential as antibacterial agents.

## 4.4 Nucleotides in chemotherapy

As discussed in the preceding sections, many of the currently used antiviral drugs are structural analogs of nucleosides. Their biological activity as inhibitors of DNA or RNA polymerase or RNA reverse transcriptase, however, depends on their conversion to 5′-triphosphate via 5′-mono- and di-phosphates [22]. This process is catalyzed by intracellular kinases, phosphorylation to 5′-monophosphate being usually the rate-limiting step in human cells [23]. Structurally modified nucleosides are, however, phosphorylated by kinases less efficiently than canonical nucleosides [24]. Accordingly, a nucleoside analog may become rejected as a drug candidate due to too inefficient phosphorylation, as triphosphates could be an efficient polymerase or reverse transcriptase inhibitor. Administration of nucleosides as 5′-monophosphates should largely bypass the phosphorylation threshold and lead to improved biological activity. Unfortunately, nucleoside monophosphates are not able to penetrate cellular membranes due to their ionic character. Masking of the phosphate moiety with a biodegradable lipophilic protecting group, hence, offers a viable prodrug strategy with which the therapeutic potential of nucleoside analogs could be improved [25].

To be applicable, a nucleotide prodrug has to: (i) exhibit sufficient extracellular stability, (ii) be sufficiently lipophilic to allow passive diffusion through the cell membrane so that therapeutically significant intracellular concentration is achieved and

(iii) be able, after internalization, to release the parent nucleotide drug by removal of the masking group in nontoxic form. Two of the nonbridging phosphate oxygens are under physiological conditions negatively charged and have to be masked in order to obtain a neutral, lipophilic phosphotriester. Usually a so-called tripartate strategy is applied. This means that an enzyme-labile protecting group is attached to the phosphate group via a linker. An intracellular enzyme removes the actual protecting group and the remaining linker rapidly drops off from the phosphate group by a nonenzymatic mechanism [26].

The feasibility of numerous prodrug strategies has been verified by in vitro studies [27], but only a few of them have so far ended up to clinical use. The first FDA-approved prodrugs of antiviral nucleoside analogs were the bis(pivaloyloxymethyl)-protected [2-(adenine-9-yl)ethoxy]methylphosphonic acid [bis(POM)-adefovir] [28] and bis(isopropoxycarbonyloxymethyl)-protected R-([1-(adenine-9-yl)propan-2-yl]oxymethylphosphonic acid [bis(POC)-tenofovir] [29] for treatment of HBV (2002) and HIV (2001), respectively (Figure 4.6A and B). Bis(POM)-protected 1-[(2-amino-9H-purin-9-yl)methylcyclopropoxy]methylphosphonic acid (besifovir, Figure 4.6C) [30] is in clinical trials as an anti-HBV agent. All these prodrugs seem to release the phosphonate drug by the same mechanism. Intracellular carboxyesterases easily hydrolyze one of the carboxy ester bonds, triggering elimination of formaldehyde (with bisPOM, Figure 4.7A) or stepwise elimination of $CO_2$ and formaldehyde (with bisPOC, Figure 4.7B). The negative charge accumulated on the phosphonate group upon removal of the first protecting group markedly retards the action of carboxyesterase. With 2,2-disubstituted 3-acetyloxypropyl groups, for example, the retardation is more than three orders of magnitude [31]. Evidently, a phosphodiesterase rather than carboxyesterase removes the remaining POC group of bisPOC-protected nucleotides.

The most successful among more recent prodrug strategies is undoubtedly the ProTide strategy [32, 33]. The phosphate group of nucleoside 5′-monophosphate is replaced with a phosphoramidate group derived from an α-amino acid ester, usually L-alanine ester, and the remaining dissociable phosphoryl hydroxyl function is protected as an aryl ester. The stepwise release of a phosphate or phosphonate antiviral from these prodrugs is depicted in Figure 4.7C. Intracellular esterases first cleave the alaninyl ester bond. The exposed carboxy group then attacks on phosphorus displacing the aryloxy ligand. The cyclic intermediate obtained is spontaneously hydrolyzed to an open-chain structure, and the phosphoramidase enzyme cleaves the P–N bond [34]. So far one nucleoside phosphate and one phosphonate prodrug have been approved by the FDA: sofosbuvir (Figure 4.7E) [35] and tenofovir alafenamide (Figure 4.7D) [36] to treat HCV and HIV infections, respectively. Numerous ProTide drug candidates are at various phases of clinical trials [26].

Another prodrug strategy that has produced drug candidates to clinical trials is HepDirect approach based on oxidation of the phosphate protecting group by CYP3A, a cytochrome P450 enzyme in the liver [37]. The drug candidate is protected as 4-phenyl-1,3,2-dioxaphosphinane 2-oxide. Oxidation leads to hydroxylation of the benzylic carbon followed by spontaneous opening of the dioxaphosphinane ring and removal of the pro-

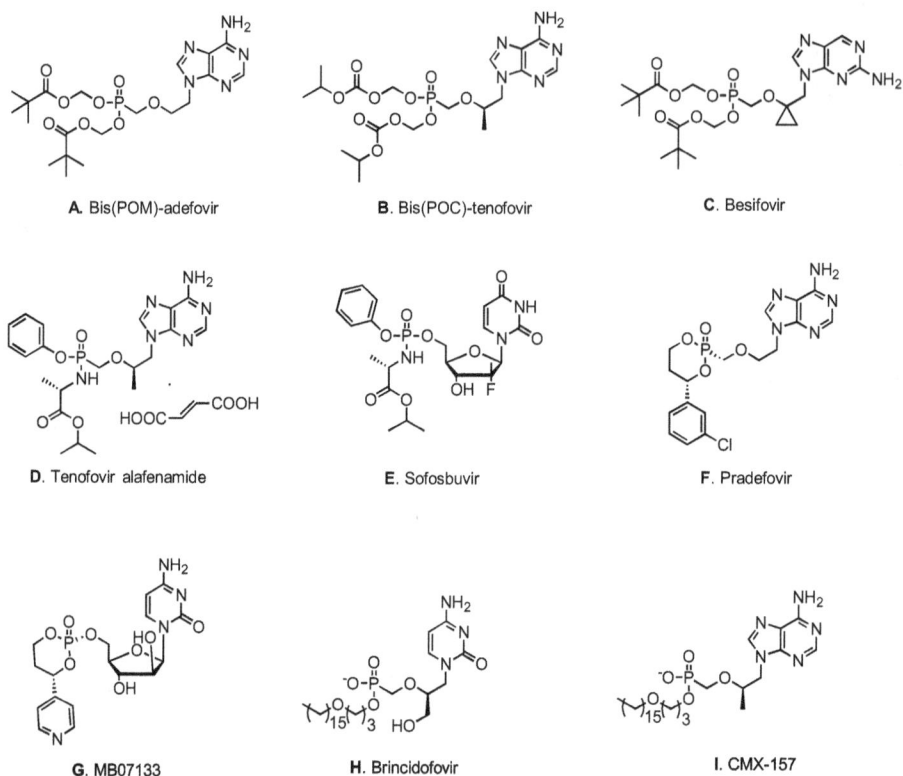

**Figure 4.6:** Nucleoside 5′-monophosphate prodrugs approved for clinical use or in late-phase clinical trials as antivirals.

tecting group by β-elimination (Figure 4.7D). Two HepDirect candidates have proceeded to clinical trials: pradefovir (Figure 4.7F) as an anti-HBV agent and MB07133 (Figure 4.7G) for the treatment of hepatocellular carcinoma.

Interestingly, two phosphonate monoesters of acyclic nucleoside analogs, brincidofovir (Figure 4.7H) [38] and CMX-157 (Figure 4.7I) [39], have proceeded to clinical trials, in spite of the negative charge on the prodrug. Owing to the long hydrophobic 3-(hexadecyloxy)propyl tail, the compounds become metabolized to free phosphonates by phospholipases. Brincidofovir exhibits broad-spectrum antiviral activity, and CMX-157 is a potential anti-HBV drug.

In addition, two thoroughly studied prodrug approaches deserve to be mentioned, namely SATE and CycloSal, even though no candidates based on these strategies are in drug development pipeline. The underlying idea of SATE (S-acyl-2-thioethyl) strategy is esterase-dependent deacylation and subsequent departure of the 2-mercaptoethyl group as ethylene sulfide (Figure 4.7E) [40]. In a variant of this approach, the acyl group is replaced with an alkylthio group. This is removed by reductive cleavage of the disulfide linkage by glutathione, the rest of the deprotection process being identical with the original SATE approach. Possibly, the potential toxicity of ethylene sulfide has retarded the progress of a drug candidate to clinical trials in spite of extensive in vitro studies.

CycloSal strategy is the only purely chemical prodrug approach. The rate of release of saligenyl group from the 5′-phosphate may be adjusted by substituents on the phenyl ring [41]. The phenolic oxygen departs more easily than the benzylic oxygen giving a phosphodiester intermediate (Figure 4.7F). The phenolic hydroxyl group then protonates the benzylic oxygen and stabilizes the departing benzylic carbocation by resonance.

The prodrug strategies for delivery of nucleoside analogs as di- or tri-phosphates have been studied less extensively than those for nucleoside monophosphates. An approach that stands out is protection of the terminal phosphate with two 4-acyloxybenzyl groups derived from long-chain fatty acids. [42–44]. The long-chain acyl groups evidently warrant sufficient cellular uptake, in spite of negative charges of the nonterminal phosphates [45]. Deacylation by intracellular esterases results in stepwise departure of the exposed 4-hydroxybenzyl groups as 4-methylenecyclohexa-2,5-dien-1-one, leaving the phosphoanhydride linkages largely intact (Figure 4.8). Some otherwise inactive nucleoside analogues have shown anti-HIV activity when introduced into the test system as this kind of prodrugs [46]. In addition, the efficiency of abacavir, an approved anti-HIV nucleoside drug, was improved when administered as a triphosphate prodrug [43].

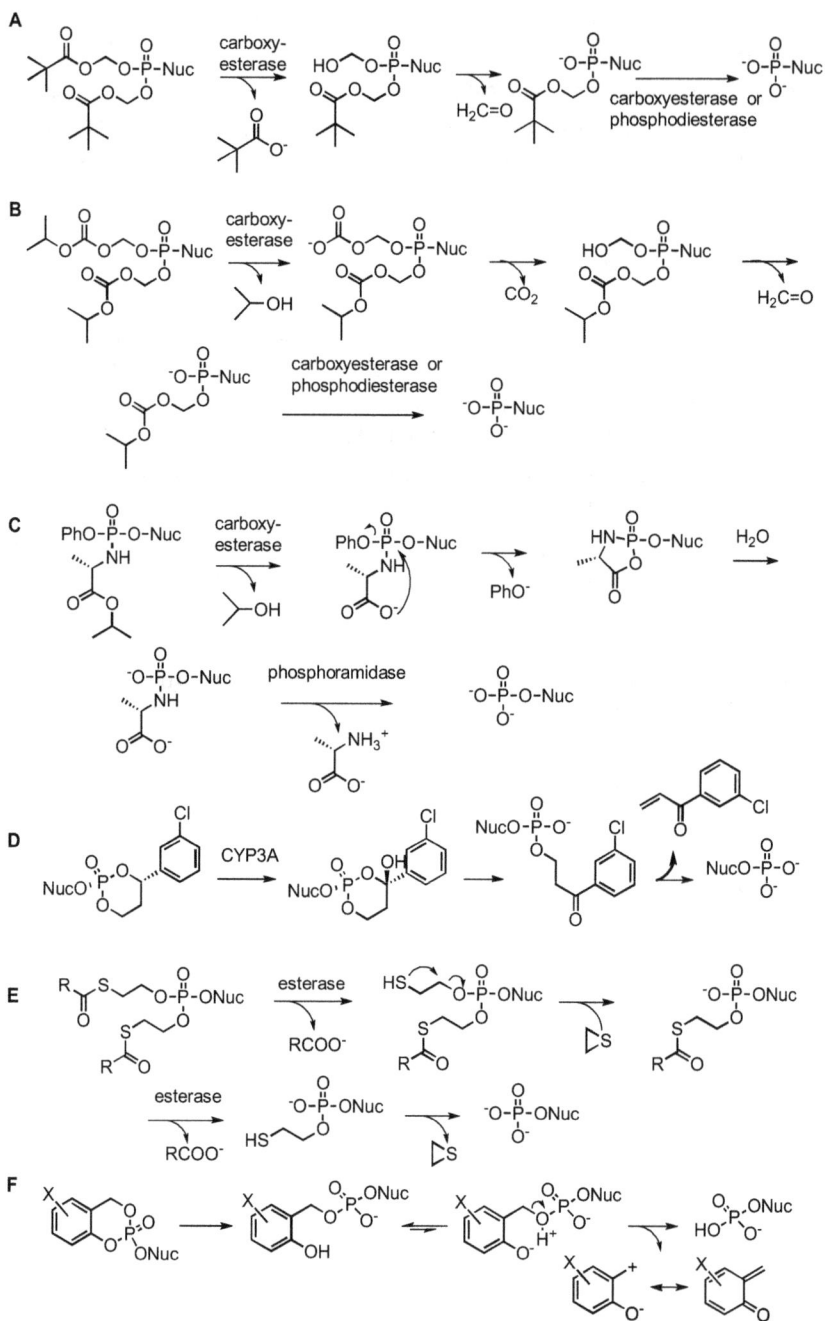

**Figure 4.7:** Mechanisms for the removal of biodegradable protecting groups from prodrugs of antiviral nucleoside 5′-monophosphates. The release of effective drugs is triggered by an intracellular esterase (A, B, C, E), by an oxidative cytochrome P450 enzyme in the liver (D) or chemically (F).

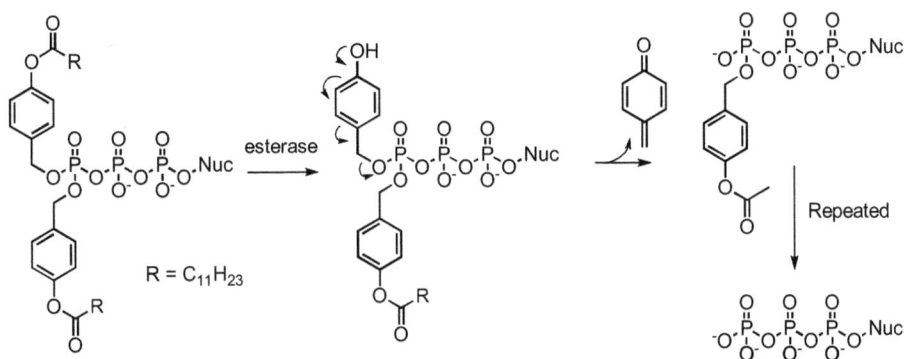

**Figure 4.8:** Esterase-triggered deprotection of bis(4-acyloxybenzyl)-protected nucleoside 5′-triphosphate.

## Further reading

Ramesh D, Vijayakumar BG, Kannan T. Advances in nucleoside and nucleotide analogues in tackling human immunodeficiency virus and hepatitis virus infections. ChemMedChem 2021, 16, 1403–1419.
Groaz E, De Clercq E, Herdewijn P. Anno 2021: Which antivirals for the coming decade. Annu Reports Med Chem 2021, 57, 49–107.

## References

[1]     De Clerc E, Li G. Approved antiviral drugs over the past 50 years. Clin Microbiol Rev 2016, 29, 695–747.
[2]     Gallo RC, Montagnier L. The discovery of HIV as the cause of AIDS. New Engl J Med 2003, 349, 2283–2285.
[3]     Ramesh D, Vijayakumar BG, Kannan T. Advances in nucleoside and nucleotide analogues in tackling human immunodeficiency virus and hepatitis virus infections. ChemMedChem 2021, 16, 1403–1419.
[4]     Graci JD, Cameron CE. Mechanisms of action of ribavirin against distinct viruses. Rev Med Virology 2006, 16, 37–48.
[5]     De Clercq E. Discovery and development of BVDU (brivudin) as a therapeutic for the treatment of herpes zoster. Biochem Pharmacol 2004, 68, 2301–2315.
[6]     Schaeffer HJ, Beauchamp L, de Miranda P, Elion GB, Bauer DJ, Collins P. 9-(2-Hydroxyethoxymethyl) guanine activity against viruses of the herpes group. Nature 1978, 272, 583–585.
[7]     De Clercq E, Field HJ. Antiviral prodrugs-the development of successful prodrug strategies for antiviral chemotherapy. Br J Pharmacol 2006, 147, 1–11.
[8]     Groaz E, De Clercq E, Herdewijn P. Anno 2021: Which antivirals for the coming decade. Annu Reports Med Chem 2021, 57, 49–107.
[9]     Galmarini CM, Mackey JR, Dumontet C. Nucleoside analogues and nucleobases in cancer treatment. Lancet Oncol 2002, 3, 415–424.
[10]   Mirza AZ. Advancement in the development of heterocyclic nucleosides for the treatment of cancer. A review. Nucleosides Nucleotides & Nucleic Acids 2019, 38, 836–857.

[11]   Damaraju VL, Damaraju S, Young JD, Baldwin SA, Mackey J, Sawyer MB, Cass CE. Nucleoside anticancer drugs: The role of nucleoside transporters in resistance to cancer chemotherapy. Oncogene 2003, 22, 7524–7536.

[12]   Koczor CA, Torres RA, Lewis W. The role of transporters in the toxicity of nucleoside and nucleotide analogs. Expert Opin Drug Metab Toxicol 2012, 8, 665–676.

[13]   O'Connell KM, Hodgkinson JT, Sore HF, Welch M, Salmond GP, Spring DR. Combating multidrug-resistant bacteria: Current strategies for the discovery of novel antibacterials. Angew Chem Int Ed 2013, 52, 10706–10733.

[14]   Yssel AEJ, Vanderleyden J, Steenackers HP. Repurposing of nucleoside- and nucleobase-derivative drugs as antibiotics and biofilm inhibitors. J Antimicrob Chemother 2017, 72, 2156–2170.

[15]   Landini P, Antoniani D, Burgess JG, Nijland R. Molecular mechanisms of compounds affecting bacterial biofilm formation and dispersal. Appl Microbiol Biotechnol 2010, 86, 813–823.

[16]   Vermes A, Guchelaar H, Dankert J. Flucytosine: A review of its pharmacology, clinical indications, pharmacokinetics, toxicity and drug interactions. J Antimicrob Chemother 2000, 46, 171–179.

[17]   Gieringer JH, Wenz AF, Just HM, Daschner FD. Effect of 5-fluorouracil, mitoxantrone, methotrexate, and vincristine on the antibacterial activity of ceftriaxone, ceftazidime, cefotiam, piperacillin, and netilmicin. Chemotherapy 1986, 32, 418–424.

[18]   Attila C, Ueda A, Wood T. 5-Fluorouracil reduces biofilm formation in Escherichia coli K-12 through global regulator AriR as an antivirulence compound. Appl Microbiol Biotechnol 2009, 82, 525–533.

[19]   Greenstein RJ, Su L, Haroutunian V, Shahidi A, Brown ST. On the action of methotrexate and 6-mercaptopurine on M. avium subspecies paratuberculosis. PLoS One 2006, 2, e161 (5 pages).

[20]   Casado JL, Valdezate S, Calderon C, Navas E, Frutos B, Guerrero A, Martinez-Beltran J. Zidovudine therapy protects against Salmonella bacteremia recurrence in human immunodeficiency virus-infected patients. J Infect Dis 1999, 179, 1553–1556.

[21]   Sandrini MP, Clausen AR, On SL, Aarestrup FM, Munch-Petersen B, Piskur J. Nucleoside analogues are activated by bacterial deoxyribonucleoside kinases in a species-specific manner. J Antimicrob Chemother 2007, 60, 510–520.

[22]   De Clercq E. Strategies in the design of antiviral drugs. Nat Rev Drug Discovery 2002, 1, 13–25.

[23]   Jordheim LP, Durantel D, Zoulim F, Dumontet C. Advances in the development of nucleoside and nucleotide analogues for cancer and viral diseases. Nat Rev Drug Discovery 2013, 12, 447–464.

[24]   Stein DS, Moore KH. Phosphorylation of nucleoside analog antiretrovirals: A review for clinicians. Pharmacotherapy 2001, 21, 11–34.

[25]   Pradere U, Garnier-Amblard EC, Coats SJ, Amblard F, Schinazi RF. Synthesis of nucleoside phosphate and phosphonate prodrugs. Chem Rev 2014, 114, 9154–9218.

[26]   Thornton PJ, Kadri H, Miccoli A, Mehellou Y. Nucleoside phosphate and phosphonate prodrug clinical candidates. J Med Chem 2016, 59, 10400–10410.

[27]   Poijärvi-Virta P, Lönnberg H. Prodrug approaches of nucleotides and oligonucleotides. Curr Med Chem 2006, 13, 3441–3465.

[28]   Srivastva DN, Farquhar D. Bioreversible phosphate protective groups: Synthesis and stability of model acyloxymethyl phosphates. Bioorg Chem 1984, 12, 118–129.

[29]   Arimilli MN, Kim CU, Dougherty J, Mulato AS, Oliyai R, Shaw JP, Cundy KC, Bischofberger N. Synthesis, in vitro biological evaluation and oral bioavailability of 9-[2-(phosphonomethoxy)propyl]adenine (PMPA) prodrugs. Antiviral Chem Chemother 1997, 8, 557–564.

[30]   Choi JR, Cho DG, Roh KY, Hwang JT, Ahn S, Jang HS, Cho WY, Kim KW, Cho YG, Kim J, Kim YZA. Novel class of phosphonate nucleosides. 9-[(1-Phosphonomethoxycyclopropyl)methyl]guanine as a potent and selective anti-HBV agent. J Med Chem 2004, 47, 2864–2869.

[31]   Ora M, Taherpour S, Linna R, Leisvuori A, Hietamäki E, Poijärvi-Virta P, Beigelman L, Lönnberg H. Biodegradable protections for nucleoside 5′-monophosphates: Comparative study on the

removal of O-acetyl and O-acetyloxymethyl protected 3-hydroxy-2,2-bis(ethoxycarbonyl)propyl groups. J Org Chem 2009, 74, 4992–5001.

[32] Mehellou Y, Balzarini J, McGuigan C. Aryloxy phosphoramidate triesters: A technology for delivering monophosphorylated nucleosides and sugars into cells. ChemMedChem 2009, 4, 1779–1791.

[33] Mehellou Y, Rattan HC, Balzarini J. The ProTide prodrug technology: From the concept to the clinic. J Med Chem 2018, 61, 2211–2226.

[34] Murakami E, Tolstykh T, Bao HY, Niu CR, Steuer HMM, Bao DH, Chang W, Espiritu C, Bansal S, Lam AM, Otto MJ, Sofia MJ, Furman PA. Mechanism of activation of PSI-7851 and its diastereoisomer PSI-7977. J Biol Chem 2010, 285, 3437–3447.

[35] Sofia MJ, Bao D, Chang W, Du J, Nagarathnam D, Rachakonda S, Reddy PG, Ross BS, Wang P, Zhang HR, Bansal S, Espiritu C, Keilman M, Lam AM, Steuer HM, Niu C, Otto MJ, Furman PA. Discovery of a beta-D-2'-deoxy-2'-alphafluoro-2'-beta-C-methyluridine nucleotide prodrug (PSI-7977) for the treatment of hepatitis C virus. J Med Chem 2010, 53, 7202–7218.

[36] Chapman H, Kernan M, Prisbe E, Rohloff J, Sparacino M, Terhorst T, Yu R. Practical synthesis, separation, and stereochemical assignment of the PMPA pro-drug GS-7340. Nucleosides Nucleotides Nucleic Acids 2001, 20, 621–628.

[37] Erion MD, Bullough DA, Lin CC, Hong Z. Hepdirect prodrugs for targeting nucleotide-based antiviral drugs to the liver. Curr Opin Invest Drugs 2006, 7, 109–117.

[38] Ciesla SL, Trahan J, Wan WB, Beadle JR, Aldern KA, Painter GR, Hostetler KY. Esterification of cidofovir with alkoxyalkanols increases oral bioavailability and diminishes drug accumulation in kidney. Antiviral Res 2003, 59, 163–171.

[39] Painter GR, Almond MR, Trost LC, Lampert BM, Neyts J, De Clercq E, Korba BE, Alder KA, Beadle JR, Hostetler KY. Evaluation of hexadecyloxypropyl-9-R-[2-(phosphonomethoxy)propyl]- adenine, CMX157, as a potential treatment for human immunodeficiency virus type 1 and hepatitis B virus infections. Antimicrob Agents Chemother 2007, 51, 3505–3509.

[40] Puech F, Gosselin G, Lefebvre I, Pompon A, Aubertin A-M, Kirn A, Imbach J-L. Intracellular delivery of nucleoside monophosphates through a reductase-mediated activation process. Antiviral Res 1993, 22, 155–174.

[41] Meier C. cycloSal phosphates as chemical Trojan horses for Intracellular nucleotide and glycosylmonophosphate delivery – Chemistry meets biology. Eur J Org Chem 2006, 0, 1081–1102.

[42] Jessen HJ, Schulz T, Balzarini J, Meier C. Bioreversible protection of nucleoside diphosphates. Angew Chem Int Ed 2008, 47, 8719–8722.

[43] Weising S, Sterrenberg W, Schols D, Meier C. Synthesis and antiviral evaluation of TriPPPro-AbacavirTP, TriPPPro-CarbovirTP, and their 1',2'-cis-disubstituted analogues. ChemMedChem 2018, 13, 1771–1778.

[44] Zhao C, Weber S, Schols D, Balzarini J, Meier C. Prodrugs of γ-alkyl-modified nucleoside triphosphates: Improved inhibition of HIV reverse transcriptase. Angew Chem Int Ed 2020, 59, 22063–22071.

[45] Gollnest T, de Oliveira TD, Schols D, Balzarini J, Meier C. Lipophilic prodrugs of nucleoside triphosphates as biochemical probes and potential antivirals. Nature Commun 2015, 6, 8716 (15 pages).

[46] Gollnest T, de Oliveira TD, Rath A, Hauber I, Schols D, Balzarini J, Meier C. Membrane-permeable triphosphate prodrugs of nucleoside analogues. Angew Chem Int Ed 2016, 55, 5255–5258.

# 5 Oligonucleotides: synthesis

## 5.1 Synthesis of oligodeoxyribonucleotides

### 5.1.1 Preparation of building blocks

Oligonucleotides are linear polymers of nucleosides linked to each other invariably through 3',5'-phosphodiester linkages. The length of the chain may vary from just a few nucleotides up to 100 units. Oligonucleotides used as tools in cell biology or drug development typically fall in the range from 15 to 30 nucleotides.

Although oligonucleotides may be synthesized by a convergent strategy in solution, they usually are prepared by stepwise coupling on a solid support. Most often, the 3'-terminal nucleoside is attached through 3'-O onto a support via an appropriate linker. The next nucleoside is then coupled to the 5'-OH of the support-anchored nucleoside. Both P(III) and P(V) chemistry may be applied, but P(III) chemistry is nowadays almost exclusively used in solid-supported synthesis. Phosphoramidite chemistry [1] predominates as the coupling method, but H-phosphonate chemistry [2, 3] and P(V)chemistry [4] *are* also used for special purposes.

Before 2'-deoxynucleosides are subjected to phosphitylation or phosphorylation to obtain P(III) or P(V) building blocks, respectively, their 5'-OH and primary amino functions have to be protected. The most common approach is to trimethylsilylate the sugar hydroxyl functions, acylate the base moiety amino group, remove the silyl protections and introduce 4,4'-dimethoxytrityl (DMTr) group regioselectively to 5'-O. As indicated in Figure 5.1, 2'-deoxyadenosine is usually protected with benzoyl, 2'-deoxyguanosine with isobutyryl and 2'-deoxycytidine with benzoyl or acetyl groups. Thymidine is normally used without base protection.

The appropriately protected nucleosides are converted to phosphoramidite building blocks either by treating with 2-cyanoethyl-*N,N*-diisopropylphosphorochloridite in MeCN in the presence of Et$_3$N [5, 6] or with 1,1'-(2-cyanoethoxyphosphanediyl)bis(1H-1,2,4-triazole) in the same solvent (Figure 5.2) [7]. In the latter case, the first triazole ligand is displaced with the nucleoside and the second with silylated diisopropylamine. The phosphitylation agents employed are commercially available, but they may also be obtained by replacing two of the chloro substituents of PCl$_3$ either with 2-cyanoethanol and diisopropyl amine, in this order, or with 1H-1,2,4-triazole.

The reactivity of a phosphoramidite building block largely depends on the size of the alkyl substituents on nitrogen. If exceptionally high reactivity is needed, *N,N*-diethylphosphoramidite can be used. *N,N*-Dimethyl analog is already too unstable for practical purposes. As discussed below in more detail, methyl, allyl or 2-chlorophenyl groups are sometimes used instead of 2-cyanoethyl.

Nucleoside phosphoramidites may be purified by column chromatography on dried silica gel. To warrant the stability of the phosphoramidite moiety, it is advisable

https://doi.org/10.1515/9783111325637-005

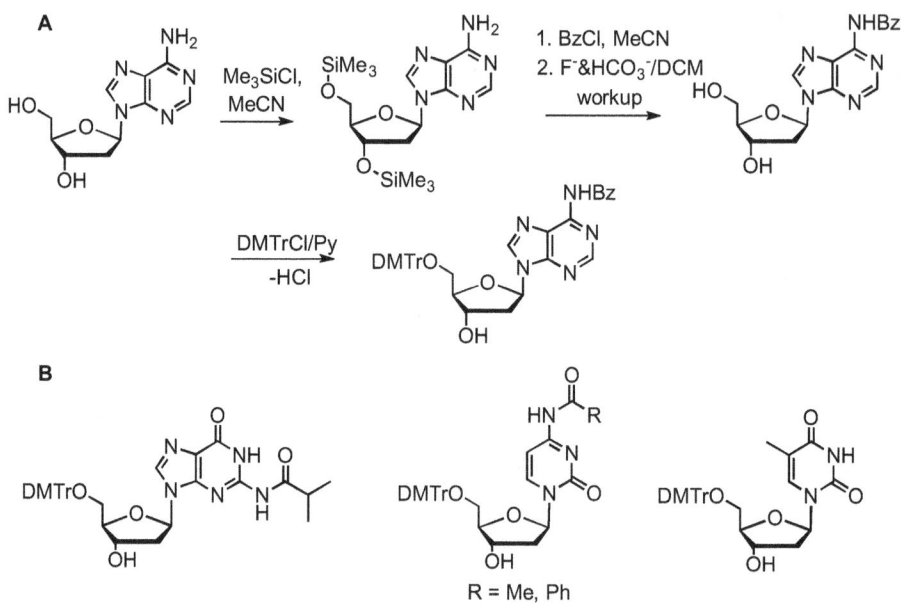

**Figure 5.1:** (A) Protection of 2′-deoxyadenosine for the synthesis of building blocks by 3′-O phosphitylation or phosphorylation. (B) 5′-O-(4,4′-Dimethoxytrityl) $N^2$-isobutyryl-2′-deoxyguanosine, $N^4$-acetyl/benzoyl-2′-deoxycytidine and thymidine used for phospitylation/phosphorylation.

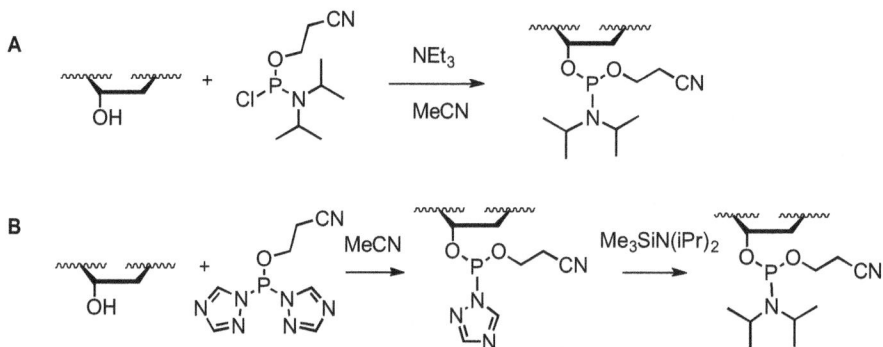

**Figure 5.2:** Alternative syntheses of phosphoramidite building blocks.

to maintain 1–2% of $Et_3N$ in the eluent. As long as no acidic impurities are present, phosphoramidites are stable compounds. Even N,N-diethylphosphoramidites withstand silica gel chromatography. The purity of phosphoramidites may be assessed by $^{31}P$ NMR spectroscopy. As the P(III) atom is chiral in nucleoside phosphoramidites, it displays two $^{31}P$ signals at about 149 ppm referring to the two diastereomers of the compound. The hydrolysis products of phosphoramidites, H-phosphonates, possibly present as impurity, display peaks at 8 and 10 ppm. Nucleoside phosphoramidites are stable compounds with a prolonged shelf-life when stored as powders under anhydrous conditions in the absence of air at temperature below 4 °C. They withstand mild basic conditions, but in the presence of even mild acids, phosphoramidites perish rapidly.

As mentioned above, *H*-phosphonate chemistry is used for special purposes. The most convenient way for conversion of protected nucleosides to *H*-phosphonates is treatment with excess of diphenyl *H*-phosphonate in pyridine, followed by hydrolysis of the phenyl *H*-phosphonate diester to *H*-phosphonate monoester (Figure 5.3A) [8]. Several other phosphitylating agents, e.g., salicoyl phosphorochloridite (Figure 5.3B), pyrophosphate in pyridine and $PCl_3$ in the presence of imidazole and $Et_3N$, are viable [9, 10].

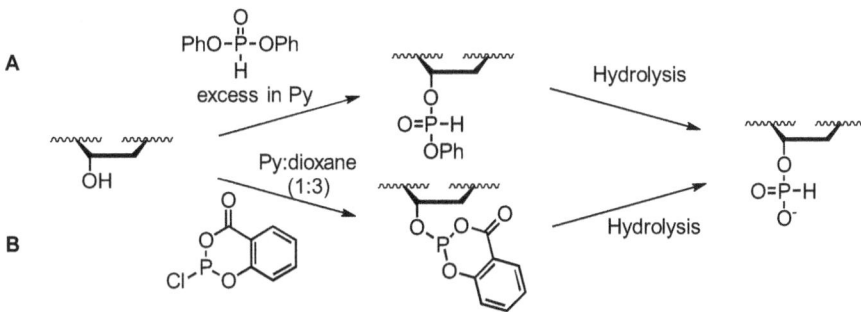

**Figure 5.3:** Alternative syntheses of H-phosphonate building blocks.

On using P(V) chemistry, three different coupling chemistries are available, each of which making use of different kinds of building block. The syntheses of these building blocks are depicted in Figure 5.4. 3′-(2-Chlorophenyl phosphate) is simply obtained by the treatment of the appropriately protected nucleoside with 2-chlorophenyl phosphorobis(1,2,4-triazolide) in MeCN, followed by hydrolysis in aqueous pyridine that contains triethyl amine (Figure 5.4A) [11]. Synthesis of 3′-(1-oxido-4-methoxypicolinyl phosphate) is more complicated (Figure 5.4B). 3′-(2,5-Dichloropbenyl phosphate) is first prepared essentially in the way described above for 3′-(2-chlorophenyl phosphate). 1-Oxido-4-methoxypicolinyl alcohol is then coupled to this phosphodiester by 2,4,6-triisopropylbenzenesulfonyl chloride/1-methylimidazole activation in pyridine. Removal of 2,5-dichlorophenol with DBU in aqueous MeCN gives the desired diester [12]. The third alternative is 3′-(benzotriazolyl 2-chlorophenyl phosphate) that is obtained by the reaction of protected nucleoside in dioxane with a prefabricated benzotriazolyl 2-chlorophenyl phosphorochloridate (Figure 5.4C) [13].

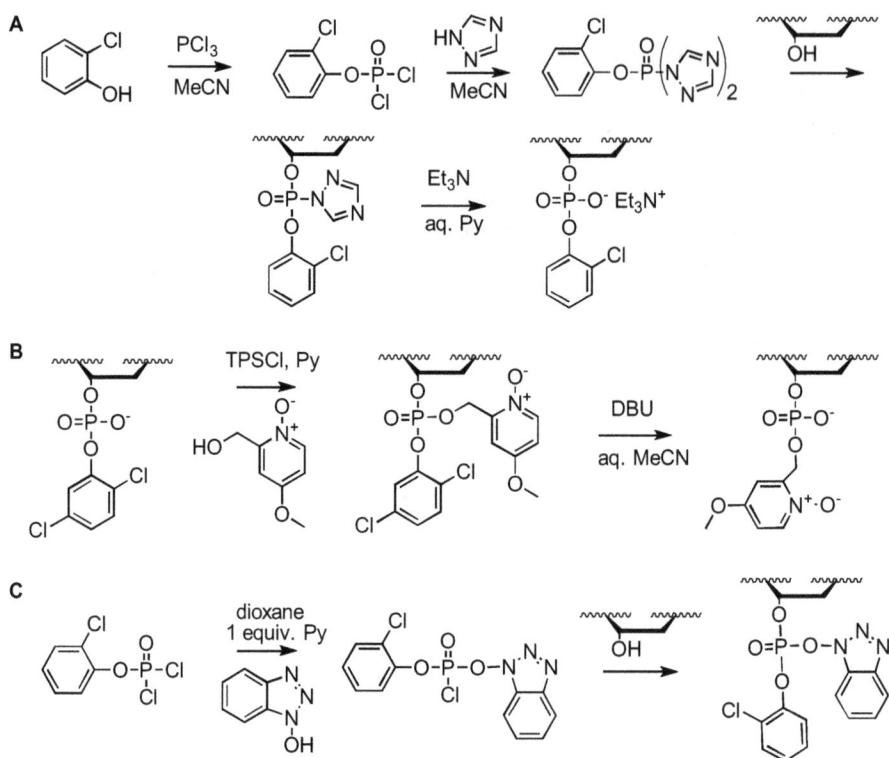

**Figure 5.4:** Preparation of building blocks for the synthesis of oligonucleotides by various P(V) chemistries.

### 5.1.2 Coupling chemistries

An activator is needed to couple one of the building blocks indicated in Figures 5.2–5.4 to the 5'-OH of an appropriately protected nucleoside. 3'-O-Phosphoramidites (Figure 5.2) are usually activated with azoles, the $pK_a$ values of which fall between 4.1 and 5.2 (Figure 5.5). In other words, they are weak acids that are sufficiently strong to protonate the departing diisopropylamino ligand in polar aprotic solvents. Without assistance by protonation, diisopropylamide ion is too unstable to serve as a leaving group. The most basic atom in phosphoramidite is the phosphorus atom, but the P-protonated species reacts only reluctantly with nucleophiles such as alcohols or azoles [14, 15]. According to molecular modeling, P-protonation shortens and strengthens the P-N bond, whereas N-protonation lengthens and weakens it [16]. In other words, the N-protonated species resembles an adduct of phosphenium ion and a neutral amine. One might envisage that the reaction initially is dissociative in nature, but reaches transition state only when the developing products, that is, the amide anion and phosphenium cation, are stabilized by protonation and nucleophilic attack, respectively. The nucleophile and acid participate in the reaction in a concerted manner, but the proton transfer is in the transition state more advanced than the bond formation between the nucleophile and phosphorus atom (Figure 5.6) [15, 17]. The activator has a dual role serving both as a proton donor and a nucleophilic catalyst. Displacement of diisopropylamine with azole activator is the rate-limiting step of the coupling process. Displacement of the azole ligand by alcohol, such as 5'-OH, is a much faster reaction. The released azolide ion precipitates as the diisopropylammonium salt, which means that a stoichiometric amount of the activator is needed.

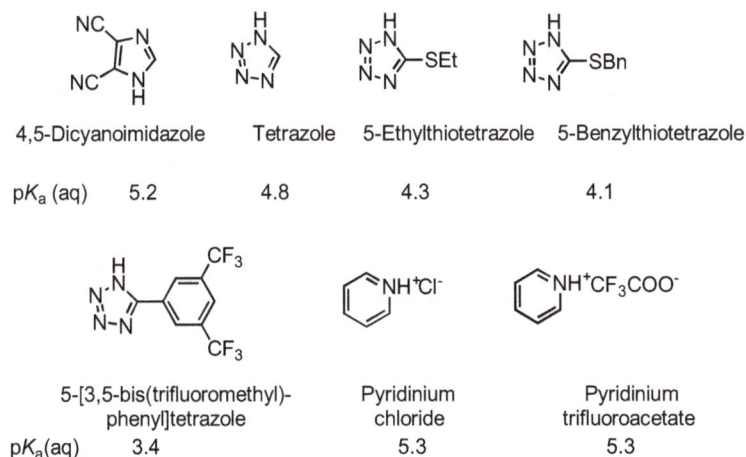

| | | | |
|---|---|---|---|
| 4,5-Dicyanoimidazole | Tetrazole | 5-Ethylthiotetrazole | 5-Benzylthiotetrazole |
| p$K_a$ (aq)    5.2 | 4.8 | 4.3 | 4.1 |

| | | |
|---|---|---|
| 5-[3,5-bis(trifluoromethyl)-phenyl]tetrazole | Pyridinium chloride | Pyridinium trifluoroacetate |
| p$K_a$(aq)    3.4 | 5.3 | 5.3 |

**Figure 5.5:** Activators for coupling of phosphoramidites with alcohols.

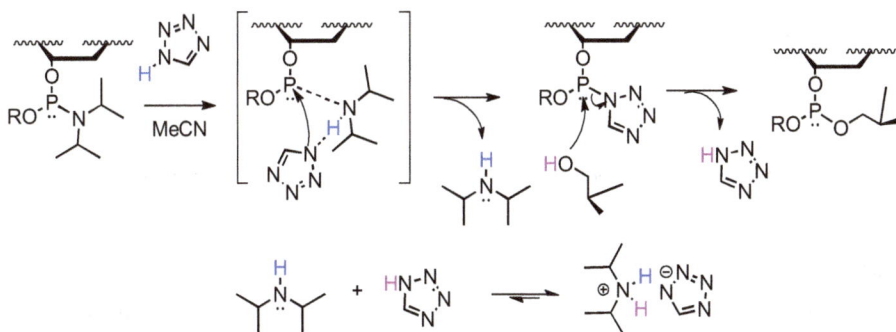

**Figure 5.6:** Mechanism of phosphoramidite coupling.

As discussed above, the efficiency of activation depends on both the acidity of activator and nucleophilicity of its conjugate base. 5-Ethylthiotetrazole and 5-benzylthiotetrazole, having aqueous p$K_a$ values 0.5 and 0.7 units lower than unsubstituted tetrazole, are more efficient activators than tetrazole [18]. More acidic activators cannot be used since they tend to remove the 5'-O-DMTr protecting group during coupling. 5-[3,5-Bis(trifluoromethyl)phenyl]tetrazole, for example, is already too acidic and can only be used for special purposes. The efficiency of an azole activator can also be increased by using more nucleophilic azoles. 5,6-Dicyanoimidazole is most widely used among such activators. Other examples are pyridinium chloride and pyridinium trifluoroacetate.

The coupling reaction yields a phosphite triester that has to be oxidized to a phosphate triester before next coupling step because the phosphite ester is too acid-labile to withstand the acidolytic removal of the commonly used 5'-ODMTr group. Most common oxidizing agent is aqueous iodine under basic conditions, that is, in a solution containing $I_2$ (0.2 g), THF (4 mL), $H_2O$ (2 mL) and 2,6-lutidine (1 mL) (Figure 5.7A) [19]. The other oxidation methods include (i) *tert*-butylperoxide in toluene or bis(trifluoromethyl)peroxide and catalytic amount of trimethylsilyltriflate in DCM [20], (ii) a mixture of NBS, DMSO and MeCN [21], (iii) 1,1-dihydroperoxycyclododecane in a mixture of DCM and EtOAc [22] and (iv) (1S)-(+)-(10-camphorsulfonyloxaziridine) in MeCN [23] and 2-butanone peroxide in DCM [24].

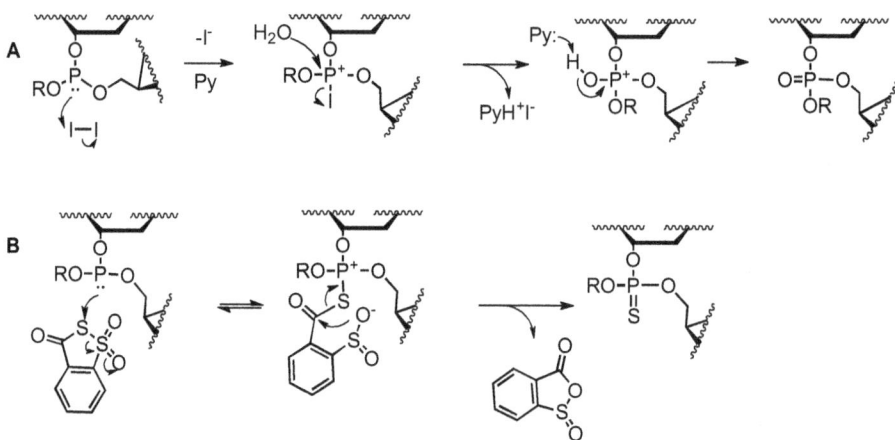

**Figure 5.7:** Oxidation (A) and sulfurization (B) of phosphite triester to phosphate triester or phosphorothioate triester, respectively.

Besides oxidation to phosphate triesters, phosphite triesters may be sulfurized to phosphorothioate triesters. The most commonly used sulfur transfer agents are listed in Table 5.1. As an example, the mechanism of sulfur transfer by the so-called Beaucage reagent (3H-1,2-benzodithiole-3-one 1,1-dioxide) is depicted in Figure 5.7B. A related mechanism is utilized by all the reagents listed. Reducing inorganic salts, such as $Na_2S_2O_3$, may be added to suppress desulfurization during cleavage from support and deprotection [32].

The most commonly used activator for H-phosphonate coupling is pivaloyl chloride in a mixture of pyridine and MeCN [10]. Pivaloyl chloride when used in moderate

**Table 5.1:** Alternative sulfurization agents for oxidation of phosphite triesters to phosphorothioate triesters.

| Sulfurization agent | Structure | Solvent | References |
|---|---|---|---|
| Beaucage reagent (3*H*-1,2-benzodithiole-3-one 1,1-dioxide) | | MeCN | [25] |
| DDTT (*N,N*-dimethyl-*N′*-(3-thioxo-3*H*-1,2,4-dithiazol-5-yl)methanimidamide) | | Py | [26] |
| PADS (phenylacetyl disulfide) | | Py | [27] |
| EDITH (3-ethoxy-1,2,4-dithiazoline-5-one) | | MeCN | [28] |
| DDD (diethyldithiocarbonate disulfide) | | MeCN | [29] |
| DTD (tetraethylthiuram disulfide) | | Py | [30] |
| XH (xanthene hydride) | | MeCN/Py | [31] |

excess (2–5 equiv.) first acylates the H-phosphonate monoester and the pivaloyloxy anion becomes subsequently displaced by pyridine. Attack of the 5′-OH of another appropriately protected nucleoside on the phosphorus atom with concomitant departure of pyridine completes the reaction (Figure 5.8A). Besides pivaloyl chloride, adamantane-1-carbonyl chloride, 2-chloro-5,5-dimethyl-1,3,2-dioxaphosphinane 2-oxide and bis (2-oxooxazolidin-3-yl)phosphinic chloride are used as activators [10]. Two reactions tend to compete with the coupling: double acylation of the H-phosphonate monoester and acylation of the 5′-OH that should attack on the phosphorus atom. While the latter reaction inevitably prevents coupling, nucleoside bis(acyl) phosphite still gives the correct product, but substantially more slowly than the H-phosphonate-carboxylic acid anhydride. Accordingly, the excess of activator, mixing time before the addition of 5′-OH nucleoside the contact with solid support, and solvent composition should be carefully optimized to reach the nearly quantitative coupling efficiency [10].

H-phosphonate diesters are much more stable than phosphite triesters under mildly acidic conditions. No decomposition has been observed in 20 h in DCM-containing 2% dichloroacetic acid, that is., under conditions often used for the removal of 5′-ODMTr protection in solid-supported synthesis [33]. That is why oxidation to phosphate diester or sulfurization to phosphorothioate diester is not necessary after each coupling, but can be carried out after completion of the chain assembly. Less than half an hour treatment with 2% iodine in pyridine–water (98:2 v/v) is generally applied [10]. To obtain a phosphorothioate oligomer, the oligomer is treated with $S_8$ in pyridine for 20 h.

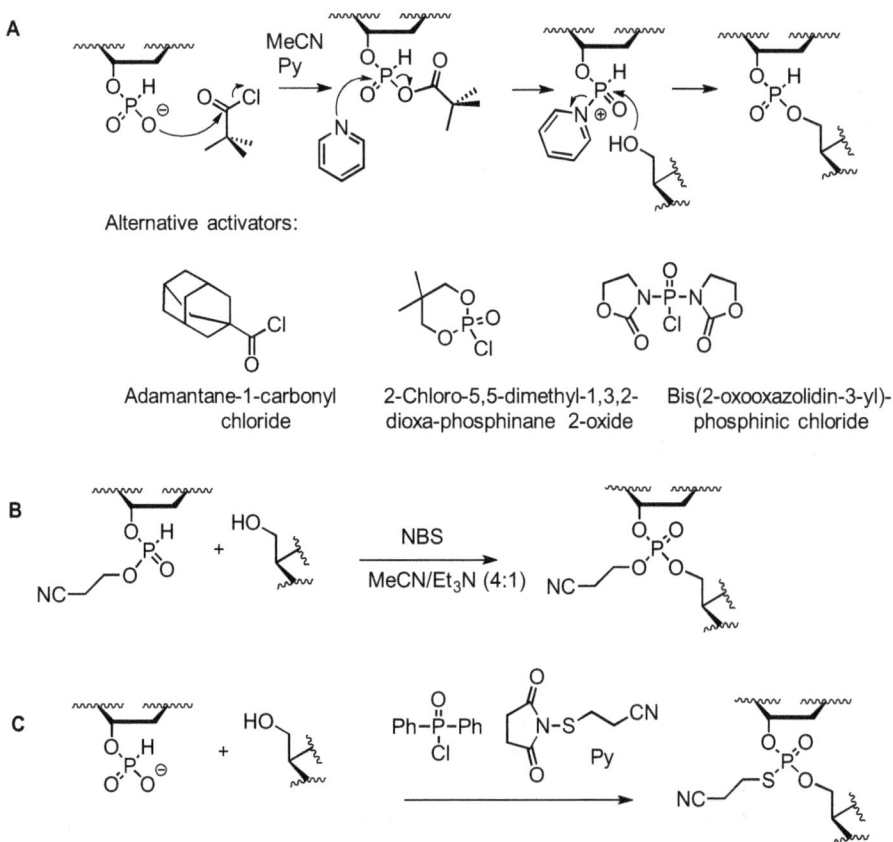

**Figure 5.8:** Mechanism of H-phosphonate coupling using pivaloyl chloride as an activator (A). Oxidative H-phosphonate coupling yielding 2-cyanoethyl-protected phosphodiester (B) [34] or phosphorothioate (C) [35].

Coupling and oxidation/sulfurization can also be carried out in one pot. For example, 2-cyanoethyl-protected dinucleoside-3′,5′-diesters that most often are obtained by phosphoramidite coupling and subsequent oxidation can be prepared in a single step by H-phosphonate chemistry using a nucleoside 3′-(2-cyanoethyl-H-phosphonate) as a building block and NBS in a 4:1 mixture of MeCN and Et$_3$N as an activator (Figure 5.8B) [34]. The corresponding phosphorothioate triesters have, in turn, be synthesized by a four-component oxidative coupling, using diphenylphosphinic chloride and N-(2-

cyanoethylthio)pyrrolidine-2,5-dione as an activator and sulfur transfer agent, respectively (Figure 5.8C) [35].

Oligonucleotide synthesis was originally developed on the basis of P(V) chemistry, first by the so-called phosphodiester and later by phosphotriester strategy [4]. In the early version of phosphodiester chemistry, a nucleoside 3'-monophosphate was coupled with the aid of an activator to the 5'-*O* of a support-anchored growing oligonucleotide chain. The product is negatively charged phosphodiester oligomer that is laborious to purify. In the next generation approach, phosphotriester chemistry, a nucleoside 3'-arylphosphate was used as a building block instead of 3'-monophosphate. Oligonucleotide is, hence, obtained in a fully protected form as a neutral phosphotriester oligomer. The coupling times, however, are longer than on using P(III) chemistries. Phosphotriester chemistry is nowadays seldom used in solid-phase synthesis and is often regarded as outdated. Nevertheless, for certain purposes it still may be the method of choice, as discussed later in this chapter, and hence, worth knowing. Three different approaches are available. First, a 3'-(2-chlorophenyl phosphate) diester is activated with arylsulfonyl chloride or azolide in the presence of 1-methylimidazole (Figure 5.9A) [36]. Second, a 3'-(1-oxido-4-methoxypicolinyl phosphate), a building block that bears a nucleophilic functionality, may be used [12]. The 1-oxido function allows intramolecular cycli-

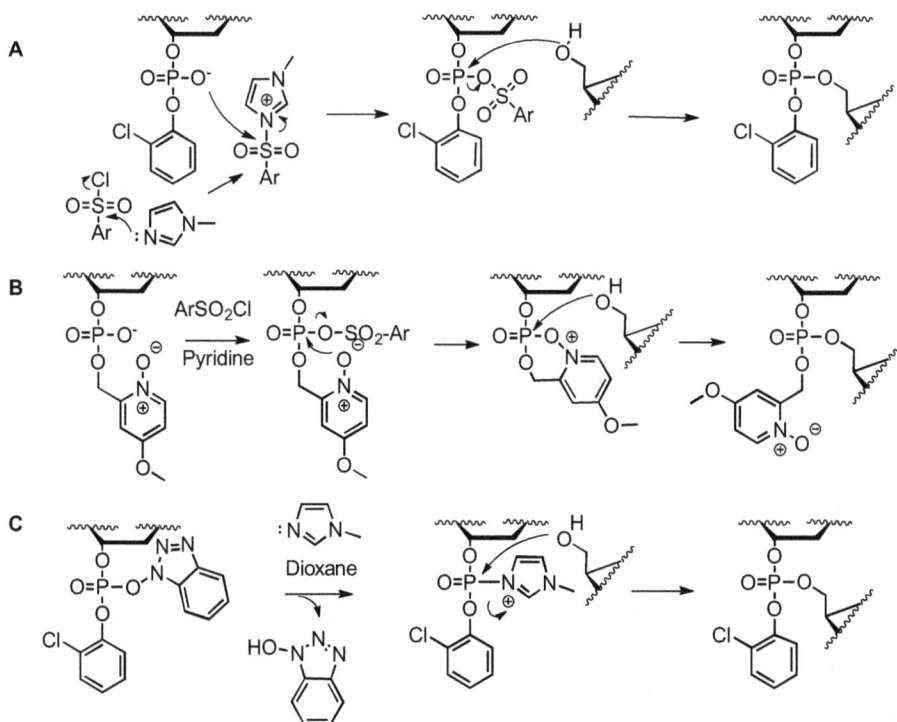

**Figure 5.9:** Alternative P(V) coupling chemistries: A [36], B [12] and C [13].

zation by attack on the phosphorus atom and this markedly facilitates the subsequent coupling (Figure 5.9B). The third coupling alternative utilizes a prefabricated or in situ generated 3'-(1-hydroxybenzotriazolyl 2-chlorophenyl phosphate in the presence of 1-methyl imidazole for coupling (Figure 5.9C) [13].

### 5.1.3 Global protecting group strategies

The global protecting group strategies for the synthesis of oligodeoxyribonucleotides by the phophoramidite, H-phosphonate and P(V) strategies are rather similar. The temporary 5'-O-protecting group that is removed in every coupling cycle is acid-labile, whereas the base and phosphate moiety protecting groups are base-labile (Figure 5.10). Usually, a base-labile linker is used to attach the 3'-nucleoside to the support, but linkers cleavable under orthogonal conditions are also available. In contrast to phosphoramidite and phosphotriester chemistry, no phosphate protecting groups are required in H-phosphonate chemistry.

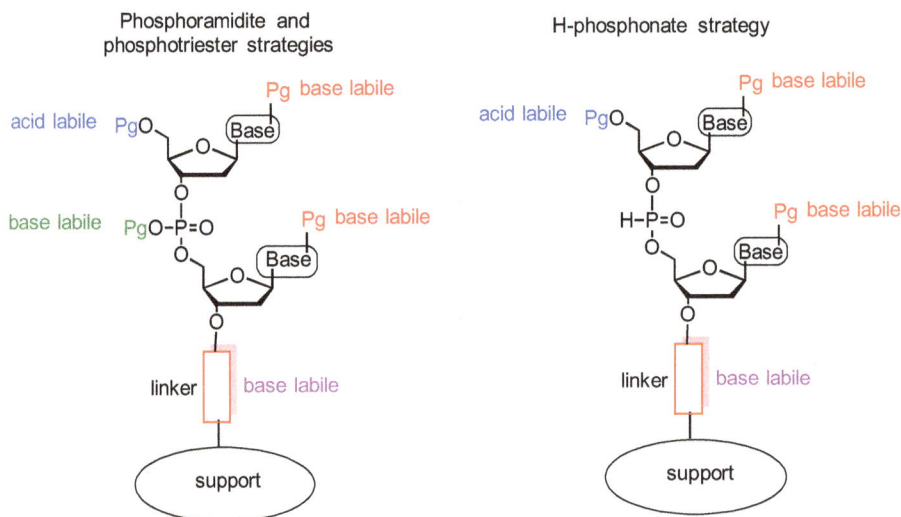

**Figure 5.10:** Global protecting group strategies for solid supported synthesis of oligodeoxyribonucleotides by various coupling chemistries.

The coupling cycles of phosphoramidite and H-phosphonate chemistry are outlined in Figures 5.11 and 5.12, respectively. The first two steps in both strategies consist of the removal of 5'-O protecting group from the support-bound nucleoside and coupling of the next building block. The uncoupled 5'-OH groups are then capped to prevent the formation of failure sequences. The conventional method is acetylation with $Ac_2O$ in the presence of 1-methylimidazole, but more efficient phosphitylation with a di-O-alkyl phosphoramidite is also used. On applying phosphoramidite chemistry, the phosphite triester is then subjected to oxidation before the next coupling cycle, whereas using H-phosphonate chemistry the oxidation takes place only after completion of the entire chain assembly. For the same reason, the coupling cycle of the phosphotriester chemistries is simpler than that of the phosporamidite chemistry; no oxidation step is required. An example is given in Figure 5.13.

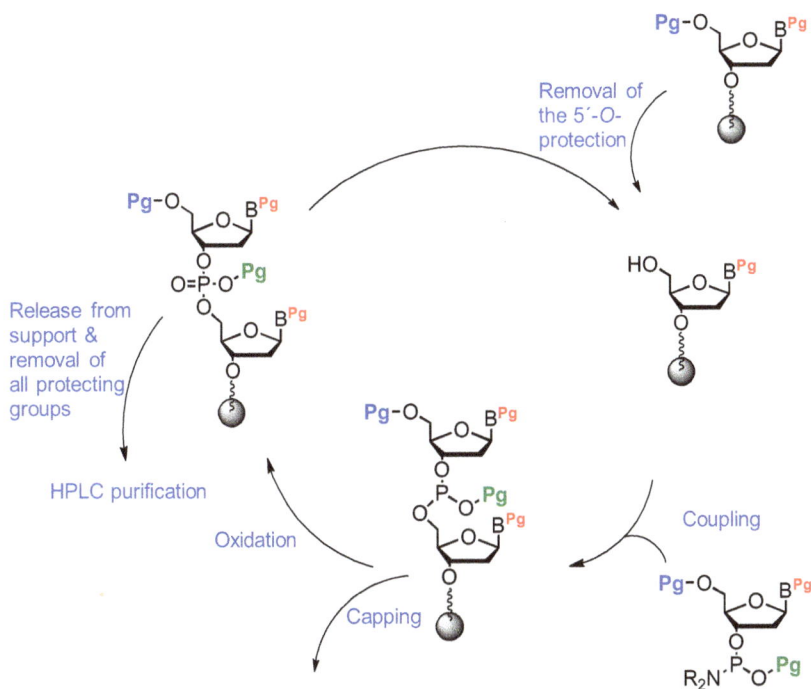

**Figure 5.11:** Coupling cycle for the solid-phase synthesis of oligodeoxyribonucleotides by the phosphoramidite approach.

## 5.1.4 Solid supports

Solid-phase synthesis of oligonucleotides in small scale is traditionally carried out on a controlled pore glass (CPG) support. The average size of the pores varies from 500 to 3,000 Angstroms according to the length of the oligonucleotide synthesized. The smallest pore-sized supports are used for the preparation of oligomers shorter than 50 nucleotides and the largest pore-sized supports for the synthesis of 200-mers. The surface of CPG is covalently coated with long aminoalkyl chains (LCAA-CPG) to which the 3'-terminal nucleoside is conjugated. The surface of the material is first treated with (3-aminopropyl)-triethoxysilane to give aminopropyl CPG. The aminopropyl arm may be

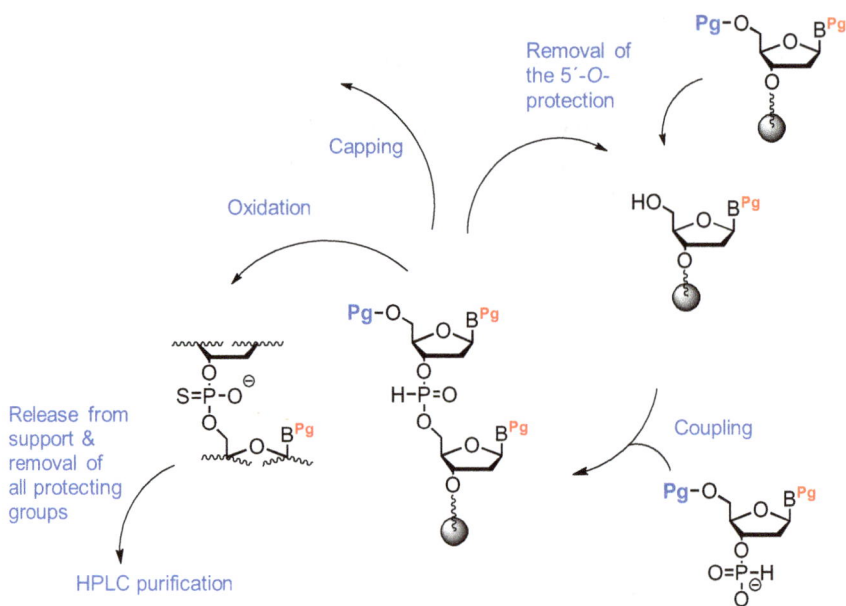

**Figure 5.12:** Coupling cycle for the solid-phase synthesis of oligodeoxyribonucleotides by the H-phosphonate approach.

further extended to result in long chain aminoalkyl (LCAA) CPG. More recently, low-swelling, highly cross-linked macroporous polystyrene that contains 60% divinylbenzene as a branching unit (MPPS), has gained popularity in small-scale synthesis. With these nonswelling glass or polystyrene supports, the loading of oligonucleotide remains below 50 μmol g$^{-1}$. Up to seven times higher loadings can be obtained with porous divinylbenzene cross-linked swelling polystyrene supports [37].

## 5.1.5 Linkers

The most common linker is succinyl linker. The 3′-OH of an appropriately protected nucleoside is esterified with succinic anhydride and immobilized to an amino functionalized support with the aid of an activator, which often is a combination of 1-(2-mesitylenesulfonyl)-3-nitro-1H-1,2,4-triazolide and 1-methylimidazole or diisopropylcarbodiimide and N-hydroxysuccinimide (Figure 5.14). Instead of succinic acid, several other dicarboxylic acids may be used as more base-labile linkers. These include in the order of increasing lability malonic [38], diglycolic (2,2′-oxydiacetic acid) [39], oxalic [40] and hydroquinone-O,O′-diacetic acid (2,2′-[1,4-phenylenebis(oxy)]diacetic acid) [41] (Figure 5.15). All these can be cleaved by an acyl substitution mechanism with aqueous ammonia, and the last one, known as Q-linker with dilute potassium carbonate in

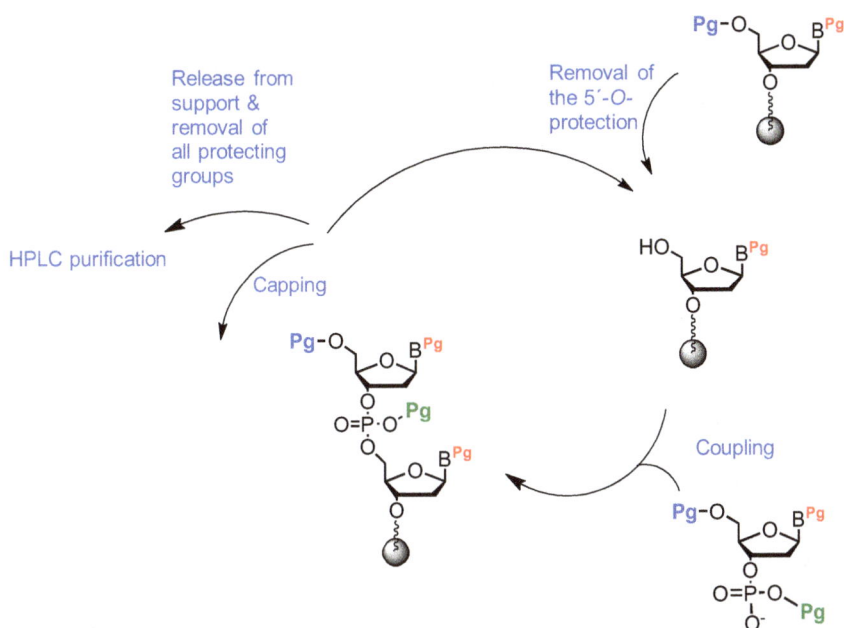

**Figure 5.13:** Coupling cycle for the solid-phase synthesis of oligodeoxyribonucleotides by the phosphotriester approach.

MeOH or neat $Et_3N \cdot 3HF$. In addition, a number of linkers are cleavable under orthogonal conditions: 4-oxoheptane-1,7-dioic acid linker with hydrazinium acetate in pyridine [42], that is, by a mechanism used for the removal of levulinoyl protecting groups. Hydrazine first forms imine with the carbonyl carbon (C4) and the terminal amino group displaces the 3'-terminal nucleoside by concomitant cyclization to 4,5-dihydropyridazin -3(2H)-one. Silyl linkers are cleaved with fluoride ion under neutral pH [43], disulfide linker reductively with concomitant departure of 2-mercaptomethyl group [44], 1,3-dithian-2-methoxycarbonyl linker (Dmoc-linker) by $NaIO_4$ oxidation to highly base-labile disulfoxide [45] and linkers containing a photolabile 2-nitrobenzyl unit by UV irradiation [46].

In case the purpose is to prepare a set of oligonucleotides containing diverse 3'-terminal nucleosides, the amino-functionalized support may be derivatized with a so-called universal linker to avoid the immobilization of each 3'-terminal nucleoside separately to the support. On using such a universal support, the 3'-terminal nucleoside is attached by normal phosphoramidite coupling. Therefore, the linker must be so designed that the oligonucleotide is on ammonolysis released as a 3'-alcohol and not a 3'-phosphate. Two such linkers are commercially available (Figure 5.16). With both of them, the key structural feature is an esterified hydroxyl group in α-position to the 3'-terminal phosphate of the oligonucleotide. This hydroxyl group is exposed upon ammonolysis. In one case (Figure 5.16A), an optimally oriented hydroxyl group performs

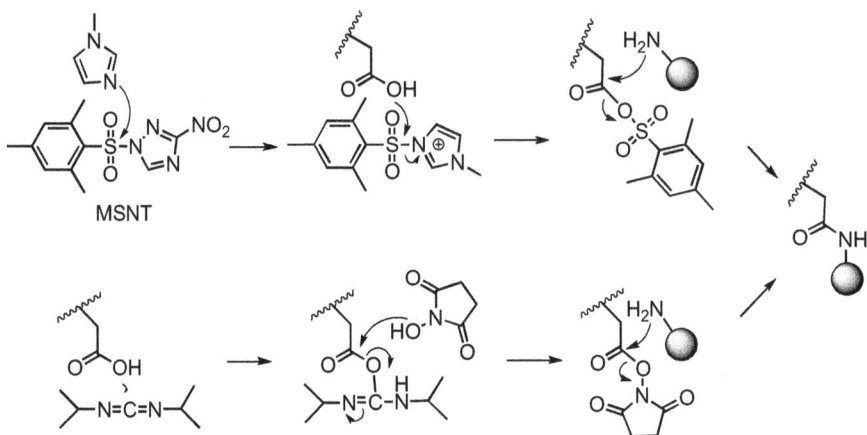

**Figure 5.14:** Immobilization of carboxylic acid linkers to aminoalkylated supports.

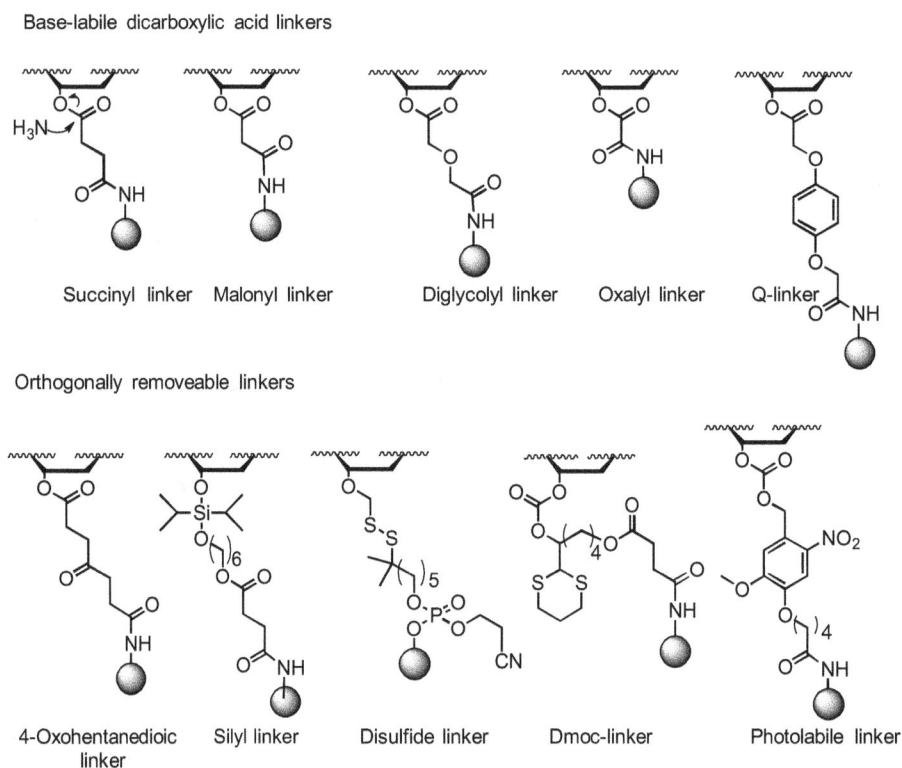

Base-labile dicarboxylic acid linkers

Succinyl linker  Malonyl linker  Diglycolyl linker  Oxalyl linker  Q-linker

Orthogonally removeable linkers

4-Oxohentanedioic linker  Silyl linker  Disulfide linker  Dmoc-linker  Photolabile linker

**Figure 5.15:** Base-labile and orthogonally removable linkers for oligonucleotide synthesis.

a hydroxide-ion-catalyzed intramolecular attack on the phosphorus atom [47]. In the other case (Figure 5.16B), the attack has been argued to be catalyzed by the neighboring amide function [48]. In both cases, a highly unstable dianionic phosphorane is formed and broken down by cleavage of the 3'-O of the oligonucleotide.

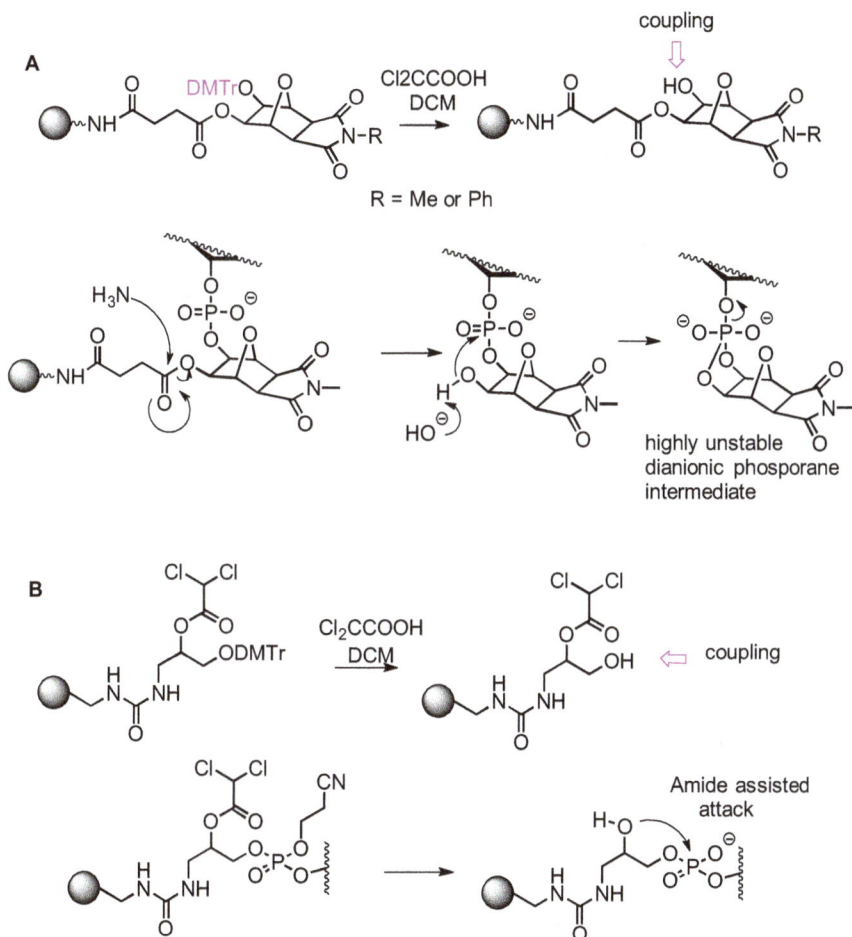

**Figure 5.16:** Universal linkers for synthesis of oligonucleotides.

### 5.1.6 Phosphate protecting groups

As discussed above, the internucleosidic phosphodiester linkages have to be kept protected on using phosphoramidite or phosphotriester coupling chemistry. Internucleosidic H-phosphonate linkages, in turn, are neutral and so stable that no additional protection is needed. 2-Cyanoethyl group is almost invariably used for phosphate protection in phosphoramidite chemistry. The alternatives are methyl and allyl groups. The 2-cyanoethyl groups are eliminated as acrylonitrile during the treatment with aqueous ammonia that is carried out after completion of the chain assembly to release the oligonucleotide from the support and to remove all protecting groups. The methyl protections are removed from phosphate groups either with disodium-2-carbamoyl-2-cyanoethylene-1,1-dithiolate (N2S2) in DMF [49] or with a mixture of thiophenol, $Et_3N$ and dioxane [50]. Allyl groups are removed orthogonally by treatment with a mixture of $Pd_2(dba)_3$–$CHCl_3$ complex, $Ph_3P$, butylamine and formic acid at 50 °C for 0.5–1 h [51]. Normal succinyl linker withstands this treatment. Conventional ammonolysis can also be used if 2% 2-mercaptoethanol is added [23].

The most commonly used phosphate protecting group in phosphotriester chemistry is 2-chlorophenyl group. This group is removed before global ammonolytic deprotection with (E)-2-nitrobenzaldehyde oxime or (E)-picolinaldehyde oxime [52]. The 1-oxido-4-alkoxy-2-picolyl groups (cf. Figure 5.9B) are, in turn, removed with piperidine or triethylammonium thiophenate [12].

### 5.1.7 Base moiety protecting groups

As indicated in Figure 5.1, the standard base moiety protections are $N^6$-benzoyl for Ade, $N^2$-isobutyryl for Gua, $N^4$-benzoyl or $N^4$-acetyl for Cyt [4]. Thymidine usually is used unprotected. In case more base-labile groups are needed, phenoxyacetyl or methoxyacetyl groups can be used. The acyl protecting groups are removed after chain assembly by acyl substitution in concentrated aqueous ammonia (Figure 5.17A). $N,N$-Dimethylformamidine protection is a frequently used alternative for benzoyl group [53]. The group is removed hydrolytically at high pH (Figure 5.17B). The advantage compared to benzoyl protection is that acid-catalyzed depurination occurring as a side reaction upon removal of the 5′-O protecting group is markedly retarded.

For the synthesis of oligonucleotides containing base-sensitive modifications or conjugate groups, orthogonally removable base moiety protecting groups are required (Figure 5.17C). The obvious alternatives are silyl and allyl protecting groups. 2-(Trimethylsilyl)ethoxycarbonyl groups can be removed with $ZnBr_2$ [54]. Allyloxycarbonyl groups are, in turn, removed with $Pd^0$ in THF (e.g., $Pd_2(dba)_3 \cdot CHCl_3$, $Ph_3P$, HCOOH, $BuNH_2$; washing with $N,N$-diethylthiocarbamate) [51]. Base-sensitive oligonucleotides have been synthesized by this strategy on a photolabile linker. All the protecting groups are removed in a single step with $Pd^0$ in THF (e.g., $Pd_2(dba)_3 \cdot CHCl_3$, $Ph_3P$, HCOOH,$BuNH_2$; washing with $N,N$-diethylthiocarbamate). Still one interesting possibility is offered by 4-pentenoyl protections that may be removed by NIS in aqueous THF [55]. Owing to the essential role of $I^+$ in cleavage of the 4-pentenoyl groups, the conventional oxidation of phosphite triesters with aqueous $I_2$ has to be replaced with *tert*-butyl hydroperoxide oxidation in an organic solvent. Recently, 1-(1,3-dithian-2-yl)-1-methylethoxycarbonyl (Dmoc) groups has been introduced for this purpose [45]. The group withstands the removal of 2-cyanoethyl groups from phosphodiester linkages with DBU in MeCN, but is converted hydrolytically unstable by oxidation to disulfoxide with aqueous NaIO$_4$. It should be noted that all these orthogonally removable groups are aimed at being used together with an orthogonally cleavable phosphate protecting group and a linker that allows the base-sensitive modification of conjugate group remain unchanged upon release from the support.

Photolabile protecting groups should in principle allow synthesis of base-sensitive oligonucleotides. The limitation is that nucleobases are prone to photochemical transformation reactions, which limits the intensity of UV irradiation that can be used for deprotection. 2,2-Bis(2-nitrophenyl)ethoxycarbonyl group has, however, been successfully used for the purpose on a photolabile support [56]. Oligonucleotides of limited length have also been synthesized without any base moiety protections [57]. The conventional *N,N*-diisopropylphosphoramidites are converted to more *O*-selective species by in situ displacement of the diisopropylamino group with 1-hydroxybenzotriazole or its 6-trifluoromethyl derivative. More recently, oxazaphospholidine chemistry (discussed in Section 5.3) has been reported to allow oligonucleotide synthesis without base moiety protections [58].

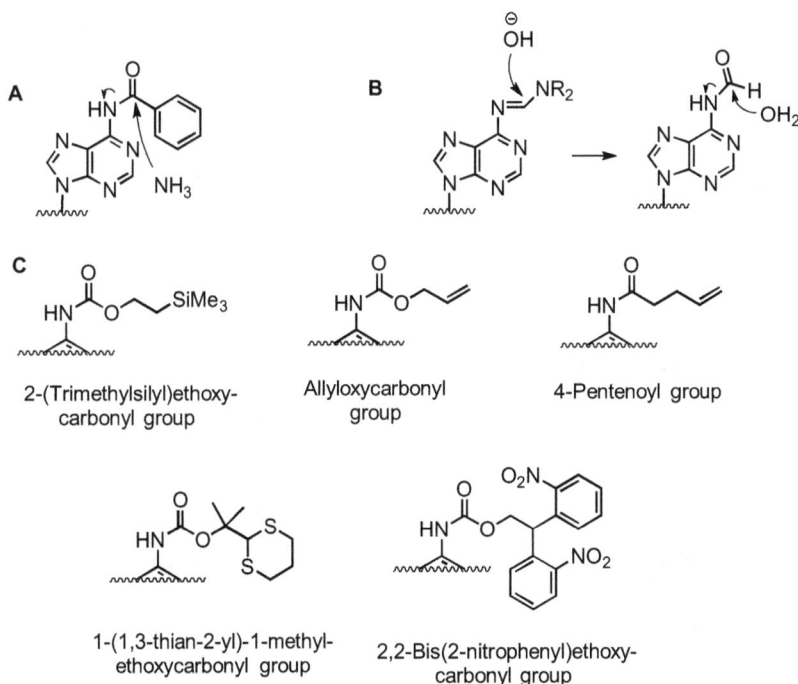

2-(Trimethylsilyl)ethoxy-carbonyl group

Allyloxycarbonyl group

4-Pentenoyl group

1-(1,3-thian-2-yl)-1-methyl-ethoxycarbonyl group

2,2-Bis(2-nitrophenyl)ethoxy-carbonyl group

**Figure 5.17:** (A) Removal of acyl protecting groups by acyl substitution with ammonia, (B) removal of amidine-type protecting groups by base-catalyzed hydrolysis and (C) structures of orthogonally removable base moiety protecting groups.

### 5.1.8 5′-O-Protecting groups

DMTr group is almost invariably used as the temporary 5′-O-protecting group and is re-moved with di- or tri-chloroacetic acid in DMF. The advantageous features of this group include convenient synthesis of building blocks due to regioselective introduction to the primary 5′-OH and release as an orange carbocation, which allows easy quantification of the coupling efficiency in automated solid-phase synthesis. The DMTr carbocation is exceptionally long-lived and hence alkylating. This does not cause any problem in ma-chine-assisted solid-phase synthesis, but in batch-type solution-phase synthesis scav-angers have to be used. Triethylsilane, triisopropylsilane, ethanedithiol, anisol, pyrene and furan are most frequently utilized for this purpose. A more serious problem is that removal of DMTr by acid treatment tends to be accompanied by depurination. This side reaction is particularly noteworthy with acyl-protected purine bases. As discussed in Chapter 2, acyl protection accelerates depurination by one order of magnitude com-pared to unprotected or amidine-protected purine nucleosides. Upon ammonolytical de-protection of the assembled sequence, the chain is cleaved by β- and δ-elimination at the site of depurination (Figure 5.18A). The resulting truncated sequences fortunately are rather easy to remove. In place of DMTr group, three to four times more acid-labile pixyl groups (Figure 5.18B) have been used [59, 60].

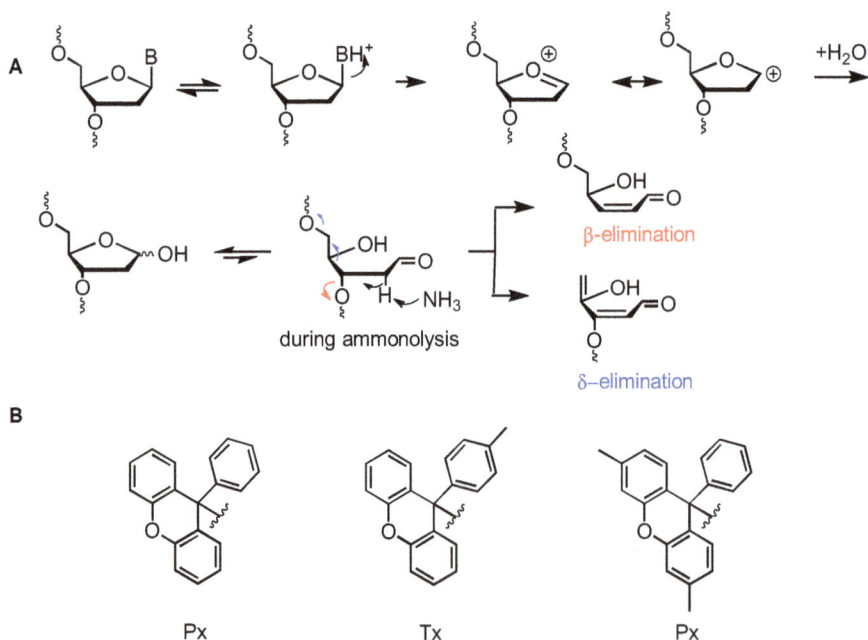

**Figure 5.18:** (A) Formation of apurinic sites in oligodeoxyribonucleotides and (B) structures of acid-labile pixyl groups used as alternatives for trityl groups.

### 5.1.9 Trimeric building blocks

Oligonucleotides are sometimes assembled from trimeric building blocks that refer to the three base codes of amino acids. Such approach is especially popular in combinatorial synthesis of oligonucleotide pools [61] that are used for protein mutagenesis [62] and generation of antibody libraries [63] because the problem of frame shift in protein synthesis is avoided. The trimeric blocks can be synthesized on a soluble support in fully protected form, that is, ready for phosphitylation [64, 65]. Accordingly, they are compatible with the normal solid-phase phosphoramidite synthesis. The internucleosidic phosphodiester linkages, however, may need to be protected with methyl [64] or 2-chlorophenyl groups [65], but this does not necessitate any changes in the conventional protocols. The coupling efficiency depends on the base composition. On the average, it is lower than with monomeric phosphoramidites. In the best cases, the overall efficiency is close to that obtained by stepwise coupling of monomers on a solid support. The advantage is that formation of $n - 1$ sequences that are difficult to remove is eliminated.

### 5.1.10 Tandem oligonucleotide synthesis

Efficiency of solid-supported oligonucleotide synthesis can be increased by assembling several different sequences as a single support-bound chain of oligonucleotides that are attached to each other by a cleavable linker. Among numerous linkers used for the purpose, simple 2,2′-sulfonyldiethanol linker seems to be simplest [66]. The linker is introduced as a DMTr-protected phosphoramidite and removed during conventional ammonolytic cleavage/deprotection step. It is compatible with both DNA and RNA synthesis.

### 5.1.11 Deprotection and purification

A two-step procedure is usually applied to deprotection of support-bound ODNs. Washing with $Et_3N$ in MeCN (1:1, v/v) is first carried out to remove 2-cyanoethyl protections without cleavage of the oligomer from support [67]. The purpose is to prevent alkylation of base moieties by acrylonitrile under the conditions of ammonolysis. The release from support and deprotection of the base moieties is then achieved by treatment with concentrated aqueous ammonia or a mixture of methylamine and ammonia [68] or with gaseous ammonia or methylamine [69]. Solid support is removed by filtration and solvent by evaporation. After this, two different protocols are used (Figure 5.19A). Either, the oligonucleotide still bearing the 5′-O-DMTr group is purified by HPLC on a C18 column, precipitated with EtOH and treated with aqueous NaOAc (10 mM, pH 3.0) to remove 5′-O-DMTr and convert the oligomer to a sodium salt. Finally 3 M NaAc is added to precipitate the oligomer as sodium salt with EtOH [70]. The alternative approach consists of removal of 5'-O-DMTr, ion-exchange HPLC and lyophilization [37].

In addition to these conventional purification methods, polymer catching [71] and temporary immobilization to a solid phase [72] have been introduced as alternative purification techniques. The underlying idea of the polymer-catching approach is that in the last step of the solid-supported assembly of the ODN, an abasic nucleoside unit having 2'-O protected with a fluoride ion labile TBDMSi group is introduced as an O-ethyl phosphoramidite. Accordingly, this terminal unit becomes coupled to the 2-cyanoethyl-protected ODN sequence via a phosphotriester linkage that withstands the removal of 2-cyanoethyl phosphate protections. The 5'-terminal unit additionally bears a 5'-linked 5-oxohexyl tail that allows easy immobilization to an aminoxy-dervatized support (Figure 5.19B) [71]. After completion of otherwise normal chain assembly, the oligonucleotide is released to solution and the protecting groups, except the protections of the 5'-terminal unit, are removed by ammonolysis. The construct is bound to an aminoxy support, which allows the removal of shorter truncated sequences by washing. The fluoride labile 2'-OTBDMSi group is removed and the exposed 2'-OH attacks under basic conditions on the neighboring phosphotriester center displacing the 5'-linked ODN. On applying purification by covalent immobilization to a solid support, the 5'-terminal building block bears a diisopropylsiloxyl ether tether with a terminal ketone structure (Figure 5.19C) [72]. Treatment with a solid support bearing aminoxy tethers results in the attachment of the full-length sequence to the support, leaving truncated sequences and wastes in solution. The oligonucleotide is then released by cleavage of the silyl ether with TBAF in DMSO.

Removal of truncated sequences does not, however, guarantee homogeneity of the oligonucleotide since base modifications may also take place during the synthesis (Figure 5.20). To reliably detect the presence of such modifications, enzymatic digestion to nucleosides, usually with a mixture of snake venom phosphodiesterase and alkaline phosphatase, followed by HPLC MS analysis is required. The source of modifications includes repeated oxidations, repeated capping and final deprotection. $N^2$-Isobutyrylguanine may become aminated to 2,6-diamonopurine or 2-amino-6-(methylamino)purine, depending on whether the final deprotection is done with ammonia or methylamine. $O^6$ is first displaced with 1-methylimidazole during capping and this with ammonia or methylamine during the final global deprotection [73]. The $N^2$-isobutyryl group may additionally become replaced with acetyl group during capping. $N^4$-Benzoylcytosine may be oxidized to uracil and, on using methylamine for final deprotection, converted to 4-methylcytosine. $N^6$-Benzoyladenine undergoes oxidation to hypoxanthine and is converted during capping and ammonolysis to 4-(5-acetyl-4-methyl-6-oxo-1,6-dihydropyrimidin-2-yl)-5-amino -1H-imidazole, appearing as an $n + 98$ amu impurity in ms [74].

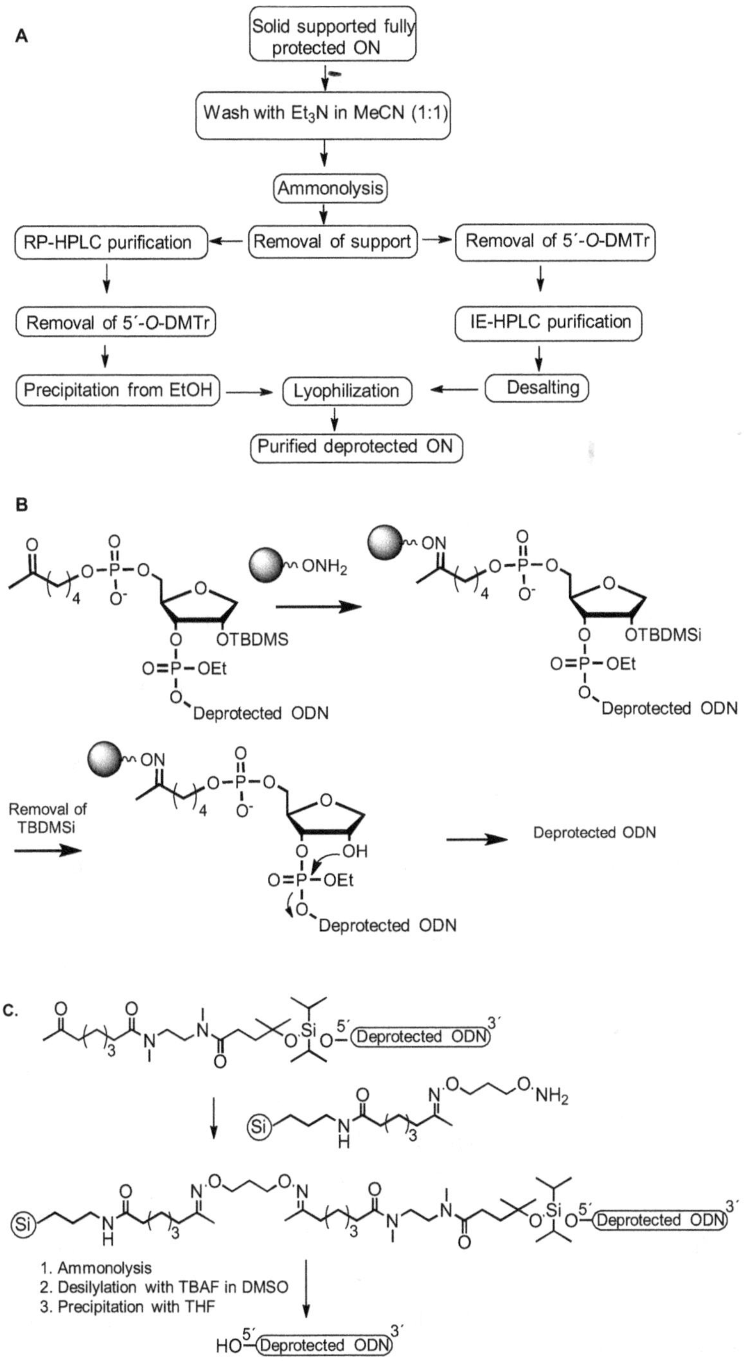

**Figure 5.19:** (A) Conventional purification of oligonucleotides, (B) purification by polymerization [71] and (C) purification by catching to a solid support [72].

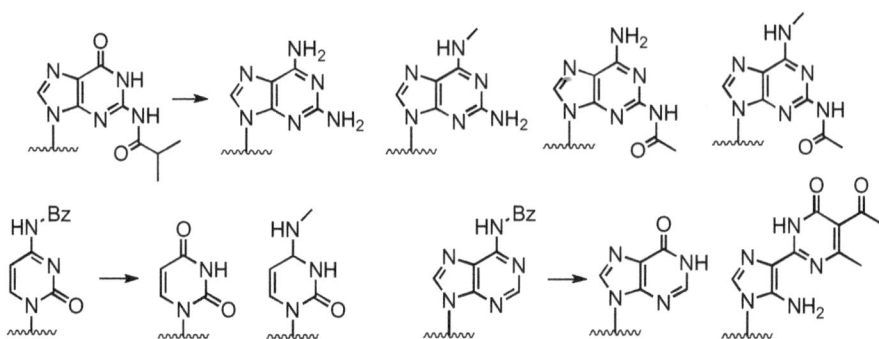

**Figure 5.20:** Base modifications reported to take place during solid-supported synthesis of oligonucleotides.

## 5.2 Synthesis of oligoribonucleotides

Oligoribonucleotides are more difficult to synthesize than oligodeoxyribonucleotides. The additional hydroxyl group at C2′ must be kept protected and this forms a steric hindrance for coupling reaction. The longer coupling time increases the risk of side reactions. In addition, the 2′-O protection must be fully stable under the conditions used to remove the phosphate protecting groups. Unprotected 2′-OH attacks extremely fast on the neighboring phosphotriester, resulting in either chain cleavage or isomerization of the 3′,5′-linkage to a 2′,5′-linkage, as indicated in Figure 5.21 [75]. The reaction is hydroxide-ion-catalyzed already at pH > 3 and very fast even under neutral conditions, the half-life being of the order of seconds. Accordingly, a negative charge has to be generated on the phosphate group before the removal of the 2′-O-protecting group. Formation of sequences containing a 2′,5′-linkage obviously form a formidable challenge for purification. Hence, selection of a proper protecting group is of outmost importance.

The most widely applied 2′-O-protecting groups are fluoride ion-sensitive *tert*-butyldimethylsilyl [76] (TBDMS) and triisopropylsilyloxymethyl [77] (TOM) groups. Phosphoramidite building blocks protected with these groups are commercially available. TBDMS group is introduced regioselectively to 2′-O by initial masking of the 3′- and 5′-hydroxy functions of a base moiety-protected nucleoside with a cyclic *tert*-butylsilylene protection (Figure 5.22A). While this group is also fluoride labile, it may be selectively removed with HF in a mixture of pyridine and DCM leaving the 2′-O-TBDMS group intact [78, 79]. 2′-O-TOM-protection is, in turn, obtained by chromatographic separation of the mixture of 2′-O and 3-O-protected nucleosides. This works well with TOM protection, which does not tend to migrate during the separation process [77], in contrast to TBDMS group. To obtain the mixture of regioisomers, 5′-O-DMTr-protected nucleoside is activated in situ as a 2′,3′-dibutylstannylene acetal (Figure 5.22B) [80]. The use of equiva-

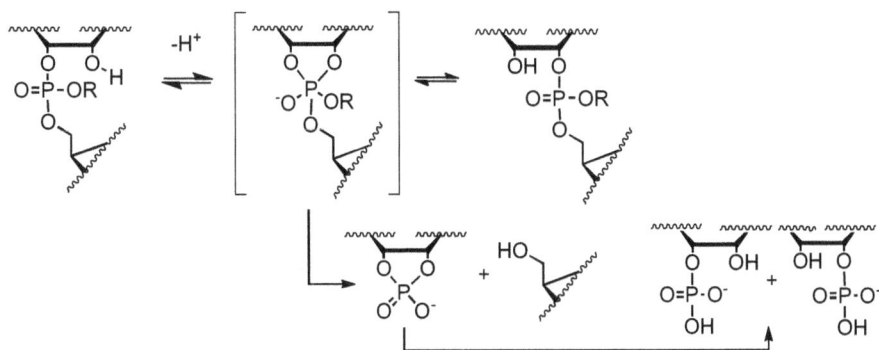

**Figure 5.21:** Hydroxide-ion-catalyzed isomerization and cleavage of internucleosidic phophotriester linkages.

lent amount of tin, however, is problematic from the point of view of biological applications. Thorough removal of this toxic metal is needed.

In addition to TBDMS and TOM groups, numerous orthogonally removable 2'-O-protecting groups have been introduced and utilized in solid-phase synthesis of oligoribonucleotides. Representative examples of these are shown in Table 5.2. The most common strategy for their introduction is initial masking of the 3'- and 5'OH with a cyclic 1,1,3,3-tetraisopropyl-1,3-disiloxane-1,3-diyl group [81]. The 2'-O-protecting group may then be introduced in a number of different manners. A protecting group that deserves special attention is the 1,1-dioxidothiomorpholine-4-carbonothioyl group (entry 3 in Table 5.2) that is compatible with conventional acyl-protected nucleobases and 2-cyanoethyl-protected phosphate linkages [82]. After chain assembly, all protecting groups are removed in a single step by treatment with neat ethylenediamine. The group is introduced by thiocarbonylation of the 2'-OH of 3',5'-di-O-1,1,3,3-tetraisopropyldisiloxane-1,3-diyl-protected nucleoside with 1,1'-thiocarbonyldiimidazole, followed by treatment with thiomorpholine 1,1-dioxide (Figure 5.22C). A more frequently used approach for 2'-O-protection is, however, conversion to 2'-O-methylthiomethyl ether by Pummerer reaction and subsequent displacement of the methylthio group by NIS or $Br_2$ activation (Figure 5.22D). Removal of the 1,1,3,3-tetraisopropyl-1,3-disiloxane-1,3-diyl protection by fluoride ion and regioselective introduction of DMTr protection to 5'-OH affords the nucleoside ready for phosphitylation. This or a closely related approach has been used to prepare 2'-O-(2-cyanoethoxymethyl) (entry 4) [83], 2'-O-(tert-butyldithiomethyl) (entry 5) [84], 2'-O-(iminooxymethyl ethyl propanoate) (entry 6) [85], 2'-O-(2-cyano-2,2-dimethylethanimine-N-oxymethyl) (entry 7) [86] and 2'-O-(4-(N-dichloroacetyl-N-methylamino)benzyloxymethyl [87] (entry 8)-protected nucleosides. 2'-O-(2-Cyanoethyl) group (entry 9) [88] has been introduced by direct alkylation of 3',5'-di-O-(1,1,3,3-tetraisopropyl-1,3-disiloxane-1,3-diyl)-protected nucleoside and 2'-O-[2-(4-tolylsulfonyl)ethoxymethyl] [89] protection (entry 10) by 2',3'-

**A**

1. tBu₂Si(OTf)₂, DMF, 0 °C
2. Imidazole, 0-25 oC

TBDMSCl

1. HF/Py/DCM, 0 oC
2. DMTrCl, Py

**B**

Bu₂SnCl₂, iPr₂NEt
DCE

(iPr)₃SiOCH₂Cl
80 oC

+

Chromatographic
separation

**C**

TIPDS-Cl₂
Py

Im₂CS

1. NH4F,MeOH
2. DMTrCl, Py

**D**

Ac₂O, AcOH
DMSO

ROH
NIS/TfOH
DCE

1. NH₄F,MeOH
2. DMTrCl, Py

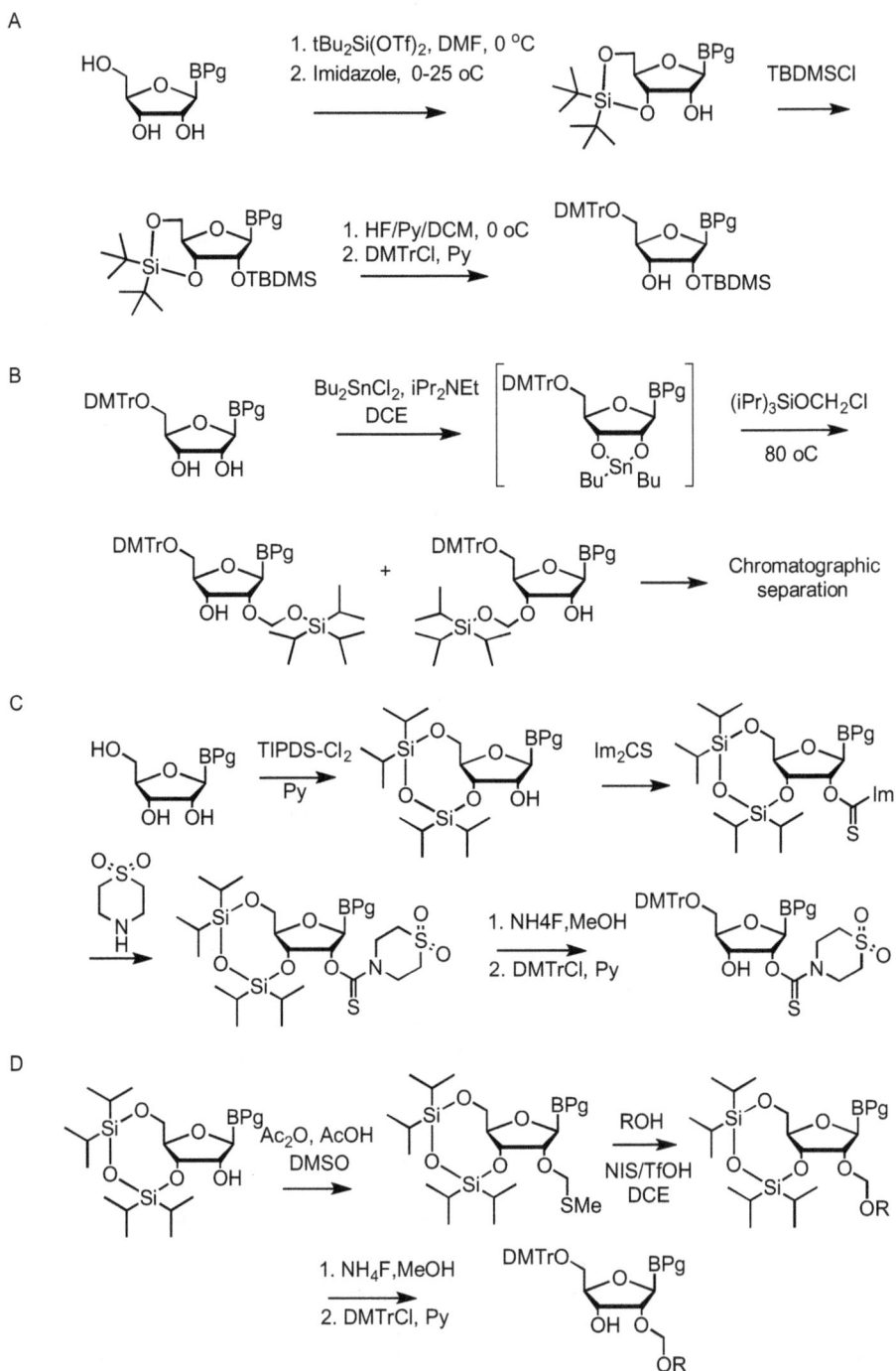

**Figure 5.22:** Alternative approaches for the synthesis of 2′-O-protected ribonucleosides.

**Table 5.2:** 2′-O-Protecting groups used in the synthesis of oligoribonucleotides.

| Entry | 2′-O-Protecting group | Conditions for removal | References |
|---|---|---|---|
| 1 | | 1 M TBAF in THF for 4 h | [76] |
| 2 | | 1 M TBAF in THF for 14 h | [77] |
| 3 | | Neat ethylenediamine (2 h at RT) | [82] |
| 4 | | 0.5 M TBAF in DMSO, containing 0.5% MeNO$_2$ as an acrylonitrile scavenger (5 h at RT) | [83] |
| 5 | | 1,4-Dithiothreitol (DTT) or tris(2-carboxyethyl) phosphine (TCEP) at pH 7.6 and 55 °C. | [84] |
| 6 | | 1. 1 M aqueous NaOH (3 h at 37 °C) 2. Neutralization with AcOH and evaporation to dryness 3. Decarboxylation with 0.5 M TBACl in DMSO/AcOH 9:1 (3 h at 65 °C) | [85] |
| 7 | | 0.5 M TBAF in DMSO (24–48 h at 55 °C | [86] |
| 8 | | 1. Concentrated aqueous NH$_3$ (10 h at 55 °C) 2. 0.1 M AcOH (15–40 min at 90 °C) | [87] |
| 9 | | 1 M TBAF in THF-containing 5% PrNH$_2$ (15 h at RT) | [88] |
| 10 | | 1 M TBAF in THF-containing 10% PrNH$_2$ and 1% bis(2-mercaptoethyl) ether (20 h at RT) | [89] |

dibutylstannylane activation (Figure 5.22B). The conditions required to remove all these protecting groups at the end of synthesis are indicated in Table 5.2.

The overall strategy for the solid-phase synthesis of oligoribonucleotides is very similar to that for their 2′-deoxy counterparts (Figure 5.11). The 2′-O-protecting groups usually tolerate both the removal of 2-cyanoethyl groups from phosphodiester linkages with 50% Et$_3$N in MeCN and the subsequent release from the support and removal of base moiety protecting groups with a 3:1 mixture of aq. ammonia and EtOH or even with aqueous ammonia. Utilization of acetyl and phenoxyacetyl groups instead of benzoyl and isobutyryl groups may be necessary to avoid a prolonged treatment. Two of the 2'-aminooxymethyl-derived protecting groups (entries 6 and 8), however, require a two-step removal. In case of the 2'-O-(iminooxymethyl ethyl propanoate) protection (entry 6), the ester linkage is first hydrolyzed with NaOH [85]. The remaining carboxy group still prevents departure of the aminooxymethyl group. When the neutralized and dried oligonucleotide is treated with TBACl in aqueous DMSO, the 2'-O-protecting group is cleaved as $CO_2$, MeCN and $CH_2O$. Likewise, the 4-(2,2-dichloro-N-methylacetamido)benzyloxymethyl group (entry 8) is upon ammonolysis deacylated to 4-(N-methylamino)benzyloxy]methyl group [84] that is cleaved under mildly acidic conditions by successive departure of N-methyliminoquinone methide ($H_2C = C_6H_4 = NH^+Me$) and formaldehyde.

Besides the 2'-O-protecting groups discussed above, still one group is worth mentioning. 2'-O-Azidomethyl-protected ribonucleoside 3'-(4-methoxy-1-oxido-2-picolyl) phosphates have been successfully utilized in solid-phase synthesis by phosphtriester approach [90]. The azidomethyl group well tolerates the removal of 4-methoxy-1-oxido-2-picolyl group with 1 M LiI in MecN and ammonolysis with 28% aqueous ammonia. Reduction to hydrolytically labile aminomethyl group with $MePPh_2$ in 4:1 dioxane water mixture then removes the 2'-OH protection in 2 h. It should be noted that this protecting group is not compatible with the phosphoramidite chemistry, owing to intramolecular Staudinger reaction [91] between the azido and phosphoramidite groups to an iminophosphorane.

As mentioned above, the coupling times tend to be longer in the synthesis of oligoribonucleotides than in the synthesis of oligodeoxyribonucleotides. For this reason, acidic activators, such as 5-benzylthiotetrazole [92], 5-ethylthiotetrazole [93] and N-phenylimidazolium triflate [94], are generally used. Acidic activators can be used because ribonucleosides are depurinated less readily than 2'-deoxyribonucleosides. After completion of chain assembly, oligoribonucleotides are deprotected and released from the support in principle the same way as their 2'-deoxy counterparts, but under somewhat milder conditions. Conventionally, a 3:1 mixture of aqueous ammonia and EtOH [95] or a 1:1 mixture of $MeNH_2$ and aqueous ammonia [96] is used. The extensively used 2'-O-TBDMS and 2'-O-TOM groups well tolerate these treatments.

Short oligoribonucleotides are routinely purified by the "DMTr-on" approach. The hydrophobic DMTr group facilitates chromatographic separation of the full-length product from shorter sequences. With sequences up to 100 nucleotides, temporary immobilization to a solid support may be a method of choice since chromatographic separation turns rather time consuming on approaching 100-mer ORN sequences. Figure 5.23 shows a recent example [97]. The synthesis is carried out on a CPG support bearing a photolabile linker. The 5'-terminal OH of the otherwise fully protected RNA (TBDMSi for 2'-OH, 2-cyanoethyl for phosphates) is tagged with a 1,2,4,5-tetrazine tail. The construct is released photochemically from the support and immobilized through the tetrazine

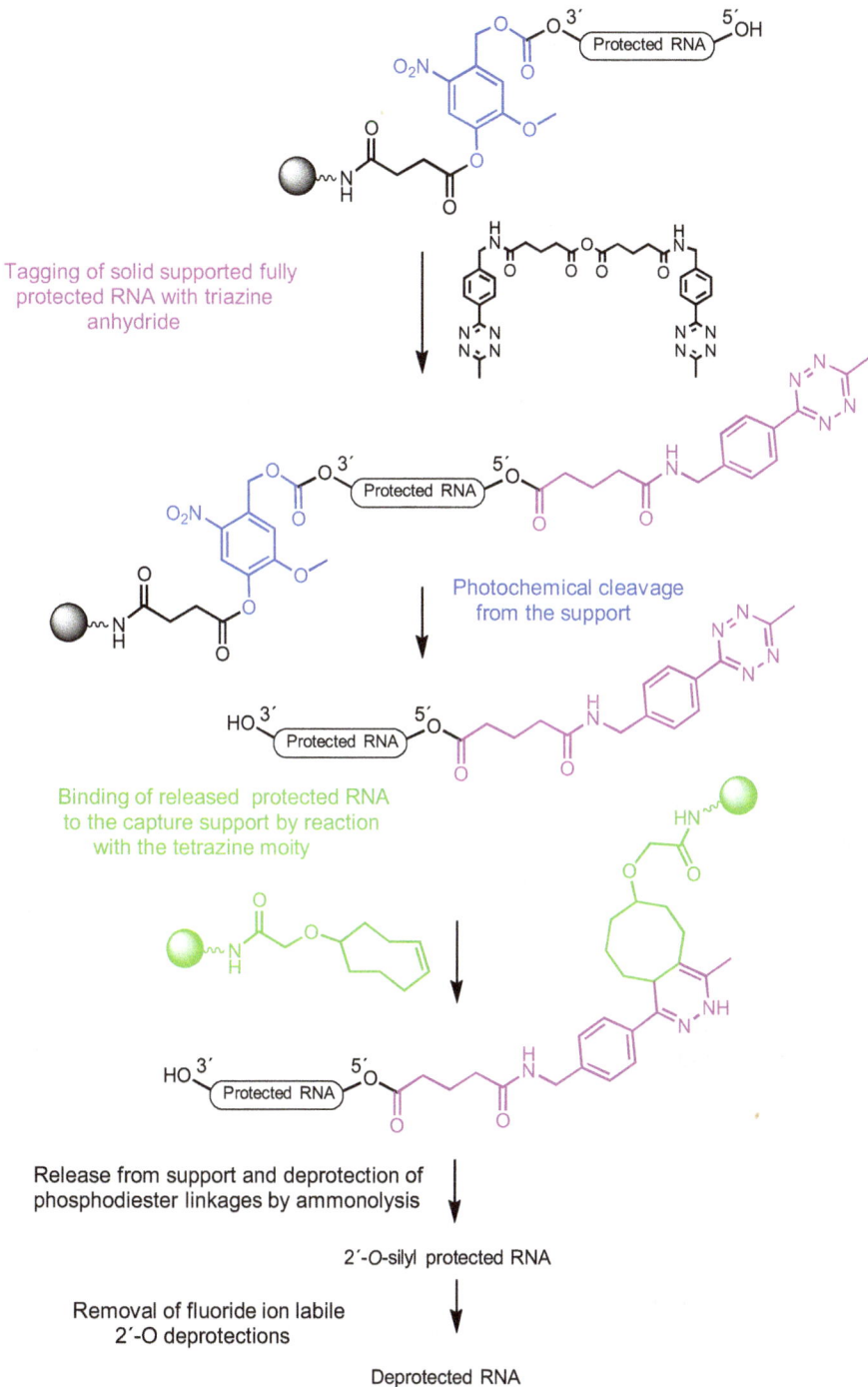

Tagging of solid supported fully protected RNA with triazine anhydride

Photochemical cleavage from the support

Binding of released protected RNA to the capture support by reaction with the tetrazine moity

Release from support and deprotection of phosphodiester linkages by ammonolysis

2′-O-silyl protected RNA

Removal of fluoride ion labile 2′-O deprotections

Deprotected RNA

Figure 5.23: Purification of long ORN sequences by binding to a catcher support.

unit to a *trans*-cyclooctene derivatized support. Oligomers that do not bear the tetrazine tag remain in solution and are washed away. Finally, the ORN is released from the support by ammonolysis, which also removes the phosphate protection. Removal of 2'-O-TBDMSi protections by fluoride ion then yields the fully deprotected ORN.

## 5.3 Enzymatic synthesis of oligonucleotides

### 5.3.1 Synthesis of oligodeoxyribonucleotides

Enzymatic synthesis is often used to introduce base and/or sugar-modified nucleosides in ODNs [98–100]. In many cases, this could be done by chemical synthesis as well, but synthesis of a phosphoramidite building block is avoided on using an enzymatic approach. Conversion to 5'-triphosphate is still needed. As regards base modifications, the preferred sites of substitution are C5 of pyrimidines and C7 of 7-deazapurines. Substituents at these sites do not severely hinder binding to polymerase and become oriented in the major groove upon duplex formation (cf. Section 8.1) [101]. In addition, 2-substituted purine [102] and $N^4$-substituted cytosine nucleotides [103] can be incorporated enzymatically. The enzymes used for the purpose are thermostable polymerases [104, 105], the same enzymes that are commonly used for enzymatic amplification of DNA by polymerase chain reaction [106], and even more appropriate polymerases have been obtained by directed evolution [107, 108]. Nucleotides containing a sugar modification are usually poor substrates for polymerases. Engineered polymerases are needed for their incorporation [109]. Among phosphate modifications, only $S_P$-α-phosphorothioate-modified dNTPs are accepted as substrates [110].

An extensively used method for enzymatic synthesis of long sequences is primer extension on a chemically synthesized template. In practice, a short chemically synthesized primer is hybridized with a template conjugated to biotin. It should be noted that when one of the canonical nucleotides is replaced by its modified analog, the analog becomes incorporated at each site opposite of the complementary base within the template. After completion of the enzymatic polymerization, the template is removed at elevated temperature or high pH by streptavidin-coated magnetic beads that bind biotin (Figure 5.24A) [111]. Alternatively, approaches related to amplification can be used [112]. Both of these approaches, however, suffer from the limitation that firm binding of primer to the template is needed. Consequently, short modified oligonucleotides (<15 nucleotides) cannot be prepared. To obtain such oligomers, a more complicated methodology known as nicking enzyme amplification reaction has been developed [113]. The primer contains a recognition site for a nicking endonuclease close to its 3'-terminus. When the polymerase has added 8–20 nucleotides to the primer, the nicking enzyme becomes able to recognize its binding site and cleaves the primer at its 3'-terminus. The product of the polymerase reaction still is too short to stay hybridized and its release into solution exposes the primer-template hybrid to next cycle of polymerization.

A more extensively used method for the preparation of short modified ODNs is based on 3′-elongation of a very short primer by terminal deoxynucleotidyl transferase (TdT) [114]. This process does not require any template and the primer can be as short as a trimer having the 5′-terminus phosphorylated. Accordingly, it looks at a first glance a perfect candidate for enzymatic synthesis of even regular unmodified oligonucleotides. One complicating fact, however, is markedly different incorporation rates of various dNTPs, the order of decreasing reactivity being G > C > A > T [115]. To avoid introduction of more than one nucleotide per coupling cycle, the 3′-OH of the NTP must be protected (Figure 5.24B). After removing the excess of coupled NTP and the enzyme, the 3′-$O$-protecting group is removed and then the next nucleotide can be coupled [116]. Unfortunately, the 3′-$O$-protecting group inevitably lowers the efficiency of enzymatic incorporation. 3′-$O$-(2-Cyanoethoxy)methyl [117], 3′-$O$-allyl [118], 3′-$O$-NH$_2$ [119] and 3′-$O$-azidomethyl [120] protections have been introduced for the template-dependent polymerization and the same groups are viable candidates also for template-independent incorporation by TdT. In spite of the apparent simplicity of this enzymatic approach, several thresholds still exist on the way to an automated synthesis that could compete with conventional phosphoramidite chemistry [115].

## 5.3.2 Synthesis of oligoribonucleotides

As regards synthesis of oligoribonucleotides, enzymatic polymerization on a DNA template plays a role when sequences longer than 100 nucleotides are prepared or the sequence contains base-modified nucleotides. Bacteriophage T7 RNA polymerase is usually used for the synthesis of both unmodified and modified RNA sequences. This enzyme does not require a separate primer, but initiates the synthesis at a specific promoter region within the DNA template [121]. T7 RNA polymerase also accepts a broad spectrum of modified NTPs as substrates [122]. In addition, the wild type has been by enzyme engineering modified to accept NTPs that the wild type does not recognize [123]. The other useful enzymes include SFM19 mutant of Taq DNA polymerase [124], and E664K and Y409G mutants of Tgo DNA polymerase [125] that allow the synthesis of several 2′-modified oligomers.

Oligoribonucleotides have also been synthesized by T4 RNA ligase using nucleoside 3′,5′-bisphosphates as monomeric building blocks [126]. The synthesis proceeds from the 5′- to 3′-terminus. After attachment of the bisphosphate to 3′-OH of the 5′-anchored primer, 5′-phosphate the bisphosphonate is removed with alkaline phosphatase and the next ligation is carried out. The efficiency of ligation is low, and the approach has never developed to the level of a workable protocol.

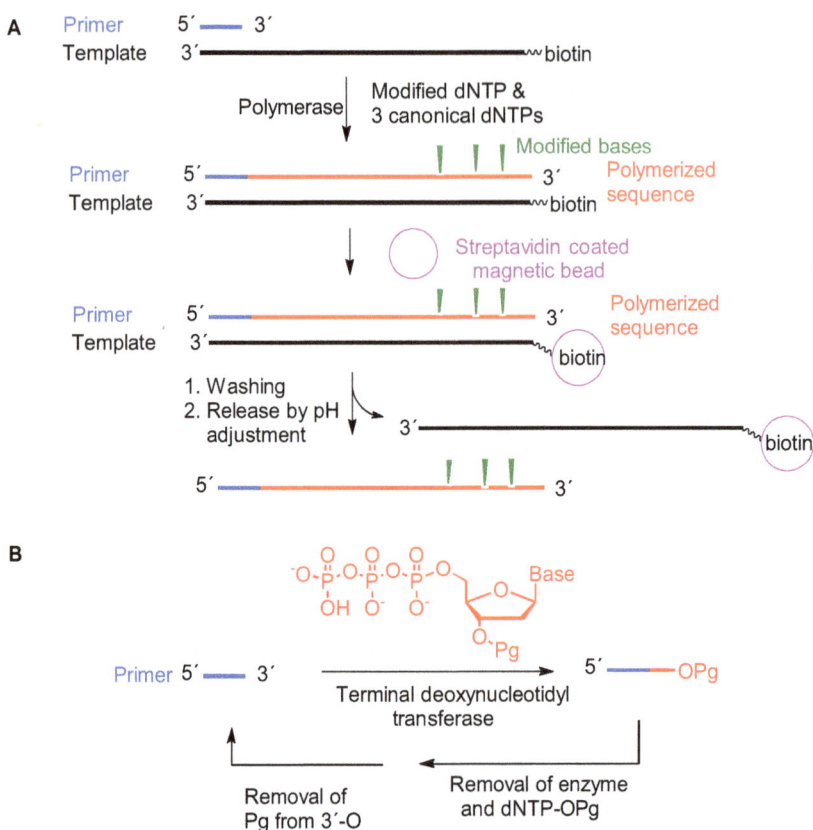

**Figure 5.24:** (A) Principle of synthesis by primer extension, making use of streptavidin-coated magnetic beads for the removal of template [110] and (B) principle of chain elongation by terminal deoxynucleotidyl transferase-catalyzed stepwise incorporation of 3′-protected nucleoside 5′-triphosphates [114].

## 5.4 Synthesis of phosphodiester-modified oligonucleotides and their congeners

### 5.4.1 Synthesis of stereoregular phosphorothioate oligonucleotides

Oxidation of phosphite triesters to phosphorothioates with the aid of sulfur transfer agents (Section 4.1) is not a stereoselective process but yields both $R_P$ and $S_P$ diastereomers in comparable amount. Since the stereochemistry of internucleosidic phosphorothioate linkages may influence on the stability of double helix formed with an unmodified phosphodiester oligonucleotide, preparation of stereopure phosphorothioate oligomers has been a subject of considerable interest since 1980s. The pioneering solid-supported synthesis of stereoregular phosphorothioate oligomers was based on chro-

matographically separated pure diastereoisomers of 5'-O-DMTr-nucleoside 3'-O-(2-thio
-1,3,2-oxathiaphospholane) synthons [127]. DBU promoted attack of the 5'-OH of a support-
bound oligomer on the phosphorus atom of this synthon in a mixture of MeCN and pyridine resulted in opening of the 1,3,2-oxathiaphospholane ring by PS-bond cleavage and
concomitant departure of the 2-mercaptoethyl group as ethylenesulfide (Figure 5.25).

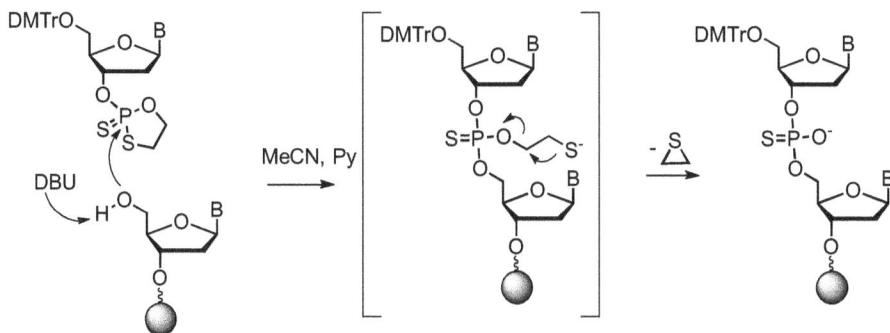

**Figure 5.25:** The 2-thio-1,3,2-oxathiaphospholane approach for the synthesis of stereoregular
oligonucleotide phosphorothioates [127].

Later, bicyclic 1,3,2-oxazapholidines have been used as synthons [128]. They are obtained by kinetically controlled reaction of appropriately protected nucleoside with
5-phenyl substituted bicyclic 1,3,2-oxazapholidine chloride having the phenyl substituent *cis*- to the pyrrolo ring (Figure 5.26). The reaction is highly stereoselective yielding only the *trans*-isomer, that is, the nucleoside and 5-phenyl substituent are bound on
opposite sides of the oxazapholidine ring. The $R_P$ and $S_P$ enantiomers formed in
comparable amount are separated chromatographically. The stereopure building blocks
are then used to assemble stereopure oligodeoxyribonucleotides as outlined in Figure 5.26. A special feature of this approach is usage of N-cyanomethylpyrrolidinium salt
as coupling activator. The purpose is to minimize the nucleophilicity of the activator and
utilize it only as a proton donor because nucleophilic attack on phosphorus could lead to
racemization around the phosphorus atom [129]. Otherwise, the coupling cycle closely resembles of conventional phosphoramidite chemistry. N,N'-Dimethylthiuram disulfide is
used as the sulfurization agent. Stereo-defined oligoribonucleotides have been prepared
similarly by using 2'-O-(2-cyanoethoxymethyl)ribonucleoside 3'-(1,3,2-oxazapholidine)
monomers [130]. Phosphorothioate ODNs-containing both $R_P$- and $S_P$-linkages, alternating
in a predesigned manner, have been obtained by a rather similar procedure using $R_P$
and $S_P$ stereoisomers of nucleoside 3'-O-(3-methyl-3-phenyl-[1,3,2]oxazapholidine)s to
obtain $S_P$- and $R_P$-linkages, respectively [131]. Phenoxyacetic anhydride as a capping reagent acylated the pyrrolidine nitrogen, resulting in spontaneous removal of the chiral
auxiliary upon subsequent sulfurization with S-cyanoethyl methylthiosulfonate.

Synthesis of building blocks:

Base moiety protections:

Coupling cycle:

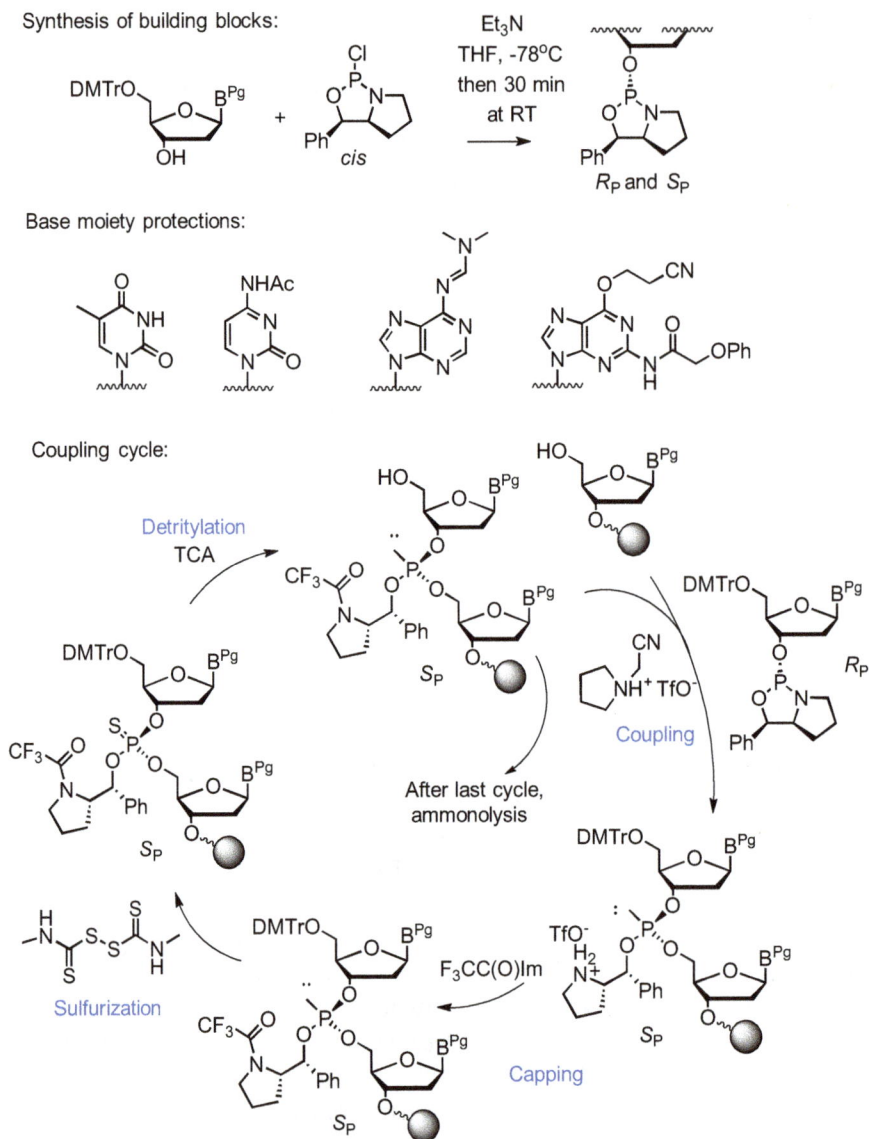

**Figure 5.26:** The 1,3,2-oxazaphospholidine approach for the synthesis of stereoregular oligonucleotide phosphorothioates [128].

In addition to the P(III) chemistries discussed above, a competitive P(V) alternative has been developed [132]. The key synthon is obtained by reacting either (−)- or (+)-limonene oxide with bis(pentafluorophenyl hydrogen phosphorotetrathioate (Figure 5.27). The 3′-OH of an appropriately protected nucleoside displaces the pentafluorothiophenyl ligand giving a stereopure phosphorodithioate. DBU-assisted attack of 5′-OH of the support bound sequence on the phosphorus atom then results in displacement of the limo-nene-derived ligand by 100% inversion. By this approach any combination of stereo-meric phosphorothioate linkages ($P_S$, $P_R$, $P_{racemic}$) can be conveniently introduced. Figure 5.27 shows the coupling cycle. The reported efficiency is comparable to that of P(III) protocols [133]. Stereoregular methylphosphonate oligomers are obtained by the same technique [134].

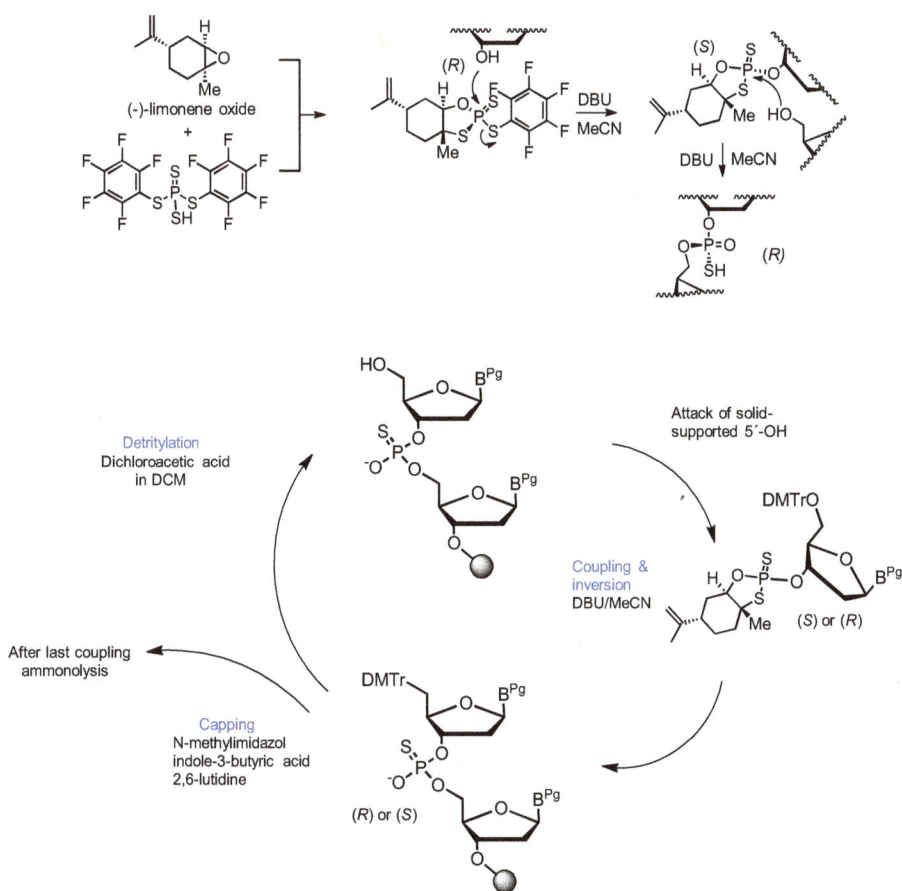

**Figure 5.27:** Synthesis of a stereopure ($R_P$)-dinucleoside-3′,5′-phosphoromonothioate and the coupling cycle for synthesis of stereoregular phosphorothioate oligonucleotides [132].

## 5.4.2 Synthesis of boranephosphonate oligonucleotides

Borane phosphonate-linked oligonucleotides have received interest since they seem to mimic DNA in biological processes [135]. Short borane phosphonate oligomers are obtained by consecutive silylation and boronation of *H*-phosphonate oligonucleotides bearing no base moiety protections (Figure 5.28A) [136]. Higher yields and longer oligomers are, however, obtained by phosphoramidite chemistry using silyl-protecting group for the base moieties and 5′-OH [137]. The coupling cycle is depicted in Figure 5.28B. O-Methyl phosphoramidites and di-tert-butylisobutylsilyl base protections withstand boronation with BH₃. Trimethylphosphite borane serves as an efficient trityl scavenger. Stereopure $R_P$- and $S_P$-boranephosphonate oligonucleotides are obtained by the 1,3,2-oxazaphospholidine approach described above (Figure 5.26), when acid labile protections for base moieties are used [138].

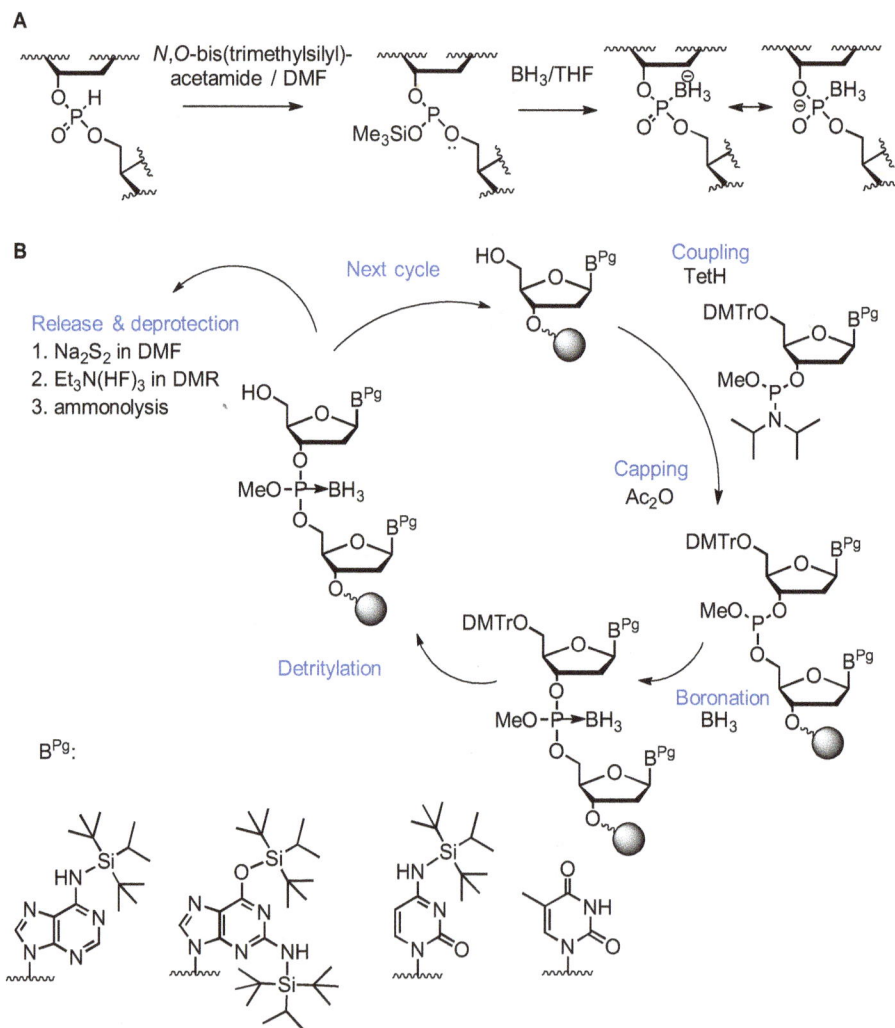

**Figure 5.28:** Synthesis of short boranephosphonate oligonucleotides without base-moiety protection (A) [136] and long oligomers by making use of base moiety silylation (B) [137].

Chimeric ODNs-containing boranophosphate and phosphorothioate linkages in addition to phosphodiester bonds have been assembled on a solid support from *H*-boranophosphonate, *H*-phosphonothioate and *H*-phosphonate-derived 3'-*O*-{3-(4-methoxyphenyl)-[1,3,2]oxazaphospholidine} monomers. The coupling cycle consisted of *N*-cyanomethylpyrrolidinium salt promoted condensation of an appropriate building block with solid supported 5'-OH and subsequent detritylation. The primary amino functions of dCtd and dAdo were 4-methoxybenzyloxycarbonyl protected and $O^6$ of dGuo trimethylsilylethyl-protected. After completion of the chain assembly, the *H*-phosphonate diester linkages were subjected to oxidation, base moiety deprotection and release from support [139, 140].

Boranephosphonate diesters react with pyridine and tertiary amines when oxidized with iodine. A zwitterionic covalent B–N$^+$ bond is formed [141]. This reaction is noteworthy since chimeric oligonucleotides-containing pyridinium boranephosphonate linkages in selected sites have showed enhanced cellular uptake [142]. In fact, boranophosphate ODN serves as a versatile precursor for various oligonucleotides containing phosphate-modified linkages [143]. Boranophosphodiester reacts in pyridine with pivaloyl chloride giving an acyl phosphite intermediate that allows conversion by electrophilic or nucleophilic substitution to a phosphorothioate, phosphoroalkylamidate, phosphotriester, phosphorothioate triester or phosphorothioatealkylamidate linkage.

### 5.4.3 Synthesis of phosphorodiamidate and thiophosphoramidate morpholino (PMO) oligomers

Phosphorodiamidate morpholino oligomers are analogs of oligonucleotides having nucleobases attached to C2 of 6-hydroxymethylmorpholines that are linked to each other via phosphrodiamidate linkages (see Figure 5.29). These oligomers form stable duplexes with nucleic acids and have shown promise as sequence-selective therapeutics that modify RNA splicing. In fact, three morpholino oligomers have already been approved for clinical use (see Section 11.3).

In spite of extensive pharmacological interest and first solution phase syntheses in mid-1990s [144], workable procedures for the synthesis of morpholino oligomers on solid support have been published only recently [145, 146]. Preparation of morfolino building blocks is rather straightforward: the C2'-C3' bond of a base-protected ribonucleoside is cleaved by oxidation with sodium periodate followed by treatment of the resulting dialdehyde with ammonia and reduction with sodium cyanoborohydride (Figure 5.29A) [144]. The ring nitrogen is tritylated and the hydroxymethyl group phosphorylated with *N,N*-dimethylphosphorodichloridate. Longer than 20-mer PMOs have been assembled in 20% yield from these building blocks on a solid support. The reactions shown to work in solution were largely used, but all the steps were carefully optimized [145]. The main problem in construction of a workable solid support synthesis is that the reactions of PMO synthesis are rather slow. Hence, finding the optimal conditions on solid support is crucial. The coupling cycle is described in Figure 5.29B. The support was prepared by immobilizing the O-terminal N-tritylated Morpholino unit to an amino functionalized support via a succinyl linker. Trityl protection was removed with a mixture of MeSO$_3$H (0.5%) and TCA (3%) in DCM. Coupling of the exposed ring nitrogen to the dichlorophosphorylated building block was then catalyzed by 5-(ethylthio)-1H-tetrazole and unreacted nitrogens were capped by acetylation. Conventional ammonolysis removed the base moiety protections and released the oligomer from the support. Fmoc-protection was also successfully used in place of trityl protection [145]. Using very similar chemistry, a fully automated flow-based oligonucleotide synthesizer has been developed. The flow-type approach has been reported to shorten the coupling times by more than one order of magnitude [146].

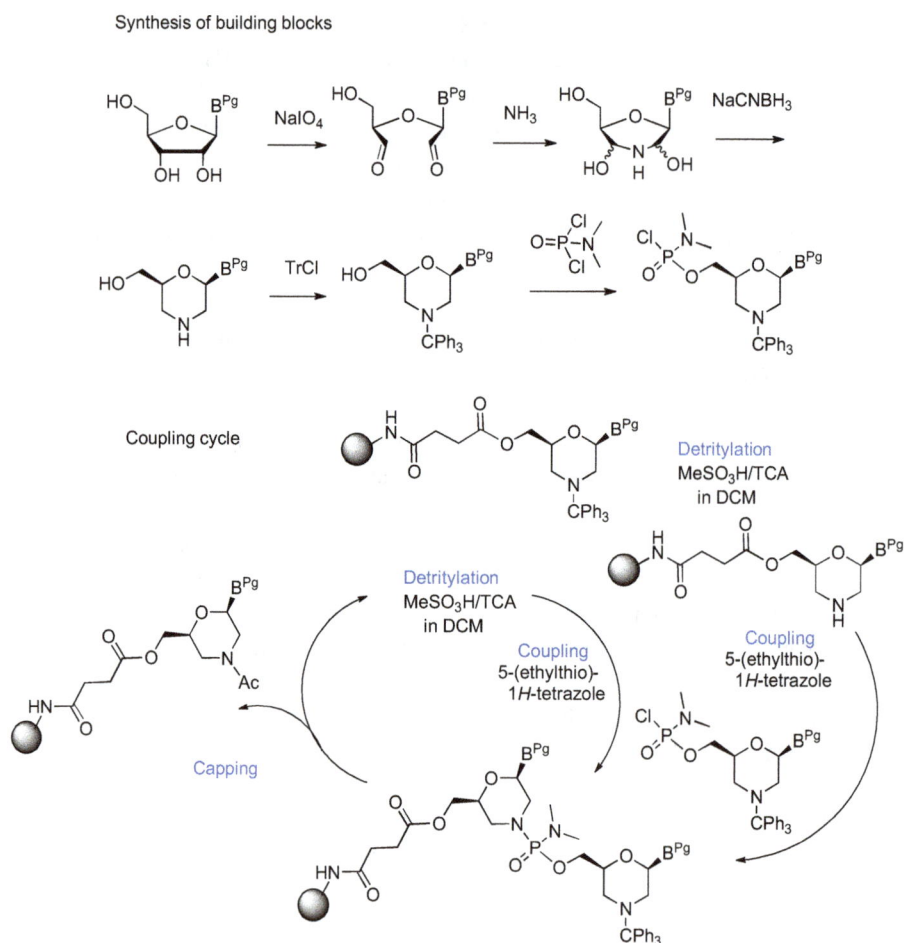

**Figure 5.29:** Synthesis of building blocks (A) and coupling cycle for their solid-supported assembly (B) to phosphordiamidate morpholino oligomers (145).

Recently, a promising solution-phase synthesis based on H-phosphonate chemistry has been reported for phosphordiamidate morpholino oligomers [147]. Phosphonium-type condensing agents commonly used in peptide synthesis efficiently accelerate condensing of an H-phosphonate monomer with morpholino ring nitrogen, giving an H-phosphonamidate linkage that, when treated with dimethylamine, is converted to N,N-dimethylamino phosphorodiamidate. The attractive features of this approach are short coupling time and applicability to convergent synthesis strategies.

Thiophosphoramidate morpholino oligomers have been obtained by phosphoramidite coupling as shown in Figure 5.30 [148]. The approach is quite flexible allowing introduction of nucleosides or their modified analogs at any site within the oligomer, a property of interest in optimizing the affinity, biological stability and toxicity of oligonucleotide therapeutics.

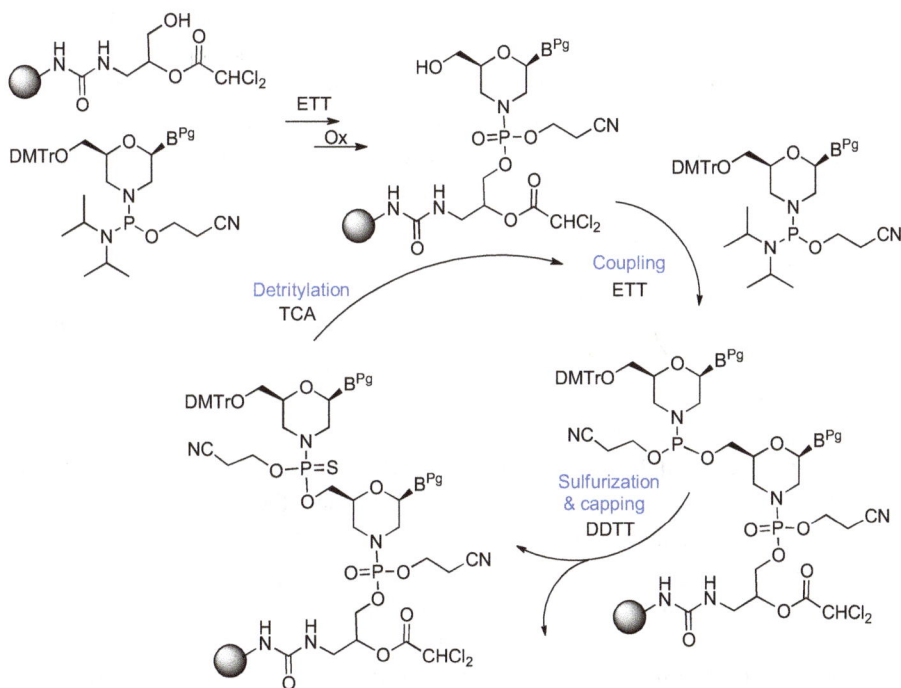

**Figure 5.30:** Synthesis of thiophosphoramidate morpholino oligomers [148].

## 5.5 Solution-phase synthesis of oligonucleotides

### 5.5.1 Convergent synthesis

Although solid-supported phosphoramidite chemistry currently is the method of choice for preparation of oligonucleotides, not only in lab but even in kilogram scale production, solution-phase synthesis has still retained some interest as a method for large-scale production. In fact, the interest has recently increased. The underlying idea is to reduce the excess of building block and activator required for quantitative coupling to get rid of the cost of solid support and to scale up the synthesis to be compatible with existing infrastructure of production. Additionally, the possibility for thorough characterization after each coupling is a desirable feature of solution-phase synthesis.

Oligonucleotides may be assembled in solution either in a convergent manner (from monomers, to dimers, tetramers, octamers, etc.) or in a stepwise manner by making use of a soluble support or recyclable solid-supported reagents. Preparation of Vitravene offers a good example of the possibilities of a conventional convergent strategy [149]. Vitravene is a 21-mer phosphorothioate oligodeoxyribonucleotide approved by FDA in 1988 for the treatment of cytomegalovirus retinitis in immunocompromised patients. The drug was withdrawn in 2006.

The base sequence of Vitravene is 5'-GCG TTT GCT CTT CTT CTT GCG-3'. Accordingly, the sequence contains four different trimers, which were prepared by two consecutive oxidative H-phosphonate couplings, as outlined in Figure 5.31 [149]. The 3'-$O$-levulinoyl group was removed and the exposed hydroxyl function was phosphitylated. To obtain a hexamer, a detritylated trimer was coupled to the 3'-(H-phosphonate) group of a 5'-tritylated trimer. This kind of stepwise chain elongation by trimers was continued until an 18-mer was obtained. In the last coupling, a 3'-$O$-levulinoylated trimer was used. The coupling efficiency remained high, around 94%, throughout the synthesis.

A more recent example is offered by synthesis of an 18-mer phosphorothioate ODN by conventional phosphoramidite chemistry [150]. Four tetra/pentameric blocks were assembled in a stepwise manner on a 3'-$O$-(*tert*-butydiphenylsilyl)-protected 3'-terminal nucleoside. Base moiety protections were conventional and 4,5-dicyanoimidazole was used as activator. The fluoride ion-sensitive 3'-$O$-protection of each assembled block was removed and the exposed bydroxy function was phosphitylated. The first (3'-terminal) block was then coupled to uridine bearing a large hydrophobic protecting group at 3'-$O$. Sulfurization with xanthane hydride was performed and the 5'-DMTr protection was removed. This coupling cycle was repeated until all the blocks were coupled. The hydrophobic 3'-$O$-protecting group played a role in purification by precipitation after each coupling. The fact that the 18-mer ODN was prepared in kilogram scale lends support for potential of convergent approach in large-scale synthesis.

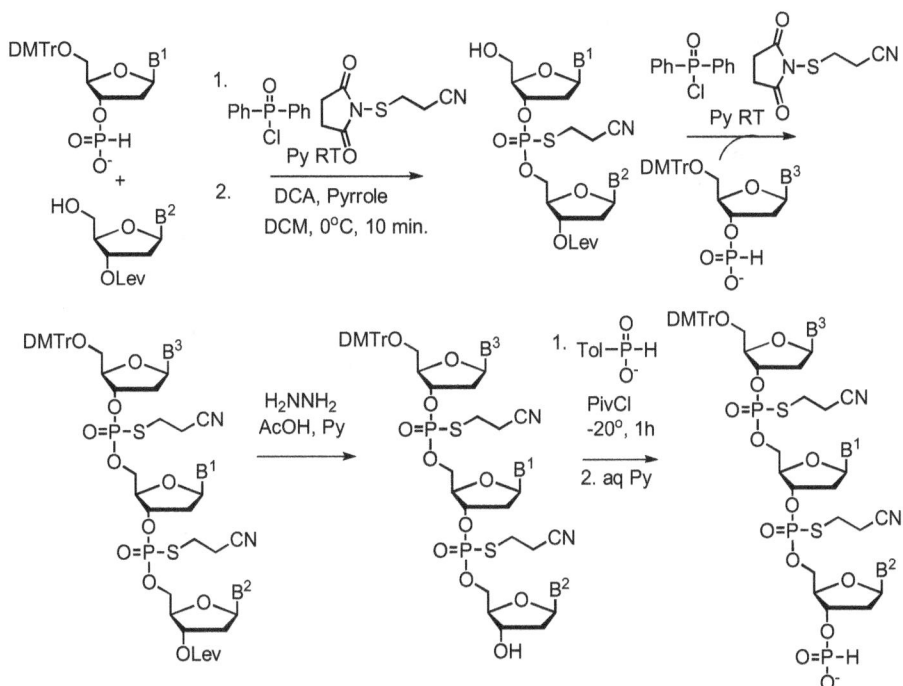

Base moiety protections:

Gua: $N^2$-iBu, $O^6$-(2,5-di-Cl-Ph; Cyt: $N^4$-Bz; Thy: $O^4$-Ph

**Figure 5.31:** Solution-phase synthesis of trimeric blocks used for stepwise assembly of 18-mer Vitravene by essentially the same H-phosphonate chemistry [149].

### 5.5.2 Synthesis on a soluble support

Synthesis of oligonucleotides on a soluble support has received rather extensive interest, but so far no serious attempt on a truly large scale has been reported. The main difference between the synthesis on a solid and soluble support lies in the removal of reagents and wastes after coupling and 5′-O deprotection. While in solid-supported synthesis this takes place by simple washing, in soluble support synthesis more laborious techniques, such as precipitation, extraction, chromatography or nanofiltration, have to be applied. The pioneering studies in this field were based on utilization of a polyetyhyleneglycol support (PEG support in Figure 5.32) that precipitates from Et$_2$O. Conventional phosphotriester chemistry with amino-acylated 3′-(o-chlorophenyl phosphate) building blocks was utilized (Figure 5.33) [151]. Each coupling cycle contained three precipitations, viz. after coupling, capping and removal of 5′-O-DMTr protection, which made the synthesis rather laborious. In spite of this, one coupling cycle could be carried out in 5 h, the coupling yield ranging from 90 to 95%. Comparable results were obtained with

3'-(*o*-chlorophenyl benzotriazol-1-yl phosphate) building blocks [152]. These building blocks were more recently used for the synthesis of pentameric oligonucleotides [153] on a branched tetrakis-*O*-[4-(azidomethyl)phenyl]pentaerythritol-derived support (tetrapodal support in Figure 5.32) [154]. The support precipitates quantitatively from MeOH. It has later been used for stereo-controlled assembly of tetranucleotide phosphorothioate shortmers from limonene-derived oxathiaphospholane sulfide building blocks

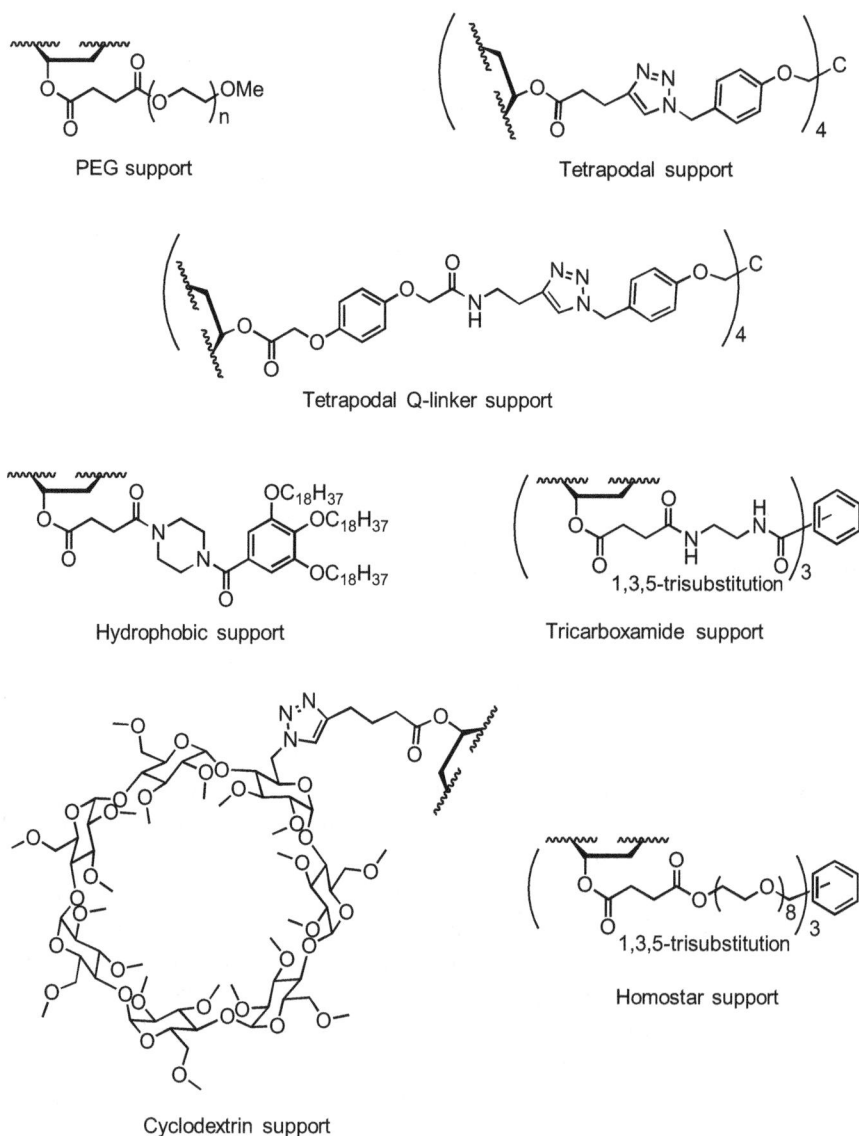

PEG support

Tetrapodal support

Tetrapodal Q-linker support

Hydrophobic support

Tricarboxamide support

Cyclodextrin support

Homostar support

**Figure 5.32:** Soluble supports used in the synthesis of oligonucleotides.

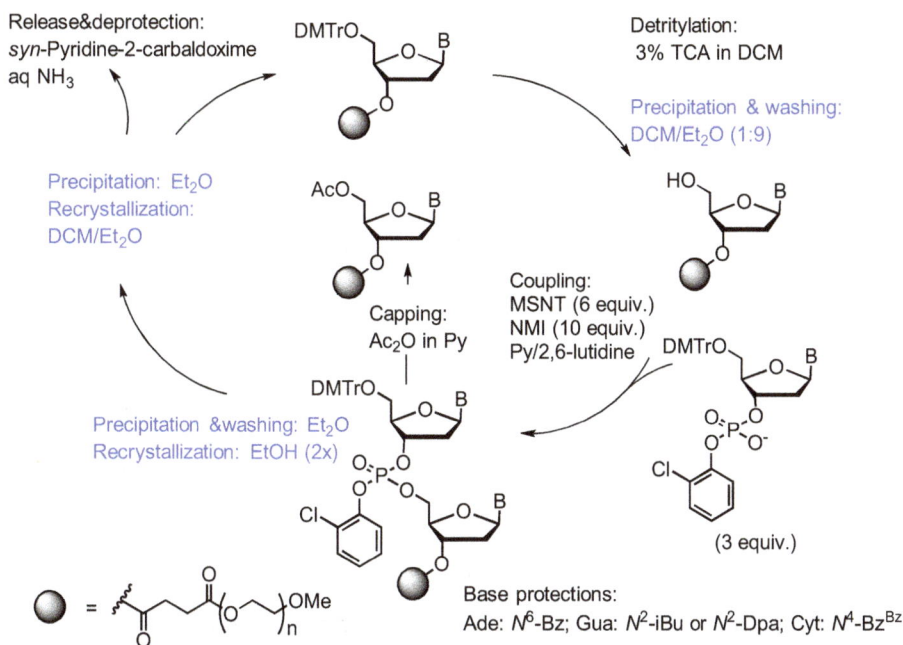

**Figure 5.33:** Synthesis of ODNs on a PEG support by the phosphotriester chemistry [151].

by P(V) chemistry [155]. Although the coupling efficiency in phosphotriester chemistry is somewhat lower than in phosphoramidite chemistry, a clear advantage is avoidance of oxidation step that markedly simplifies the coupling cycle.

PEG and tetrapodal supports have been used for the synthesis ODNs by P(III) coupling chemistry in spite of more complicated coupling cycle [154, 156]. Synthesis on the PEG support is depicted as an example in Figure 5.34. Four precipitations were carried out in each coupling cycle: after coupling, capping, oxidation with *tert*-butyl hydroperoxide and detritylation. The overall yield was comparable to that obtained by the phosphotriester chemistry. Phosphorothioate oligonucleotides were obtained similarly, the only difference being replacement of the *tert*-butyl hydroperoxide oxidation with tetraethylthiuram disulfide sulfurization [157].

Studies on soluble support synthesis by *H*-phosphonate chemistry are scarce but noteworthy. The building blocks, 3′-(2-cyanoethyl)-*H*-phosphonate diesters, were coupled oxidatively on a PEG support using NBS as an activator (Figure 5.35) [34]. Two precipitations were carried out in each coupling cycle: one after coupling and the other after 5′-detritylation. The average coupling efficiency in the synthesis of a 10-mer was 98% on using 2.5 equiv. of the 3′-*H*-phosphonate building block.

Extraction offers still one alternative method for the separation of the growing oligonucleotide chain from small molecule reagents and wastes (Figure 5.36) [158]. The oligonucleotide is kept so hydrophobic by appropriate protecting groups that

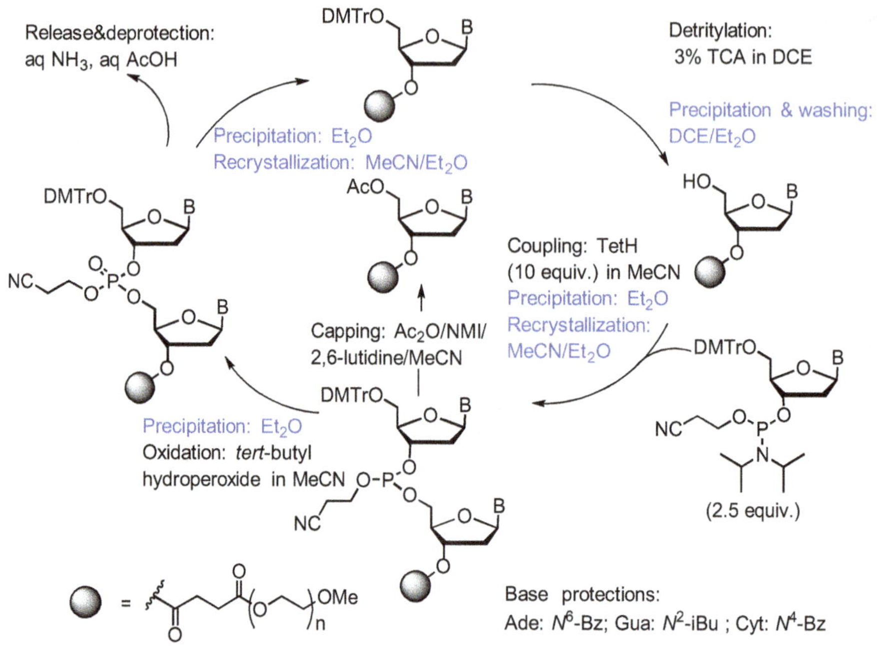

**Figure 5.34:** Synthesis of oligodeoxyribonucleotides on a PEG support by the phosphoramidite chemistry [157].

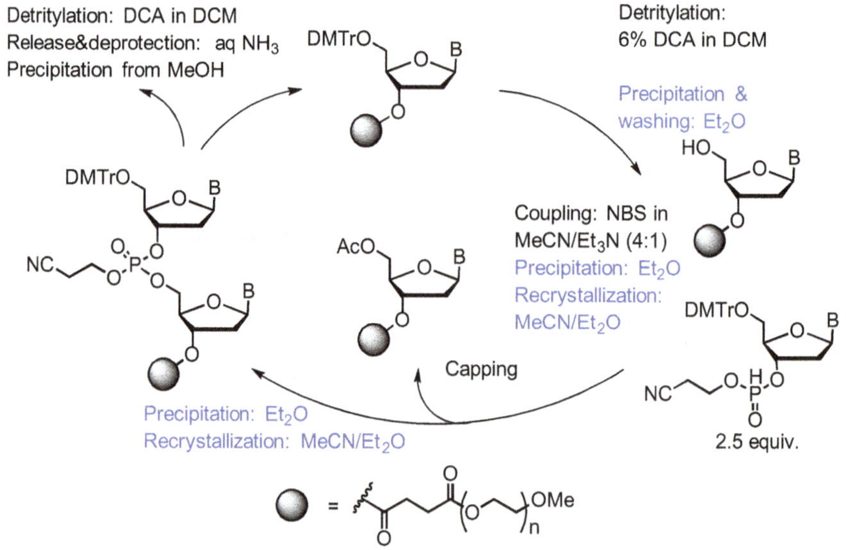

**Figure 5.35:** Synthesis of oligodeoxyribonucleotides on a PEG support by the oxidative H-phosphonate chemistry [34].

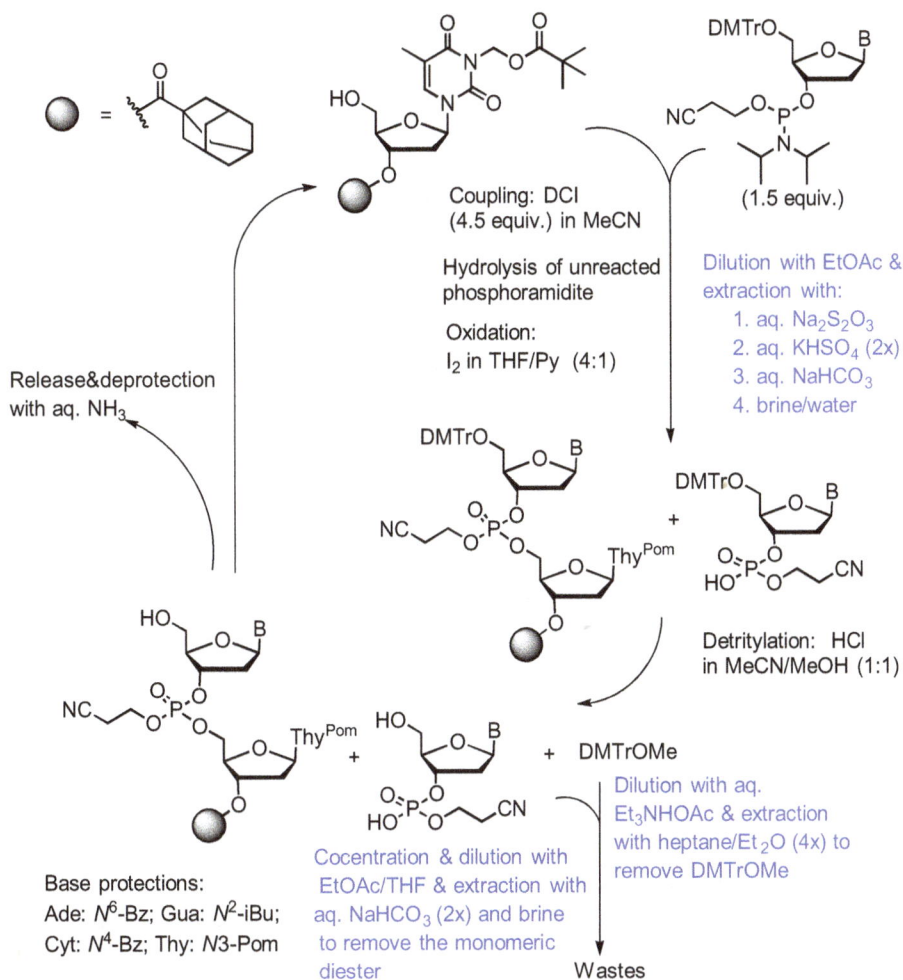

**Figure 5.36:** Synthesis of oligodeoxyribonucleotides on an adamantylcarbonyl support by making use of phosphoramidite chemistry and extractive removal of wastes [158].

most of the small molecules can be removed by water extraction. The hydrophobicity should, however, be only moderate since the DMTr methyl ether produced upon detritylation in a MeOH/MeCN mixture is removed by extraction with a very nonpolar solvent. Each coupling cycle involves 12 extractions. A fully protected hexamer has been obtained in 67% yield on using 1.5 equiv. of phosphoramidite blocks, which corresponds to 93% efficiency per coupling cycle.

Chromatographic separation as a viable, though somewhat tedious, separation method for soluble support oligonucleotide synthesis has been attempted in combination with tripodal $N^1,N^3,N^5$-tris(2-aminoethyl)benzene-1,3,5-tricarboxamide (tricarboxamide support in Figure 5.32) as a support [159]. Conventional phosphoramidite strategy with tetrazole activation and *tert*-butyl hydoperoxide oxidation was used. Two chromatographic gel permeation separations in MeOH were carried out in each coupling cycle: after coupling and detritylation. A 10-mer was assembled with an average efficiency of 87% per coupling cycle. Another example is synthesis of short oligodeoxyribonucleotides on a permethylated β-cyclodextrin support (cyclodextrin support in Figure 5.32) [160]. 5'-*O*-(1-Methoxy-1-methylethyl)-2'-deoxyribonucleoside 3'-phosphoramidites were used, and hence 5'-*O* deprotection in acidic methanol produced dimethyl acetal of acetone that could be removed by evaporation. Only one chromatographic separation (after the coupling step) in each coupling cycle was needed.

Last, but by no means least, nanofiltration in organic solvent has been introduced as a novel paradigm of solution-phase synthesis [161, 162]. In other words, the small molecular reagents and wastes are separated from the growing oligonucleotide on a soluble support by passing the mixture through a membrane that is permeable for the small molecules. 1,3,5-Tris(hydroxymethyl)benzene bearing three octa(ethylenelycol) chains was used as the support (homostar in Figure 5.32). Two filtrations through a polybenzimidazole-based membrane were carried out in each coupling cycle: after coupling and detritylation. A 9-mer 2'-*O*-methyl oligoribonucleotide phosphorothioate was prepared. The yield of the first coupling cycles was low, increasing gradually from 75 to 90%, but remained high (90–95%) after the fourth cycle. As with all the other methods developed, the approach is interesting but not yet ready for routine use.

Soluble support synthesis of ORNs has been studied to a much less extent than the synthesis of ODNs. A 21-mer oligoribonucleotide has been assembled in high yield by phosphoramidite chemistry from 5'-*O*-DMTr 2'-*O*-TBDMS-protected building blocks on a hydrophobic 4-oxo-4-{4-[3,4,5-tris(octadecyloxy)benzoyl]piperazin-1-yl}butanoyl support [163]. The support is soluble in THF, CHCl$_3$ and DCM but insoluble in MeOH and MeCN. Couplings were carried out with 1.5–2.0 equiv. in a 1:10 mixture of MeCN and DCM by 5-(benzylthio)-1*H*-tetrazole activation. The phosphite ester obtained was oxidized by the addition of 2-butanone peroxide in DCM immediately after coupling. Each coupling cycle contained only two precipitations from MeOH, one after coupling/oxidation and the other after detritylation. No capping step was included. The 21-mer was obtained in fully protected form in 46% yield, which means that the average yield per coupling cycle must be as high as 98%.

Short ORNs have been synthesized on the tetrapodal support, but unusual protecting groups were used to reduce hydrophobicity of the building blocks in order to warrant quantitative precipitation from MeOH. Accordingly, the 5'-*O*-DMTr group was replaced with 1-methoxy-1-methylethyl acetal protection and the 2'-*O*-TBDMS group with 2-cyanoethyl ethyl group [164]. When commercially available 2'-*O*-TBDMS 5'*O*-DMTr-protected building blocks were used, the coupling and oxidation steps were followed by precipitation from water, acidolytic detritylation and chromatographic purification in each coupling cycle [165].

Studies on soluble-supported synthesis of oligonucleotides are commonly motivated by the development of an industrial-scale synthesis. Several of the methodologies discussed above may, however, be useful for laboratory synthesis up to gram scale since no special equipment is required. Future will show whether one of them will really be used in industrial scale.

## 5.5.3 Synthesis with solid-supported reagents

Experience on usage of solid-supported reagents in oligonucleotide synthesis is rather limited. Only two serious attempts have been published, one by using phosphoramidite chemistry [166] and the other by *H*-phosphonate chemistry [167]. By phosphoramidite

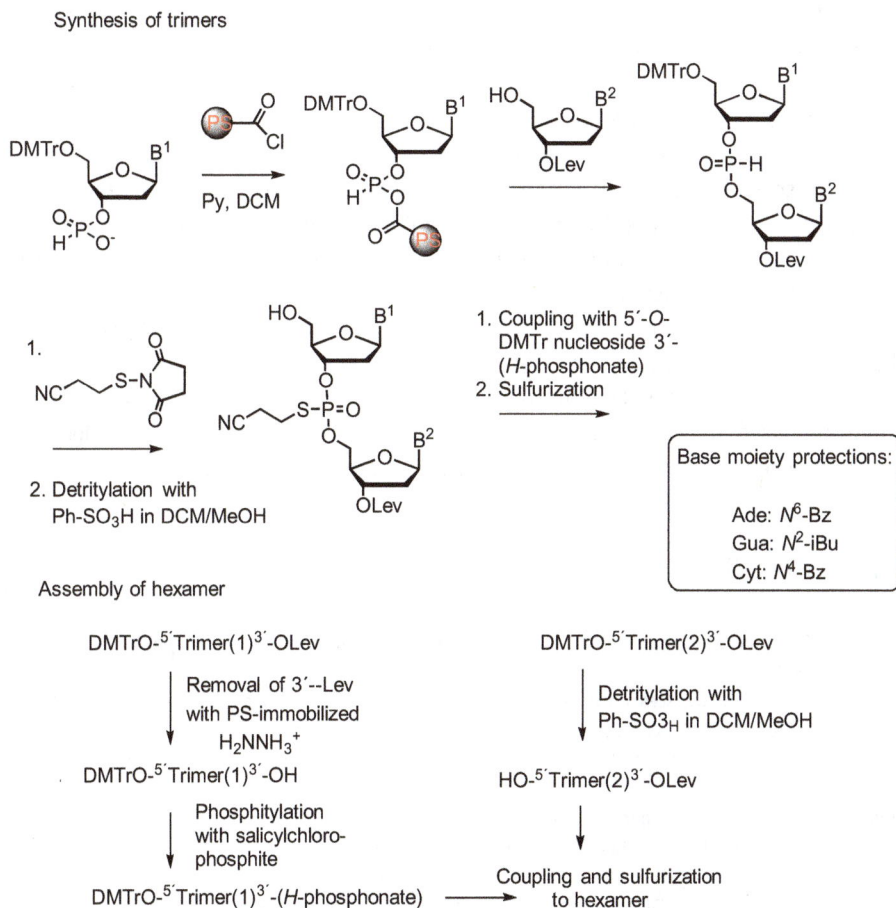

**Figure 5.37:** Synthesis of a hexameric phosphorothioate in solution by H-phosphonate chemistry with the aid of solid-supported reagents [167].

chemistry, only trimeric phosphodiester and phosphorothioate ODNs were prepared. The coupling of 5'-O-DMTr-protected nucleoside 3'phosphoramidite (1.5 equiv.) with a 3'-O-Lev nucleoside was promoted with excess (10 equiv.) of polyvinylpyridinium tosylate in MeCN. The unreacted phophoramidite was hydrolyzed to 2-cyanoethyl *H*-phosphonate and the crude product mixture was oxidized with polystyrene-anchored quaternary ammonium periodate of sulfurized with polystyrene-anchored quaternary ammonium tetrathionate. Finally, the 3'-O-Lev group was removed with hydrazinium polystyrene-sulfonate. The exposed hydroxyl function was phosphitylated with 2-cyanoethyl-*N,N,N'*, *N'*-tetraisopropylphosphoramidite using polyvinylpyridinium chloride as a catalyst. All polymer-supported reagents could be removed by filtration. Unfortunately, attempts to fully remove the residual water from the polyvinylpyridinium tosylate failed, which complicated the synthesis by leading to the hydrolysis of phosphoramidites to *H*-phosphonate diesters.

By *H*-phosphonate chemistry, a phosphorothioate hexamer was synthesized applying a convergent strategy (Figure 5.37). A 5'-O-DMTr-nucleoside 3'-(*H*-phosphonate) (1.2 equiv.) was first activated with a polystyrene-supported acid chloride (4 equiv.). Attack of a 3'-O-Lev nucleoside (1 equiv.) on the phosphorus atom of a supported-bound mixed carboxylic *H*-phosphonate anhydride then released the *H*-phosphonate diester in solution. Evidently, the unreacted *H*-phosphonate remained bound to the resin. The *H*-phosphonate dimer was immediately oxidized to a phosphorothioate dimer by the addition of *N*-(2-cyanoethylthio)succinimide. The 5'-O-DMTr group was then removed acidolytically into solution and the 5'-O-deprotected dimer was coupled to another 5'-O-DMTr-nucleoside 3'-(*H*-phosphonate). To achieve coupling of two trimers, the 3'-O-Lev group was removed with resin-immobilized hydrazinium ion and the exposed 3'-OH was phosphitylated in solution phase with salicylchlorophosphite. Another trimer prepared in a similar manner was subjected to detritylation instead of delevulinoylation. These two trimers were finally coupled and sulfurized to a hexamer.

## Further reading

Abramova T. Frontiers and approaches to chemical synthesis of oligodeoxyribonucleotides. Molecules 2013, 18, 1063–1075.

Beaucage SL, Iyer RP. Advances in the synthesis of oligonucleotides by the phosphoramidite approach. Tetrahedron 1992, 48, 2223–2311.

Flamme M, McKenzie LK, Sarac I, Hollenstein M. Chemical methods for the modification of RNA. Methods 2019, 161, 64–82.

Guga P, Koziolkiewicz M. Phosphorothioate nucleotides and oligonucleotides – Recent progress in synthesis and application. Chem Biodiv 2011, 8, 1642–1681.

Hocek M. Enzymatic synthesis of base-functionalized nucleic acids for sensing, crosslinking, and modulation of protein–DNA binding and transcription. Acc Chem Res 2019, 52, 1730–1737.

Lönnberg H. Synthesis of oligonucleotides on a soluble support. Beilstein J Org Chem 2017, 13, 1368–1387.

McKenzie LK, El-Khoury R, Thorpe JD, Damha MJ, Hollenstein M. Recent progress in non-native nucleic acid modifications. Chem Soc Rev 2021, 50, 5126–5164.

Nurminen E, Lönnberg H. Mechanisms of the substitution reactions of phosphoramidites and their congeners. J Phys Org Chem 2004, 17, 1–17.

Reese CB. Oligo- and polynucleotides: 50 years of chemical synthesis. Org Biomol Chem 2005, 3, 3861–3868.

Roy S, Caruthers M. Synthesis of DNA/RNA and their analogs via phosphoramidite and H-phosphonate chemistries. Molecules 2013, 18, 14268–14284.

Sarac I, Hollenstein M. Terminal deoxynucleotidyl transferase in the synthesis and modification of nucleic acids. ChemBioChem 2019, 20, 860–871.

Stawinski J, Strömberg R. Di- and oligonucleotide synthesis using H-phosphonate chemistry. Methods Mol Biol 2005, 288, 81–100.

Wei X. Coupling activators for the oligonucleotide synthesis via phosphoramidite approach. Tetrahedron 2013, 69, 3615–3637.

Virta P. Solid-phase synthesis of base-sensitive oligonucleotides. Arkivoc 2009, (iii), 54–83.

# References

[1] Beaucage SL, Caruthers MH. Deoxynucleoside phosphoramidites – A new class of key intermediates for deoxypolynucleotide synthesis. Tetrahedron Lett 1981, 22, 1859–1862.

[2] Garegg PJ, Lindh I, Regberg T, Stawinski J, Strömberg R, Hendrichson C. Nucleoside H-phosphonates. III. Chemical synthesis of oligodeoxyribonucleotides by the hydrogenphosphonate approach. Tetrahedron Lett 1986, 27, 4051–4054.

[3] Froehler BC, Ng PG, Matteucci MD. Synthesis of DNA via deoxynucleoside H-phosphonate intermediates. Nucleic Acids Res 1986, 14, 5399–5407.

[4] Reese CB. Oligo- and poly-nucleotides: 50 years of chemical synthesis. Org Biomol Chem 2005, 3, 3851–3868.

[5] MacBride LJ, Caruthers MH. An investigation of several deoxynucleoside phosphoramidites useful for synthesizing deoxyoligonucleotides. Tetrahedron Lett 1983, 24, 245–248.

[6] Sinha ND, Biernat J, Koester H. β-Cyanoethyl N,N-dialkylamino/N-morpholinomonochloro phosphoamidites, new phosphitylating agents facilitating ease of deprotection and work-up of synthesized oligonucleotides. Tetrahedron Lett 1983, 24, 5843–5846.

[7] Fourrey J-L, Varenne J. Improved procedure for the preparation of deoxynucleoside phosphoramidites: arylphosphoramidites as new convenient intermediates for oligodeoxynucleotide synthesis. Tetrahedron Lett 1984, 25, 4511–4514.

[8] Jankowska J, Sobkowski M, Stawinski J, Kraszewski A. Studies on aryl H-phosphonates. I. An efficient method for the preparation of deoxyribo- and ribonucleoside 3′-H-phosphonate monoesters by transesterification of diphenyl H-phosphonate. Tetrahedron Lett 1994, 35, 3355–3358.

[9] Marugg JE, Tromp M, Kuyl-Yeheskiely E, van der Marel GA, van Boom JH. A convenient and general approach to the synthesis of properly protected d-nucleoside-3′-hydrogenphosphonates via phosphite intermediates. Tetrahedron Lett 1986, 27, 2661–2664.

[10] Stawinski J, Strömberg R. Di- and oligonucleotide synthesis using H-phosphonate chemistry. Methods Mol Biol 2005, 288, 81–100.

[11] Chattopadhyaya JB, Reese CB. Some observations relating to phosphorylation methods in oligonucleotide synthesis. Tetrahedron Lett 1979, 20, 5059–5062.

[12] Efimov A, Buryakova AA, Dubey IY, Polushin NN, Chakhmakhcheva OG, Ovchinnikov YuA. Application of new catalytic phosphate protecting groups for the highly efficient phosphotriester oligonucleotide synthesis. Nucleic Acids Res 1986, 14, 6525–6540.

[13] De Vroom E, Fidder A, Marugg JE, van der Marel GA, Van Boom JH. Use of a 1-hydroxybenzotriazole activated phosphorylating reagent towards the synthesis of short RNA fragments in solution. Nucleic Acids Res 1986, 14, 5885–5900.

[14] Nurminen EJ, Mattinen JK, Lönnberg H. Protonation of phosphoramidites. The effect on nucleophilic displacement. J Chem Soc Perkin Trans 2000, 2, 0, 2238–2240.

[15] Nurminen EJ, Mattinen JK, Lönnberg H. Nucleophilic and acid catalysis in phosphoramidite alcoholysis. J Chem Soc Perkin Trans 2001, 2, 2159–2165.

[16] Korkin AA, Tsvetkov EN. Theoretical investigation of the interaction between the HNX-YHM molecules and electron acceptors – the ab initio study of the effect of protonation on the P-N bond in $PH_2NH_2$. Bull Soc Chim Fr 1988, 0, 335–338.

[17] Nurminen E, Lönnberg H. Mechanisms of the substitution reactions of phosphoramidites and their congeners. J Phys Org Chem 2004, 17, 1–17.

[18] Wei X. Coupling activators for the oligonucleotide synthesis via phosphoramidite approach. Tetrahedron 2013, 69, 3615–3637.

[19] Mateucci MD, Caruthers MH. Synthesis of deoxyoligonucleotides on a polymer support. J Am Chem Soc 1981, 103, 3185–3191.

[20] Hayakawa Y, Uchiyama M, Noyori R. Nonaqueous oxidation of nucleoside phosphites to the phosphates. Tetrahedron Lett 1986, 27, 4191–4194.

[21] Uzagare MC, Padiya KJ, Salunkhe MM, Sanghvi YS. NBS–DMSO as a nonaqueous nonbasic oxidation reagent for the synthesis of oligonucleotides. Bioorg Med Chem Lett 2003, 13, 3537–3540.

[22] Saneyoshi H, Miyata K, Seio K, Sekine M. 1,1-Dihydroperoxycyclododecane as a new, crystalline non-hygroscopic oxidizer for the chemical synthesis of oligodeoxyribonucleotides. Tetrahedron Lett 2006, 47, 8945–8947.

[23] Manoharan M, Lu Y, Casper MD, Just G. Allyl group as a protecting group for internucleotide phosphate and thiophosphate linkages in oligonucleotide synthesis: facile oxidation and deprotection conditions. Org Lett 2000, 2, 243–246.

[24] Kataoka M, Hattori A, Okino S, Hyodo M, Asano M, Kawai R, Hayakawa Y. Ethyl(methyl)dioxirane as an effficient reagent for the oxidation of nucleoside phosphites into phosphates under nonbasic anhydrous conditions. Org Lett 2001, 3, 85–88 (correction in Org Lett 2001, 3, 2939).

[25] Lyer RP, Egan W, Regan JB, Beaucage SL. 3H-1,2-Benzodithiole-3-one 1,1-dioxide as an improved sulfurizing reagent in the solid-phase synthesis of oligodeoxyribonucleoside phosphorothioates. J Am Chem Soc 1990, 112, 1253–1254.

[26] Guzaev AP. Reactivity of 3H-1,2,4-dithiazole-3-thiones and 3H-1,2-dithiole-3-thiones as sulfurizing agents for oligonucleotide synthesis. Tetrahedron Lett 2011, 52, 434–437.

[27] Cheruvallath ZS, Wheeler PD, Cole DL, Ravikumar VT. Use of phenylacetyl disulfide (PADS) in the synthesis of oligodeoxyribonucleotide phosphorothioates. Nucleosides, Nuicleotides 1999, 18, 485–492.

[28] Xu Q, Barany G, Hammer RP, Musier-Forsyth K. Efficient introduction of phosphorothioates into RNA oligonucleotides by 3-ethoxy-1,2,4-dithiazoline-5-one (EDITH). Nucleic Acids Res 1996, 24, 3643–3644.

[29] Eleuteri A, Cheruvallath ZS, Capaldi DC, Cole DL, Ravikumar VT. Synthesis of dimer phosphoramidite synthons for oligodeoxyribonucleotide phosphorothioates using diethyldithiocarbonate disulfide as an efficient sulfurizing reagent. Nucleosides Nucleotides 1999, 18, 1803–1807.

[30] Vu H, Hirschbein BL. Internucleotide phosphite sulfurization with tetraethylthiuram disulfide. Phosphorothioate oligonucleotide synthesis via phosphoramidite chemistry. Tetrahedron Lett 1991, 32, 3005–3008.

[31] Tang J-Y, Han Y, Tang JX, Zhang Z. Large-scale synthesis of oligonucleotide phosphorothioates using 3-amino-1,2,4-dithiazole-5-thione as an efficient sulfur-transfer reagent. Org Process Res Dev 2000, 4, 194–198.

[32] Habuchi T, Terao Y, Utsugi M. A robust method using reducing inorganic salts for preventing the desulfurization of phosphorothioate oligonucleotides during the cleavage and deprotection step. Tetrahedron Let 2023, 133, 154843 (4 pages).

[33] Heinonen P, Winqvist A, Sanghvi Y, Strömberg R. Studies on the stability of dinucleoside H-phosphonates. Nucleosides Nucleotides & Nucleic Acids 2003, 22, 1387–1289.

[34] Padiya KJ, Salunkhe MM. Large scale, liquid phase oligonucleotide synthesis by alkyl H-phosphonate approach. Bioorg Med Chem 2000, 8, 337–342.

[35] Reese CB, Yan HJ. Solution phase synthesis of ISIS 2922 (Vitravene) by the modified H-phosphonate approach. J Chem Soc Perkin Trans 2002, 1, 0, 2619–2633.

[36] Reese CB, Zhang P-Z. Phosphotriester approach to the synthesis of oligonucleotides: a reappraisal. J Chem Soc Perkin Trans 1993, 1, 0, 2291–2301.

[37] Catani M, De Luca C, Alcantara JMG, Manfredini N, Perrone D, Marchesi E, Weldon R, Müller-Späth T, Cavazzini A, Morbidelli M, Sponchioni M. Oligonucleotides: current trends and innovative applications in the synthesis, characterization, and purification. Biotechnol J 2020, 15, 1900226 (14 pages).

[38] Guzaev A, Lönnberg H. A novel solid support for synthesis of 3′-phosphorylated chimeric oligonucleotides containing internucleosidic methyl phosphotriester and methylphosphonate linkages. Tetrahedron Lett 1997, 38, 3989–3992.

[39] Mullah B, Livak K, Andrus A, Kenney P. Efficient synthesis of double dye-labeled oligodeoxyribonucleotide probes and their application in a real time PCR assay. Nucleic Acids Res 1998, 26, 1026–1031.

[40] Alul RH, Singman CN, Zhang GR, Letsinger RL. Oxalyl-CPG: a labile support for synthesis of sensitive oligonucleotide derivatives. Nucleic Acids Res 1991, 19, 1527–1532.

[41] Pon RT, Yu S. Hydroquinone-O,O′-diacetic acid ('Q-linker') as a replacement for succinyl and oxalyl linker arms in solid phase oligonucleotide synthesis. Nucleic Acid Res 1997, 25, 3629–3635.

[42] Leisvuori A, Poijärvi-Virta P, Virta P, Lönnberg H. 4-Oxoheptanedioic acid: an orthogonal linker for solid-phase synthesis of base-sensitive oligonucleotides. Tetrahedron Lett 2008, 49, 4119–4121.

[43] Routledge A, Wallis MP, Ross KC, Fraser W. A new deprotection strategy for automated oligonucleotide synthesis using a novel silyl-linked solid support. Bioorg Med Chem Lett 1995, 5, 2059–2064.

[44] Semenyuk A, Kwiatkowski M. A base-stable dithiomethyl linker for solid-phase synthesis of oligonucleotides. Tetrahedron Lett 2007, 48, 469–472.

[45] Shahsavari S, Eriyagama DNAM, Halami B, Begoyan V, Tanasova M, Chen J, Fang S. Electrophilic oligodeoxynucleotide synthesis using dM-Dmoc for amino protection. Beilstein J Org Chem 2019, 15, 1116–1128.

[46] Venkatesan H, Greenberg MM. Improved utility of photolabile solid phase synthesis supports for the synthesis of oligonucleotides containing 3′-hydroxyl termini. J Org Chem 1996, 61, 525–529.

[47] Guzaev AP, Manoharan M. A conformationally preorganized universal solid support for efficient oligonucleotide synthesis. J Am Chem Soc 2003, 125, 2380–2381.

[48] Azhayev A, Antopolsky M, Tennila TM, Mackie H, Randolph JB. Advancements in oligonucleotide synthesis. GEN 2005. 25.

[49] Dahl BJ, Bjergarde K, Henriksen L, Dahl O. A higly reactive odourless substitute for thiophenol/triethylamine as a deprotection reagent in synthesis of oligonucleotides and their analogues. Acta Chem Scand 1990, 44, 639–641.

[50] Daub GW, van Tamelen EE. Synthesis of oligoribonucleotides based on the facile cleavage of methyl phosphotriester intermediates. J Am Chem Soc 1977, 99, 3526–3528.

[51]   Hayakawa Y, Wakabayashi S, Kato H, Noyori R. The allylic protection method in solid-phase oligonucleotide synthesis. An efficient preparation of solid-anchored DNAoligomers. J Am Chem Soc 1990, 112, 1691–1696.

[52]   Reese CB, Zard L. Some observations relating to the oximate ion promoted unblocking of oligonucleotide aryl esters. Nucleic Acids Res 1981, 9, 46114–4626.

[53]   McBride LJ, Kierzek R, Beaucage SL, Caruthers MH. Amidine protecting groups for oligonucleotide synthesis. J Am Chem Soc 1986, 108, 2040–2048.

[54]   Ferreira F, Morvan F. Silyl protecting groups for oligonucleotide synthesis removed by a ZnBr$_2$ treatment. Nucleosides, Nucleotides, Nucleic Acids 2005, 24, 1009–1013.

[55]   Iyer RP, Yu D, Habus I, Ho NH, Johnson S, Decling T, Jiang Z, Zhou W, Xie J, Agraval S. N-pent-4-enoyl (PNT) group as a universal nucleobase protector: applications in the rapid and facile synthesis of oligonucleotides, analogs, and conjugates. Tetrahedron 1997, 53, 2731–2750.

[56]   Alvarez K, Vasseur -J-J, Beltran T, Imbach J-L. Photocleavable protecting groups as nucleobase protections allowed the solid-phase synthesis of base-sensitive SATE-prooligonucleotides. J Org Chem 1999, 64, 6319–6328.

[57]   Ohkubo A, Ezawa Y, Seio K, Sekine M. O-Selectivity and utility of phosphorylation mediated by phosphite triester intermediates in the N-unprotected phosphoramidite method. J Am Chem Soc 2004, 126, 10884–10896.

[58]   Kakuta K, Kasahara R, Sato K, Wada T. Solid-phase synthesis of oligodeoxynucleotides using nucleobase N-unprotected oxazaphospholidine derivatives bearing a long alkyl chain. Org Biomol Chem 2023, 21, 7580–7592.

[59]   Reese CB, Yan H. Alternatives to the 4,4′-dimethoxytrityl (DMTr) protecting group. Tetrahedron Lett 2004, 45, 2567–2570.

[60]   Tram K, Sanghvi YS, Yan H. Further optimization of detritylation in solid-phase oligodeoxyribonucleotide synthesis. Nucleosides Nucleotides & Nucleic Acids 2011, 30, 12–19.

[61]   Arunachalam TS, Wichert C, Appela B, Müller S. Mixed oligonucleotides for random mutagenesis: best way of making them. Org Biomol Chem 2012, 10, 4641–4650.

[62]   Sondek J, Shortle D. A general strategy for random insertion and substitution mutagenesis: substoichiometric coupling of trinucleotide phosphoramidites. Proc Natl Acad Sci USA 1992, 89, 3581–3585.

[63]   Fellouse FA, Esaki K, Birtalalan S, Raptis D, Cancasdi VJ, Koide A, Jhurani P, Vasser M, Wiesmann C, Kossiakoff AA, Koide S, Sidhu SS. High-throughput generation of synthetic antibodies from highly functional minimalist phage-displayed libraries. J Mol Biol 2007, 373, 924–940.

[64]   Suchsland R, Appel B, Virta P, Müller S. Synthesis of fully protected trinucleotide building blocks on a disulphide-linked soluble. RSC Adv 2021, 11, 3892–3896.

[65]   Kungurtsev V, Lönnberg H, Virta P. Synthesis of protected 2′-O-deoxyribonucleotides on a precipitative soluble support: a useful procedure for the preparation of trimer phosphoramidites. RSC Adv 2016, 6, 105428–105432.

[66]   Saraya JS, O'Flaherty DK. A facile and general tandem oligonucleotide synthesis methodology for DNA and RNA. ChemBioChem 2024, 25, e202300870 (6 pages).

[67]   Capaldi DC, Gaus H, Krotz AH, Arnoold J, Carty RL, Moore MN, Scozzari AN, Lowery K, Cole DL, Ravikumar VT. Synthesis of high-quality antisense drugs. Addition of acrylonitrile to phosphorothioate oligonucleotides: adduct characterization and avoidance. Org Process Res Develop 2003, 7, 832–838.

[68]   Reddy MP, Hanna NB, Farooqui F. Ultrafast cleavage and deprotection of oligonucleotides synthesis and use of C$^{Ac}$ derivatives. Nucleosides Nucleotides 1997, 16, 1589–1598.

[69]   Boal JH, Wilk A, Harindranath N, Max EE, Kempel T, Beaucage SL. Cleavage of oligodeoxyribonucleotides from controlled-pore glass supports and their rapid deprotection by gaseous amines. Nucleic Acids Res 1996, 24, 3115–3117.

[70]  Krotz AH, McElroy B, Scozzari AN, Cole DL, Ravikumar VT. Controlled detritylation of antisense oligonucleotides. Org Process Res Develop 2003, 7, 47–52.

[71]  Grajkowski A, Cawrse BM, Takahashi M, Beaucage SL. An improved process for the release of synthetic DNA sequences from a solid-phase capture support. Tetrahedrol Lett 2022, 106, 154077 (5 pages).

[72]  Grajkowski A, Cieslak J, Beaucage SL. Solid-phase purification of synthetic DNA sequences. J Org Chem 2016, 81, 6165–6175.

[73]  Rodriguez AA, Cedillo I, Mowery BP, Gaus HJ, Krishnamoorthy SS, McPherson AK. Formation of the $N^2$-acetyl-2,6-diaminopurine oligonucleotide impurity caused by acetyl capping. Bioorg Med Chem Lett 2014, 24, 3243–3246.

[74]  Rodriguez AA, Cedillo I, McPherson AK. Conversion of adenine to 5-amino-4-pyrimidinylimidazole caused by acetyl capping during solid phase oligonucleotide synthesis. Bioorg Med Chem Lett 2016, 26, 3468–3471.

[75]  Mikkola S, Lönnberg T, Lönnberg H. Phosphodiester models for cleavage of nucleic acids. Beilstein J Org Chem 2018, 14, 803–837.

[76]  Usman N, Ogilvie KK, Jiang M-Y, Cedergren RJ. Automated chemical synthesis of long oligoribonucleotides using 2′-O-silylated ribonucleoside 3′-O-phosphoramidites on a controlled-pore glass support: synthesis of a 43-nucleotide sequence similar to the 3′-half molecule of *Escherichia coli* formylmethionine tRNA. J Am Chem Soc 1987, 109, 7845–754.

[77]  Pitsch S, Weiss PA, Jenny L, Stutz A, Wu X. Reliable chemical synthesis of oligoribonucleotides (RNA) with 2′-O-[(triisopropylsilyl)oxy]methyl(2′-O-tom)-protected phosphoramidites. Helv Chim Acta 2001, 84, 3773–3795.

[78]  Serebryany V, Beigelman L. An efficient preparation of protected ribonucleosides for phosphoramidite RNA synthesis. Tetrahedron Lett 2002, 43, 1983–1985.

[79]  Ching SM, Tan WJ, Chua KL, Lam Y. Synthesis of cyclic di-nucleotidic acids as potential inhibitors targeting diguanylate cyclase. Bioorg. Med. Chem 2010, 18, 6657–6665.

[80]  Wagner D, Verheyden JPH, Moffatt JG. Preparation and synthetic utility of some organotin derivatives of nucleosides. J Org Chem 1974, 39, 24–30.

[81]  Markiewicz W. Tetraisopropyldisiloxane-1,3-diyl, a group for simultaneous protection of 3′-hydroxy and 5′-hydroxy functions of nucleosides. J Chem Res S 1979, 0, 24–25.

[82]  Dellinger DJ, Timar Z, Myerson J, Sierzchala AB, Turner J, Ferreira F, Kupihár Z, Dellinger G, Hill KW, Powell JA, Sampson JR, Caruthers MH. Streamlined process for the chemical synthesis of RNA using 2′-O-thionocarbamate-protected nucleoside phosphoramidites in the solid phase. J Am Chem Soc 2011, 133, 11540–11556.

[83]  Shiba Y, Masuda H, Watanabe N, Ego T, Takagaki K, Ishiyama K, Ohgi T, Yano J. Chemical synthesis of a very long oligoribonucleotide with 2-cyanoethoxymethyl (CEM) as the 2′-O-protecting group: structural identification and biological activity of a synthetic 110-mer precursor-microRNA candidate. Nucleic Acids Res 2007, 35, 3287–3296.

[84]  Semenyuk A, Földesi A, Johansson T, Estmer-Nilsson C, Blomgren P, Brännvall M, Kirsebom LA, Kwiatkowski M. Synthesis of RNA using 2′-O-DTM protection. J Am Chem Soc 2006, 128, 12356–12357.

[85]  Takahashi M, Grajkowski A, Cawrse BM, Beaucage SL. Innovative 2′-O-imino-2-propanoate-protecting group for effective solid-phase synthesis and 2′-O-deprotection of RNA sequences. J Org Chem 2021, 86, 4944–4956.

[86]  Cieslak J, Ausin C, Grajkowski A, Beaucage SL. The 2-cyano-2,2-dimethylethanimine-N-oxymethyl group for 2′-hydroxyl protection of ribonucleosides in the solid-phase synthesis of RNA sequences. Chem Eur J 2013, 19, 4623–4632.

[87] Cieslak J, Grajkowski A, Kauffman JS, Duff RJ, Beaucage SL. The (4-N-dichloroacetyl-N-methylamino) benzyloxymethyl group for 2′-hydroxyl protection of ribonucleosides in solid-phase synthesis of oligoribonucleotides. J Org Chem 2008, 73, 2774–2783.

[88] Saneyoshi H, Ando K, Seio K, Sekine M. Chemical synthesis of RNA via 2′-O-cyanoethylated intermediates. Tetrahedron 2007, 63, 11195–11203.

[89] Zhou C, Honcharenko D, Chattopadhyaya J. 2-(4-Tolylsulfonyl)ethoxymethyl (TEM) – A new 2′-OH protecting group for solid supported RNA synthesis. Org Biomol Chem 2007, 5, 333–343.

[90] Efimov VA, Aralov AV, Klykov VN, Chakhmakhcheva OG. Synthesis of RNA by the rapid phosphotriester method using azido-based 2′-protecting groups. Nucleosides, Nucleotides, Nucleic Acids 2009, 28, 846–865.

[91] Lin FL, Hoyt HM, Halbeek HV, Bergman RG, Bertozzi CR. Mechanistic investigations of the Staudinger ligation. J Am Chem Soc 2005, 127, 2686–2695.

[92] Welz R, Müller S. 5-(Benzylmercapto)-1H-tetrazole as activator for 2′-O-TBDMS phosphoramidite building blocks in RNA synthesis. Tetrahedron Lett 2002, 43, 795–797.

[93] Sproat B, Colonna F, Mullah B, Tsou D, Andrus A, Hampel A, Vinayak R. An efficient method for the isolation and purification of oligoribonucleotides. Nucleosides Nucleotides 1995, 14, 255–273.

[94] Hayakawa Y, Kawai R, Hirata A, Sugimoto J-I, Kataoka M, Sakakura A, Hirose M, Noyori R. Acid/azole complexes as highly effective promoters in the synthesis of DNA and RNA oligomers via the phosphoramidite method. J Am Chem Soc 2001, 123, 8165–8176.

[95] Ogilvie KK, Usman N, Nicoghosian K, Cedergren RJ. Total chemical synthesis of a 77-nucleotide-long RNA sequence having methionine-acceptance activity. Proc Natl Acad Sci USA 1988, 85, 5764–5768.

[96] Reddy MP, Farooqui F, Hanna NB. Methylamine deprotection provides increased yield of oligoribonucleotides. Tetrahedron Lett 1995, 36, 8929–8932.

[97] He M, Wu X, Mao S, Haruehanroengra P, Khan I, Sheng J, Royzen M. Bio-orthogonal chemistry enables solid phase synthesis and HPLC and gel-free purification of long RNA oligonucleotides. Chem Commun 2021, 57, 4263–4266.

[98] Hollenstein M. Nucleoside triphosphates – Building blocks for the modification of nucleic acids. Molecules 2012, 17, 13569–13591.

[99] Hocek M Synthesis of base-modified 2′-deoxyribonucleoside triphosphates and their use in enzymatic synthesis of modified DNA applications in bioanalysis and chemical biology. J Org Chem 2014, 79, 9914–9921.

[100] Kuwahara M, Sugimoto N. Molecular evolution of functional nucleic acids with chemical modifications. Molecules 2010, 15, 5423–5444.

[101] Jäger S, Rasched G, Kornreich-Leshem H, Engeser M, Thum O, Famulok M. A versatile toolbox for variable DNA functionalization at high density. J Am Chem Soc 2005, 127, 15071–15082.

[102] Matyasovsky J, Perlíkova P, Malnuit V, Pohl R, Hocek M. 2-Substituted dATP derivatives as building blocks for polymerase-catalyzed synthesis of DNA modified in the minor groove. Angew Chem Int Ed 2016, 55, 15856–15859.

[103] Jakubovska J, Tauraite D, Birstonas L, Meskys R. N⁴-Acyl-2′-deoxycytidine-5′-triphosphates for the enzymatic synthesis of modified DNA. Nucleic Acids Res 2018, 46, 5911–5923.

[104] Sawai H, Nagashima J, Kuwahara M, Kitagata R, Tamura T, Matsui I. Differences in Substrate Specificity of C(5)-Substituted or C(5)-Unsubstituted Pyrimidine Nucleotides by DNA Polymerases from Thermophilic Bacteria, Archaea, and Phages. Chem Biodiversity 2007, 4, 1979–1995.

[105] Kuwahara M, Takano Y, Kasahara Y, Nara H, Ozaki H, Sawai H, Sugiyama A, Obika S. Study on suitability of KOD DNA polymerase for enzymatic production of artificial nucleic acids using base/sugar modified nucleoside triphosphates. Molecules 2010, 15, 8229–8240.

[106] Saiki RK, Gelfand DH, Stoffel S, Scharf SJ, Higuchi R, Horn GT, Mullis KB, Erlich HA. Primer-directed enzymatic amplification of DNA with a thermostable DNA polymerase. Science 1988, 239, 487–491.

[107] Staiger N, Marx A. A DNA polymerase with increased reactivity for ribonucleotides and C5-Modified deoxyribonucleotides. ChemBioChem 2010, 11, 1963–1966.

[108] Ramsay N, Jemth A-S, Brown A, Crampton N, Dear P, Holliger P. CyDNA: synthesis and replication of highly Cy-Dye substituted DNA by an evolved polymerase. J Am Chem Soc 2010, 132, 5096–5104.

[109] Pinheiro VB, Taylor AI, Cozens C, Abramov M, Renders M, Zhang S, Chaput JC, Wengel J, Peak-Chew SY, McLaughlin SH, Herdewijn P, Holliger P. Synthetic genetic polymers capable of heredity and evolution. Science 2012, 336, 6271–6284.

[110] Röthlisberger P, Hollenstein M. Aptamer chemistry. Adv Drug Deliv Rev 2018, 134, 3–21.

[111] Brazdilova P, Vrabel M, Pohl R, Pivonkova H, Havran L, Hocek M, Fojta M. Ferrocenylethynyl derivatives of nucleoside triphosphates: synthesis, incorporation, electrochemistry, and bioanalytical applications. Chem Eur J 2007, 13, 9527–9533.

[112] Kuwahara M, Nagashima J, Hasegawa M, Tamura T, Kitagata R, Hanawa K, Hososhima S, Kasamatsu T, Ozaki H, Sawai H. Systematic characterization of 2′-deoxynucleoside- 5′-triphosphate analogs as substrates for DNA polymerases by polymerase chain reaction and kinetic studies on enzymatic production of modified DNA. Nucleic Acids Res 2006, 34, 5383–5394.

[113] Menova P, Raindlova V, Hocek M. Scope and limitations of the nicking enzyme amplification reaction for the synthesis of base modified oligonucleotides and primers for PCR. Bioconjugate Chem 2013, 24, 1081–1093.

[114] Sarac I, Hollenstein M. Terminal deoxynucleotidyl transferase in the synthesis and modification of nucleic acids. ChemBioChem 2019, 20, 860–871.

[115] Jensen MA, Davis RW. Template-independent enzymatic oligonucleotide synthesis (TiEOS): its history, prospects, and challenges. Biochemistry 2018, 57, 1821–1832.

[116] Bollum FJ. Oligodeoxyribonucleotide-primed reactions catalyzed by calf thymus polymerase. J Biol Chem 1962, 237, 1945–1949.

[117] Földesi A, Keller A, Stura A, Zigmantas S, Kwiatkowski M, Knapp D, Engels JW. The fluoride cleavable 2-(cyanoethoxy)methyl (CEM) group as reversible 3′-O-terminator for DNA sequencing-by-synthesis – synthesis, incorporation, and cleavage. Nucleosides Nucleotides Nucleic Acids 2007, 26, 271–275.

[118] Ju J, Kim DH, Bi L, Meng Q, Bai X, Li Z, Li X, Marma MS, Shi S, Wu J, Edwards JR, Romu A, Turro NJ. Four-color DNA sequencing by synthesis using cleavable fluorescent nucleotide reversible terminators. Proc Natl Acad Sci USA 2006, 103, 19635–19640.

[119] Hutter D, Kim MJ, Karalkar N, Leal NA, Chen F, Guggenheim E, Visalakshi V, Olejnik J, Gordon S, Benner SA. Labeled nucleoside triphosphates with reversibly terminating aminoalkoxyl groups. Nucleosides Nucleotides Nucleic Acids 2010, 29, 879–895.

[120] Modified template-independent enzymes for polydeoxynucleotide synthesis. 2016, US20160108382A1.

[121] Yin YW, Steitz TA. Structural basis for the transition from initiation to elongation transcription in T7 RNA polymerase. Science 2002, 298, 1387–1395.

[122] Milisavljevic N, Perlikova P, Pohl R, Hocek M. Enzymatic synthesis of base-modified RNA by T7 RNA polymerase. A systematic study and comparison of 5-substituted pyrimidine and 7-substituted 7-deazapurine nucleoside triphosphates as substrates. Org Biomol Chem 2018, 16, 5800–5807.

[123] Flamme M, McKenzie LK, Sarac I, Hollenstein M. Chemical methods for the modification of RNA. Methods 2019 May, 15, 161, 64–82.

[124] Chen TJ, Hongdilokkul N, Liu ZX, Adhikary R, Tsuen SS, Romesberg FE. Evolution of thermophilic DNA polymerases for the recognition and amplification of C2′-modified DNA. Nat Chem 2016, 8, 557–563.

[125] Cozens C, Pinheiro VB, Vaisman A, Woodgate R, Holliger P. A short adaptive path from DNA to RNA polymerases. Proc Natl Acad Sci 2012, 109, 8067–8072.

[126] England TE, Uhlenbeck OC. Enzymatic oligoribonucleotide synthesis with T4 RNA ligase. Biochemistry 1978, 17, 2069–2076.

[127] Stec WJ, Grajkowski A, Koziolkiewicz M, Uznanski B. Novel route to oligo(deoxyribonucleoside phosphorothioates). Stereocontrolled synthesis of P-chiral oligo(deoxyribonucleoside phosphorothioates). Nucleic Acids Res 1991, 19, 5883–5888.

[128] Oka N, Yamamoto M, Sato T, Wada T. Solid-phase synthesis of stereoregular oligodeoxyribonucleoside phosphorothioates using bicyclic oxazaphospholidine derivatives as monomer units. J Am Chem Soc 2008, 130, 16031–16037.

[129] Oka N, Wada T, Saigo K. An oxazaphospholidine approach for the stereocontrolled synthesis of oligonucleoside phosphorothioates. J Am Chem Soc 2003, 125, 8307–8317.

[130] Nukaga Y, Yamada K, Ogata T, Oka N, Wada T. Stereocontrolled solid-phase synthesis of phosphorothioate oligoribonucleotides Using 2′-O-(2-cyanoethoxymethyl)nucleoside 3′-O-oxazaphospholidine Monomers. J Org Chem 2012, 77, 7913–7922.

[131] Iwamoto N, Butler DCD, Svrzikapa N, Mohapatra S, Zlatev I, Sah DWY, Meena, Standley SM, Lu G, Apponi LH, Frank-Kamenetsky M, Zhang JJ, Vargeese C, Verdine GL. Control of phosphorothioate stereochemistry substantially increases the efficacy of antisense oligonucleotides. Nat Biotechnol 2017, 35, 845–851.

[132] Knouse KW, deGruyter JN, Schmidt MA, Zheng B, Vantourout JC, Kingston C, Mercer SE, Mcdonald IM, Olson RE, Zhu Y, Hang C, Zhu J, Yuan C, Wang Q, Park P, Eastgate MD, Baran PS. Unlocking P(V): reagents for chiral phosphorothioate synthesis. Science 2018, 361, 1234–1238.

[133] Huang Y, Knouse KW, Qiu S, Hao W, Padial NM, Vantourout JC, Zheng B, Mercer SE, Lopez-Ogalla J, Narayan R, Olson RE, Blackmond DG, Eastgate MD, Schmidt MA, McDonald IM, Baran PS. A P(V) platform for oligonucleotide synthesis. Science 2021, 373, 1265–1270.

[134] Xu D, Rivas-Bascon N, Padial NM, Knouse KW, Zheng B, Vantourout JC, Schmidt MA, Eastgate MD, Baran PS. Enantiodivergent formation of C–P bonds: synthesis of P-chiral phosphines and methylphosphonate oligonucleotides. J Am Chem Soc 2020, 142, 5785–5792.

[135] Sergueev DS, Ramsey Shaw B. H-Phosphonate approach for solid-phase synthesis of oligodeoxyribonucleoside boranophosphates and their characterization. J Am Chem Soc 1998, 120, 9417–9427.

[136] Sergueev DS, Sergueeva ZA, Ramsey Shaw B. Synthesis of oligonucleoside boranophosphates via an H-phosphonate method without nucleobase protection. Nucleosides, Nucleotides, Nucleic Acids 2001, 20, 789–795.

[137] Roy S, Olesiak M, Shang S, Caruthers MH. Silver nanoassemblies constructed from boranephosphonate DNA. J Am Chem Soc 2013, 135, 6234–6241.

[138] Hara RI, Saito T, Kogure T, Hamamura Y, Uchiyama N, Nukaga Y, Iwamoto N, Wada T. Stereocontrolled synthesis of boranophosphate DNA by an oxazaphospholidine approach and evaluation of its properties. J Org Chem 2019, 84, 7971–7983.

[139] Sato K, Imai H, Shuto T, Hara IR, Wada T. Solid-phase synthesis of phosphate/boranophosphate chimeric DNAs using the H=phosphonate−H-boranophosphonate method. J. Org. Chem 2019, 84, 15032–15041.

[140] Takahashi Y, Sato K, Wada T. Solid-phase synthesis of boranophosphate/phosphorothioate/ phosphate chimeric oligonucleotides and their potential as antisense oligonucleotides. J Org Chem 2022, 87, 3895–3899.

[141] Golebiewska J, Stawinski J. Reaction of boranephosphonate diesters with pyridines or tertiary amines in the presence of iodine: synthetic and mechanistic studies. J Org Chem 2020, 85, 4312–4323.

[142] Roy S, Paul S, Roy M, Kundu R, Monfregola L, Caruthers MH. Pyridinium boranephosphonate modified DNA oligonucleotides. J Org Chem 2017, 82, 1420–1427.

[143] Takahashi Y, Kakuta K, Namioka Y, Igarashi A, Sakamoto T, Hara RI, Sato K, Wada T. Synthesis of P-modified DNA from boranophosphate DNA as a precursor via accyl phosphite intermediates. J Org Chem 2023, 88, 10617–10631.

[144] Summerton J, Weller D. Morpholino antisense oligomers: design, preparation and properties. Antisense Nucleic Acid Drug Dev 1997, 7, 187–195.

[145] Kundu J, Ghosh A, Ghosh U, Das A, Nagar D, Pattanayak S, Ghose A, Sinha S. Synthesis of phosphorodiamidate morpholino oligonucleotides using trityl and Fmoc chemistry in an utaomated oligo synthesizer. J Org Chem 2022, 87, 9466–9478.

[146] Li C, Callahan AJ, Simon MD, Totaro KA, Mijalis AJ, Phadke K-S, Zhang G, Hartrampf N, Schissel CK, Zhou M, Zong H, Hanson GJ, Loas A, Pohl NLB, Verhoeven DE, Pentelute BL. Fully automated fast-flow synthesis of antisense phosphorodiamidate morpholino oligomers. Nat Commun 2021, 12, 4396 (8 pages).

[147] Tsurusaki T, Sato K, Imai H, Hirai K, Takahashi D, Wada T. Convergent synthesis of phosphorodiamidate morpholino oligonucleotides (PMOs) by the H-phosphonate approach. Scientific Reports 2023, 13, 12576 (10 pages).

[148] Langner HK, Jastrzebska K, Caruthers MH. Synthesis and characterization of thiophosphoramidate Morpholino oligonucleotides and chimeras. J Am Chem Soc 2020, 142, 16240–16253.

[149] Reese CB, Yan HJ. Solution phase synthesis of ISIS 2922 (Vitravene) by the modified H-phosphonate approach. Chem Soc Perkin Trans 2002, 1, 2619–2633.

[150] Zhou X, Kiesman WF, Yan W, Jiang H, Antia FD, Yang J, Fillon YA, Xiao L, Shi X. Development of kilogram-scale convergent liquid-phase synthesis of oligonucleotides. J Org Chem 2022, 87, 2087–2110.

[151] Bonora GM, Scremin CL, Colonna FP, Garbesi A. HELP (high efficiency liquid phase) new oligonucleotide synthesis on soluble polymeric support. Nucleic Acids Res 1990, 18, 3155–3159.

[152] Colonna FP, Scremin CL, Bonora GM. Large scale H.E.L.P. synthesis of oligodeoxynucleotides by the hydroxybenzotriazole phosphotriester approach. Tetrahedron Lett 1991, 32, 3251–3254.

[153] Kungurtsev V, Virta P, Lönnberg H. Synthesis of short oligodeoxyribonucleotides by phosphotriester chemistry on a precipitative tetrapodal support. Eur J Org Chem 2013, 0, 7886–7890.

[154] Kungurtsev V, Laakkonen J, Gimenez Molina A, Virta P. Solution-phase synthesis of short oligo-2′-deoxyribonucleotides by using clustered nucleosides as a soluble support. Eur J Org Chem 2013, 0, 6687–6693.

[155] Rosenqvist P, Saari V, Pajuniemi E, Gimenez Molina A, Ora M, Horvath A, Virta P. Stereo-controlled liquid phase synthesis of phosphorothioate oligonucleotides on a soluble support. J Org Chem 2023, 88, 10156–10163.

[156] Bonora GM, Biancotto G, Maffini M, Scremin CL. Large scale, liquid phase synthesis of oligonucleotides by the phosphoramidite approach. Nucleic Acids Res 1993, 21, 1213–1217.

[157] Scremin CL, Bonora CM. Liquid phase synthesis of phosphorothioate oligonucleotides on polyethylene glycol support. Tetrahedron Lett 1993, 34, 4663–4666.

[158] de Koning MC, Ghisaidoobe ABT, Duynstee HI, Ten Kortenaar PBW, Filippov DV, van der Marel GA. Simple and efficient solution-phase synthesis of oligonucleotides using extractive work-up. Org Process Res Dev 2006, 10, 1238–1245.

[159] Wör R, Köster H. The use of liquid phase carriers for large scale oligonucleotide synthesis in solution via phosphoamidite chemistry. Tetrahedron 1999, 55, 2957–2972.

[160] Molina AG, Kungurtsev V, Virta P, Lönnberg H. Acetylated and methylated β-cyclodextrins as viable soluble supports for the synthesis of short 2′-oligodeoxyribo-nucleotides in solution. Molecules 2012, 17, 12102–12120.

[161] Gaffney PRJ, Kim JF, Valtcheva IB, Williams GD, Anson MS, Buswell AM, Livingston AG. Liquid-phase synthesis of 2′-methyl-RNA on a homostar support through organic-solvent nanofiltration. Chem Eur J 2015, 21, 9535–9543.

[162] Kim JF, Gaffney PRJ, Valtcheva IB, Williams G, Buswell AM, Anson MS, Livingston AG. Organic solvent nanofiltration (OSN); new technology platform for liquid-phase oligonucleotide synthesis (LPOS) Org Process Res Dev 2016, 20, 1439–152.

[163] Kim S, Matsumoto M, Chiba K. Liquid-phase RNA synthesis by using alkyl-chain-soluble support. Chem Eur J 2013, 19, 8615–8620.

[164] Molina AG, Jabgunde AM, Virta P, Lönnberg H. Solution phase synthesis of short oligoribonucleotides on a precipitative tetrapodal support. Beilstein J Org Chem 2014, 10, 2279–2285.

[165] Molina AG, Jabgunde AM, Virta P, Lönnberg H. Assembly of short oligoribonucleotides from commercially available building blocks on a tetrapodal soluble support. Curr Org Synth 2015, 12, 202–207.

[166] Dueymes C, Schönberger A, Adamo I, Navarro A-E, Meyer A, Lange M, Imbach J-L, Link F, Morvan F, Vasseur -J-J. High-yield solution-phase synthesis of di- and trinucleotide blocks assisted by polymer-supported reagents. Org Lett 2005, 7, 3485–3488.

[167] Adamo I, Dueymes C, Schönberger A, Navarro A-E, Meyer A, Lange M, Imbach J-L, Link F, Morvan F, Vasseur -J-J. Solution-phase synthesis of phosphorothioate oligonucleotides using a solid supported acyl chloride with *H*-phosphonate chemistry. Eur J Org Chem 2006, 0, 436–448.

# 6 Oligonucleotides: reactions

## 6.1 Cleavage and isomerization of RNA phosphodiester linkages: small molecular models

### 6.1.1 Cleavage by Brönsted acids and bases

Oligoribonucleotides are stable molecules under physiological conditions in the absence of any catalyst. The half-life for the cleavage at pH 7 and 25 °C is of the order of 10 years [1], but acids, bases and metal ions markedly accelerate the reaction. The reaction is initiated by an intramolecular attack of 2'-OH on the phosphorus atom giving a pentacoordinated species, pentaoxyphosphorane, that can be either an intermediate having a finite lifetime, or only a transitions state. The phosphorane intermediate (or transition state) may undergo breakdown by departure of the 5'-O-linked nucleoside, giving a 2',3'-cyclic phosphate that is rapidly hydrolyzed to a mixture of 2'- and 3'-monophosphates by the attack of water on phosphorus (Figure 6.1) [2]. In case the pentaoxyphosphorane has a sufficient lifetime, it may additionally undergo an endo-cyclic PO bond fission, leading to isomerization of the 3',5'-phosphodiester bond to a 2',5'-bond.

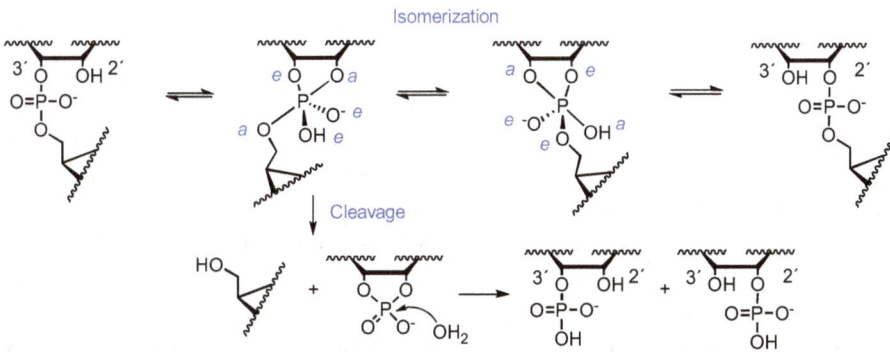

**Figure 6.1:** Cleavage and isomerization of the phosphodiester linkages of oligoribonucleotides.

The pentaoxyphosphorane intermediate is a trigonal bipyramid having two apical and three equatorial ligands, as already discussed in Section 3.4. The attacking nucleophile always adopts an apical position within the intermediate and departure from the intermediate is possible only through an apical position. This regularity is known as one of the Westheimer's rules [3]. According to the same rules, electronegative ligands prefer apical position, whereas electron rich ligands, anionic ligands in particular, are locked to an equatorial position. Among the two oxygen atoms of the five-membered ring formed upon the attack of 2'-OH on phosphorus, one must always be

https://doi.org/10.1515/9783111325637-006

apical and the other equatorial. A phosphorane having a finite life-time may undergo a conformational change known as Berry pseudorotation [4]. Two of the equatorial ligands take an apical position, while the apical ligands turn equatorial. One of the equatorial ligands remains equatorial. The phosphorane intermediate in Figure 6.1 is depicted to undergo Berry pseudorotation: the initially apical 2'-O and 5'-O turn equatorial, while the initially equatorial 3'-O and nonbridging phosphoryl OH take an apical position, enabling departure of 3'-O.

The intermediate additionally contains two nonbridging OH ligands. The protonation state of these largely determines the lifetime of the phosphorane intermediate. The first dissociation takes place around pH 8 and the second at pH > 14 [5]. In other words, neutral and monoanionic phosphoranes coexist in the physiological pH range. If a dianionic phosphorane is initially formed, it either decomposes immediately or undergoes rapid thermodynamically favored protonation to the more stable monoanion.

As mentioned above, the cleavage of RNA phosphodiester linkages are subject to catalysis by both acids and bases. Figure 6.2 shows the pH-rate profile for the cleavage and isomerization of 3',5'-UpU, the shortest possible model of RNA, at 90 °C [6]. The rather complex shape of the pH-rate profiles refers to existence of several ionic forms through which the reactions may proceed. These are shown in Figure 6.2. The pH regions where the reaction predominantly proceeds through a given ionic form are indicated by colored bars. Figure 6.3, in turn, records the percentage contributions of various partial reactions of cleavage and isomerization as a function of pH.

The phosphodiester linkages are negatively charged at pH > 1, which dramatically retards the attack of 2'-OH on phosphorus. Neutralization of the charge by protonation facilitates the nucleophilic attack to such an extent that both the cleavage and isomerization turn acid catalyzed at pH < 4. Evidently, a rapid pre-equilibrium protonation of the phosphodiester linkage is followed by the attack of 2'-OH on the phosphorus atom with concomitant, possibly water mediated, transfer of proton to the unprotonated nonbridging oxygen of the developing phosphorane intermediate. The neutral phosphorane is sufficiently stable to pseudorotate, which enables isomerization (Figure 6.4). These reactions through the neutral phosphorane predominate at pH 2–4 (purple code).

Reaction order in hydronium ion concentration, however, gradually increases from 1 at pH 4 to 2 at pH 1, and levels off toward 1 on passing the $pK_a$ value 0.7 of the phosphodiester linkage [6]. In other words, a reaction through a doubly protonated monocationic phosphodiester takes over under very acidic conditions. This second-order reaction proceeds by rate-limiting attack of 2'-OH on a monocationic phosphodiester giving a monocationic phosphorane that undergoes rapid protolytic equilibration between the bridged oxygens. A protonated PO bond then rapidly cleaves. The ratio of the fission of $PO^{2'}$-, $PO^{3'}$- and $PO^{5'}$-bonds is 1.0:1.2:0.4 (Figure 6.5). One should also bear in mind that depurination competes with phosphoester reactions at pH 2–4. Since depurination is first-order in hydronium ion concentration over the whole acidity range, the competition in less significant in more acidic solutions.

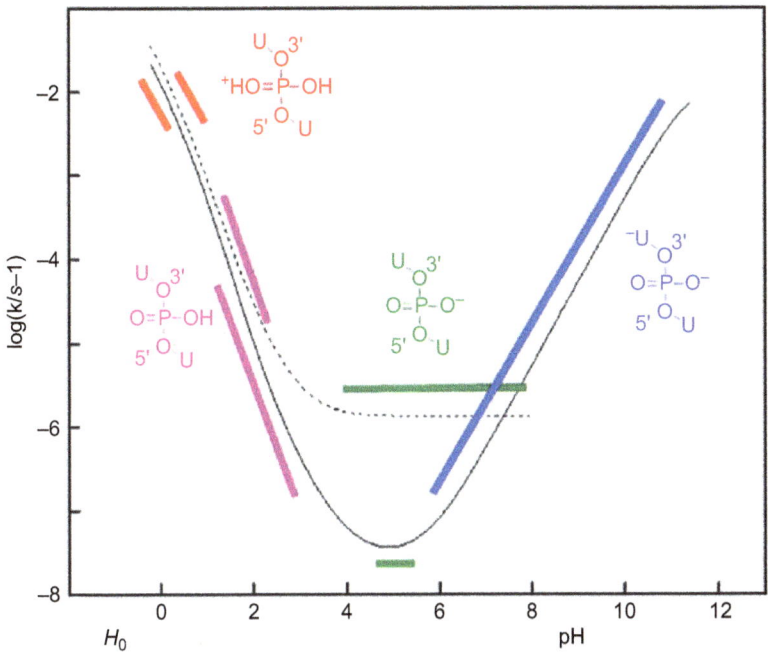

**Figure 6.2:** pH-rate profile for the cleavage (solid line) and isomerization (dotted line) of uridyl-3',5'-uridine at 90 °C [6].

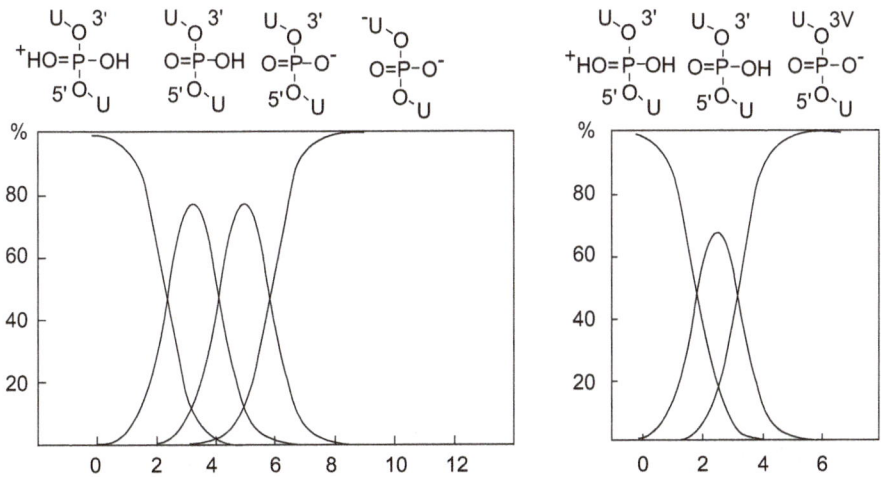

**Figure 6.3:** Contributions of various partial reactions to the overall rate of cleavage (A) and isomerization (B) of RNA phosphodiester linkages. The data refers to cleavage of uridyl-3',5'-uridine at 90 °C [6].

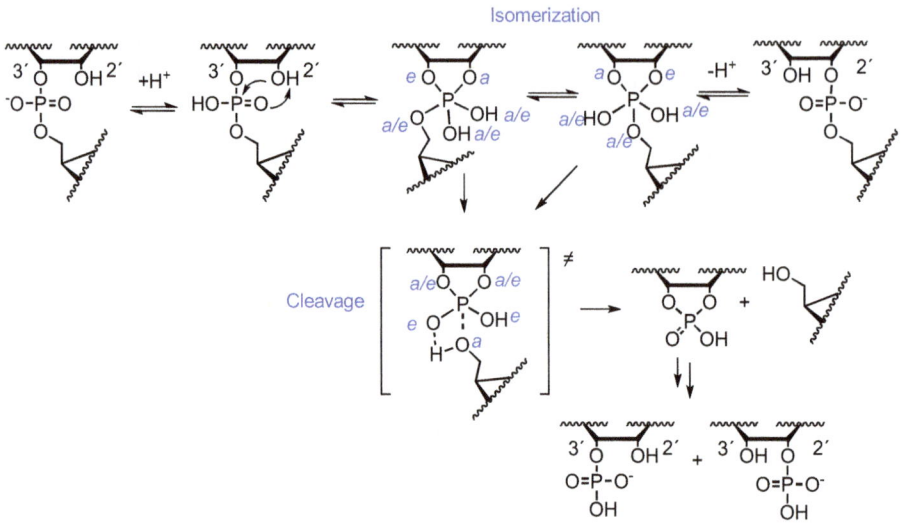

**Figure 6.4:** Mechanism for the cleavage and isomerization of internucleosidic 3′,5′-phosphodiester linkages of oligoribonucleotides through a neutral phosphorane intermediate. These reactions predominate at pH 2–4 (purple code in Figure 6.2).

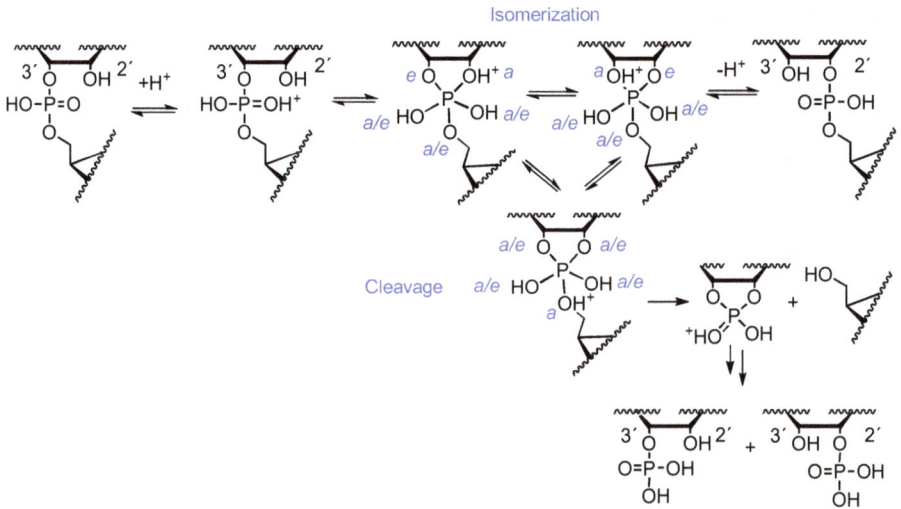

**Figure 6.5:** Mechanism for the cleavage and isomerization of internucleosidic 3′,5′-phosphodiester linkages of oligoribonucleotides through a monocationic phosphorane intermediate. These reactions predominate at pH < 1 (red code in Figure 6.2).

In the pH range 4–6, isomerization is pH-independent and more than one order of magnitude faster than cleavage. According to DFT calculations, the phosphorane monoanion is sufficiently stable to pseudorotate [7], but it is also possible that pseudorotation takes place by kinetically invisible consecutive protonation and deprotonation. The same calculations suggest that breakdown of the monoanionic phosphorane intermediate to a 2′,3′-cyclic phosphate is much slower than endocyclic fission to a 2′,5′- or 3′,5′-acyclic phosphodiester. pH-independent cleavage is experimentally observed, but only over a narrow pH region around pH 5 (Figure 6.6) [6]. The departing 5′-O most likely accepts a proton from phosphorane OH ligand concerted with the PO bond rupture [8]. This proton transfer facilitates the exocyclic PO bond cleavage by destabilizing the phosphorane and stabilizing the leaving group. Since the endocyclic cleavage is by 10 kcal mol$^{-1}$ more facile than the exocyclic cleavage, only the 2′-O and 3′-O may depart as oxyanions, and they become protonated only after the transition state.

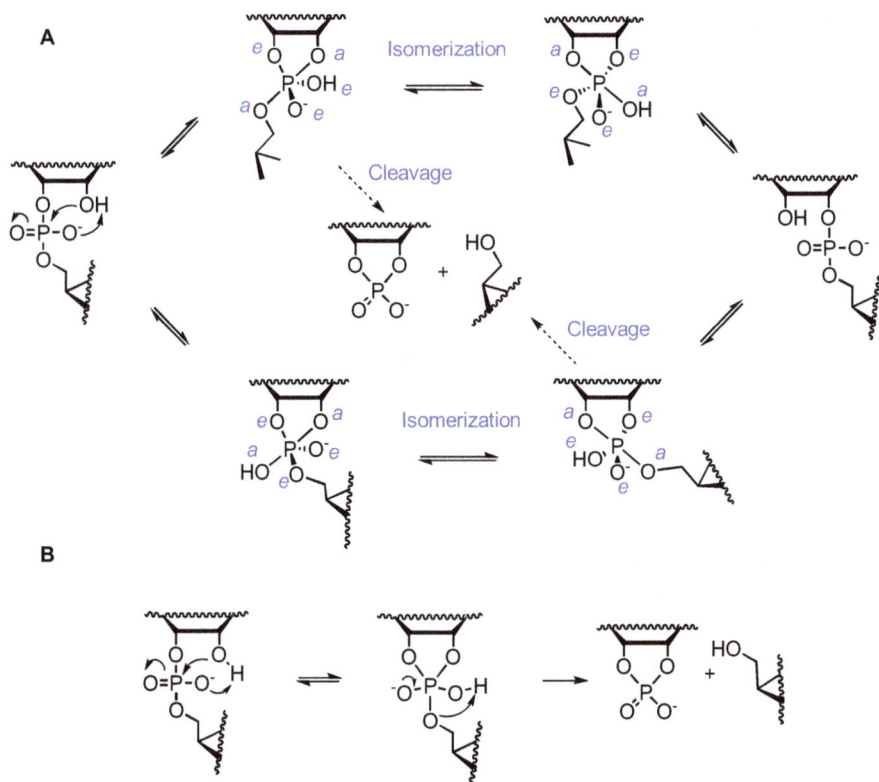

**Figure 6.6:** pH- and buffer-independent isomerization (A) and cleavage (B) of RNA phosphodiester linkages (green code in Figure 6.2).

Hydroxide ion-catalyzed cleavage becomes the dominant reaction at pH > 7. The reaction is first-order in hydroxide ion concentration at pH 7–11 and starts then to level off to pH independence on approaching the $pK_a$ value 12.6 of 2'-OH [6]. Accordingly, 2'-OH is deprotonated in a rapid pre-equilibrium stage, the resulting oxyanion attacks on the phosphorus atom, and the dianionic phosphorane obtained undergoes rate-limiting breakdown by departure of the 5'-O as oxyanion (Figure 6.7) [9]. No isomerization occurs indicating that the phosphorane dianion, although being still an intermediate with a finite life-time, is too unstable to undergo isomerization via a kinetically invisible protonation to monoanion [10–12]. The transition state is late, which means that the formation of the P–O2' and the fission of the P–O5' bond are both considerably advanced.

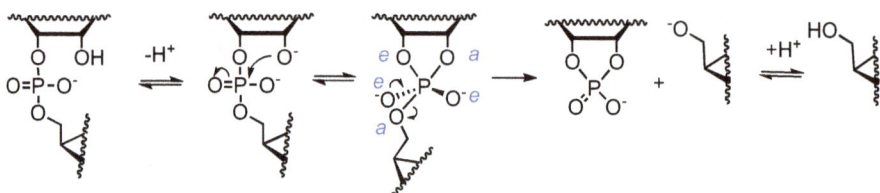

**Figure 6.7:** Hydroxide ion-catalyzed cleavage of RNA phosphodiester linkages (blue code in Figure 6.2).

The cleavage of RNA phosphodiester linkages is additionally subject to buffer catalysis. The reaction has received interest as a model of enzymatic cleavage by RNases, above all RNAse A. The buffer catalysis, however, is rather inefficient. That is why the kinetic measurements have been performed at high concentrations of imidazole or morpholine buffers, which has made elimination of salt and cosolute effects highly challenging. Two mechanisms appear to operate in parallel. One of them is general base-catalyzed attack of 2'-OH on phosphorus, giving a dianionic phosphorane that breaks down without any kinetically visible catalysis (Figure 6.8A). This reaction plays a role in basic buffers. The second reaction is general acid-catalyzed cleavage. According to Breslow et al. [13, 14], the reaction actually takes place by a specific acid/general base mechanism: rapid pre-equilibrium protonation of the phosphodiester linkage is followed by general base-catalyzed attack of 2'-OH on the phosphorus (Figure 6.8B). The dianionic phosphorane obtained breaks down without any catalysis. The reaction, hence, occurs via a minor tautomer of the phosphodiester linkage. According to Kirby et al. [15], the general acid-catalyzed cleavage proceeds via the major tautomer, that is, by rapid pre-equilibrium formation of a monoanionic phosphorane and subsequent general acid-catalyzed fission of the P-O$^{5'}$ bond (Figure 6.8C).

Besides general base-catalyzed isomerization (Figure 6.8A), general acid-catalyzed isomerization also occurs: O3' of the monoanionic phosphorane is protonated by the general acid instead of O5' leading to O3'-P bond rupture. The general acid catalysis of isomerization, however, is much weaker than that of the cleavage, which makes elimination of salt and cosolute effects challenging. Regardless of whether the cleavage reaction takes place by Breslow's or Kirby's mechanism, one thing is indisputable. Buffer catalysis markedly influences on the partition of the phosphorane intermediate to cleavage and isomerization products. With 5'-TTUTT-3', for example, the buffer independent isomerization is at pH 5.85 (80 °C, $I = 0.5$ M) almost 30 times as fast as cleavage, but at buffer concentration 0.7 M, cleavage is the predominant reaction, being 4 times as fast as isomerization.

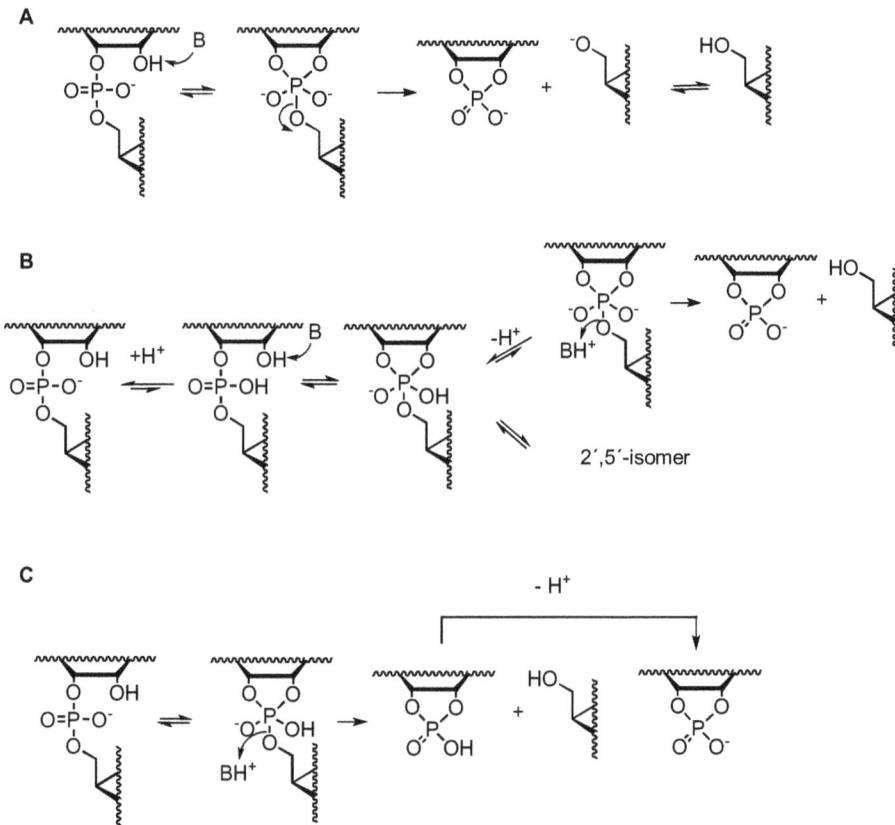

**Figure 6.8:** Mechanisms proposed [13, 15] for the buffer-catalyzed cleavage of RNA phosphodiester linkages by imidazole and morpholine buffers.

## 6.1.2 Cleavage by multifunctional catalysts

Imidazole, guanidine and primary amino group are the amino acid side chain functionalities that often play role in enzyme-catalyzed reactions of RNA [16]. The chemical basis for the role of imidazole has been elucidated by the mechanistic studies of buffer catalysis discussed above. Guanidine and primary amino groups are both present in protonated guanidinium and ammonium form at physiological pH. Accordingly, both reduce electron density in their vicinity, not only inductively through bonds but also electrostatically through space. In addition, they may serve as weak general acids. Mechanistically most interesting is, however, the ability of guanidine group to shuttle a proton through various tautomeric forms [17]. To learn more of the efficiency of this process, a model system has been constructed by anchoring a 2,4-diamino-1,3,5-triazine core in the vicinity of the phosphodiester bond of 3′,5′-UpU with the aid of two cyclen-functionalized side chains that in the presence of Zn²⁺ ion bind tightly to the uracil bases [18]. The triazine core accelerates the cleavage of the phosphodiester linkage by two orders of magnitude at pH 6. A mechanistic interpretation is given in Figure 6.9. The triazine core first abstracts a proton from 2′-OH and donates another proton to nonbridging phosphoryl oxygen. In other words, it strengthens the attacking nucleophile by deprotonation and stabilizes the developing phosphorane intermediate by protonation. In addition to formation of the phosphorane intermediate, the triazine core also able facilitates its breakdown, now by mediating a proton transfer from the phosphorane hydroxyl to the departing 5′-O.

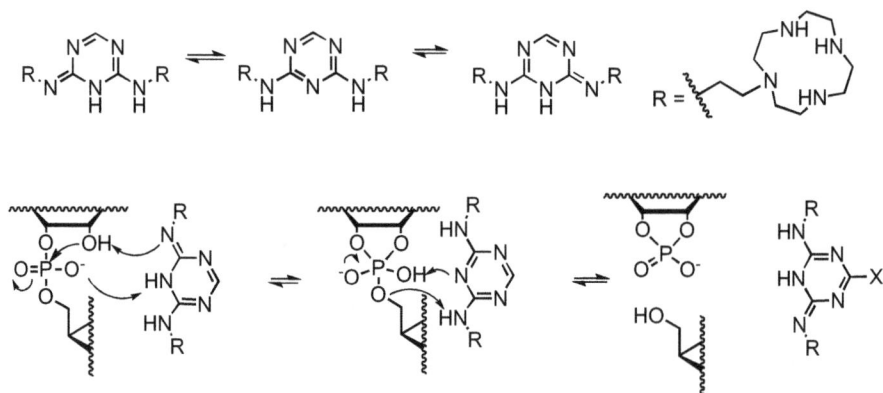

**Figure 6.9:** Cleavage of RNA phosphodiester linkages by 2,4-diamino-1,3,5-triazine anchored in the vicinity of the scissile bond [18].

Efficient guanidine-based cleaving agents for RNA have been obtained by attaching two guanidine groups to the upper rim of calix[4]arene having the lower rim substituted with long alkoxy chains (Figure 6.10A). At low millimolar concentrations of the agent, the cleavage of dinucleoside-3′,5′-monophosphate in 80% aqueous DMSO is accelerated up to 10⁴-fold compared to the hydroxide ion-catalyzed background reaction [19]. The catalytic efficiency is highest at pH 10.4, that is, under conditions where only one of the guanidine groups is protonated. The protonated guanidinium group most likely binds to the phosphate group and facilitates the attack of 2′-OH on phosphorus. The attack is additionally subject to general base catalysis by the unprotonated guanidine group. More recent studies with analogous cleaving agents bearing four guanidine or arginine groups at the upper rim have evidenced efficient cleavage in aqueous solution on using small molecular RNA and DNA models, 2-hydroxypropyl 4-nitrophenyl phosphate and bis-(4-nitrophenyl) phosphate [20]. Hydrophobic aggregation has been shown to play a key role. Evidently the catalytic functions of monomers undergo within an aggregate a structural preorganization that improves catalytic efficiency compared to nonaggregated monomers. Within aggregate, intermolecular interactions between monomeric units may play a role not present in solution phase.

Another indication of catalytic efficiency of guanidinium group is offered by arginine rich peptide structures that as peptide-oligonucleotide conjugates allow sequence-selective cleavage of target RNA [21]. An illustrative example of such a cleaving agent is decapeptide H[LRLRGLRLRG]OH that contains four catalytic arginine residues. The intervening leucine residues also are essential, likely owing to increasing hydrophobicity of the microscopic environment. The sequence recognizing oligonucleotide sequence was designed such that a 5-nt bulge of RNA was formed upon hybridization as the target site for the peptide moiety. The postulated mechanism involves participation of two guanidinium group (Figure 6.10B).

Tris[2-(benzimidazol-2-yl)ethyl]amine (Figure 6.10C) is still one efficient guanidine-based cleaving agent [22]. It tends to aggregate in aqueous solution, but tethering to an oligodeoxyribonucleotide prevents aggregation, affording a rather efficient sequence-selective cleaving agent that exhibits the half-life of 12–17 h for cleavage of oligoribonucleotides at 37 °C and 1.5 µM concentration of the cleaving agent conjugate [23]. An even slightly higher efficiency has been reported for a PNA conjugate of diethylenetriamine (Figure 6.10D) [24] (for the structure of PNA, see Section 11.4). Quaternized 1,4-diazabicyclo[2,2,2]octane of histidine (Figure 6.10E), in turn, serves as a sequence-independent cleaving agent that promotes the cleavage within single stranded regions of large RNA molecules [25]. The cleaving activity is comparable to that of the conjugates described above.

### 6.1.3 Reactions of internucleosidic phosphotriesters

Alkylation of a nonbridging phosphoryl oxygen has a dramatic effect on the reactivity of the phosphodiester linkage. At pH < 1, the influence still is modest; the $O$-methylated phosphodiester linkage is cleaved and isomerized roughly speaking as readily as the monoprotonated (neutral) linkage. At pH > 2, the situation is different. Isomerization turns hydroxide ion-catalyzed already at pH > 2 and shows first-order dependence on hydroxide ion concentration at higher pH. The cleavage, in turn, remains pH-independent in the pH region 2–7, and becomes then hydroxide ion-catalyzed [26]. Accordingly, the half-life for isomerization at pH 7 is of the order of seconds, while the half-life for cleavage is approximately 30 h at 25 °C. To prevent isomerization, the neighboring 2'-OH must be kept protected always when the negative charge of the phosphodiester linkage is somehow masked.

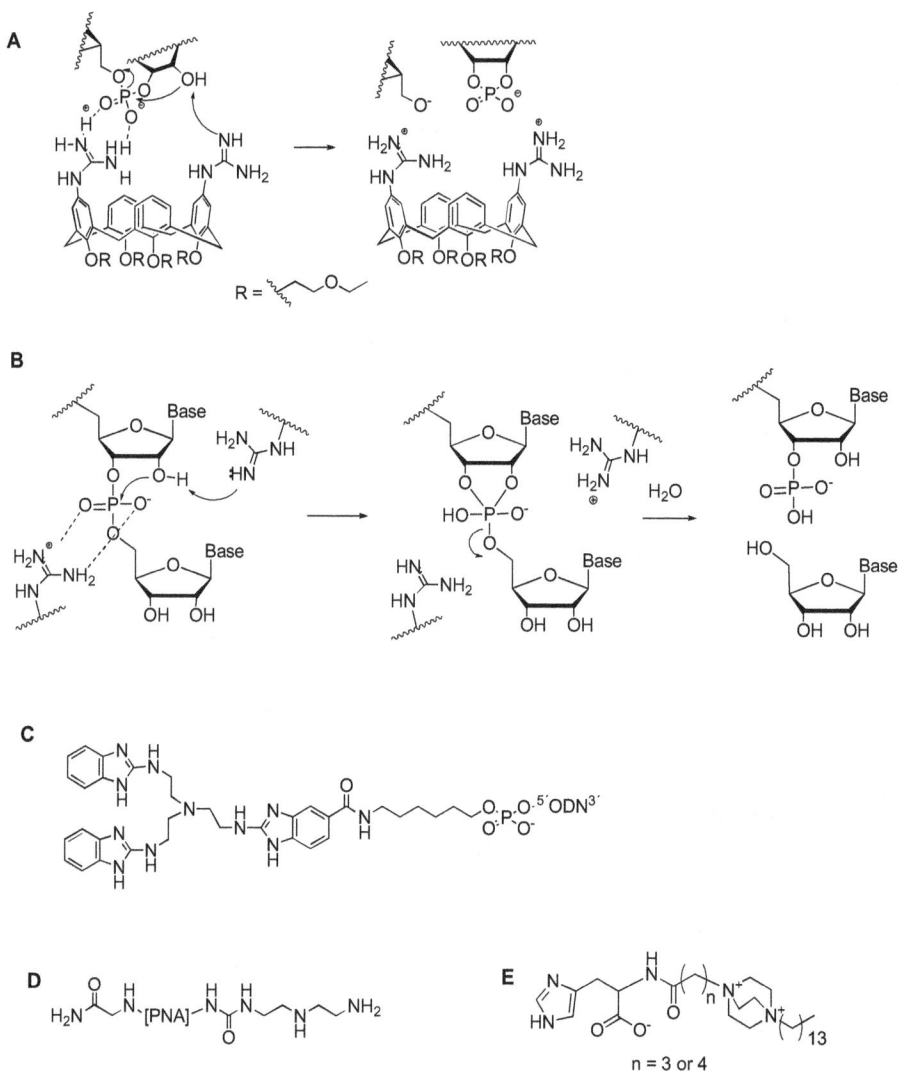

**Figure 6.10:** Nonmetallic cleaving agents for RNA phosphodiester linkages: (A) mechanism of cleavage by bis(guanidine)calix[4]arene [19], (B) mechanism of cleavage by arginine rich peptide oligonucleotide conjugates [21], (C and D) sequence-selective artificial ribonucleases [22–24] and (E) nonselective artificial ribonuclease [25].

## 6.2 Dinucleoside(3′,5′)monophosphates as RNA models: interplay between conformation and intramolecular stacking

Dinucleoside(3′,5′)monophosphates (NpN) are exceptionally interesting nucleotides since they are the shortest possible mimics of nucleic acids. While it is known that 5′-phosphorylation of nucleosides moderately weakens intermolecular stacking [27, 28] and has minor effects on the conformational equilibria [29], the interplay between the conformation and intramolecular stacking of NpNs is of particular interest, because it sheds light to the behavior of single stranded nucleic acids in solution at the nearest neighbor level. Although NpNs are conformationally highly flexible in aqueous solution, the tendency for intramolecular stacking of the base moieties is still clearly detectable [29, 30]. Table 6.1 records the percentage of NpNs estimated by NMR analysis to be in a stacked conformation. The striking feature is poor stacking tendency of uridine compared, not only to the purine nucleosides, but also to cytidine. The overall conformation of the stacked form evidently largely depends on the mutual orientation of the stacked bases. Stacking results in constraint that favors N-type (3′-endo) ring puckering. The difference between the 3′-endo population of NpN and monomeric Np and pN is particularly noticeable with purine nucleosides. Upon dimerization, the 3′-endo population is increased from 30–40% to 50–60%. The confor-

**Table 6.1:** Percentage of dinucleoside(3′,5′)monophosphates ($N^1pN^2$) in a stacked form by NMR analysis (A), percentage of $N^1$ (in parentheses $N^1p$) in 3′-endo form (B), percentage of $N^2$ ($pN^2$) (C), percentage of $N^1pN^2$ in stacked form by molecular dynamics (D) and percentage of $N^1pN^2$ in stacked form by potential mean force calculations (E).

| $N^1pN^2$ | % of $N^1pN^2$ stacked by NMR[a] | % of $N^1(N^1p)$ in 3′-endo form[a] | % of $N^2(pN^2)$ in 3′-endo form[a] | % of $N^1pN^2$ stacked by MD[b] | % of $N^1pN^2$ stacked by PMF[c] |
|---|---|---|---|---|---|
| ApA | 38 ± 2 | 58 (31) | 61 (40) | 85 | 49 |
| ApG | 25 ± 2 | 49 (31) | 51 (37) | 87 | 69 |
| ApC | 38 ± 2 | 64 (31) | 75 (55) | 75 | 28 |
| ApU | 34 ± 3 | 57 (31) | 59 (46) | 74 | 61 |
| GpA | 30 ± 2 | 55 (33) | 56 (40) | 90 | 33 |
| GpC | 45 ± 4 | 71 (33) | 79 (55) | 67 | 68 |
| GpU | 27 ± 5 | 49 (33) | 62 (46) | 73 | 52 |
| CpC | 35 ± 1 | 74 (60) | 69 (55) | 29 | 4 |
| CpA | 24 ± 2 | 71 (60) | 57 (40) | 47 | 91 |
| CpG | 23 ± 3 | 71 (60) | 57 (37) | 52 | 31 |
| CpU | 33 ± 2 | 66 (60) | 64 (46) | 29 | 5 |
| UpU | 8 ± 5 | 56 (56) | 53 (46) | 44 | 2 |
| UpA | 15 ± 3 | 53 (56) | 51 (40) | 47 | 6 |
| UpG | 10 ± 5 | 54 (56) | 50 (37) | 68 | 74 |
| UpC | 18 ± 2 | 62 (56) | 63 (55) | 66 | 6 |

[a]From Refs. [29, 30] at 20 °C, [b] from Ref. [31], [c] from Ref. [32] at 5 Å distance between the glycosylated nitrogens of N1 and N2.

mation around the C4′–C5′ and C5′–O bonds of the 5′-linked nucleoside does not, in turn, appreciably differ from that in monomeric pN. The O3′–P–O5′ bond system, hence, is highly flexible allowing the series of conformational events required to minimize the constraints caused by intramolecular stacking [29].

The results obtained by theoretical calculations differ considerably from those based on NMR analysis. According to calculations by classical molecular dynamics [31], the stacked form is more favored than reported on the basis of NMR analysis. More importantly, even the order of stacking efficiency among various dinucleotide monophosphates is different. In particular, the stacking tendency of cytidine is comparable rather to stacking of uridine than to stacking of purine nucleosides (Table 6.1). The results of potential of mean force calculations also argue against efficient stacking of cytidine [32], but otherwise the consistency between these two theoretical approaches is modest. The values given in Table 6.1 refer to the free energy of unstacking along the reaction coordinate at 5 Å distance of the glycosylated nitrogen atoms (N1 with pyrimidines, N9 with purines).

## 6.3 Cleavage of oligoribonucleotides: the effect of base sequence and secondary structure on the cleavage rate

The chain length appears to have only a minor effect on the average stability of RNA phosphodiester bonds. On average, the phosphodiester bonds in poly(U) are cleaved in aqueous alkali as rapidly as 3′,5′-UpU [33]. The acid-catalyzed cleavage and isomerization reactions of polyU are, in turn, one order of magnitude faster than with 3′,5′-UpU. Acid catalysis also somewhat favors the reactions of nonterminal phosphodiester bonds, while the alkaline cleavage is not subject to similar selectivity.

Although there seems to be no fundamental difference between phosphodiester cleavage of polyU and 3′,5′-U̇pU, polyU is not in every respect a proper model for RNA. The stacking interaction between uracil bases is weak, if any, compared to stacking of adjacent purine bases. This does not cause marked difference in hydrolytic stability of various 3′,5′-NpNs, but the situation is different with RNA oligomers. With both synthetic ORNs [34] and natural RNA oligomers [35, 36], linkages that consist of a 3′-linked pyrimidine and 5′-linked purine nucleoside are often exceptionally labile. Nevertheless, one should note that all such linkages are not exceptionally labile. The identity of the 3′- and 5′-linked nucleosides does not alone determine the hydrolytic stability of a given phosphodiester linkage. Studies with chimeric ORNs, 5′-GGGUAN|AAGUGC-3′, where N is an unsubstituted ribonucleoside and all the other nucleosides are 2′-O-methylated, suggest that the cleavage rates may differ by two orders of magnitude. With these dodecamers, 15-fold accelerations and 15-fold retardations compared to the cleavage of a similar tetramer model have been observed at pH 8.5 [37]. It appears that even rather remote stacking interactions still affect stacking at the cleavage site. Stacking geometry of neighboring bases within an oligomer differs from the optimal geometry present in 3′,5′-NpN monomer [38]. Accordingly, stacking of each base pair must compromise with the stacking tendency of other base pairs making the stacking sequence dependent [39, 40].

Another factor that markedly affects hydrolytic stability is secondary structure. Above all, the bonds within double helical regions are much more stable than those within single strands. Base-stacking in double helix favors the so-called "adjacent" geometry, where the angle $O^{2'}$–P–$O^{5'}$ is around 60°. A prerequisite for the cleavage, in turn, is that both the attacking $O^{2'}$ and departing $O^{5'}$ are able to adopt an apical position within the phosphorane intermediate/transition state. Either, the scissile bond should initially take an "in-line" geometry with the $O^{2'}$–P–$O^{5'}$ angle being 180°, or the phosphorane intermediate having the $O^{5'}$ in equatorial position should pseudorotate. The double helical structure resists both of these conformational changes, hence, stabilizing the phosphodiester bonds against both cleavage and isomerization [41].

The reactivity of a particular phosphodiester linkage within a hairpin loop or bulge may differ, but not dramatically, from the reactivity of the same bond within a linear strand. The relative cleavage rates of a 3′,5′-CpA bond in various sites within tetra- to hepta-loops of otherwise 2′-O-methylated hairpins vary from 0.78 to 8.7 compared to the same bond within a linear chain [42]. Similar results have been obtained with pentanucleotide bulges [43].

## 6.4 Cleavage of phosphate-modified analogs of oligoribonucleotides

### 6.4.1 Phosphorothioates and phosphorothiolates

Phosphorothioate oligonucleotides are extensively used as antisense oligonucleotides in chemotherapy (cf. Chapter 11). In addition, phosphorothioates [44], as well as 3′- or 5′-S-phosphorothiolate oligonucleotides [45] are often utilized in mechanistic studies of enzyme catalysis. The chirality of phosphorothioate linkage allows examination of the stereochemical requirements of the enzyme catalysis, whereas replacement of bridging oxygen with sulfur markedly alters the leaving group property of 3′- and 5′-oxygens and/or the geometry of the attack of 2′-OH on phosphorus. Kinetic studies with thiosubstituted analogs of dinucleoside-3′,5′-monophosphates have provided background information useful for interpretation of the results of enzymatic studies [46].

Thiosubstitution of one of the nonbridging phosphoryl oxygen atoms creates a new stereogenic center in dinucleoside monophosphates, as already discussed in Chapter 3 $R_P$ and $S_P$ diastereomers are obtained (Figure 6.11). There fortunately is a convenient method for assignment of the stereochemistry. Snake venom phosphodiesterase cleaves the $R_P$ but not the $S_P$ linkage [47].

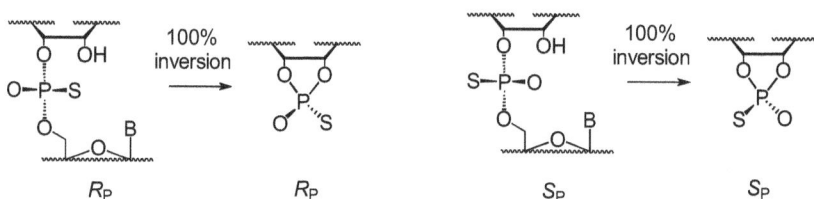

**Figure 6.11:** Stereochemistry of the cleavage of internucleosidic phosphorothioate linkages.

Table 6.2 records the thioeffects, $k_O/k_S$, for the cleavage and isomerization of the internucleosidic phosphorothioate linkages, and for the hydrolysis of their immediate cleavage product, 2′,3′-cyclic phosphorothioate. The nonbridging thiosubstitution has only a minor effect on the stability of the phosphodiester linkage at pH > 8, i.e., under conditions where the reaction is hydroxide ion-catalyzed. The thioeffect, $k_O/k_S$, is 1.3 and 0.8 for the $R_P$ and $S_P$ diastereomer, respectively [48]. Reactions proceed with 100% inversion. The transition state is late as with the oxygen counterparts. In fact, the $\beta_{lg}$ values for the cleavage of 3′-alkyl phosphates [49] and phosphorothiotes [50] are equal within the limits of experimental errors. No conversion to a 2′,5′-isomer takes place. DFT calculations suggest that compared to oxygen, sulfur atom in principle stabilizes the dianionic phosphorane intermediate [51]. This does not, however, lead to enhanced cleavage rate since the inherent stabilization is counterbalanced by weaker solvation of the sulfur containing phosphorane. With dithioate, the thioeffect is somewhat more marked, $k_O/k_S = 2.8$ [52].

**Table 6.2:** Thioeffects, $k_O/k_S$, for the cleavage and isomerization of the internucleosidic phosphorothioate linkages.

| Reaction | $R_P$ | $S_P$ | dithio | Ref. |
|---|---|---|---|---|
| **Cleavage** | | | | |
| Hydroxide ion-catalyzed | 1.3 | 0.8 | 2.8 | [48, 52] |
| pH-independent | 0.1 | 0.3 | 0.2 | [52, 53] |
| Acid-catalyzed (via neutral phosphorane) | 0.9 | 0.4 | 0.6 | [52, 53] |
| Acid-catalyzed (via phosphorane monocation) | 22 | 45 | | [53] |
| **Isomerization** | | | | |
| pH-independent | 5 | 7 | 8 | [52, 53] |
| Acid-catalyzed (via neutral phosphorane) | 9 | 9 | 4 | [52, 53] |
| Acid-catalyzed (via phosphorane monocation) | 80 | 60 | | [53] |
| Hydrolysis of 2′,3′-cyclicphosphorothioate | | | | |
| Hydroxide ion-catalyzed | 2.1 | 1.4 | | [54] |
| **pH-independent** | 3.4 | 1.7 | | [54] |
| Acid-catalyzed (via neutral phosphorane) | 16 | 11 | | [54] |

The situation turns more complicated at pH < 8. Reactions occurring via a monoanionic phosphorane gradually start to predominate. This allows isomerization and, additionally, hydrolytic desulfurization. Hydrogen sulfide is $10^5$ times less basic than hydroxide ion and competes as a leaving group with sugar hydroxyl functions. Desulfurization actually predominates in the pH range 5–7, representing 80% of the disappearance of the starting material. As regards the cleavage and isomerization reactions, thiosubstitution favors the cleavage and retards isomerization. The thioeffects, $k_O/k_S$, for the cleavage being 0.1 ($R_P$), 0.3 ($S_P$) and 0.2 (dithioate) [52, 53]. Thioeffects for the isomerization are 5 ($R_P$), 7 ($S_P$) and 8 (dithioate).

In the pH range 2–4, the reaction takes place via neutral phosphorane, as with their oxygen counterparts. Desulfurization predominates and isomerization is retarded by the thiosubstitution. The thioeffects for the cleavage are 0.9 ($R_P$), 0.4 ($S_P$), 0.6 (dithioate) and for the isomerization 9 (both $R_P$ and $S_P$) and 4 (dithioate). At pH < 2, the behavior of phosphoro-mono- and -di-thioates clearly differs from that of their oxygen analogs. While phosphodiesters tend to react via a monocationic ionic form and, hence, the reaction order in hydronium ion activity approaches 2, this is not the case with phosphorothioates. That is why the thioeffects at pH < 2 are very prominent: for cleavage 22 ($R_P$) and 45 ($S_P$), and for isomerization 80 ($R_P$) and 60 ($S_P$).

Thioeffects on hydrolysis of the immediate cleavage product, 2′,3′-cyclic phosphorothioate, are also of interest, in particular for the mechanistic studies of ribozyme catalysis (cf. Section 10.2). The hydroxide ion-catalyzed and pH-independent reactions are, however, rather insensitive to sulfur substitution. Thioeffects of the hydroxide ion-catalyzed hydrolysis are 2.1 ($R_P$) and 1.4 ($S_P$) [54], and those for the pH-independent hydrolysis 3.4 ($R_P$) and 1.7 ($S_P$). Under acidic conditions, desulfurization severely competes with hydrolysis that is markedly slower than with its oxygen analog: $k_O/k_S = 16$ ($R_P$) and 11 ($S_P$).

Replacing a bridging oxygen atom with sulfur has a dramatic influence on the hydrolytic stability of the phosphodiester linkages. 5′-Thio substitution accelerates the hydroxide ion-catalyzed cleavage from 4 to 5 orders of magnitude [55, 56], consistent with 5 orders of magnitude higher acidity of sulfides compared to alcohols. The large rate accelerating effect of 3′-thio substitution is more difficult to explain. The hydroxide ion-catalyzed cleavage is accelerated by two to three orders of magnitude [57, 58]. Since sulfur is a less electronegative element than oxygen, it most likely reduces the electron density at phosphorus less than oxygen. In other words, thio substitution should retard rather than accelerate the attack of 2′-oxyanion on phosphorus. Evidently, better solvation of sulfur plays a role. QM/MM calculations suggest that the thiophosphorane intermediate/transition state is stabilized by solvation compared to its pentaoxy counterpart and the barrier for its breakdown is lowered [59]. The pH independent isomerization at pH 3–5 is 50 times as fast as with the oxygen counterpart [58]. The acid-catalyzed isomerization and cleavage, in turn, exhibit only modest thio effects.

### 6.4.2 Phosphoramidates

Substitution of O3' with nitrogen makes ORN stable toward nucleases, while it still hybridizes with DNA and RNA as selectively and efficiently as unmodified RNA [60]. Under neutral and alkaline conditions, the nonenzymatic stability is very similar to that of unmodified ORN, but under acidic conditions the phosphoramidate oligomers are hydrolyzed three orders of magnitude more stable than their unmodified counterparts [61]. The predominant reaction under neutral conditions is the attack of 2'-hydroxy on phosphorus atom concerted with proton transfer from O2' to a nonbridging phosphoryl oxygen. The pentacoordinated intermediate obtained breaks down by P–O5' bond cleavage concerted with proton transfer to the departing O5'. The resulting terminal nucleoside O2',N3'-cyclic phosphoramidate finally undergoes kinetically invisible hydrolysis to terminal 3'-amino-3'-deoxynucleoside 2'-monophosphate (Figure 6.12) [61].

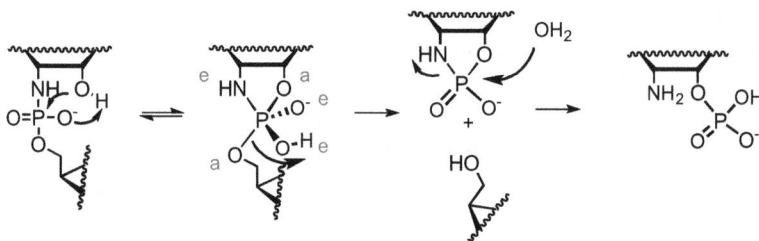

**Figure 6.12:** pH-independent cleavage of RNA phosphoramidate linkage [61].

### 6.4.3 C5'-phosphonates

Substitution of O5' with carbon prevents cleavage of the internucleosidic phosphonate linkage, but isomerization of the O3'–P–C5' phosphonate linkages to O2'–P–C5' linkage still takes place. Over a wide pH range from 5 to 9, pH-independent isomerization is the only reaction taking place [62].

## 6.5 Metal ion-promoted cleavage of oligoribonucleotides

### 6.5.1 Mechanism of cleavage by metal ion complexes

Metal ions have been known to promote the cleavage of RNA phosphodiester bonds since the early studies of Dimroth et al. in 1959 [63]. A few exceptions apart, metal ions accelerate only cleavage, not isomerization of phosphodiester bonds [64]. Alkaline metal ions are not catalytically active. Among divalent ions, $Cu^{2+}$ and $Zn^{2+}$ exhibit greatest rate accelerations. Trivalent lanthanide ions are much more efficient cleaving agents, but the active species evidently is a multinuclear hydroxide complex formed on approaching the pH in alkaline region where precipitation of the complex takes place [65–67].

Precipitation of metal ions as hydroxide ion complexes may be prevented by binding the metal ion to a neutral organic ligand that forms a stable complex. Terpyridine, bipyridine and polyazamacrocycles, above all 1,4,7-triazacyclononane and 1,5,9-triazacyclododecane, are often used for the purpose. Lanthanide ions bind weakly to neutral ligands and, hence, ligands with encapsulating side arms have to be used, although formation of catalytically most active multinuclear species may be prevented [68]. Ligands with negatively charged donor atoms tend to abolish the catalytic activity.

The efficiency of metal ion catalysis can be increased by binding two metal ions to a ligand that locks them at an appropriate distance to each other. Complexes A [69], B [70] and C [71] in Figure 6.13 are examples of such dinuclear complexes. The catalytic activity of a mononuclear complex, in turn, may be increased by an appropriately situated H-bond donor, like the amino groups in complex D. Complex D [72] is almost 100-fold as efficient cleaving agents as its counterpart E devoid of amino groups. Nevertheless, the difference between the catalytic activity of di- and mononuclear complexes, such as the $Zn^{2+}$ complexes A and F, is even more prominent [69].

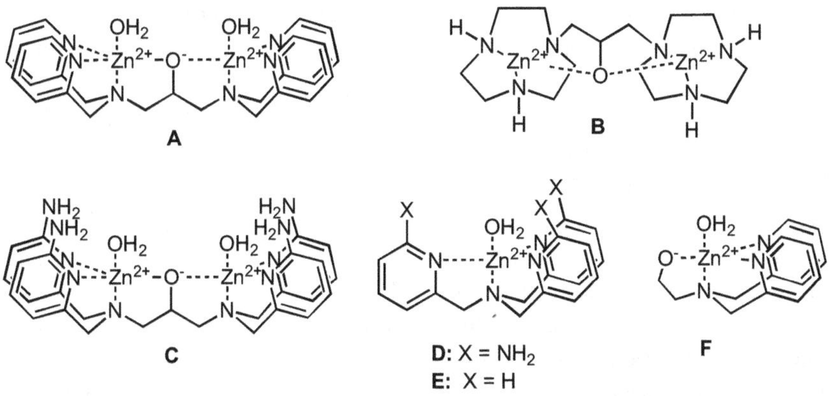

**Figure 6.13:** Examples of highly active $Zn^{2+}$ complexes for the cleavage of phosphodiester linkages of oligoribonucleotides.

Complexes B and C in Figure 6.13 are the most active dinuclear cleaving agents described so far. Half-life for the cleavage of 3′,5′-UpU at 2 mM concentration of B is about one week at pH 7.0 and 25 °C [73]. With complex C, the half-life is reduced to 7 h at 1 mM catalyst concentration at pH 6.5 and 25 °C [74]. Complex C additionally catalyzes the isomerization of 3′,5′-UpU to 2′,5′-UpU, although much less efficiently than the cleavage [75]. The cleavage is accelerated by 6 order of magnitude and isomerization by two orders of magnitude. This is the only reported example of metal ion catalytic isomerization of RNA phosphodiester bond. Additionally, only isomerization of phosphorothioate linkage is known to be subject to weak catalysis by $Zn^{2+}$ [76].

Much higher catalytic activities have been reported for metal ion complexes when 2-hydroxypropyl $p$-nitrophenyl phosphate (HpNP) is used as a simplified model of dinucleoside-3′,5′-monophosphates, often in methanol or ethanol. One should, however, bear in mind that $p$-nitrophenol with $pK_a = 7.15$ is overwhelmingly better leaving group than a 5′-linked nucleoside, the $pK_a$ value of which is 6 or 7 orders of magnitude higher. Replacement of the solvent water with methanol or ethanol still dramatically accelerates the cleavage of this simple RNA model [77]. Main reason for the impressive acceleration is that the catalyst binds more tightly to the phosphoester.

Dependence of the observed rate constant on pH is for metal ion-promoted cleavage either sigmoidal of bell-shaped. Phosphodiester evidently participates as a monoanion. In case the catalyst (metal aquo ion) undergoes only one deprotonation, the rate profile is sigmoid. In case the catalyst undergoes two consecutive deprotonations to a dianionic species, the rate-profile is bell-shaped [78]. Several mechanisms have been proposed to explain these observations, but no one of them has gained general acceptance. The common feature of all the mechanisms is deprotonation of 2′-OH by metal ion or its deprotonated aqua ligand. Metal ion decreases the electron density of phosphorus atom by binding to nonbridging phosphoryl oxygen and it may, in principle, increase the electron density of 2′-$O$ by displacing the dissociable proton (Figure 6.14A) [79]. This mechanism mainly receives support by studies with HpNP. Instead of direct coordination to 2′-$O$, a phosphate bound metal ion may deprotonate 2′-OH by its hydroxide ligand concerted with formation of the P–O2′ bond (Figure 6.14B) [80]. With dinuclear complexes, double Lewis acid activation takes place through binding of a metal ion to both nonbridging phosphoryl oxygens, but only one of them participates in deprotonation of 2′-OH. The metal ion can also serve merely as an electrophile decreasing the electron density on the phosphorus atom. With dinuclear complexes, a more efficient double Lewis acid activation takes place [73]. The attacking nucleophile is, depending on pH, 2′-OH or 2′-oxyanion, but the metal ion does not participate in proton abstraction (Figure 6.14C). Finally, a general acid catalysis has received considerable support (Figure 6.14D). According to this mechanism, a dianionic phosphorane intermediate is formed in a rapid pre-equilibrium step. The otherwise highly unstable intermediate is stabilized by binding of metal aquo ion to one of the nonbridging oxygens, in case of a dinuclear catalyst to both nonbridging oxygens. The metal aquo ion then facilitates the departure of the 5′-linked nucleoside by donating a proton from aqua ligand to the departing 5′-$O$. The observations that lend support for this mechanism include rather modest influence of basicity of the leaving group on the cleavage rate [74, 81], acceleration of isomerization by dinuclear complexes [74], primary $^{18}O$ isotope effect for the departure of the 5′-linked nucleoside [82], and kinetic solvent isotope effect ($k_H/k_D = 2.7$) that is consistent with rate-limiting proton transfer [74].

**A.** Nucleophilic catalysis

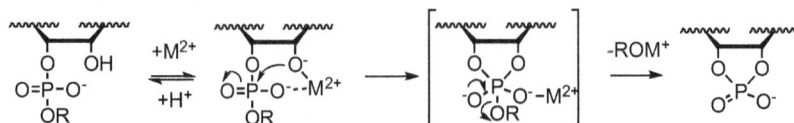

**B.** General base catalysis

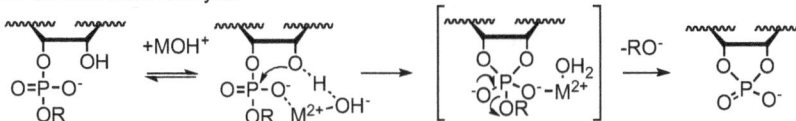

**C.** Elecrophilic catalysis

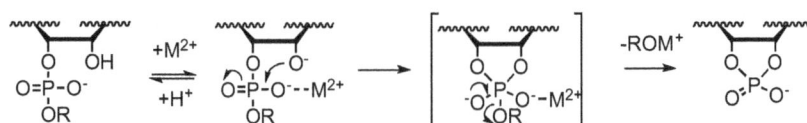

**D.** General acid catalysis

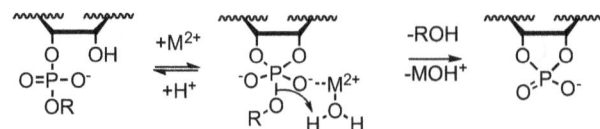

**Figure 6.14:** Alternative mechanisms for the metal ion-promoted cleavage of RNA phosphodiester linkages.

### 6.5.2 Metal ion-promoted cleavage of oligoribonucleotides

The presence of several phosphodiester linkages in oligoribonucleotides allows formation of metal ion chelates by binding to two phosphate groups. Dinucleoside diphosphates bearing a terminal 3′-monophosphate group, for example, are cleaved 100-times as fast as dinucleoside monophosphates devoid this functionality [83]. The most obvious explanation for this rate enhancement is simultaneous binding of metal ion to both of the negatively charged phosphate groups. Studies with Up(Tp)$_n$-type oligomers have shown that the rate-accelerating effect is only moderately decreased on increasing the length of the thymidine 3′-phosphates from 1 to 8 [84]. Most likely back-folding of the chain allows interaction between the 3′-terminal monophosphate group and the 5′-terminal phosphodiester linkage and this process is mediated by additional metal ion(s) binding to the intervening phosphodiester centers. The rate-deceleration with the increasing distance between the monophosphate group and the cleavage site is more marked in case the intervening phosphodiester linkages are replaced with neutral methylphosphonate linkages [85].

Bridging of two phosphodiester linkages may also be a source of rate acceleration, but the acceleration is much more modest than that caused by bridging with a terminal monophosphate group. Consistent with this view, phosphodiester bonds of poly(U) are cleaved by the 1,5,9-triazacyclododecane complex of $Zn^{2+}$ 11 times as fast as the same bond in 3′,5′-UpU [86]. The distribution between terminal and nonterminal bond cleavages is statistical and the products bear a terminal 2′,3′-cyclic phosphate groups as a detectably stable intermediate. The same chelate also cleaves the phosphodiester bonds within bulges approximately as readily as those within linear strands [87]. The cleavage is somewhat faster within a large than within a small bulge, but even a single nucleotide bulge still allows cleavage. The double helical stem remains intact. The cleavage within tetra- to hepta-nucleotide hairpin loops is likewise approximately as fast as within a linear chain [88].

### 6.5.3 Nucleobase-selective cleavage

The 1,5,9-triazacyclododecane complex of $Zn^{2+}$ that efficiently cleaves ORN phophodiester linkages, is additionally able to recognize uracil base. The zinc ion binds to deprotonated N3, and two of the secondary amino groups are H-bonded to the carbonyl oxygens [89]. This has allowed utilization of a tris($Zn^{2+}$-)azacrown complex as a cleaving agent exhibiting moderate base moiety selectivity (Figure 6.15) [90]. Two of the $Zn^{2+}$ (azacrown) side arms bind to two adjacent bases within the chain and the third one serves as a catalyst cleaving the intervening phosphodiester bond. In addition to uracil, guanine is also recognized, although less efficiently than uracil [91]. Accordingly, the cleavage rate decreases at low concentration of the cleaving agent in the order UpU > UpN ≈ NpU > GpG ≫ NpN (N = G,A,C) [92].

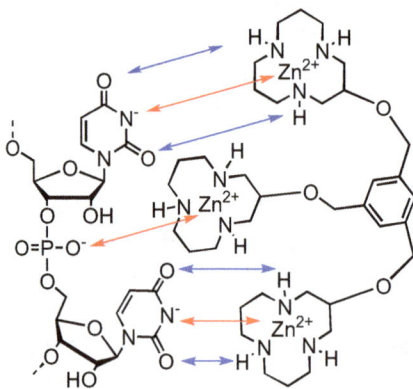

**Figure 6.15:** Chemical basis for the base moiety-selective cleavage of RNA phosphodiester linkages [90].

### 6.5.4 Artificial ribonucleases

Real sequence selectivity for the metal ion-promoted cleavage is achieved by attaching a metal ion complex to an oligomer that recognizes the base sequence of the target RNA by hybridization. Usually 2'-deoxyribonucleotide (ODN), 2'-*O*-methyloligoribonucleotide (2'-*O*-Me-ORN) or a peptide nucleic acid (PNA) is used for the purpose. This kind of conjugates, called artificial ribonucleases, can be used for tailoring of large RNA molecules in vitro [93]. Their development has often been motivated by highly selective degradation of mRNA, but several barriers still exist on the way to in vivo applications of this approach.

The early studies on the artificial ribonucleases were largely based on lanthanide ion complexes. Among them, 5'-ODN conjugates derived from $Dy^{3+}$ complex of texaphyrin [94] (Figure 6.16A) and $Eu^{3+}$ complex of a pyridine cyclophane [95] (Figure 6.16B) turned out to be most efficient. They cleaved the target at the 3'-side of the first and third unpaired nucleotide. The half-lives at pH 7.5 and 37 °C were 2 and 4 h, respectively. When these complexes were attached in an intrachain position of the sequence-recognizing ODN, the release of conjugates from the cleaved target was facilitated and the cleavage showed turnover [96, 97]. Since metal ions cannot efficiently cleave phosphodiester bonds within a double helical region, the recognizing sequence is usually planned in such a way that a bulge is formed at the cleavage site upon hybridization.

Besides lanthanide complexes, $Cu^{2+}$ and $Zn^{2+}$ complexes have been used for site specific cleavage of RNA. The most efficient $Cu^{2+}$-based construct consists of two catalytic moieties, that is, terpyridine conjugates of 2'-*O*-methyl-ORNs attached to each other via a flexible linker (Figure 6.16C) [98]. The half-life for cleavage by this dinuclear $Cu^{2+}$ complex conjugate is 2–3 h at pH 7.4 and 45 °C. Intrachain $Cu^{2+}$ complex of 2,9-dimethyl-5-aminophenentroline (Figure 6.16D) is approximately as efficient, the half-life being 5 h at pH 7.5 and 37 °C [99]. The role of the methyl substituents is essential, since they prevent dimerization of the $Cu^{2+}$ complex. The cleaving activity of the corresponding $Zn^{2+}$ conjugate is half of that of the $Cu^{2+}$ complex [100].

Even more efficient sequence-selective catalysts have been obtained by using PNA (peptide nucleic acid) for sequence recognition instead of ODN (Figure 6.16E) [101, 102]. With intrachain $Cu^{2+}$-(2,9-dimethyl-5-aminophenantroline) conjugates that target three or four nucleotide bulges, less than half an hour half-lives has been achieved at 37 °C. $Zn^{2+}$-(dimethyl-dipyridophenazine conjugates) show comparable or even slightly higher cleaving activity [103]. Both conjugates also show turnover in excess of RNA. These evidently are the most efficient artificial ribonucleases described so far.

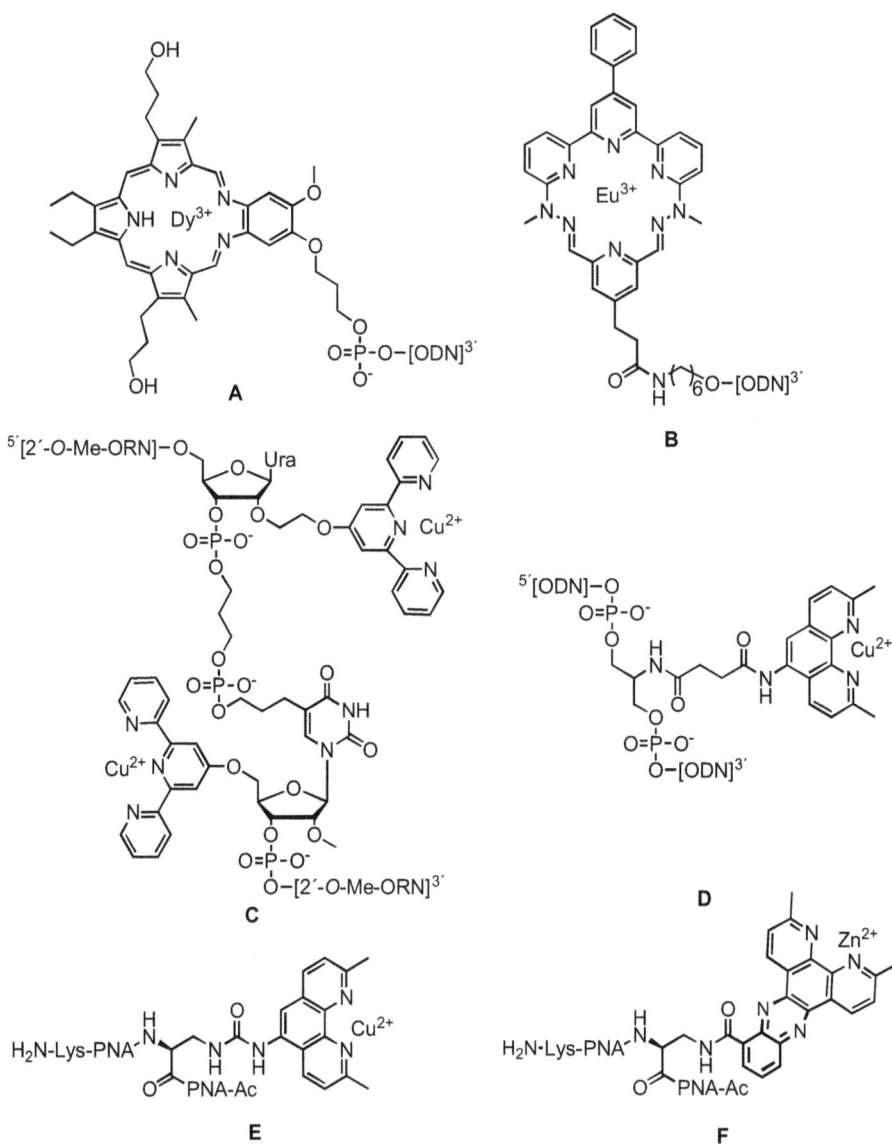

**Figure 6.16:** Sequence-selective artificial ribonucleases: A [94], B [95], C [98], D [99], E [101] and F [103].

## 6.5.5 Functionalized nanoparticles as artificial ribonucleases

Gold nanoparticles coated with ω-mercaptoalkyl-functionalized 1,4,7-triazacyclononane azacrowns constitute an interesting group of artificial metallonucleases. These particles in the presence of $Zn^{2+}$ efficiently cleave HpNP, di(ribonucleoside)-3′,5′-monophosphates and even DNA (Figure 6.17A) [104]. The dependence on the concentration of $Zn^{2+}$ is sigmoid, strongly suggesting that two $Zn^{2+}$ chelates operatively participates in the transition state. Similar $Cu^{2+}$ nanozymes even exhibit modest enantioselectivity: particles coated with the azacrown conjugate B in Figure 6.17 cleaves 3′,5′-UpU somewhat more efficiently than particles bearing the enantiomeric conjugates C [105]. The polarity of the microenvironment in the vicinity of the azacrown complex plays an important role. The catalytic activity was significantly enhanced on increasing the length of the alkyl tether between the gold surface and the catalytic $Zn^{2+}$ azacrown complex (Figure 6.17D) [106]. Recent kinetic studies on cleavage of uridine 3′-alkylphosphates have shown that simultaneous interaction of two $Zn^{2+}$ azacrown complexes with the scissile phosphodiester linkage is a prerequisite for fast cleavage. The moderately negative $\beta_{lg}$-value has been interpreted to suggest that neither the nucleophilic attack of 2′-OH on phosphorus nor the cleavage of the P–O5′ bond is alone rate-limiting, but formation of the O2′-P bond is more advanced than the fission of the P–O5′ bond. [107].

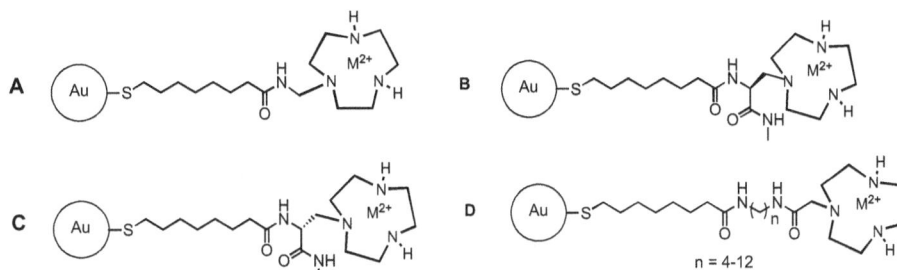

Figure 6.17: Gold nanoparticle cleaving agents: A [104], B and C [105], D [106].

## 6.6 Cleavage of oligodeoxyribonucleotide phosphodiester linkages

### 6.6.1 Cleavage by Brönsted acids and bases

Phosphodiester linkages of oligodeoxyribonucleotides (ODNs) are exceptionally stable. In fact, it is not known how stable they are, since other degradative processes are faster. The most reliable estimate that is based on hydrolysis of dineopentyl phosphate is 30 million years at pH 7 and 25 °C [108]. Most likely, the 5′-linked nucleoside departs more rapidly by C5′–O bond cleavage than by P–O5′ cleavage. Hydrolysis of dimethyl phosphate, for comparison, proceeds 99% by nucleophilic attack on carbon instead of phosphorus [109]. One should also bear in mind that much faster depurination leads to opening of sugar ring and the aldehyde form of the sugar enables chain cleavage by elimination, as discussed below in more detail [110].

Two mechanisms may be envisaged for the cleavage of DNA phosphodiester bonds on the basis of pH independence of the hydrolysis dineopentyl phosphate monoanion at pH 7–12: attack of water on the phosphorus atom concerted with proton transfer to a nonbridging oxygen (Figure 6.18A), or attack of hydroxide ion on a neutral phosphodiester linkage (Figure 6.18B). Both reactions lead to formation of a pentaoxyphosphorane monoanion. Computational calculations prefer the former alternative, and lend support for a finite life-time of the monoanionic phosphorane [111].

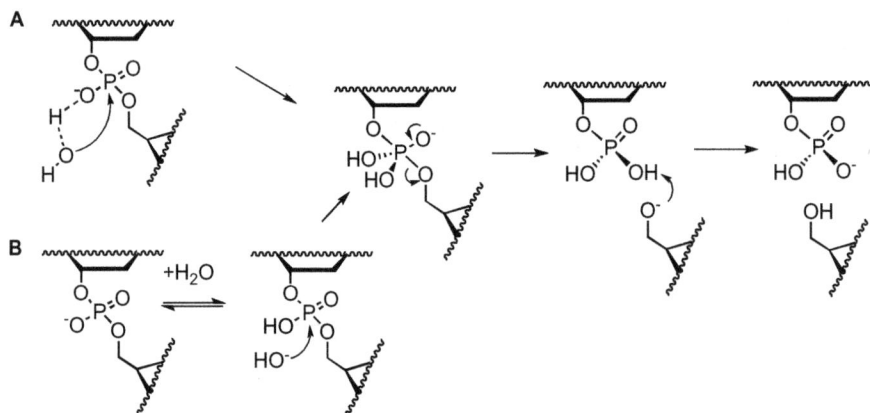

**Figure 6.18:** Alternative mechanisms for the cleavage of oligodeoxyribonucleotide phosphodiester linkages. mechanism A is preferred on the basis of theoretical calculations [111].

The fastest reaction of ODNs is hydrolysis of the *N*-glycosidic bond of purine nucleosides, depurination (cf. Section 2.6). Consequently, abasic sites are formed and these induce strand breaks *via* β- and δ-elimination. In other words, either C3′–O or C5′–O bond is cleaved by concomitant formation of a 2′,3′-double bond or 2′,3′- and 4′,5′-double bonds as discussed in Section 5.1 (cf. Figure 5.18A). While depurination is accelerated by acids, the elimination step is catalyzed by bases, and hence, even these eliminations are slow under neutral conditions.

## 6.6.2 Cleavage promoted by metal ions

Among the numerous metal ions studied, only $Ce^{4+}$ [112, 113], $Zr^{4+}$ [114], $Th^{4+}$ [115] and $Co^{3+}$ [116] cleave ODNs or DNA at a rate useful for practical purposes. Cleavage by $Ce^{4+}$ is fastest, the half-life of TpT being 4.3 h in 10 mM aqueous $Ce(NH_4)_2(NO_3)_6$ at pH 7.0 and 50 °C. Two hydroxide-bridged $Ce^{4+}$ ions evidently result in double Lewis acid activation by binding to the nonbridging phosphoryl oxygen and an additional hydroxide ligand serves as a nucleophile displacing either the 3′- or 5′-linked nucleoside (Figure 6.19A) [117]. The reaction continues by dephosphorylation of the resulting 3′- and 5′-monophosphates.

Interestingly, addition of $PrCl_3$ into the mixture ($[PrCl_3] = 5$ mM) still accelerates the cleavage by one order of magnitude [118]. The enhanced catalysis has been attributed to cooperative action of two $Ce^{4+}$ ions and one $Pr^{3+}$ ion, as depicted in (Figure 6.19B). The $Ce^{4+}$ ions still result in the double Lewis acid activation, while the hydroxide ligand of the $Pr^{3+}$ ion, bridged between the $Ce^{4+}$ ions, serves as an intracomplex nucleophile. This mode of action is advantageous since the hydroxide ligand of $Pr^{3+}$ is a better nucleophile than the hydroxide ligand of $Ce^{4+}$.

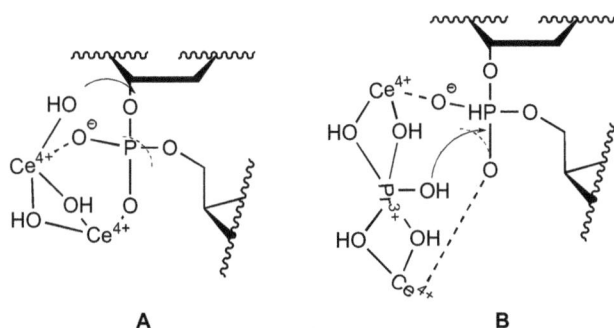

**Figure 6.19:** Mechanisms proposed for the $Ce^{4+}$ ion-promoted cleavage of DNA phosphodiester linkages [117, 118].

The $Ce^{4+}$-promoted cleavage has been converted sequence selective with the aid of two ethylenediamine-*N,N,N′,N′*-tetrakis(methylenephosphonic acid) conjugated oligodeoxyribonucleotide probes, one of which is a 3′- and the other a 5′-conjugate (Figure 6.20A) [119]. The base sequences of the conjugates are such that the conjugate groups become upon hybridization situated opposite to the desired cleavage site. $Ce^{3+}$ ions become bound to the phosphonic acid ligands and are oxidized in the presence of atmospheric oxygen to catalytically active $Ce^{4+}$. Based on this and several other $Ce^{4+}$ based techniques, useful methods for site-selective manipulation of large genomes have been developed [120]. Oligodeoxyribonucleotides are also cleaved by redox-active metal ions, especially by their dinuclear complexes. These reactions proceed, however, by C–O rather than P–O bond scission, initiated by abstraction of hydrogen atom from either the sugar or base moiety [121]. Gold nanoparticles coated with $Zn^{2+}$-binding ligand, has been shown to cleave efficiently plasmid DNA [104] (Figure 6.20B). The mechanism evidently resembles that of metallonuclease enzymes: a plasmid phosphodiester linkage is anchored via $Zn^{2+}$-azacrown complex and the guanidinium group facilitates by transition state stabilization the attack of terminal serine residue on the phosphate group.

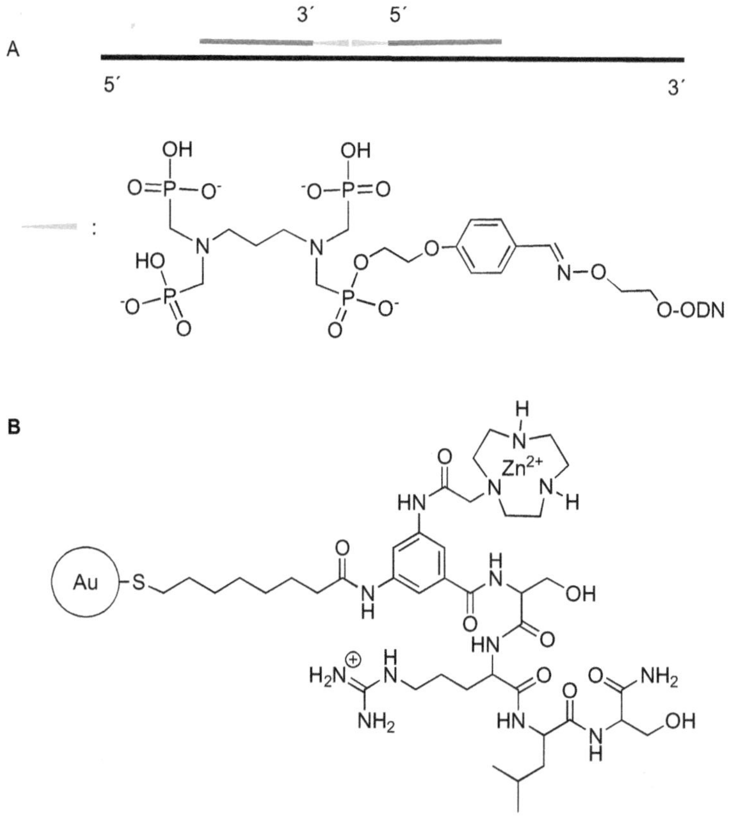

**Figure 6.20:** (A) Site-selective Ce$^{4+}$-promoted cleavage of DNA [119] and (B) gold nanoparticle cleaving agent shown to cleave DNA [104].

## Further reading

Bevilacqua PC, Harris ME, Piccirilli JA, Gaines C, Ganguly A, Kostenbader K, Ekesan S, York DM. An ontology for facilitating discussion of catalytic strategies of RNA-cleaving enzymes. ACS Chem Biol 2019, 14, 1068–1076.

Brown RS, Lu Z-L, Liu CT, Tsang WY, Edwards DR, Neverov AA. Dinuclear Zn(II) catalysts as biomimics of RNA and DNA phosphoryl transfer enzymes: Changing the medium from water to alcohol provides enzyme-like rate enhancements. J Phys Org Chem 2010, 23, 1–15.

Emilsson GM, Nakamura S, Roth A, Breaker RR. Ribozyme speed limits. RNA 2003, 9, 907–918.

N-S L, Frederiksen JK, Piccirilli JA. Synthesis, properties, and applications of oligonucleotides containing an RNA dinucleotide phosphorothiolate Linkage. Acc Chem Res 2011, 44, 1257–1269.

Lönnberg H. Structural modifications as tools in mechanistic studies of the cleavage of RNA phosphodiester linkages. Chem Rec 2022, 22, e202200141 (17 pages).

Mancin F, Prins LJ, Pengo P, Pasquato L, Tecilla P, Scrimin P. Hydrolytic metallo-nanozymes: From micelles and vesicles to gold nanoparticles. Molecules 2016, 21, 1014–1032.

Mikkola S, Lönnberg T, Lönnberg H. Phosphodiester models for cleavage of nucleic acids. Beilstein J Org Chem 2018, 14, 803–837.

Morrow JR. Artificial ribonucleases. Adv Inorg Biochem 1994, 9, 41–74.

Niittymäki T, Lönnberg H. Artificial ribonucleases. Org Biomol Chem 2006, 4, 15–25.

Oivanen M, Kuusela S, Lönnberg H. Kinetics and mechanisms for the cleavage and isomerization of the phosphodiester bonds of RNA by Brønsted acids and bases. Chem Rev 1998, 98, 961–990.

Perrault DM, Anslyn EV. Unifying the current data on the mechanism of cleavage–transesterification of RNA. Angew Chem Int Ed 1997, 36, 432–450.

# References

[1]     Kaukinen U, Lyytikäinen S, Mikkola S, Lönnberg H. The reactivity of phosphodiester bonds within linear single-stranded oligoribonucleotides is strongly dependent on the base sequence. Nucleic Acids Res 2002, 30, 468–474.

[2]     Oivanen M, Kuusela S, Lönnberg H. Kinetics and mechanisms for the cleavage and isomerization of the phosphodiester bonds of RNA by Brønsted acids and bases. Chem Rev 1998, 98, 961–990.

[3]     Westheimer FH. Pseudo-rotation in the hydrolysis of phosphate esters. Acc Chem Res 1968, 1, 70–78.

[4]     Berry RS. Correlation of rates of intramolecular tunneling processes, with application to some group V compounds. J Chem Phys 1960, 32, 933–938.

[5]     Lopez X, Schaefer M, Dejaegere A, Karplus M. Theoretical evaluation of $pK_a$ in phosphoranes: Implications for phosphate ester hydrolysis. J Am Chem Soc 2002, 124, 5010–5018.

[6]     Järvinen P, Oivanen M, Lönnberg H. Interconversion and phosphoester hydrolysis of 2',5'- and 3',5'-dinucleoside monophosphates: Kinetics and mechanisms. J Org Chem 1991, 56, 5396–5401.

[7]     Lopez CS, Faza ON, Gregersen BA, Lopez X, de Lera AR, York DM. Pseudorotation of natural and chemically modified biological phosphoranes: Implications in RNA catalysis. ChemPhysChem 2004, 5, 1045–1049.

[8]     Yang Y, Cui Q. Does water play an important role in phosphoryl transfer reactions? Insight from theoretical study of a model reaction in water and tert-butanol. J Phys Chem B 2009, 113, 4930–4939.

[9]     Harris ME, Dai Q, Gu H, Kellerman DL, Piccirilli JA, Anderson VE. Kinetic isotope effects for RNA cleavage by 2'-O-transphosphorylation: Nucleophilic activation by specific base. J Am Chem Soc 2010, 132, 11613–11621.

[10]    Lönnberg H, Strömberg R, Williams A. Compelling evidence for a stepwise mechanism of the alkaline cyclisation of uridine 3'-phosphate esters. Org Biol Chem 2004, 2, 2165–2167.

[11]    Ye J-D, Li N-S, Dai Q, Piccirilli JA. The mechanism of RNA strand scission: An experimental measure of the Brønsted coefficient, $\beta_{nuc}$. Angew Chem Int Ed 2007, 46, 3714–3717.

[12]    Huang M, York DM. Linear free energy relationships in RNA transesterification: Theoretical models to aid experimental interpretations. Phys Chem Chem Phys 2014, 16, 15846–15855.

[13]    Breslow R, Dong SD, Webb Y, Xu R. Further studies on the buffer-catalyzed cleavage and isomerization of uridyluridine. Medium and ionic strength effects on catalysis by morpholine, imidazole, and acetate buffers help clarify the mechanisms involved and their relationship to the mechanism used by the enzyme ribonuclease and by a ribonuclease mimic. J Am Chem Soc 1996, 118, 6588–6600.

[14]    Anslyn E, Breslow R. On the mechanism of catalysis by ribonuclease: Cleavage and isomerization of the dinucleotide UpU catalyzed by imidazole buffers. J Am Chem Soc 1989, 111, 4473–4482.

[15]  Beckmann C, Kirby AJ, Kuusela S, Tickle DC. Mechanisms of catalysis by imidazole buffers of the hydrolysis and isomerisation of RNA models. J Che Soc Perkin Trans 1998, 2, 0, 573–581.

[16]  Salvio R. The guanidinium unit in the catalysis of phosphoryl transfer reactions: From molecular spacers to nanostructured supports. Chem Eur J 2015, 21, 10960–10971.

[17]  Perreault DM, Cabell LA, Anslyn EV. Using guanidinium groups for the recognition of RNA and as catalysts for the hydrolysis of RNA. Bioorg Med Chem 1997, 5, 1209–1220.

[18]  Lönnberg TA, Helkearo M, Jancso A, Gajda T. Mimics of small ribozymes utilizing a supramolecular scaffold. Dalton Trans 2012, 41, 3328–3338.

[19]  Salvio R, Cacciapaglia R, Mandolini L, Sansone F, Casnati A. Diguanidinocalix[4]arenes as effective and selective catalysts of the cleavage of diribonucleoside monophosphates. RSC Advances 2014, 4, 34412–34416.

[20]  Lisi D, Vezzoni CA, Casnati A, Sansone F, Salvio R. Intra- and intermolecular cooperativity in the catalytic activity of phosphodiester cleavage by self-assembled systems based on guanidinylated calix[4]arenes. Chem Eur J 2023, 29, e202203213 (8 pages).

[21]  Amirloo B, Staroseletz Y, Yousaf S, Clarke DJ, Brown T, Aojula H, Zenkova MA, Bichenkova EV. "Bind, cleave and leave": Multiple turnover catalysis of RNA cleavage by bulge–loop inducing supramolecular conjugates. Nucleic Acids Res 2022, 50, 651–673.

[22]  Scheffer U, Strick A, Ludwig V, Peter S, Kalden E, Göbel MW. Metal-free catalysts for the hydrolysis of RNA derived from guanidines, 2-aminopyridines, and 2-aminobenzimidazoles. J Am Chem Soc 2005, 127, 2211–2217.

[23]  Gnaccarini C, Sascha P, Scheffer U, Vonhoff S, Klussmann S, Göbel MW. Site-specific cleavage of RNA by a metal-free artificial nuclease attached to antisense oligonucleotides. J Am Chem Soc 2006, 128, 8063–8067.

[24]  Verheijen JC, Deiman BALM, Yeheskiely E, Van der Marel GA, Van Boom JH. Efficient hydrolysis of RNA by a PNA – Diethylenetriamine adduct. Angew Chem Int Ed 2000, 39, 369–372.

[25]  Kuznetsova IL, Zenkova MA, Gross HJ, Vlassov VV. Enhanced RNA cleavage within bulge-loops by an artificial ribonuclease. Nucleic Acids Res 2005, 33, 1201–1212.

[26]  Kosonen M, Lönnberg H. General and specific acid/base catalysis of the hydrolysis and interconversion of ribonucleoside 2′- and 3′-phosphotriesters: Kinetics and mechanism of the reactions of 5′-O-pivaloyluridine 2′- and 3′-dimethylphosphates. J Chem Soc Perkin Trans 1995, 2, 0, 1203–1209.

[27]  Scheller KH, Hofstetter F, Mitchell PR, Prijs P, Sigel H. Macrochelate formation in monomeric metal ion complexes of nucleoside 5′-triphosphates and the promotion of stacking by metal ions. Comparison of the self-association of purine and pyrimidine 5′-triphosphates using proton nuclear magnetic resonance. J Am Chem Soc 1981, 103, 247–260.

[28]  Mitchell PR, Sigel H. A proton nuclear-magnetic-resonance study of self-stacking in purine and pyrimidine nucleosides and nucleotides. Eur J Biochem 1978, 88, 149–154.

[29]  Lee C-H, Ezra FS, Kondo NS, Sarma RH, Danyluk SS. Conformational properties of dinucleoside monophosphates in solution: Dipurines and dipyrimidines. Biochemistry 1976, 15, 3627–3639.

[30]  Ezra FS, Lee C-H, Kondo NS, Danyluk SS, Sarma RH. Conformational properties of purine-pyrimidine and pyrimidine-purine dinucleoside monophosphates. Biochemistry 1977, 16, 1977–1987.

[31]  Jafilan S, Klein L, Hyun C, Florián J. Intramolecular base stacking of dinucleoside monophosphate anions in aqueous solution. J Phys Chem B 2012, 116, 3613–3618.

[32]  Norberg J, Nilsson L. Stacking free energy profiles for all 16 natural ribodinucleoside monophosphates in aqueous solution. J Am Chem Soc 1995, 117, 10832–10840.

[33]  Kuusela S, Lönnberg H. Hydrolysis and isomerization of internucleosidic phosphodiester bonds of polyuridylic acid: Kinetics and mechanism. J Chem Soc Perkin Trans 1994, 2, 0, 2109–2113.

[34]  Bibillo A, Figlerowicz M, Ziomek K, Kierzek R. The nonenzymatic hydrolysis of oligoribonucleotides. VII. Structural elements affecting hydrolysis. Nucleosides Nucleotides Nucleic Acids 2000, 19, 977–994.

[35]  Bibillo A, Figlerowicz M, Kierzek R. The non-enzymatic hydrolysis of oligoribonucleotides VI. The role of biogenic polyamines. Nucleic Acids Res 1999, 27, 3931–3937.

[36]  Vlassov VV, Zuber G, Felden B, Behr J-P, Giege R. Cleavage of tRNA with imidazole and spermine imidazole constructs: A new approach for probing RNA structure. Nucleic Acids Res 1995, 23, 3161–3167.

[37]  Kaukinen U, Lyytikäinen S, Mikkola S, Lönnberg H. The reactivity of phosphodiester bonds within linear single-stranded oligoribonucleotides is strongly dependent on the base sequence. Nucleic Acids Res 2002, 30, 468–474.

[38]  Hobza P, Sponer J. Structure, Energetics, and Dynamics of the Nucleic Acid Base Pairs: Nonempirical Ab Initio Calculations. Chem Rev 1999, 99, 3247–3276.

[39]  Kaukinen U, Venäläinen T, Lönnberg H, Peräkylä M. The base sequence dependent flexibility of lonear single-stranded oligoribonucleotides correlates with the reactivity of the phosphodiester bond. Org Biomol Chem 2003, 1, 2439–2447.

[40]  Kaukinen U, Lönnberg H, Peräkylä M. Stabilisation of the transition state phosphodiester bond cleavage within linear single-stranded oligoribonucleotides. Org Biomol Chem 2004, 2, 66–73.

[41]  Usher DA, McHale AH. Hydrolytic stability of helical RNA; a selective advantage for the natural 3',5'-bonds. Proc Natl Acad Sci 1976, 73, 1149–1153.

[42]  Zagorowska I, Mikkola S, Lönnberg H. Hydrolysis of phosphodiester bonds within hairpin loops in buffer solutions: The effect of secondary structure on the inherent reactivity of RNA phosphodiester bonds. Helv Chim Acta 1999, 82, 2105–2111.

[43]  Kaukinen U, Bielecki L, Mikkola S, Adamiak RW, Lönnberg H. The cleavage of phosphodiester bonds within small RNA bulges in the presence and absence of metal ion catalysts. J Chem Soc Perkin Trans 2 2001, 0, 1024–1031.

[44]  Frederiksen JK, Piccirilli JA. Identification of catalytic metal ion ligands in ribozymes. Methods 2009, 49, 148–166.

[45]  Li N-S, Frederiksen JK, Piccirilli JA. Synthesis, properties, and applications of oligonucleotides containing an RNA dinucleotide phosphorothiolate Linkage. Acc Chem Res 2011, 44, 1257–1269.

[46]  Ora M, Lönnberg T, Lönnberg H. From Nucleic Acid Sequences to Molecular Medicine. Erdman VA, Barciszewski J. ed., Berlin Heidelberg: Springer Verlag, 2012, 47–65.

[47]  Burgers PMJ, Eckstein F. Diastereomers of 5'-O-adenosyl 3'-O-uridyl phosphorothioate: Chemical synthesis and enzymatic properties. Biochemistry 1979, 18, 592–596.

[48]  Almer H, Strömberg R. Intramolecular transesterification in thiophosphate-analogues of an RNA-dimer. Tetrahedron Lett 1991, 32, 3723–3726.

[49]  Kosonen M, Yousefi-Salakdeh E, Strömberg R, Lönnberg H. Mutual isomerization of uridine 2'- and 3'-alkylphosphates and cleavage to a 2',3'-cyclic phosphate: The effect of the alkyl group on the hydronium- and hydroxide-ion-catalyzed reactions. J Chem Soc Perkin Trans 1997, 2, 0, 2661–2666.

[50]  Ora M, Hanski A. Stepwise mechanism of hydroxide Ion catalyzed cyclization of uridine 3'-thiophosphates. Helv Chim Acta 2011, 94, 1563–1574.

[51]  Liu Y, Gregersen BA, Hengge A, York DM. Transesterification thio effects of phosphate diesters: Free energy barriers and kinetic and equilibrium isotope effects from density-functional theory. Biochemistry 2006, 45, 10043–10053.

[52]  Ora M, Järvi J, Oivanen M, Lönnberg H. Hydrolytic reactions of the phosphorodithioate analogue of uridylyl(3,5')uridine: Kinetics and mechanisms for the cleavage, desulfurization, and isomerization of the internucleosidic linkage. J Org Chem 2000, 65, 2651–2657.

[53]  Oivanen M, Ora M, Almer H, Strömbrg R, Lönnberg H. Hydrolytic reactions of the diastereomeric phosphoromonothioate analogs of uridylyl(3',5')uridine: Kinetics and mechanism for

desulfurization, phosphoester hydrolysis and transesterification to the 2′,5′-isomers. J Org Chem 1995, 60, 5620–5627.

[54] Ora M, Oivanen M, Lönnberg H. Hydrolysis and desulfurization of the diastereomeric phosphoromonothioate analogs of uridine 2′,3′-cyclic monophosphate. J Org Chem 1996, 61, 3951–3955.

[55] Liu X, Reese CB. Uridylyl-(3′,5′)-(5′-thiouridine). An exceptionally base-labile di-ribonucleoside phosphate analogue. Tetrahedron Lett 1995, 36, 3413–3416.

[56] Thomson JB, Patel BK, Jimenez V, Eckart K, Eckstein F. Synthesis and properties of diuridine phosphate analogues containing thio and amino modifications. J Org Chem 1996, 61, 6273–6281.

[57] Weinstein LB, Earnshaw DJ, Cosstick R, Cech TR. Synthesis and characterization of an RNA dinucleotide containing a 3′-S-phosphorothiolate linkage. J Am Chem Soc 1996, 118, 10341–10350.

[58] Elzagheid MI, Oivanen M, Klika KD, Jones BCNM, Cosstick R, Lönnberg H. Hydrolytic reactions of 3′-deoxy-3′-thioinosylyl-(3′, 5′)-uridine; an RNA dinucleotide containing a 3′-S-phosphorothiolate linkage. Nucleosides Nucleotides 1999, 18, 2093–2108.

[59] Gregersen BA, Lopez X, York DM. Hydrid QM/MM study of thio effects in transphosphorylation reactions: The role of solvation. J Am Chem Soc 2004, 126, 7504–7513.

[60] Matray TJ, Gryaznov SM. Synthesis and properties of RNA analogs – Oligoribonucleotide N3′→P5′ phosphoramidates. Nucleic Acids Res 1999, 27, 3976–3985.

[61] Ora M, Mattila K, Lönnberg T, Oivanen M, Lönnberg H. Hydrolytic reactions of riribonucleoside 3′,5′-(3′-N-phosphoramidates): Kinetics and mechanisms for the P–O and P–N bond Cleavage of 3′-amino-3′-deoxyuridylyl-3′,5′-uridine. J Am Chem Soc 2002, 124, 14364–14372.

[62] Oivanen M, Mikhailov SN, Padyukova NSh, Lönnberg H. Kinetics of mutual isomerization of the phosphonate analogs of dinucleoside 2′,5′- and 3′,5′-monophosphates in aqueous solution. J Org Chem 1993, 58, 1617–1619.

[63] Dimroth K, Jaenicke L, Heizel D. Die spaltung der pentose-nuckeinsäure der hefe mit bleihydroxyd. Liebigs Ann Chem 1950, 566, 206–210.

[64] Kuusela S, Lönnberg H. Metal ions that promote the hydrolysis of nucleoside phosphoesters do not enhance intramolecular phosphate migration. J Phys Org Chem 1993, 6, 347–356.

[65] Morrow JR. Artificial ribonucleases. Adv Inorg Biochem 1994, 9, 41–74.

[66] Kuusela S, Lönnberg H. Effect of metal ions on the hydrolytic reactions of nucleosides and their phosphoesters. Metal Ions Biol Systems 1996, 32, 271–300.

[67] Matsumura K, Komiyama M. Enormously fast RNA hydrolysis by lanthanide(III) ions under physiological conditions: Eminent candidates for novel tools of biotechnology. J Biochem 1997, 122, 387–394.

[68] Chin KOA, Morrow JR. RNA cleavage and phosphate diester transesterification by encapsulated lanthanide ions: Traversing the lanthanide series with lanthanum(III), europium(III), and lutetium(III) complexes of 1,4,7,10-tetrakis(2-hydroxyalkyl)-1,4,7,10-tetraazacyclododecane. Inorg Chem 1994, 33, 5036–5041.

[69] Feng G, Mareque-Rivas JC, Williams NH. Comparing a mononuclear Zn(II) complex with hydrogen bond donors with a dinuclear Zn(II) complex for catalysing phosphate ester cleavage. Chem Commun 2006, 0, 1845–1847.

[70] Iranzo O, Elmer T, Richard JP, Morrow JR. Cooperativity between metal Ions in the cleavage of phosphate diesters and RNA by dinuclear Zn(II) catalysts. Inorg Chem 2003, 42, 7737–7746.

[71] Feng G, Natale D, Prabaharan R, Mareque-Rivas JC, Williams NH. Efficient phosphodiester binding and cleavage by a ZnII complex combining hydrogen-bonding interactions and double Lewis acid activation. Angew Chem Int Ed 2006, 45, 7056–7059.

[72] Feng G, Mareque-Rivas JC, de Rosales RTM, Williams NH. A highly reactive mononuclear Zn(II) complex for phosphodiester cleavage. J Am Chem Soc 2005, 127, 13470–13471.

[73] O'Donoghue A, Pyun SY, Yang MI, Morrow JR, Richard JP. Substrate specificity of an active dinuclear Zn(II) catalyst for cleavage of RNA analogues and a dinucleoside. J Am Chem Soc 2006, 128, 1615–1621.

[74] Linjalahti H, Feng G, Mareque-Rivas JC, Mikkola S, Williams NH. Cleavage and isomerization of UpU promoted by dinuclear metal ion Complexes. J Am Chem Soc 2008, 130, 4232–4233.

[75] Korhonen H, Mikkola S, Williams NH. The mechanism of cleavage and isomerisation of RNA promoted by an efficient dinuclear Zn2+ complex. Chem Eur J 2012, 18, 659–670.

[76] Ora M, Peltomäki M, Oivanen M, Lönnberg H. Metal-ion-promoted cleavage, isomerization, and desulfurization of the diastereomeric phosphoromonothioate analogues of uridylyl(3′,5′)uridine. J Org Chem 1998, 63, 2939–2947.

[77] Brown RS, Lu Z-L, Liu CT, Tsang WY, Edwards DR, Neverov AA. Dinuclear Zn(II) catalysts as biomimics of RNA and DNA phosphoryl transfer enzymes: Changing the medium from water to alcohol provides enzyme-like rate enhancements. J Phys Org Chem 2010, 23, 1–15.

[78] Fanning AM, Plush SE, Gunnlaugsson T. Tri- and tetra-substituted cyclen based lanthanide(III) ion complexes as ribonuclease mimics: A study into the effect of log $K_a$, hydration and hydrophobicity on phosphodiester hydrolysis of the RNA-model 2-hydroxypropyl-4-nitrophenyl phosphate (HPNP). Org Biomol Chem 2015, 13, 5804–5816.

[79] Bonfa L, Gatos M, Mancin F, Tecilla P, Tonellato U. The ligand effect on the hydrolytic reactivity of Zn(II) complexes toward phosphate diesters. Inorg Chem 2003, 42, 3943–3949.

[80] Molenveld P, Engbersen JFJ, Kooijman H, Spek AL, Reinhoudt DN. Efficient catalytic phosphate diester cleavage by the synergetic action of two Cu(II) centers in a dinuclear cis-diaqua Cu(II) calix [4]arene enzyme model. J Am Chem Soc 1998, 120, 6726–6737.

[81] Mikkola S, Stenman E, Nurmi K, Yousefi-Salakdeh E, Strömberg R, Lönnberg H. The mechanism of the metal ion promoted cleavage of RNA phosphodiester bonds involves a general acid catalysis by the metal aquo ion on the departure of the leaving group. J Chem Soc, Perkin Trans 1999, 2, 0, 1619–1625.

[82] Zhang S, Gu H, Chen H, Strong E, Ollie EW, Kellerman D, Liang D, Miyagi M, Anderson VE, Piccirilli JA, York DM, Harris ME. Isotope effect analyses provide evidence for an altered transition state for RNA 2′-O-transphosphorylation catalyzed by $Zn^{2+}$. Chem Commun 2016, 0, 4462–4465.

[83] Butzow JJ, Eichhorn GL. Interaction of metal ions with nucleic acids and related compounds. XVII. On the mechanism of degradation of polyribonucleotides and oligoribinucleotides by Zn(II) ions. Biochemistry 1971, 11, 2019–2027.

[84] Kuusela S, Azhayv A, Guzaev A, Lönnberg H. The effect the 3′-terminal monophosphate group on the metal-ion-promoted hydrolysis of the phosphodiester bonds of short oligonucleotides. J Chem Soc Perkin Trans 1995, 2, 0, 1197–1202.

[85] Kuusela S, Guzaev A, Lönnberg H. Acceleration of the Zn2+ promoted phosphodiester hydrolysis of oligoribonucleotides by the 3′-terminal monophosphate group: Intrastrand participation over several nucleoside units. J Chem Soc Perkin Trans 1996, 2, 0, 1895–1899.

[86] Kuusela S, Lönnberg H. Hydrolysis and isomerization of the internucleosidic phosphodiester bonds of polyuridylic acid: Kinetics and mechanism. J Chem Soc Perkin Trans 1994, 2, 0, 2109–2113.

[87] Kaukinen U, Bielecki L, Mikkola S, Adamiak RW, Lönnberg H. The cleavage of phosphodiester bonds within small RNA bulges in the presence and absence of metal ion catalysts. J Chem Soc Perkin Trans 2001, 2, 0, 1024–1031.

[88] Zagorowska I, Kuusela S, Lönnberg H. Metal ion dependent hydrolysis of RNA phosphodiester bonds within hairpin loops. A comparative kinetic study on chimeric ribo/2′-O-methylribo oligoribonucleotides. Nucleic Acids Res 1998, 26, 3392339–6.

[89] Kimura E, Shiota T, Koike M, Shiro M, Kodama M. A zinc(II) complex of 1,5,9-triazacyclododecane ([12]aneN3) as a model for carbonic anhydrase. J Am Chem Soc 1990, 112, 5805–5811.

[90]  Wang Q, Lönnberg H. Simultaneous interaction with base and phosphate moieties modulates the phosphodiester dleavage of dinucleoside 3‘,5‘-monophosphates by dinuclear $Zn^{2+}$ complexes of di(azacrown) ligands. J Am Chem Soc 2006, 128, 10716–10728.

[91]  Wang Q, Leino E, Jancsó A, Szilágyi I, Gajda T, Hietamäki E, Lönnberg H. Zn2+ complexes of di- and tri-nucleating azacrown ligands as base moiety selective cleaving agents of RNA 3′,5′-phosphodiester bonds: Binding to guanine base. ChemBioChem 2008, 9, 1739–1748.

[92]  Laine M, Ketomäki K, Poijärvi-Virta P, Lönnberg H. Base moiety selectivity in cleavage of short oligoribonucleotides by di- and tri-nuclear Zn(II) complexes of azacrown-derived ligands. Org Biomol Chem 2009, 7, 2780–2787.

[93]  Whitney A, Gavory G, Balasubramanian S. Site-specific cleavage of human telomerase RNA using PNA-neocuproine·Zn(II) derivatives. Chem Commun 2003, 0, 36–37.

[94]  Magda D, Crofts S, Lin A, Miles D, Wright M, Sessler JL. Synthesis and kinetic properties of ribozyme analogues prepared using phosphoramidite derivatives of dysprosium(III) texaphyrin. J Am Chem Soc 1997, 119, 2293–2294.

[95]  Hall J, Hüsken D, Pieles U, Moser HE, Häner R. Efficient sequence-specific cleavage of RNA using novel europium complexes conjugated to oligonucleotides. Chem Biol 1994, 1, 185–190.

[96]  Magda D, Miller A, Sessler JL, Iverson FL. Site-ppecific hydrolysis of RNA by europium(III) texaphyrin conjugated to a synthetic oligodeoxyribonucleotide. J Am Chem Soc 1994, 116, 7439–7440.

[97]  Hall J, Hüsken D, Häner R. Towards artificial ribonucleases: The sequence-specific cleavage of RNA in a duplex. Nucleic Acids Res 1996, 24, 3522–3526.

[98]  Sakamoto S, Tamura T, Furukawa T, Komatsu Y, Ohtsuka E, Kitamura M, Inoue H. Highly efficient catalytic RNA cleavage by the cooperative action of two Cu(II) complexes embodied within an antisense oligonucleotide. Nucleic Acids Res 2003, 31, 1416–1425.

[99]  Putnam WC, Daniher AT, Trawick BN, Bashkin JK. Efficient new ribozyme mimics: Direct mapping of molecular design principles from small molecules to macromolecular, biomimetic catalysts. Nucleic Acids Res 2001, 29, 2199–2204.

[100] Putnam WC, Bashkin JK. De novo synthesis of artificial ribonucleases with benign metal catalysts. Chem Commun 2000, 0, 767–768.

[101] Murtola M, Wenska M, Strömberg R. PNAzymes that are artificial RNA restriction enzymes. J Am Chem Soc 2010, 132, 8984–8990.

[102] Luige O, Murtola M, Ghidini A, Strömberg R. Further probing of $Cu^{2+}$-dependent PNAzymes acting as artificial RNA restriction enzymes. Molecules 2019, 24, 672 (12 pages.

[103] Luige O, Bose PP, Stulz R, Steunenberg P, Brun O, Andersson S, Murtola M, Strömberg R. Zn2+-Dependent peptide nucleic acid-based artificial ribonucleases with unprecedented efficiency and specificity. Chem Commun 2021, 57, 10911–10914.

[104] Czescik J, Zamolo S, Darpre T, Rigo R, Sissi C, Pecina A, Riccardi L, De Vivo M, Mancin F, Scrimin P. A gold nanoparticle nanonuclease relying on a Zn(II) mononuclear complex. Angew Chem Int Ed 2021, 60, 1423–1432.

[105] Chen JL-Y, Pezzato C, Scrimin P, Prins LJ. Chiral nanozymes–gold nanoparticle-based transphosphorylation catalysts capable of enantiomeric discrimination. Chem Eur J 2016, 22, 7028–7032.

[106] Diez-Castellnou M, Mancin F, Scrimin P. Efficient phosphodiester cleaving nanozymes resulting from multivalency and local medium polarity control. J Am Chem Soc 2014, 136, 1158–1161.

[107] Czescik J, Mancin F, Strömberg R, Scrimin P. The mechanism of cleavage of RNA phosphodiesters by a gold nanoparticle nanozyme. Chem Eur J 2021, 27, 8143–8148.

[108] Schroeder GK, Lad C, Wyman P, Williams NH, Wolfenden R. The time required for water attack at the phosphorus atom of simple phosphodiesters and of DNA. Proc Natl Sci Acad 2006, 103, 4052–4055.

[109] Wolfenden R, Ridgway C, Young G. Spontaneous hydrolysis of ionized phosphate monoesters and diesters and the proficiencies of phosphatases and phosphodiesterases as catalysts. J Am Chem Soc 1998, 120, 833–834.

[110] Abe YS, Sasaki S. DNA cleavage at the AP site via β-elimination mediated by the AP site-binding ligands. Bioorg Med Chem 2016, 24, 910–914.

[111] Kamerlin SCL, Williams NH, Warshel A. Dineopentyl phosphate hydrolysis: Evidence for stepwise water attack. J Org Chem 2008, 73, 6960–6969.

[112] Komiyama M, Shiiba T, Kodaama T, Takeda J, Sumaoka J, Yashiro M. DNA hydrolysis by cerium(IV) does not involve either molecular oxygen or hydrogen peroxide. Chem Lett 1994, 23, 1025–1028.

[113] Takasaki BK, Chin J. Cleavage of the phosphate diester backbone of DNA with cerium(III) and molecular oxygen. J Am Chem Soc 1994, 116, 1121–1122.

[114] Ott R, Krämer R. Rapid phosphodiester hydrolysis by zirconium(IV). Angew Chem Int Ed 1998, 37, 1597–1600.

[115] Ihara T, Shimura H, Ohmori K, Tsuji H, Takeuchi J, Takagi M. Hydrolysis of nucleotides using actinoid metal ion. Chem Lett 1996, 25, 687–688.

[116] Dixon NE, Geue RJ, Lambert JN, Moghaddas S, Pearce DA, Sargeson AM. DNA hydrolysis by stable metal complexes. Chem Commun 1996, 0, 1287–1288.

[117] Komiyama M. Design of highly active Ce(IV) catalysts for DNA hydrolysis and their applications. Chem Lett 2016, 45, 1347–1355.

[118] Takeda N, Imai T, Irisawa M, Sumaoka J, Yashiro M, Shikegawa H, Komiyama M. Unprecedentedly Fast DNA Hydrolysis by the Synergism of the Cerium(IV)-Praseodymium(III) and the cerium(IV)-neodymium(III) Combinations. Chem Lett 1996, 25, 599–600.

[119] Lönnberg T, Aiba Y, Hamano Y, Miyajima Y, Sumaoka J, Komiyama M. Oxidation of an oligonucleotide-bound Ce[III]/multiphosphonate complex for site-selective DNA scission. Chem Eur J 2010, 16, 855–859.

[120] Shigi N, Sumaoka J, Komiyama M. Applications of PNA-based artificial restriction DNA cutters. Molecules 2017, 22, 1586–1600.

[121] Anjomshoa M, Amirheidari B. Nuclease-like metalloscissors: Biomimetic candidates for cancer and bacterial and viral infections therapy. Coord Chem Rev 2022, 458, 214417 (41 pages).

# 7 Oligonucleotide conjugates

## 7.1 Introduction

Oligonucleotides may by conjugation be provided with properties that they do not inherently have. Such conjugates are increasingly used as tools in cell biology research,
diagnostics and drug discovery. The list of conjugate groups used for various purposes
is long. Reporter groups, such as fluorescent dyes and spin-labels, allow sensitive detection of oligonucleotides. Chemically reactive groups may result in cross-linking with nucleic acid binding proteins or complementary nucleic acid sequences, or they may
cleave the complementary strand sequence-selectively. Intercalators stabilize double
helices and groove binders recognize them. Hydrophobic groups facilitate cellular uptake. Metal-ion-binding ligands allow transport of metal ions, which is essential for imaging techniques and even utilized in chemotherapy. Conjugations to groups that are
able to recognize a certain cell type by binding to a receptor allow targeting and facilitate internalization of oligonucleotides. For this purpose, lipid, carbohydrate, peptide,
aptamer and small molecule conjugates are used. Even the intracellular traffic of oligonucleotides may be tuned by conjugation. Conjugation is also used to prolong the half-
life of oligonucleotides in plasma.

Conjugate groups can be attached to oligonucleotides during solid-supported oligonucleotide synthesis or post-synthetically in solution. The advantage of the solid-
supported conjugation is a less laborious purification. The conjugate group is usually
used in excess to warrant quantitative conjugation. Removal of the excess on solid
support is easy. The problems arising from different solubility of oligonucleotide and
the conjugate group are also easier to handle on a solid support. The disadvantage is
that the conventional strategy of oligonucleotide assembly most likely needs revisions.
The advantage of solution phase conjugation, in turn, is independence of the protecting group strategies applied to syntheses of oligonucleotide and the conjugate group.
Especially on preparing peptide or oligosaccharide conjugates, this may well over-
compensate the extra work caused by several technically demanding purifications.

## 7.2 Reactions used for conjugation

Largely same reactions are used for conjugation in solution phase and on a solid support. Figure 7.1 records a number of frequently applied reactions. These reactions are
so selective that an unprotected oligonucleotide can be used for conjugation in highly
polar solvents. The overwhelmingly most extensively used approaches are reactions
A and B: Cu(I)-catalyzed Huisgen's 2,3-dipolar cycloaddition between azide and alkyne
[1, 2] and its copper free version with conformational strained alkynes [3]. The source
of Cu(I) is usually a Cu(II) salt, almost invariably $CuSO_4$. The Cu(II) ion is reduced to

https://doi.org/10.1515/9783111325637-007

Cu(I) in situ with sodium ascorbate in the presence of tris[(1-benzyl-1 *H*-1,2,3-triazol-4-yl) methyl]amine. This auxiliary stabilizes the Cu(I) oxidation state but does not interfere in the catalytic action of Cu(I). Reactions C-M in Table 7.1 have also been frequently used for oligonucleotide conjugation [18,19], in particular before the invention of Cu(I)-catalyzed Huisgen's reaction in 2002. Reactions N and O, in turn, are rather recent. An interesting feature of reaction N is pH-responsive reversibility: under slightly acidic conditions, formation of the *N*-methoxyoxazolidine linker between the oligonucleotide

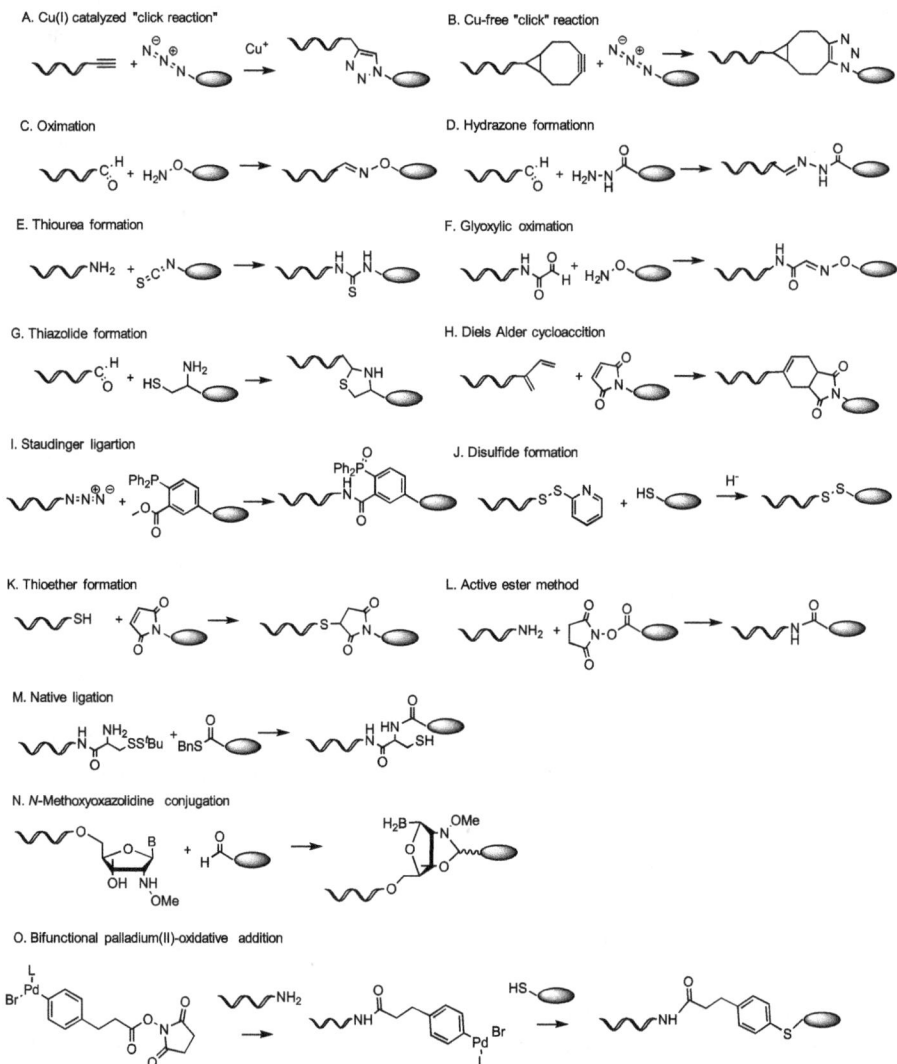

**Figure 7.1:** Reactions utilized for preparation of oligonucleotide conjugates: A [1, 2], B [3], C [4], D [5], E [6], F [7], G[8], H [9], I [10], J[11], K [12], L [13], M [14], N [15, 16] and O [17].

and conjugate group is reversible, but at neutral pH the linker is practically stable [15]. Reaction O, in turn, is the first example of an organometallic-promoted method that enables stepwise conjugation of another biomolecule to oligonucleotide [17].

## 7.3  Solid-supported synthesis of 5′-O-conjugates

Preparation of 5′-O-conjugates of oligonucleotides on a solid support is straightforward. The chain is normally assembled in the 3′- to 5′ -direction and the conjugate group may, hence, be introduced as a prefabricated phosphoramidite by an additional coupling cycle. Fluorescent dyes [20], metal chelates [21], photochemical crosslinking agents [22], bile acids [23], minor groove-binding agents [24], intercalators [25], lipids [26], and biotin [27] have been tethered in this manner to the 5′-terminus. Some examples of the phosphoramidite reagents employed are given in Figure 7.2. The conjugate group is preferably protected with base-labile groups to allow deprotection of the entire oligonucleotide conjugate in a single step. Exceptionally, hydroxy function, when available, is usually protected as a DMTr ether to enable monitoring of the coupling efficiency and/or to warrant stability during global deprotection by ammonolysis. On using phosphoramidite D in Figure 7.2, for example, an unprotected hydroxymethyl

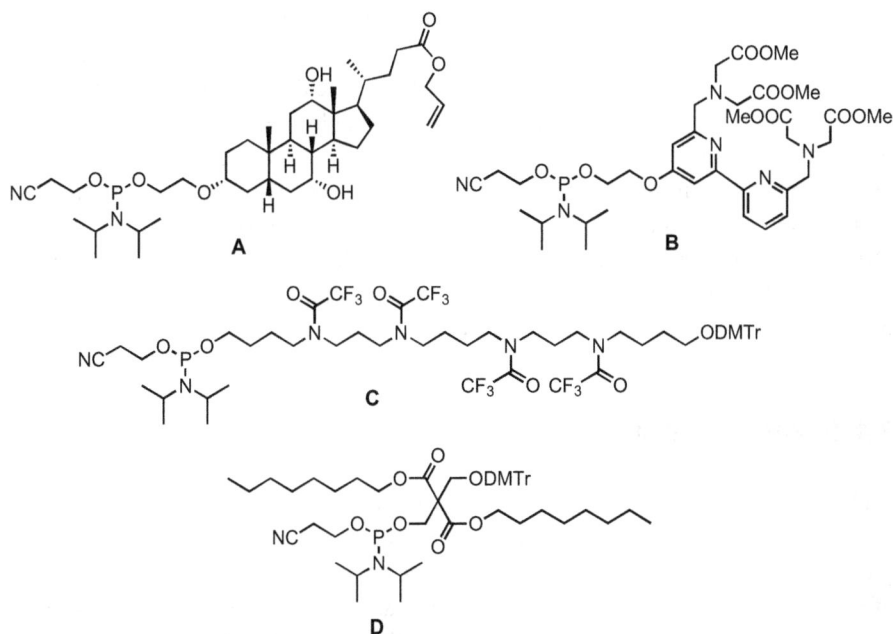

**Figure 7.2:** Phosphoramidite building blocks for introduction of bile acid (A) [23], lanthanide ion chelating (B) [28], oligoamine (C) [29] and fatty acid (D) [26] conjugate groups.

group would trigger under basic conditions breakdown of the conjugate group by retro-aldol condensation [26].

An alternative for the preparation of 5'-conjugates by coupling a prefabricated phosphoramidite is a two-step strategy: an appropriately protected linker is first coupled as a phosphoramidite to the 5'-terminus and, after removal of the protecting group on-support, conjugation is carried out. Oligonucleotide bearing a still protected linker may also be released in solution, deprotected and subjected to conjugation in solution. The most widely used linker is a 4-methoxytrityl-protected amino-linker (Figure 7.3A). The 4-methoxytrityl group can be removed under mildly acidic conditions without loss of base moiety protections or cleavage from the support. The conjugate group is then attached by one of the reactions described for amino-functionalized oligonucleotides in Figure 7.1. The amino tail can also be attached as an unprotected $\alpha,\omega$-diamine by activating the 5'-hydroxy group on-support with carbonyldiimidazole (Figure 7.3B). The alkyl carbamate linkage obtained withstands ammonolytic base moiety deprotection and cleavage from the support. On applying this technique, one should, however, bear in mind that treatment with amines may result in transamination of acyl-protected cytosine.

Instead of a nucleophilic amino linker, an electrophilic functionality may be introduced in the 5'-terminus and attacked by a nucleophile. Phosphitylation with a thioester containing phosphoramidite, for example, offers a simple method for generation an amine sensitive site at the 5'-terminus (Figure 7.3C) [30].

Azido and alkyne linkers have recently replaced the conventional amino linkers. They allow conjugation by the so-called click reaction, either by Cu(I) catalysis or in a strain-promoted manner (Figure 7.1A and B). The azido group is usually introduced by a two-step reaction via a bromide linker, owing to incompatibility of azido group and phosphoramidite chemistry (Figure 7.3D). An alternative approach is conversion of an amino linker to azido linker by treatment with fluorosulfuryl azide ($FSO_2N_3$) [31]. No metal ions are needed and unprotected amino functions of the nucleobases withstand the process. Linkers bearing a conformational strained alkyne are also used for metal-free conjugation (Figure 7.3E).

## 7.4 5'-Phosphorylation

5'-Phosphorylation of oligonucleotides is of special interest because both 5'-mono- and tri-phosphates serve as substrates in numerous enzymatic processes. Several phosphoramidite and H-phosphonate reagents have been developed that allow synthesis of 5'-monophosphates by conventional solid-phase strategies (Figure 7.4). The phosphoramidite or H-phosphonate building block that is used for the last coupling usually bears protecting groups removable by elimination or cyclization during the final ammonolysis. Oligonucleotides A [32] and B [33] in Figure 7.4, for example, expose the

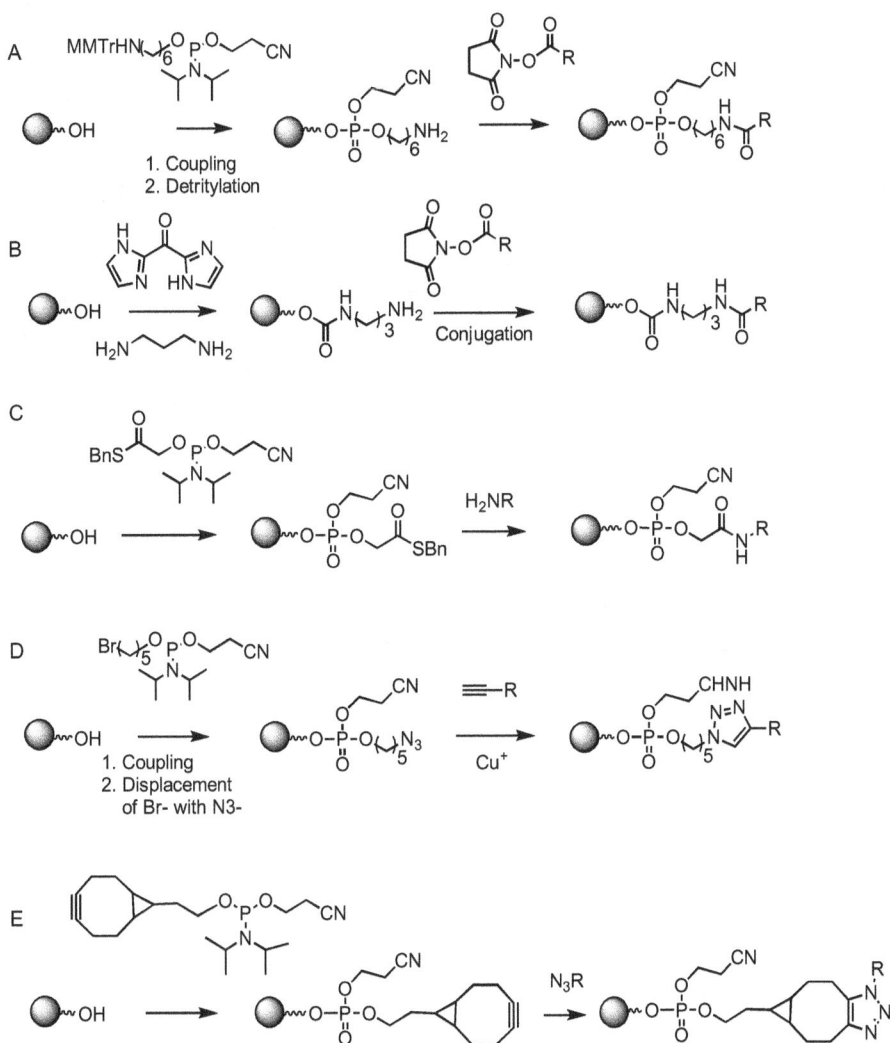

**Figure 7.3:** Examples of stepwise introduction of conjugate group to the 5′-terminus of a support-bound fully protected oligonucleotide.

terminal phosphate group by E2 elimination of the 2-cyanoethyl and 2-[(2-DMTrO-ethyl)sulfonyl]ethyl groups, respectively. With oligomer C, the thioester linkages are first cleaved by ammonia induced acyl substitution and the exposed 2-mercaptoethyl groups then depart by cyclization to ethylene sulfide [34]. With oligomer D, cleavage of the PS bond by attack of hydroxide ion on phosphorus precedes the release of ethylene sulfide [35], and with oligomer E, acid-catalyzed detritylation triggers cyclization to ethylene sulfide [36]. Detritylation of oligomer F, in turn, enables exposure of the 5′-monophosphate by a retro-aldol condensation mechanism during ammonolysis

[37]. Oligomer G differs from the others in the sense that the terminal phosphate is introduced as an H-phosphonate reagent, not as a phosphoramidite [38]. The disulfide bond is reductively cleaved, which leads to departure of ethylene sulfide. Removal by harsh ammonolysis without a reductive auxiliary is also possible. The terminal phosphate of oligomer H is introduced as phosphoramidite, the *tert*-butyl group is removed with dichloroacetic acid giving a 2-cyanoethyl H-phosphonate diester [39]. Oxidation to phosphate with iodine followed by ammonolytic removal of the 2-cyanoethyl group completes the reaction. The protecting group of oligomer I is photolabile [40]. Oligomer J is noteworthy. The terminal phosphate is introduced conventionally as phosphoramidite, and the protecting groups depart spontaneously during oxidation of the phosphite ester [41]. Oligomer K bears a thermolabile protection [42]. The 2-(methylthio)ethyl group can be removed in water (1 h, 60 °C). The remaining 2-cyanoethyl group is removed by ammonolysis.

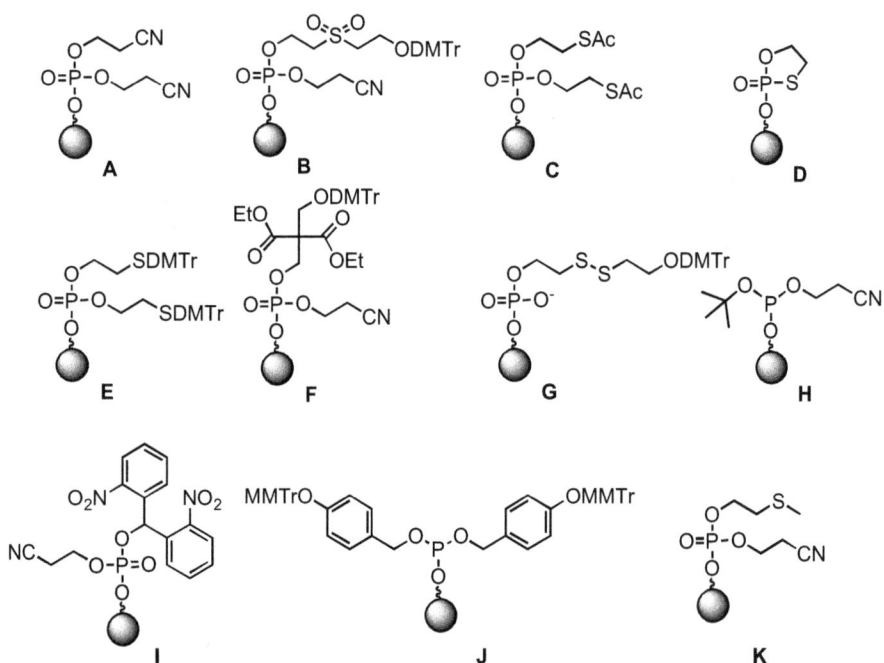

**Figure 7.4:** Conversion of 5'-O-functionalized solid-supported oligonucleotides to oligonucleotide 5'-monophosphates: A [32], B [33], C [34], D [35], E [36], F [37], G [38], H [39], I [40], J [41] and K [42].

Two approaches have been developed for the synthesis of oligonucleotide 5'-triphosphates. In both cases, the starting material is a support-bound fully protected oligonucleotide. After detritylation, 5'-OH is either H-phosphonylated with diphenyl *H*-phosphonate and hydrolyzed to 5'-*H*-phosphonate (Figure 7.5A) [43], or phosphitylated with 5-chlorosaligenyl phosphoramidite (Figure 6.5B) [44]. The H-phosphonate is then oxidized in the presence of imidazole to imidazole phosphoramidate, whereas the 5-chlorosaligenyl phosphite ester is oxidized to the corresponding phosphate ester. Attack of pyrophosphate ion then completes the triphosphate synthesis in both cases.

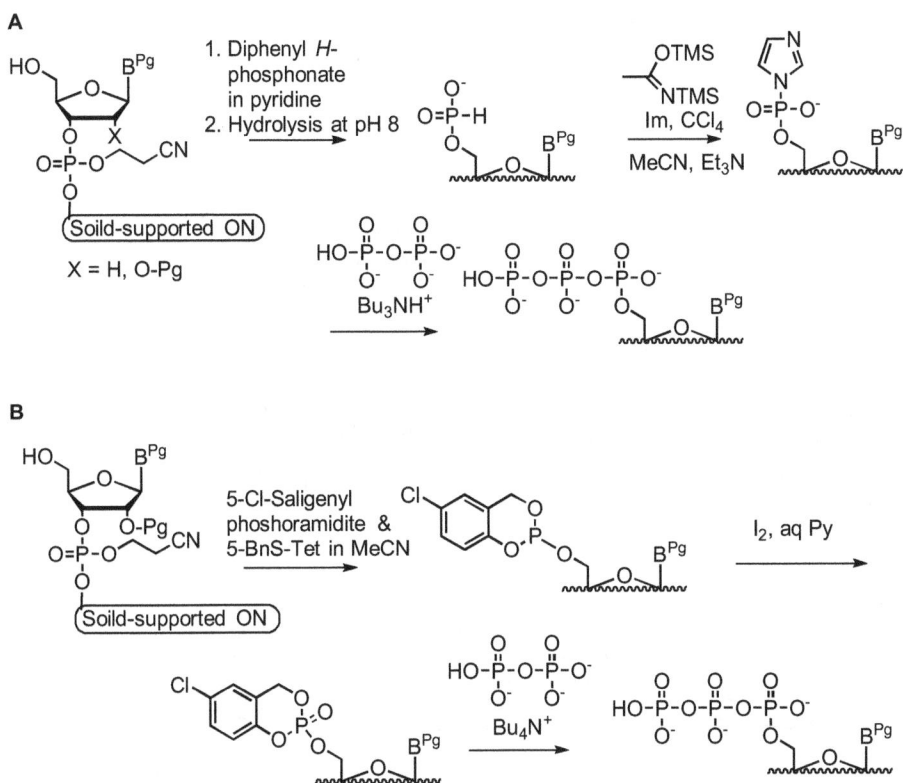

**Figure 7.5:** Synthesis of oligonucleotide 5′-triphosphates: A [43] and B [44].

## 7.5 Solid-supported synthesis of 3′-O-conjugates

Conjugation to the 3′-terminus is more complicated than 5′-conjugation, since the 3′-OH is usually attached to the support during the chain assembly. A simple but rather seldom used solution to this problem is assembly of the oligonucleotide chain in inverse direction by using 5′-phosphoramidites as synthons. The 3′-terminus may then be manipulated as discussed above for 5′-conjugation. As long as the conventional 3′ to 5′ chain assembly is used, at least three alternative approaches are available. The first one is utilization of a linker that withstands oligonucleotide synthesis but contains a linkage cleaved by the entering conjugate group upon release of the oligonucleotide from the support. Linker A in Figure 7.6, for example, contains an amine

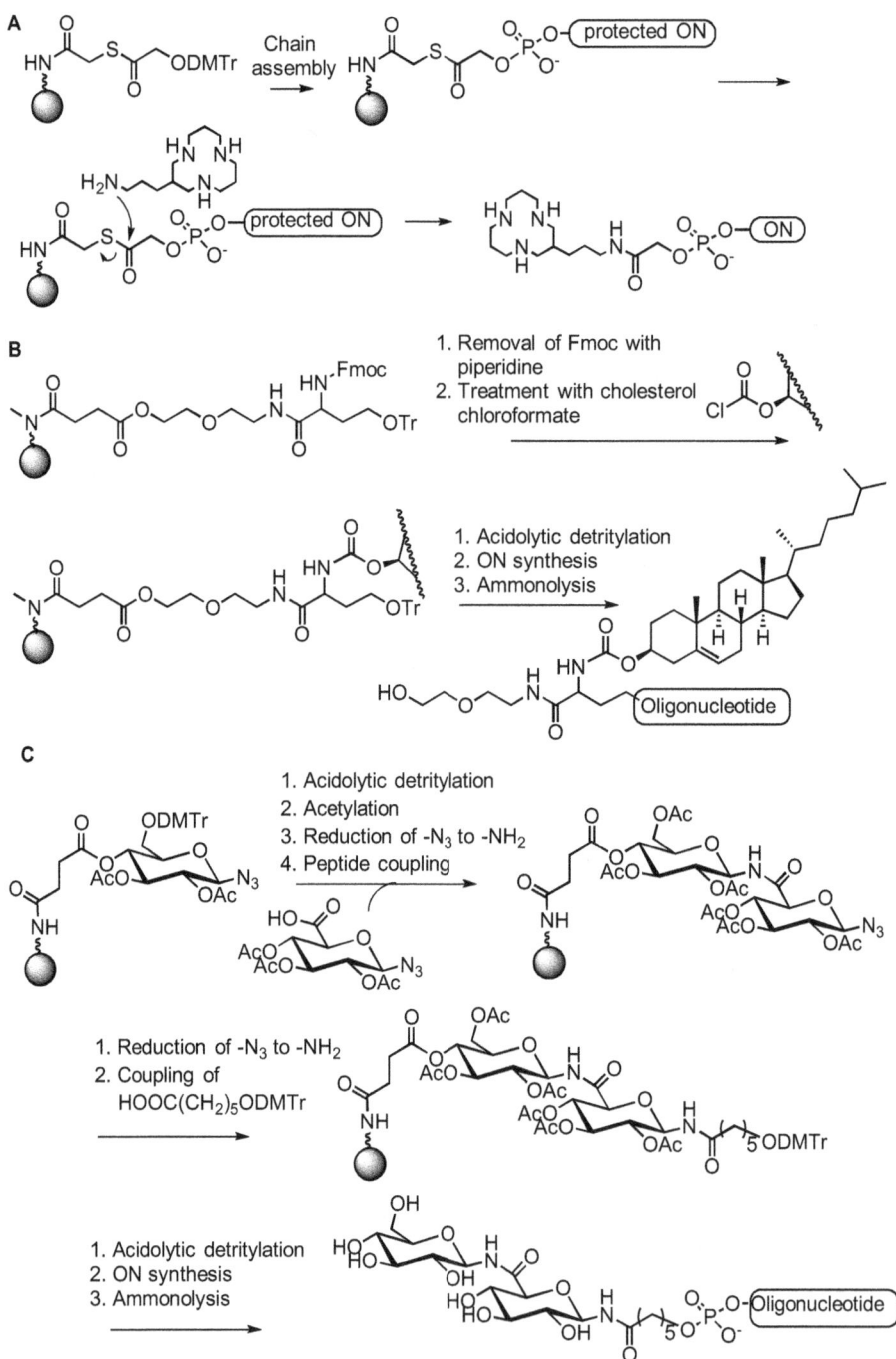

**Figure 7.6:** Alternative strategies for the synthesis of 3′-conjugates of oligonucleotides on a solid support: A [30], B [45] and C [47].

sensitive thioester linkage [30]. Alternatively, a branched orthogonally protected linker can be used that allows introduction of the conjugate group prior to or after the chain assembly. An illustrative example is given in Figure 7.6B [45]. The crucial part of the linker is homoserine, the α-amino group of which is protected with a base-labile Fmoc group and the hydroxyl function with an acid-labile trityl group. The Fmoc group is first removed with piperidine and the exposed amino group acylated with the conjugate group, in this particular case with cholesterol chloroformate. The urethane linkage formed withstands the acidic conditions required for exposure of the hydroxyl function and assembly of the oligonucleotide chain on it. This kind approach is useful for construction of peptide conjugates by consecutive peptide and oligonucleotide synthesis [46], as discussed in more detail in Section 7.8. Finally, the conjugate group and oligonucleotide may be assembled consecutively on a single linker. For example, a 3'-glycoconjugate has been prepared in this manner on a succinyl linker (Figure 7.6C) [47]. The requirement is that the conjugate group is protected with base-labile groups, removable under similar conditions as the oligonucleotide base moieties.

## 7.6 Solid-supported synthesis of intrachain conjugates

Conjugate groups may also be introduced into intrachain positions within oligonucleotides, either to base, sugar or phosphate moieties. An obvious approach is to derivatize a nucleoside with the desired conjugate group prior to its conversion to 3'-phosphoramidite and use it as a building block in any coupling cycle. Protections on the conjugate group must be base-labile. The preferred site of conjugation is C5 of pyrimidine nucleosides [48–50]. This approach has been used, for example, for decoration of oligonucleotides with sugars [51, 52]. Instead of ready-made conjugate group, the nucleosidic building block may bear an orthogonally protected functionality that is exposed for conjugation after the chain assembly or even after release in solution. 1-Aminomethyl-1,2-dideoxy-D-*erythro*-pentofuranose protected with a photolabile *N*-1-(2-nitrophenyl)ethoxycabonyl (NPEC) group serves as a recent example [53].

Instead of prefabricated nucleoside conjugates, nucleosides bearing an azido or alkynyl group are increasingly used. These groups withstand the chain assembly and allow conjugation by Cu(I)-catalyzed dipolar cycloaddition on support. Owing to incompatibility of azido substitution with phosphoramidite chemistry, the azido bearing building blocks must, however, be coupled as H-phosphonates. The couplings before and after the insertion of the azido block may be normal phosphoramidite couplings [54]. Examples are given in Figure 7.7. 5-Iodo-2'-deoxyribouridine is a convenient starting material for alkyne-functionalized nucleosides. A linker bearing two terminal triple bonds can be easily introduced by Sonogashira cross-coupling reaction [55]. After conversion of the functionalized nucleoside to phosphoramidite and incorporation into oligonucleotide chain, the remaining terminal triple bond is subjected to Cu(I)-catalyzed click reaction with an azide-functionalized molecule, such as azido coumarin dye in Figure 7.7A [56]. Alternatively, the sugar moiety may be functionalized with both alkyne an azido groups as exemplified in Figure 7.7B, where both approaches are utilized in a sequential manner [57]. Recently, thymidines that bear a methyl cyclopropene or sydnone [1,2,3-oxadiazol-5(2H)-one] group at 2'-OH have been introduced in oligonucleotides by solid-supported synthesis [58]. These groups allow post synthetic conjugation by two traceless reactions: inverse electron-demand Diels Alder reaction (methyl cyclopropene) and strain-promoted alkyne cycloaddition (sydnone).

Convertible nucleoside strategy is another viable technique for preparation of intrachain conjugates. A nucleoside bearing a good leaving group is introduced into a desired position within the chain and subjected to post synthetic displacement by a nucleophilic conjugate group on-support or in solution. This technique enables introduction of conjugate groups that do not withstand the conditions of oligonucleotide synthesis [59]. 6-Iodopurine containing ORNs can be post-synthetically transformed to their 6-alkyl-, 6-aryloxy-, 6-alkylthio- and 6-(dialylamino)-purine counterparts by post synthetic displacement of the 6-iodo substituent [60]. 2-Fluoroinosine [61], in turn, has been a precursor for guanine $N^2$-conjugates and $O^4$-(4-nitrophenyl)uracil [62] for cytosine $N^4$-conjugates. Cytosine $N^4$-conjugates can be obtained via oxidative amination of 4-thio-2'-deoxyuridine with both alkyl- and aryl-amines [63]. N3-nitrothymine [64] and 5-methyl -4-(1,2,4-triazol-1-yl)pyrimidin-2-one [65] serve as an intermediate of in preparation of N3- and C4-conjugates of thymine, respectively. Two different conjugations may be carried out when two precursors with different leaving group properties are incorporated into the chain. For example, the susceptibility of 5-methoxycarbonylmethyl- and 5-cyano-methoxycarbonylmethyl-uracil to nucleophilic attack of amines differs so much that the 5-cyanomethoxycarbonyl group can be first selectively amidated [66].

Besides nucleophilic displacements, Pd(0)-promoted cross-coupling reactions are utilized for on-support derivatization of nucleobases. 5-Iodouracil has been derivatized with terminal alkynes using Pd(PPh$_3$)$_4$ and CuI as promoters [67], and 8-bromo-$N^6$-benzoyladenine has been converted to a ferrocene conjugate by Sonogashira coupling with ferrocenyl propargylamide [68]. A closely related strategy is to utilize nucleosides masked with a photochemically removable group, such as 3,4-dimethoxy-2-nitrobenzyloxycarbonyl group. For example, the terminal amino group of a 5-(6-aminohex-1-yn-1-yl) side arm on uracil has been masked with this group and after photochemical deprotection subjected to peptide coupling [69].

2'-Hydroxyl function may be utilized for introduction of conjugate groups sequence selectively in ORNs. For this purpose, two different phosphoramidites bearing an orthogonally removable 2'-O-protecting groups are used in the solid-supported chain assembly: for example 2'-O-thiomorpholine-carbothioate as a permanent and 2'-O-tert-butyl(dimethyl)silyl as a temporary protecting group [70]. Another sequence selective post-synthetic conjugation method is based on rhodium(I)-catalyzed reaction of carbenes with unpaired guanosine [71]. In the presence of several guanosines, the de-

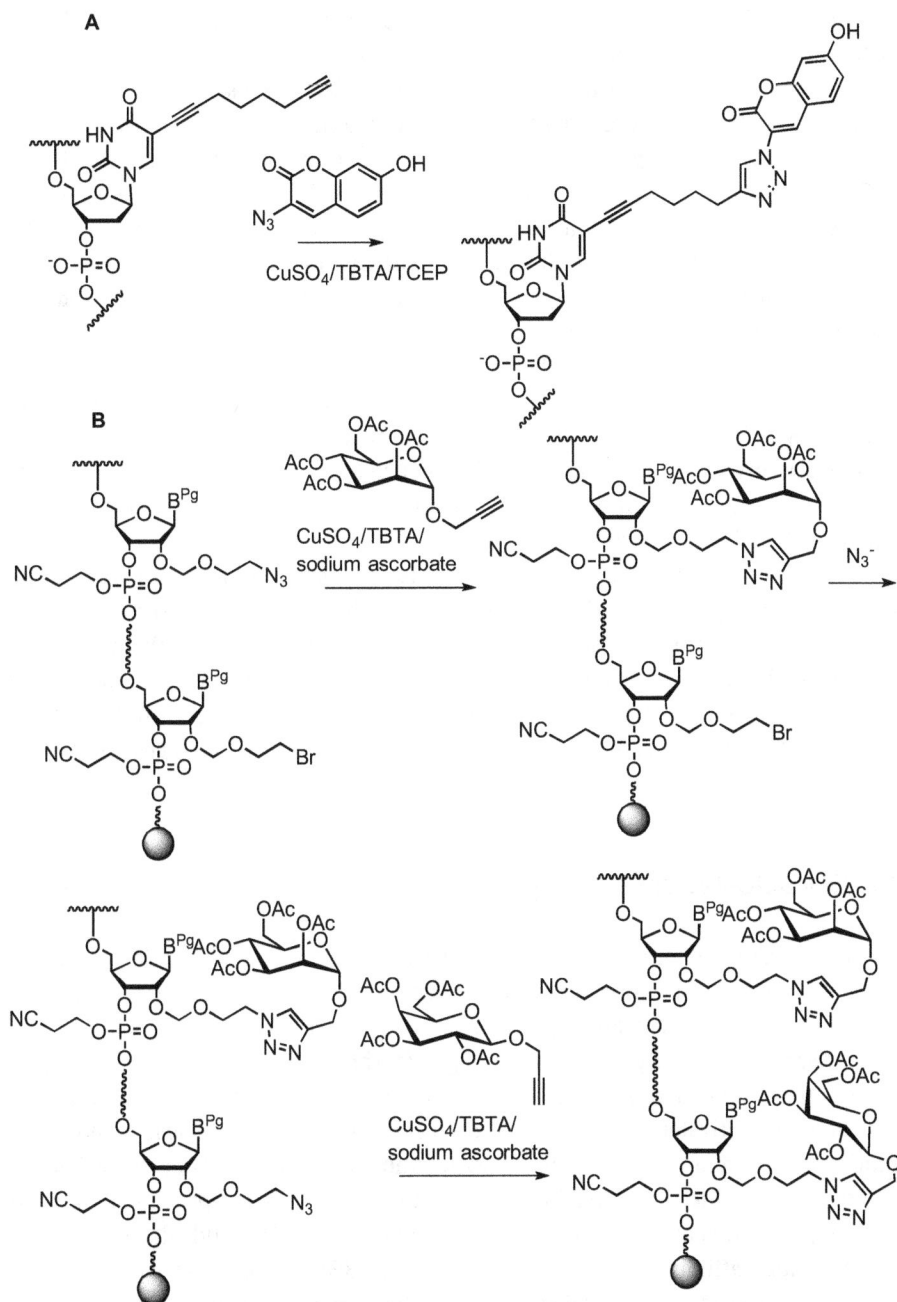

**Figure 7.7:** Utilization of Cu$^+$-catalyzed dipolar cycloaddition in preparation of oligonucleotide conjugates: A [56] and B [57].

sired one can be selected by hybridization with an otherwise complementary ODN that forces this guanosine to form a one nucleotide bulge.

Conjugation to internucleosidic phosphodiester linkages is also feasible. Nucleoside phosphoramidites bearing an appropriately protected linker or a prefabricated conjugate group may in principle be used as synthons in the chain assembly. A bulky linker or conjugate group, however, tends to reduce the coupling efficiency. That is why it may be preferable to introduce an *H*-phosphonate linkage in a desired place within the oligonucleotide chain and oxidize it with $CCl_4$ to phosphoramidate in the presence of an amine derivatized linker or conjugate group [72]. Alternatively, 2-cyanoethyl-protected phosphoramidites are used for synthesis of only those internucleosidic linkages that are aimed at being conjugated. The other linkages are introduced methyl protected. After chain assembly, the 2-cyanoethyl protections are removed with piperidine or DBU and activated with TsCl [73]. The methyl protections remain untouched. The tosyl group is then displaced with an amino-functionalized conjugate group. Finally, the methyl protections are removed with a special nucleophilic reagent (disodium 2-carbamoyl-2-cyano-ethylene-1,1-dithiolate).

A desired phosphodiester linkage may also be derivatized already during the solid-supported synthesis [74, 75]. For this purpose, the phosphite diester intermediate obtained upon coupling is reacted in the course of iodine oxidation with an alkyl sulfonyl azide. The azide functionality allows post-synthetic conjugation after completion of the chain assembly. To facilitate displacement of the phosphite protecting group during the oxidation step, the conventional 2-cyanoethyl group is replaced with a *tert*-butyl group.

## 7.7 Glycocluster conjugates

Branched linkers are required for targeting therapeutic oligonucleotides to sugar binding proteins, lectins, present on cell surface. Lectins contain several binding sites for sugar monomers. The affinity per mole of sugar units is markedly enhanced in comparison to the affinity of monovalent ligands. The phenomenon is known as glycocluster effect. In other words, the lack of strength of an individual interaction is compensated by multivalency. To ensure high affinity binding, several sugars should be attached to a common scaffold that allows simultaneous binding of all sugar ligands to the subunits of lectin. A well-known example is an asialoglycoprotein receptor in hepatocytes that requires simultaneous binding of three *N*-acetylgalactosamine ligands for high affinity interaction [76] (cf. Section 13.3). A triantennary conjugate attached to the 5'-terminus of oligonucleotide, has been shown to warrant efficient recognition the receptor and internalization of the oligonucleotide [77]. The conjugate has been prepared by an active ester method in solution; a pentafluorophenyl ester of a triantennary construct of three fully acetylated *N*-acetylgalactosmines is reacted

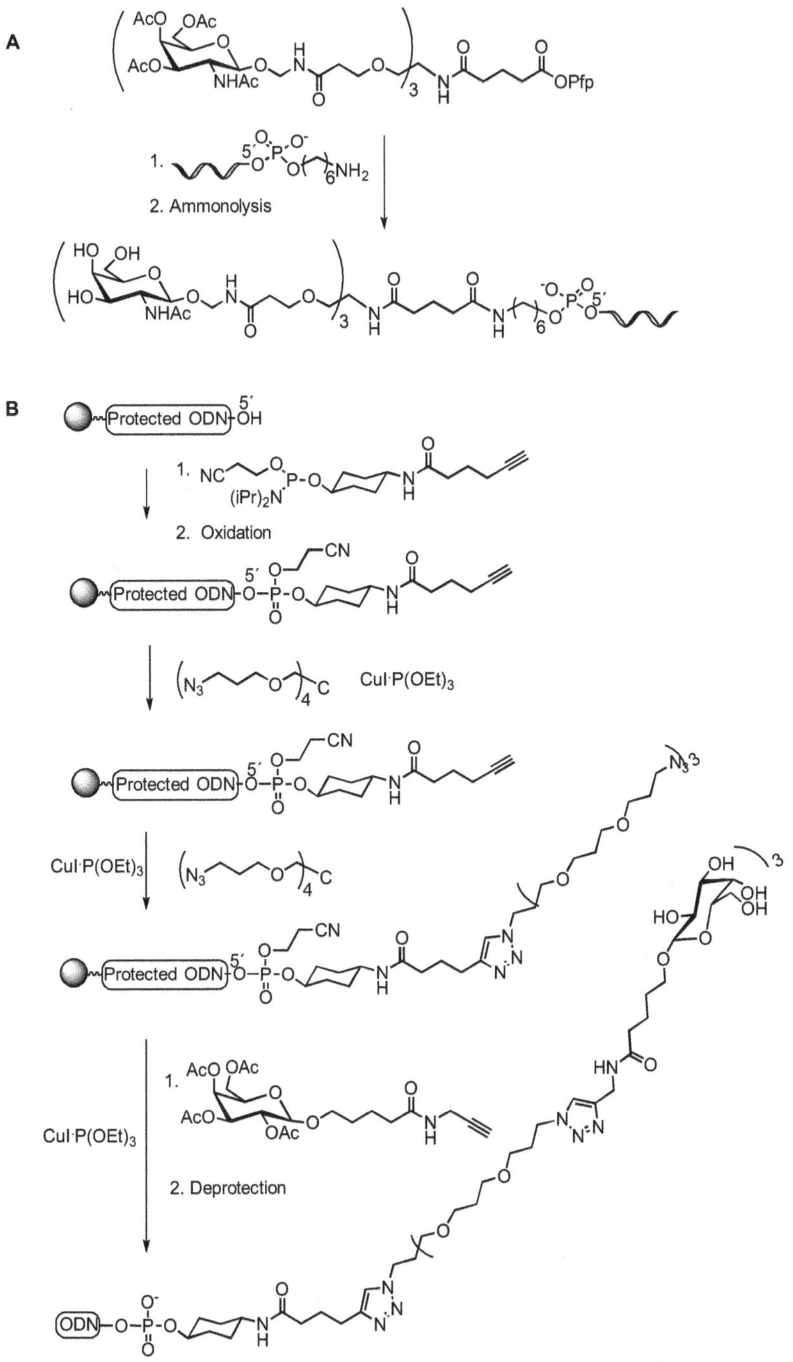

**Figure 7.8:** (A) Assembly of a triantennary oligonucleotide glycol conjugate from a prefabricated sugar cluster [77]. (B) A versatile method for the synthesis of sugar-clustered oligonucleotides on support [78].

with a 5′-aminohexyl oligonucleotide and subjected then to ammonolytic deprotection of the sugar ligands (Figure 7.8A).

A more versatile strategy for the synthesis of sugar-clustered oligonucleotides is based on sequential utilization of Cu(I)-catalyzed click reaction [78] Figure 7.8B shows an example. An alkyne-functionalized linker is first coupled as a phosphorphoramidite to the 5′-terminus of a solid-supported oligonucleotide. Click reaction with a pentaerythritol-derived tetraazide yields a triantennary azido conjugate that is subjected to another click reaction with an alkynylated sugar peracetate. Ammonolysis releases the deprotected conjugate in solution. The key step for the success of the synthesis is the Cu(I) catalyst, $CuI \cdot P(OEt)_3$, introduced in a dipolar aprotic solvent (DMF, DMSO or DMAA). The same approach allows synthesis of even nine sugar clusters, when the alkyne component of the second click reaction is a triantennary sugar cluster.

## 7.8 Peptide conjugates

Peptide conjugates of oligonucleotides are subject to wider interest than any other class of oligonucleotide conjugates. The reason is that some relatively short peptides are able to penetrate into cell, at least partly by an endocytosis-independent cytoplasmic entry [79] (cf. Section 13.2) In addition, the intracellular traffic of oligonucleotides may be influenced by nuclear localization signal peptides [80]. The most extensively used method for preparation of oligonucleotide-peptide conjugates is conjugation of a prefabricated peptide in solution to an appropriately functionalized oligonucleotide with the aid of some bio-orthogonal reaction [81]. Reactions successfully utilized include native ligation [14], disulfide formation [11], thioether formation [12], Diels-Alder cycloaddition [9], $N$-methoxyoxazolidine conjugation [15], and Cu(I)-promoted click reaction [82] (for the reactions, see Figure 7.1). The Cu(I)-promoted click reaction is exceptionally facile between a 5′-attached carbonyl activated alkyne and azido-functionalized peptide [83]. In more detail, a solid-supported 5′-(aminohexyl) tethered oligonucleotide is first acylated with 4-(propargylaminomethyl)benzoic acid and then treated overnight with the azido-functionalized peptide (2 equiv.) in the presence of $CuSO_4$ and sodium ascorbate. Ammonolysis releases and deprotects the conjugate. The $N$-methoxyoxazolidine linker differs from the others in the sense that the conjugation is reversible under slightly acidic conditions. Accordingly, the linker may be used as a pH-responsive cleavable linker [15].

Peptide oligonucleotide conjugates have also been assembled from monomeric building blocks on a single solid support, which generally necessitates modifications in the conventional protecting group strategy of peptide or oligonucleotide synthesis. Numerous approaches have described [19, 84]. The most generally applicable ones are outlined in Figure 7.9. To make the protecting group strategy of peptide synthesis compatible with that the oligonucleotide synthesis, 2-(biphenyl-4-yl)propan-2-yloxycarbonyl (Bpoc) group has been used for protection of the α-amino groups and the side chains of histidine, lysine have been acylated [85]. The α-amino groups are, hence, deprotected under the conditions normally used for 5′-*O*-detritylation and the acyl protections are removed during final ammonolysis (Figure 7.9A). After all, removal of the phosphate protecting groups from a support bound oligonucleotide does not seem to interfere with the subsequent

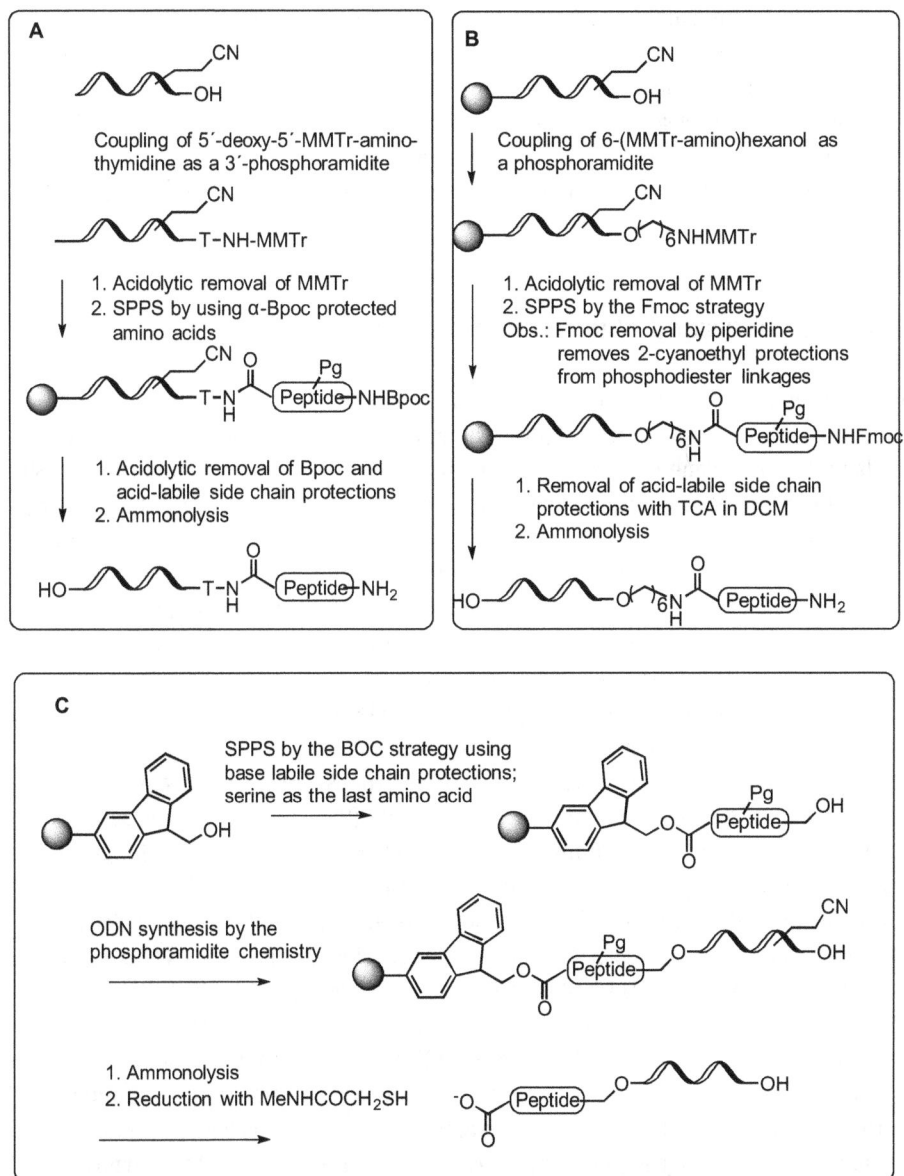

**Figure 7.9:** Alternative approaches for preparation of oligonucleotide–peptide conjugates on a single solid support: A [85], B [86] and C [87].

assembly of the peptide moiety by unmodified Fmoc PyBOP/DIPEA chemistry. Although repeated removal of the Fmoc protections from α-amino groups also removes the 2-cyanoethyl groups from the phosphodiester linkage, the peptide coupling still works (Figure 7.9B) [86]. 3′-Peptide conjugates are obtained by using a solid-supported Fmoc group as a linker [87]. The peptide moiety is first assembled by the Boc chemistry which utilizes acid labile α-amino protections, the side chain functionalities being protected with base-labile groups (Figure 7.9C). Serine or homoserine is coupled as the last amino acid, and the oligonucleotide moiety is assembled by the phosphoramidite chemistry. Ammonolysis then releases and deprotects the conjugate. A recent study shows that the conventional Boc/tBu protections of Trp, His, Arg, Asp and Glu can be removed in borate buffer at 90 °C, that is, under conditions where depurination of nucleobases is avoided [88].

## 7.9 Photoswitched oligonucleotides

Oligonucleotides bearing a photoconvertible conjugate group, that is, groups that undergo a structural change upon light irradiation, have gained popularity as research tools with which gene expression can be monitored or even controlled in vivo [89]. The light-triggered structural change can be a photolytic cleavage, crosslinking by [2 + 2] cycloaddition, photocatalytic *cis–trans* isomerization or intramolecular photocyclization [90]. Figure 7.10 shows examples of these groups as part of an ODN structure. A photoconvertible group is typically introduced during solid-supported oligonucleotide synthesis as a prefabricated nucleoside phosphoramidite, or in few cases postsynthetically to a prefabricated oligonucleotide.

Figure 7.10A shows examples of photoremovable conjugate groups. The most common is *ortho*-nitrobenzyl group (*o*NB). This group and its analogs, 1-(*ortho*-nitrophenyl) ethyl (NPE) and 2-(*ortho*-nitrophenyl)propyl (NPP) groups, are used to cage oligonucleotides by conjugation to various sites in base moieties [91]. Photocleavage is achieved by radiation in the range 345–420 nm. Recently, 6-nitropiperonyl methyl (NPM) group has gained increasing popularity because of longer wavelength of photocleavage. When incorporated into the sugar-phosphate backbone, the presence of *o*NB group enables phototriggered chain cleavage. Other common photocleavable groups are *p*-hydroxyphenacyl, thioether-enol phosphate (TEEP-OH), aryl sulfides and benzophenone [89].

Coumarin, carbazole and some vinyl derivatives constitute another set of photoconvertible groups. They undergo [2 + 2] cycloaddition with nucleobases upon irradiation at a characteristic wavelength. The reaction is, however, reversible. A 5′-terminal oligonucleotide conjugate of coumarin, for example, undergoes covalent crosslinking with thymidine residue of a complementary strand when irradiated at 350 nm (Figure 7.10B), but the reaction may reversed by 254 nm irradiation [92]. 3-Cyanovinylcarbazole nucleoside (CNVK), in turn, crosslinks with a pyrimidine base in the complementary strand at 385 nm and uncrosslinks at 312 nm without any damage in DNA [93]. Unconjugated

## A. Photo-removable groups

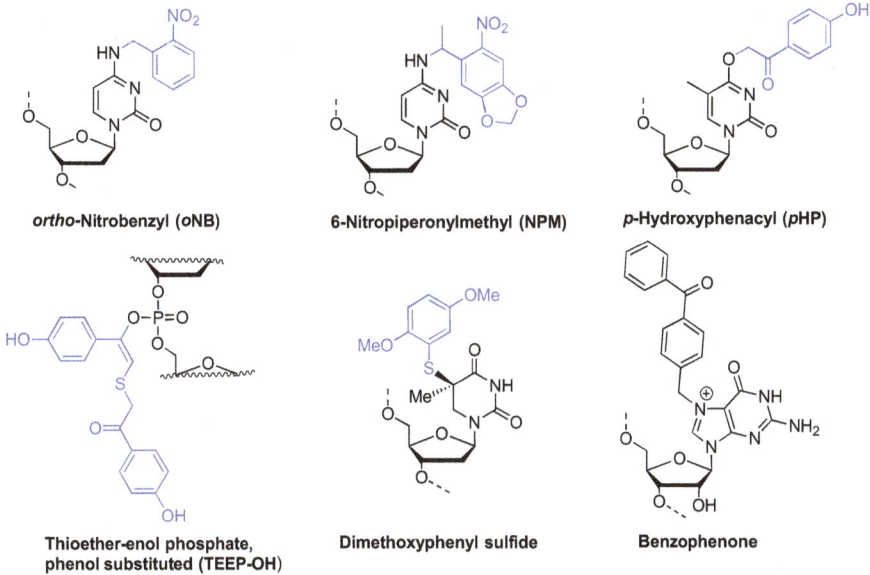

*ortho*-Nitrobenzyl (*o*NB)  6-Nitropiperonylmethyl (NPM)  *p*-Hydroxyphenacyl (*p*HP)

Thioether-enol phosphate, phenol substituted (TEEP-OH)  Dimethoxyphenyl sulfide  Benzophenone

## B. Photo-crosslinking groups

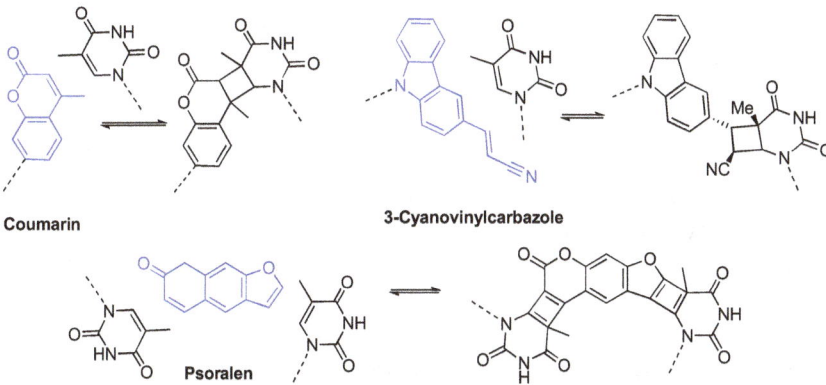

Coumarin  3-Cyanovinylcarbazole

Psoralen

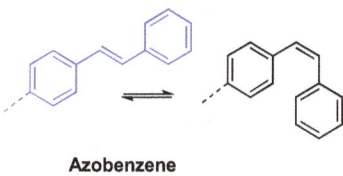

## C. Groups undergoing *cis-trans* isomerization  D. Groups undergoing spirocyclization

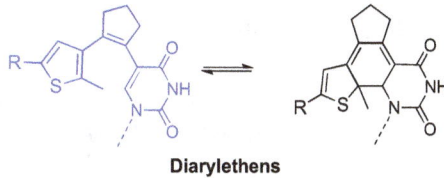

Azobenzene  Diarylethens

**Figure 7.10:** Examples of photoconvertible conjugate groups of oligonucleotides.

psoralen, a derivative of coumarin, is used for interchain crosslinking of DNA at 365 nm.

In addition to 3-cyanovinylcarbazole several other vinyl group containing conjugate groups of oligonucleotides serve as reversible photoswitches. Azobenzene is the best known example (Figure 7.10C) [94]. The planar *trans*-isomer is converted to non-planar *cis*-isomer by UV light and the process can be reversed by visible light. Other extensively used photoswitches are 8-styrylguanosine and its 8-naphthylvinyl and 8-fluorenylvinyl congeners [95].

The most common class of photoswitches that undergo photocyclization are diarylethenes derived from either 2'-deoxyuridine or 2'-deoxyadenosine (Figure 7.10D). They undergo a photochemical electrocyclic ring closure with concomitant enlargement of conjugated electron system. The visible result is that the originally colorless "open" structure turns to a colored "closed" structure. Diarylethenes are regarded as a promising class of photoswitches, the photochemical properties of which still are under active research [96].

## 7.10 Conjugation by in vitro transcription

Enzymatic in vitro transcription on a DNA template is also utilized for introduction of conjugate groups into oligonucleotides. A nucleoside bearing the desired conjugate group is converted to its 5'-triphosphate and incorporated into the oligonucleotide structure by a DNA or RNA polymerase reaction. The prerequisite naturally is that a polymerase accepting the conjugated nucleoside triphosphate as a substrate is available. Synthesis of ORN conjugates that are able to covalently crosslink with RNA-binding proteins offers a recent example. A cysteine and histidine selective binding arm, *N*-(propargyl)chloroacetamide, was first coupled to 7-iodo-7-deazaadeonosine 5'-*O*-triphosphate by Pd-promoted Sonogashira reaction. The compound turned out to be a good substrate for T7 RNA polymerase and was successfully incorporated into ORN on template [97]. Fluorescent C5-conjugates of 2'-deoxycytidine have been introduced similarly into ODNs by KOD XL DNA polymerase-catalyzed primer extension [98].

## Further reading

Bartosik K, Debiec K, Czarnecka A, Sochacka E, Leszczynska G. Synthesis of nucleobase-modified RNA oligonucleotides by post-synthetic approach. Molecules 2020, 25, 3344 (37 pages).

Fantoni NZ, El-Sagheer AH, Brown T. A Hitchhiker's Guide to Click-Chemistry with Nucleic Acids. Chem Rev 2021, 121, 7122–7154.

George JT, Srivatsan SG. Bioorthogonal chemistry-based RNA labeling technologies: Evolution and current state. Chem Commun 2020, 56, 12307–12318.

Kaila J, Raines RT. Advances in bioconjugation. Curr Org Chem 2010, 14, 138–147.

Klöcker N, Weissenboecka FP, Rentmeister A. Covalent labeling of nucleic acids. Chem Soc Rev 2020, 49, 8749–8773.

Krell K, Harijan D, Ganz D, Doll L, Wagenknecht H-A. Postsynthetic Modifications of DNA and RNA by means of copper-free cycloadditions as bioorthogonal reactions. Bioconjugate Chem 2020, 31, 990–1011.

Lönnberg H. Solid-phase synthesis of oligonucleotide conjugates useful for delivery and targeting of potential nucleic acid therapeutics. Bioconjugate Chem 2009, 20, 1065–1094.

Singh Y, Spinelli N, Defrancq E. Chemical strategies for oligonucleotide-conjugate synthesis. Curr Org Chem 2008, 12, 263–290.

Tavakoli A, Min J-H. Photochemical modifications for DNA/RNA oligonucleotides. RSC Adv 2022, 12, 6484–6507.

Ustinov AV, Stepanova IA, Duvnyakova VV, Zatsepin TS, Nozhevnikova EV, Korshun VA. Modification of nucleic acids using [3+2]-dipolar cycloaddition of azides and alkynes. Russ J Bioorg Chem 2010, 36, 401–445.

Venkatesan N, Kim BH. Peptide conjugates of oligonucleotides: Synthesis and applications. Chem Rev 2006, 106, 3712–3761.

Zatsepin TS, Romanova EA, Oretskaya TS. Nucleosides and oligonucleotides containing 2′-reactive groups: Synthesis and application. Russ Chem Rev 2004, 73, 701–733.

# References

[1]   Rostovtsev VV, Green LG, Fokin VV, Sharpless KB. A stepwise Huisgen cycloaddition process: Copper (I)-catalyzed regioselective "ligation" of azides and terminal alkynes. Angew Chem Int Ed 2002, 41, 2596–2599.

[2]   Tornoe CW, Christensen C, Meldal M. Peptidotriazoles on solid phase: [1,2,3]-triazoles by regiospecific copper(I)-catalyzed 1,3-dipolar cycloadditions of terminal alkynes to azides. J Org Chem 2002, 67, 3057–3064.

[3]   Jewetta JC, Bertozzi CR. Cu-free click cycloaddition reactions in chemical biology. Cu-free click cycloaddition reactions in chemical biology. Chem Soc Rev 2010, 39, 1272–1279.

[4]   Salo H, Virta P, Hakala H, Prakash TP, Kawasaki AM, Manoharan M, Lönnberg H. Aminooxy functuionalised oligonucleotides: Preparation, on-support derivatization, and postsynthetic attachment to polymer support. Bioconjugate Chem 1999, 10, 815–823.

[5]   Raddatz S, Mueller-Ibeler J, Kluge J, Wäss L, Burdinski G, Havens JR, Onofrey TJ, Wang D, Schweitzer M. Hydrazide oligonucleotides: New chemical modification for chip array attachment and conjugation. Nucleic Acids Res 2002, 30, 4793–4802.

[6]   Jobbagy A, Kiraly K. Chemical characterization of fluorescein isothiocyanate-protein conjugates. Biochim Biophys Acta 1966, 124, 166–175.

[7]   Singh Y, Defrancq E, Dumy P. New method to prepare peptide-oligonucleotide conjugates through glyoxylic oxime formation. J Org Chem 2004, 69, 8544–8546.

[8]   Zatsepin TS, Stetsenko DA, Arzumanov AA, Romanova EA, Gait MJ, Oretskaya TS. Synthesis of peptide-oligonucleotide conjugates with single and multiple peptides attached to 2′-aldehydes through thiazolidine, oxime, and hydrazine Linkages. Bioconjugate Chem 2002, 13, 822–830.

[9]   Marchan V, Ortega S, Pulido D, Pedroso E, Grandas A. Diels-Alder cycloadditions in water for the straightforward preparation of peptide-oligonucleotide conjugates. Nucleic Acids Res 2006, 34, e24/1–e24/9.

[10]  Bednarek C, Wehl I, Jung N, Schepers U, Bräse S. The Staudinger Ligation. Chem Rev 2020, 120, 4301–4354.

[11]  Antopolsky M, Azhayeva E, Tengvall U, Auriola S, Jääskeläinen I, Rönkkö S, Honkakoski P, Urtti A, Lönnberg H, Azhayev A. Peptide–oligonucleotide phosphorothioate conjugates with membrane translocation and nuclear localization properties. Bioconjugate Chem 1999, 10, 598–606.

[12]  Tung CH, Rudolph MJ, Stein S. Preparation of oligonucleotide-peptide conjugates. Bioconjugate Chem 1991, 2, 464–465.

[13]  Staros JV. N-Hydroxysulfosuccinimide active esters: Bis(N-hydroxysulfosuccinimide) esters of two dicarboxylic acids are hydrophilic, membrane-impermeant, protein cross-linkers. Biochemistry 1982, 21, 3950–3955.

[14]  Stetsenko DA, Gait MJ. Efficient conjugation of peptides to oligonucleotides by "native ligation". J Org Chem 2000, 65, 4900–4908.

[15]  Aho A, Sulkanen M, Korhonen H, Virta P. Conjugation of oligonucleotides to peptide aldehydes via a pH-responsive N-methoxyoxazolidine linker. Org Lett 2020, 22, 6714–6718.

[16]  Aho A, Äärelä A, Korhonen H, Virta P. Expanding the scope of the cleavable N-(methoxy)oxazolidine linker for the synthesis of oligonucleotide conjugates. Molecules 2021, 26, 490 (11 pages).

[17]  Jbara M, Rodriguez J, Dhanjee HH, Loas A, Buchwald SL, Pentelute BL. Oligonucleotide bioconjugation with bifunctional palladium reagents. Angew Chem Int Ed 2021, 60, 12109–12115.

[18]  Singh Y, Murata P, Defrancq E. Recent developments in oligonucleotide conjugation. Chem Soc Rev 2010, 39, 2054–2070.

[19]  Lönnberg H. Solid-phase synthesis of oligonucleotide conjugates useful for delivery and targeting of potential nucleic acid therapeutics. Bioconjugate Chem 2009, 20, 1065–1094.

[20]  Adamczyk M, Chan CM, Fino JR, Mattingly PG. Synthesis of 5- and 6-Hydroxymethylfluorescein Phosphoramidites. J Org Chem 2000, 65, 596–601.

[21]  Czlapinski JL, Sheppard TL. Nucleic acid template-directed assembly of metallosalen–DNA conjugates. J Am Chem Soc 2001, 123, 8618–8619.

[22]  Bendinskas KG, Harsch A, Wilson RM, Midden WR. Sequence-specific photomodification of DNA by an oligonucleotide-phenanthrodihydrodioxin conjugate. Bioconjugate Chem 1998, 9, 555–563.

[23]  Lehmann TJ, Engels JW. Synthesis and properties of bile acid phosphoramidites 5′-tethered to antisense oligodeoxynucleotides against HCV. Bioorg Med Chem 2001, 9, 1827–1835.

[24]  Robles J, Rajur SB, McLaughlin LW. A Parallel-stranded DNA triplex tethering a Hoechst 33258 analogue results in complex stabilization by simultaneous major groove and minor groove binding. J Am Chem Soc 1996, 118, 5820–5821.

[25]  Chen J-K, Weith HL, Grewal RS, Wang G, Cushman M. Synthesis of novel phosphoramidite reagents for the attachment of antisense oligonucleotides to various regions of the benzophenanthridine ring system. Bioconjugate Chem 1995, 6, 473–482.

[26]  Guzaev A, Lönnberg H. Solid support synthesis of ester linked hydrophobic conjugates of oligonucleotides. Tetrahedron 1999, 55, 9101–9116.

[27]  Pon R. A long chain biotin phosphoramidite reagent for the automated synthesis of 5′-biotinylated oligonucleotides. Tetrahedron Lett 1991, 32, 1715–1718.

[28]  Mukkala V-M, Sund C, Kwiatkowski M Spectrofluorometric method and compounds that are of value for the method, 1993 U.S. Patent 5,216,134.

[29]  Pons B, Kotera M, Zuber G, Behr J-P. Online synthesis of diblock cationic oligonucleotides for enhanced hybridization to their complementary sequence. ChemBioChem 2006, 7, 1173–1176.

[30]  Hovinen J, Guzaev A, Azhayeva E, Azhayev A, Lönnberg H. Imidazole tethered oligodeoxyribonucleotides: Synthesis and RNA cleaving activity. J Org Chem 1996, 60, 2205–2209.

[31]  Krasheninina OA, Thaler J, Erlacher MD, Micura R. Amine-to-azide conversion on native RNA via metal-free diazotransfer opens new avenues for RNA manipulations. Angew Chem Int Ed 2021, 60, 6970–6974.

[32]  Uhlmann E, Engels J. Chemical 5′-phosphorylation of oligonucleotides valuable in automated DNA synthesis. Tetrahedron Lett 1986, 27, 1023–1026.

[33] Horn T, Urdea MS. A chemical 5′-phosphorylation of oligonucleotides that can be monitored by trityl cation release. Tetrahedron Lett 1986, 27, 4705–4708.

[34] Lefebvre I, Perigaud C, Pompon A, Aubertin A-M, Girardet J-L, Kirn A, Gosselin G, Imbach J-L. Mononucleoside phosphotriester derivatives with S-Acyl-2-thioethyl bioreversible phosphate-protecting groups: Intracellular delivery of 3′-azido-2′,3′-dideoxythymidine 5′-monophosphate. J Med Chem 1995, 38, 3941–3950.

[35] Olesiak M, Krajewska D, Wasilewska E, Korczynski D, Baraniak J, Okruszek A, Stec WJ. Thiophosphorylation of biologically relevant alcohols by the oxathiaphospholane approach. Synlett 2002, 0, 967–971.

[36] Ausyn C, Grajkowski A, Cieslak J, Beaucage SL. An efficient reagent for the phosphorylation of deoxyribonucleosides, DNA oligonucleotides and their thermolytic analogues. Org Lett 2005, 7, 4201–4204.

[37] Guzaev A, Salo H, Azhayev A, Lönnberg H. A new approach for chemical phosphorylation of oligonucleotides at the 5′-terminus. Tetrahedron 1995, 51, 9375–9384.

[38] Lartia R, Asseline U. New reagent for the preparation of oligonucleotides involving a 5′-thiophosphate or a 5′-phosphate group. Tetrahedron Lett 2004, 45, 5949–5952.

[39] Meyer A, Bouillon C, Vidal S, Vasseur -J-J, Morvan F. A versatile reagent for the synthesis of 5′-phosphorylated,5′-thiophosphorylated or 5-phosphoramidate-conjugated oligonucleotides. Tetrahedron Lett 2006, 47, 8867–8871.

[40] Pradere U, Halloy F, Hall J. Chemical synthesis of long RNAs with terminal 5′-phosphate groups. Chem Eur J 2017, 23, 5210–5213.

[41] Romanucci V, Zarrelli A, Guaragna A, Di Marino C, Di Fabio G. New phosphorylating reagents for deoxyribonucleosides and oligonucleotides. Tetrahedron Lett 2017, 58, 1227–1229.

[42] Madaoui M, Meyer A, Vasseur -J-J, Morvan F. Thermolytic reagents to synthesize 5′- or 3′-mono(thio) phosphate oligodeoxynucleotides or 3′-modified oligodeoxynucleotides. Eur J Org Chem 2019, 0, 2832–2842.

[43] Zlatev I, Lavergne T, Debart F, Vasseur -J-J, Manoharan M, Morvan F. Efficient solid-phase chemical synthesis of 5′-triphosphates of DNA, RNA, and their Analogues. Org Lett 2010, 12, 2190–2193.

[44] Sarac I, Meier C. Efficient automated solid-phase synthesis of DNA and RNA 5′-triphosphates. Chem Eur J 2015, 21, 16421–16426.

[45] Stetsenko DA, Gait MJ. A convenient solid-phase method for synthesis of 3′-conjugates of oligonucleotides. Bioconjugate Chem 2001, 12, 576–586.

[46] Antopolsky M, Azhayeva E, Tengvall U, Azhayev A. Towards a general method for the stepwise solid-phase synthesis of peptide–oligonucleotide conjugates. Tetrahedron Lett 2002, 43, 527–530.

[47] D'Onofrio J, de Champdore M, De Napoli L, Montesarchio D, Di Fabio G. Glycomimetics as decorating motifs for oligonucleotides: Solid-phase synthesis, stability, and hybridization properties of carbopeptoid–oligonucleotide conjugates. Bioconjugate Chem 2005, 16, 1299–1309.

[48] Kurz A, Bunge A, Windeck A-K, Rost M, Flasche W, Arbuzova A, Strohbach D, Müller S, Liebscher J, Huster D, Herrmann A. Lipid-anchored oligonucleotides for stable double-helix formation in distinct membrane domains. Angew Chem Int Ed 2006, 45, 4440–4444.

[49] Turner JJ, Meeuwenoord NJ, Rood A, Borst P, van der Marel GA, van Boom JH. Reinvestigation into the synthesis of oligonucleotides containing 5-(β-D-glucopyranosyloxymethyl)-2′-deoxyuridine. Eur J Org Chem 2003, 0, 3832–3839.

[50] Tona R, Bertolini R, Hunziker J. Synthesis of aminoglycoside-modified oligonucleotides. Org Lett 2000, 2, 1693–1696.

[51] Matsuura K, Hibino M, Ikeda T, Kobayashi K. Construction of glycol-clusters by self-organization of site specifically glycosylated oligonucleotides and their cooperative amplification of lectin recognition. J Am Chem Soc 2001, 123, 357–358.

[52] Matsuura K, Hibino M, Ikeda T, Yamada Y, Kobayashi K. Self-organized glycoclusters along DNA: Effect of the spatial arrangement of galactoside residues on cooperative lectin recognition. Chem Eur J 2004, 10, 352–359.

[53] Martín-Nieves V, Fabrega C, Guasch M, Fernandez S, Sanghvi YS, Ferrero M, Eritja R. Oligonucleotides containing 1-aminomethyl or 1-mercaptomethyl-2-deoxy-D-ribofuranoses: Synthesis, purification, characterization, and conjugation with fluorophores and lipids. Bioconjugate Chem 2021, 32, 350–366.

[54] Kiviniemi A, Virta P, Lönnberg H. Utilization of intrachain 4′-C-azidomethylthymidine for preparation of oligodeoxyribonucleotide conjugates by click chemistry in solution and on solid support. Bioconjugate Chem 2008, 19, 1726–1734.

[55] Müggenburg F, Müller S. Azide-modified nucleosides as versatile tools for bioorthogonal labeling and functionalization. Chem Rec 2022, e202100322 (13 pages).

[56] Sirivolu VR, Chittepu P, Seela F. DNA with branched internal side chains: Synthesis of 5-tripropargylamine-dU and conjugation by an azide-alkyne double click reaction. ChemBioChem 2008, 9, 2305–2316.

[57] Kiviniemi A, Virta P, Drenichev MS, Mikhailov SN, Lönnberg H. Solid-supported 2′-oligococonjugation of oligonucleotides by azidation and click reacions. Bioconjugate Chem 2011, 22, 1249–1255.

[58] Bristiel A, Cadinot M, Pizzonero M, Taran F, Urban D, Guignard R, Guianvarc'h D. 2′-Modified thymidines with bioorthogonal cyclopropene or sydnone as building blocks for copper-free postsynthetic functionalization of chemically synthesized oligonucleotides. Bioconjugate Chem 2023, 34, 1613–1621.

[59] Bartosik K, Debiec K, Czarnecka A, Sochacka E, Leszczynska G. Synthesis of nucleobase-modified RNA oligonucleotides by post-synthetic approach. Molecules 2020, 25, 3344 (37 pages).

[60] Xie Y, Fang Z, Yang W, He Z, Chen K, Heng P, Wang B, Zhou. 6-Iodopurine as a Versatile Building Block for RNA Purine Architecture Modifications. Bioconjugate Chem 2022, 33, 353–362.

[61] Harris CM, Zhou L, Strand EA, Harris TM. New strategy for the synthesis of oligodeoxynucleotides bearing adducts at exocyclic amino sites of purine nucleosides. J Am Chem Soc 1991, 113, 4328–4329.

[62] de la Torre BG, Morales JC, Avino A, Iacopino D, Ongaro A, Fitzmaurice D, Murphy D, Doyle H, Redmond G, Eritja R. Synthesis of oligonucleotides carrying anchoring groups and their use in the preparation of oligonucleotide – Gold conjugates. Helv Chim Acta 2002, 85, 2594–2607.

[63] Wang J, Shang J, Xiang Y, Tong A. General method for post-synthetic modification of oligonucleotides based on oxidative amination of 4-thio-2′-deoxyuridine. Bioconjugate Chem 2021, 32, 721–728.

[64] Gorchs O, Hernandez M, Garriga L, Pedroso E, Grandas A, Farras J. A New Method for the Preparation of Modified Oligonucleotides. Org Lett 2002, 4, 1827–1830.

[65] Xu Y-Z, Zheng Q, Swann PF. Synthesis of DNA containing modified bases by post-synthetic substitution. Synthesis of oligomers containing 4-substituted thymine: O4-alkylthymine, 5-methylcytosine, N4-dimethylamino-5-methylcytosine, and 4-thiothymine. J Org Chem 1992, 57, 3839–3845.

[66] Shinozuka K, Kohgo S, Ozaki H, Sawai H. Multi-functionalization of oligodeoxynucleotide: A facile post-synthetic modification technique for the preparation of oligodeoxynucleotides with two different functional molecules. Chem Commun 2000, 0, 59–60.

[67] Khan SI, Grinstaff MW. Palladium(0)-catalyzed modification of oligonucleotides during automated solid-phase synthesis. J Am Chem Soc 1999, 121, 4704–4705.

[68] Beilstein AE, Grinstaff MW. Synthesis and characterization of ferrocene-labeled oligodeoxynucleotides. J Organomet Chem 2001, 637–639, 398–406.

[69] Kahl JD, Greenberg MM. Introducing structural diversity in oligonucleotides via photolabile, convertible C5-substituted nucleotides. J Am Chem Soc 1999, 121, 597–604.

[70] Krasheninina OA, Fishman VS, Lomzov AA, Ustinov AV, Venyaminova AG. Postsynthetic on-column 2′-functionalization of RNA by convenient method. Int J Mol Sci 2020, 21, 5127 (15 pages).

[71] Lee Y-H, Yu E, Park C-M. Programmable site-selective labeling of oligonucleotides based on carbene catalysis. Nat Commun 2021, 12, 1681 (10 pages).

[72] Letsinger RL, Zhang GR, Sun DK, Ikeuchi T, Sarin PS. Cholesteryl-conjugated oligonucleotides: Synthesis, properties, and activity as inhibitors of replication of human immunodeficiency virus in cell culture. Proc Natl Acad Sci 1989, 86, 6553–6556.

[73] Davis PW, Osgood SA. A new method for introducing amidate linkages in oligonucleotides using phosphoramidite chemistry. Bioorg Med Chem Lett 1999, 9, 2691–2692.

[74] Santorelli A, Gothelf KV. Conjugation of chemical handles and functional moieties to DNA during solid phase synthesis with sulfonyl azides. Nucleic Acids Res 2022, 50, 7235–7246.

[75] Hansen RA, Märcher A, Nørgaard Pedersen K, Gothelf KV. Insertion of chemical handles into the backbone of DNA during solid-phase synthesis by oxidative coupling of amines to phosphites. Angew Chem Int Ed 2023, 62, e202305373 (8 pages).

[76] Biessen AL, Vietsch H, Rump E, Fluiter K, Kuiper J, Bijsterbosch MK, Martin K, Van Berkel TJC. Targeted delivery of oligodeoxynucleotides to parenchymal liver cells in vivo. Biochem J 1999, 340, 783–792.

[77] Prakash TP, Yu J, Migawa MT, Kinberger GA, Brad Wan W, Østergaard ME, Carty RL, Vasquez G, Low A, Chappell A, Schmidt K, Aghajan M, Crosby J, Murray HM, Booten SL, Hsiao J, Soriano A, Machemer T, Cauntay P, Burel SA, Murray SF, Gaus H, Graham MJ, Swayze EE, Seth PP. Comprehensive structure–activity relationship of triantennary N-acetylgalactosamine conjugated antisense oligonucleotides for targeted delivery to hepatocytes. J Med Chem 2016, 59, 2718–2733.

[78] Farzan VM, Ulashchik EA, Martynenko-Makaev YV, Kvach MV, Aparin IO, Brylev VA, Prikazchikova TA, Maklakova SYu, Majouga AG, Ustinov AV, Shipulin GA, Shmanai VV, Korshun VA, Zatsepin TS. Automated solid-phase click synthesis of oligonucleotide conjugates: From small molecules to diverse N-acetylgalactosamine clusters. Bioconjugate Chem 2017, 28, 2599–2607.

[79] Peraro L, Kritzer JA. Emerging methods and design principles for cell-penetrant peptides. Angew Chem Int Ed 2018, 57, 11868–11881.

[80] Kubo T, Bakalova R, Zhelev Z, Ohba H, Fujii M. Controlled intracellular localization of oligonucleotides by chemical conjugation. In Taira K, Kataoka K, Niidome T, eds. Non-viral Gene Therapy. Tokyo, Japan: Springer, 2005, 187–197.

[81] Venkatesan N, Kim BH. Peptide conjugates of oligonucleotides: Synthesis and applications. Chem Rev 2006, 106, 3712–3761.

[82] Brown SD, Graham D. Conjugation of an oligonucleotide to Tat, a cell-penetrating peptide, via click chemistry. Tetrahedron Lett 2010, 51, 5032–5034.

[83] Wenska M, Alvira M, Steunenberg P, Stenberg A, Murtola M, Strömberg R. An activated triple bond linker enables 'click' attachment of peptides to oligonucleotides on solid support. Nucleic Acids Res 2011, 39, 9047–9059.

[84] Klabenkova K, Fokina A, Stetsenko D. Chemistry of peptide-oligonucleotide conjugates: A review. Molecules 2021, 26, 5420 (36 pages).

[85] Zaramella S, Yeheskiely E, Strömberg R. A method for solid-phase synthesis of oligonucleotide 5′-peptideconjugates using acid-labile α-amino protections. J Am Chem Soc 2004, 126, 14029–14035.

[86] Ocampo SM, Albericio F, Fernandez I, Vilaseca M, Eritja R. A straightforward synthesis of 5′-peptide oligonucleotide conjugates using α -Fmoc-protected amino acids. Org Lett 2005, 7, 4349–4352.

[87] Debethune L, Marchan V, Fabregas G, Pedroso E, Grandas A. Towards nucleopeptides containing any trifunctional amino acid (II). Tetrahedron 2002, 58, 6965–6978.

[88] Wang T, Cao X, Zheng Y, Chen C, Zhou L, Sun D, Fang G, Tian C. Total stepwise solid-phase synthesis of peptide–oligonucleotide conjugates using side-chain Boc/tBu protecting groups. Chem Commun 2023, 59, 5839–5842.

[89]   Wu Y, Yang Z, Lu Y. Photocaged functional nucleic acids for spatiotemporal imaging in biology. Curr Opin Chem Biol 2020, 57, 95–104.

[90]   Tavakoli A, Min J-H. Photochemical modifications for DNA/RNA oligonucleotides. RSC Adv 2022, 12, 6484–6507.

[91]   Liu Q, Deiters A. Optochemical control of deoxyoligonucleotide function via a nucleobase-caging approach. Acc Chem Res 2014, 47, 45–55.

[92]   Hagen V, Bendig J, Frings S, Eckardt T, Helm S, Reute D, Kaupp UB. Highly efficient and ultrafast phototriggers for cAMP and cGMP by Using Long-Wavelength UV/Vis-Activation. Angew Chem Int Ed 2001, 40, 1046–1048.

[93]   Yoshimura Y, Fujimoto K. Ultrafast reversible photo-cross-linking reaction: Toward in situ DNA manipulation. Org Lett 2008, 10, 3227–3230.

[94]   Beharry AA, Woolley GA. Azobenzene photoswitches for biomolecules. Chem Soc Rev 2011, 40, 4422–4437.

[95]   Ogasawara S, Maeda M. Straightforward and reversible photoregulation of hybridization by using a photochromic nucleoside. Angew Chem Int Ed 2008, 47, 8839–8842.

[96]   Kolmar T, Büllmann SM, Sarter C, Höfer K, Jäschke A. Development of high-performance pyrimidine nucleoside and oligonucleotide diarylethene photoswitches. Angew Chem Int Ed 2021, 60, 8164–8173.

[97]   Brunderova M, Krömer M, Vlková M, Hocek M. Chloroacetamide-modified nucleotide and RNA for bioconjugations and cross-linking with RNA-binding proteins. Angew Chem Int Ed 2023, 62, e202213764 (6 pages).

[98]   Spampinato A, Kuzmova E, Pohl R, Sykorova V, Vrabel M, Kraus T, Hocek M. *trans*-Cyclooctene- and bicyclononyne-linked nucleotides for click modification of DNA with fluorogenic tetrazines and live cell metabolic labeling and imaging. Bioconjugate Chem 2023, 34, 772–780.

# 8 Nucleic acids: DNA

## 8.1 Structure of DNA

### 8.1.1 Base pairing

The commonly known structure of DNA is a double helix composed of two polydeoxyribonucleotides. These two strands run in opposite directions and are H-bonded to each other forming AT and CG base pairs depicted in Figure 8.1A. These base pairs are known as Watson–Crick base pairs (WC pairs). The NH groups of nucleic acid bases serve as H-bond donors and the lone electron pairs of carbonyl-oxygens and ring-nitrogens are H-bond acceptors. The strength of one hydrogen bond varies from 8 to 12 kJ mol$^{-1}$. The CG base pair is considerably more stable than the AT pair since it is formed by three H-bonds instead of two present in the AT pair. Solvation by water molecules, however, decreases the difference to only 4 kJ mol$^{-1}$ [1]. One should, however, bear in mind that H-bonding is not the only interaction responsible for the structure and stability of the double helical structure. In fact, vertical stacking interactions between the H-bonded base pairs stabilize the double helix even more than the horizontal H-bonding [2]. From another point of view, one may argue that stacking is partially present even in DNA single strands, and hence the role of H-bonding as a driving force of double-helix formation should not be underestimated [3].

While WC base pairing is the canonical mode of H-bonding in DNA, an alternative H-bonding pattern, that is Hoogsteen base pairing depicted in Figure 8.1C, may occur in special cases. Such base pairs have been observed in triplex [4] and quadruplex DNA [5] and in complexes of DNA with proteins [6] or antibiotics [7]. They also play a role in the so-called DNA breathing, a spontaneous local conformational transition occurring within double-stranded DNA. Evidently, a temporary replacement of a WC base pair with Hoogsteen pair makes DNA structure more flexible and better adaptive to interactions with other molecules.

The order of canonical AT and GC base pairs provides the code for protein synthesis, that is, determines the order of amino acids in proteins. Interestingly, this genetic code has been extended by incorporation of unnatural artificial base pairs in the genome of bacteria [8, 9]. When the number of base pairs increases from 2 to 3, 152 new triplet codons are available. This allows introduction of an unnatural amino acid into microbe-produced protein by making use of a modified one that contains a respective anticodon [10]. Figure 8.2A shows an example of unnatural base pair, dZ-dP, with H-bonding complementarity [11]. However, H-bonding complementarity is not absolutely required, but proper size and stacking property of the base pair may be sufficient [12]. dDs-dPx in Figure 8.2B serves as an example [13].

DNA is polyanionic under physiological conditions; the p$K_a$ values of the internucleosidic phosphodiester linkages are around 1. The multiple negative charges are

https://doi.org/10.1515/9783111325637-008

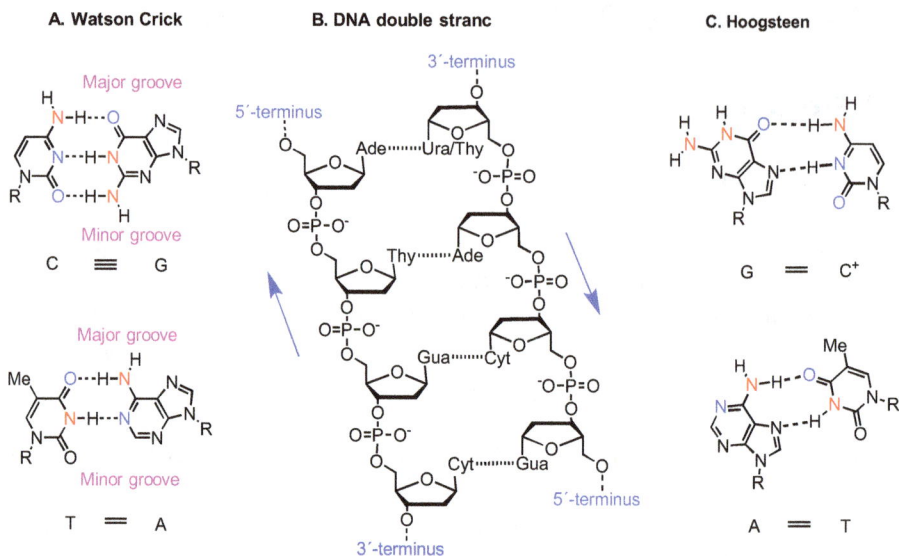

**Figure 8.1:** (A) Watson–Crick base pairs, (B) primary structure of double-stranded DNA and (C) Hoogsteen base pairs.

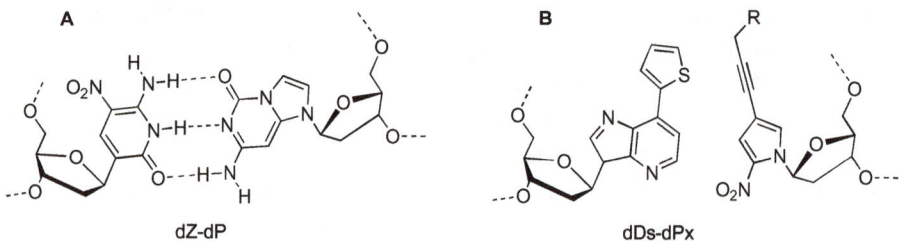

**Figure 8.2:** Examples of artificial base pairs: dZ-dP [11] and dDs-dPx [13].

largely neutralized by a dynamic cloud of monovalent cations that moves along the nucleic acid chain. Within single strands, the base moieties are up to one order of magnitude more basic than in nucleosides [14]. With double-stranded nucleic acids, the situation is less clear, but there are indications that protonated base pairs, in particular, CG pairs, may exist even in pH-range 6–7.

The WC base pair is approximately planar and perpendicular to the axis of the double helix. The pairs are isomorphous. This means that the distance between the N and O atoms is invariably 2.8 Å, the anomeric carbons are $10.6 \pm 0.2$ Å apart and the N-glycosidic bonds form a 55° angle with the line connecting the anomeric carbons. The two strands run in opposite direction, one from 5′ to 3′ and the other from 3′ to 5′ direction. Because base pairs are asymmetric, two grooves are formed on the surface of double helical cylinder, as indicated in Figure 8.1B.

## 8.1.2 Categories of DNA helices

The double-stranded helices of DNA fall in two categories: A- and B-DNA. B-DNA represents the normal DNA structure encountered in aqueous solution (or at high humidity) at low salt concentration. A-DNA, in turn, occurs at low humidity and high salt concentration. Both are anti-parallel right-handed duplexes (Figure 8.3). What does this mean? Let us think a double-stranded nucleic acid as a ladder, the right-hand side strand having the 5′-terminus up and the left-hand side strand running in the opposite direction. In case we keep the base pair at the bottom in a fixed position and start to turn the topmost base pair clockwise, we end up to a right-handed helix. If we turn counter clockwise manner we get a left-handed helix.

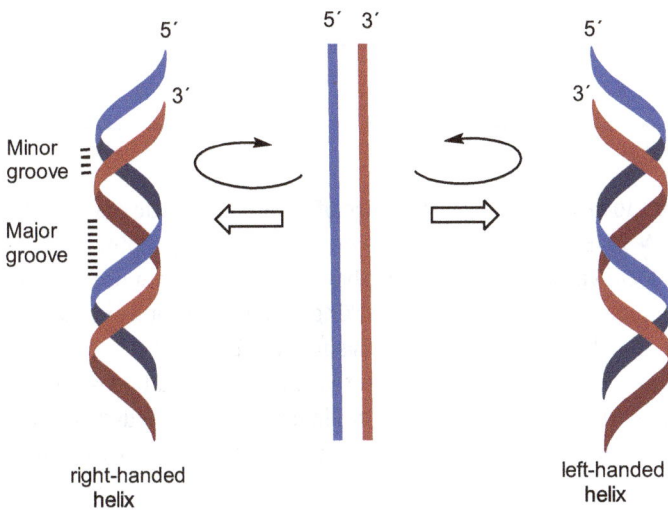

**Figure 8.3:** A schematic presentation for the meaning of right- and left-handed helices.

The conformation around the *N*-glycosidic bond of nucleosides is *anti* with both A- and B-type helices (cf. Section 1.3). The sugar rings are roughly parallel to the helix-axis and the phosphate groups are located at the outer surface of the helix cylinder. Sugar-ring puckering is different in A- and B-type helices: S in B-DNA and N in A-DNA. Distance between the base pairs is considerably longer in B-DNA (3.3 Å) than in A-DNA (2.3 Å). In A-DNA, but not in B-DNA, the base pairs are displaced 4.5 Å from the helix-axis leaving a hollow channel in the interior of the double helix. The A-type helix is therefore 3 Å wider than the B-type helix. In B-DNA the base pairs are perpendicular to the axis of double helix, while in A-DNA there is a 20° deviation from perpendicular orientation. The major groove in B-DNA is wide but shallow, in A-DNA narrow and deep. With minor groove, the situation is the opposite. The minor groove in B-DNA is narrow and deep, in A-DNA wide and shallow. The most important helical parameters for A- and B-type DNA are listed in Table 8.1 [15, 16].

**Table 8.1:** Structural parameters for A- and B-type DNA [15, 16].

|  | A-DNA | B-DNA |
|---|---|---|
| Base pairs to turn | 11 | 10.5 |
| Rise to base pair | 2.3 Å | 3.3 Å |
| Displacement of bases from the axis | 4.5 Å | Small |
| *Syn/anti*-Conformation | *Anti* | *Anti* |
| Ring puckering | N-type | S-type |
| Propellar twist within a base pair | 18° | 16° |
| Major groove | 2.7 Å wide | 12 Å wide |
|  | 13.5 Å deep | 8.8 Å deep |
| Minor groove | 11 Å wide | 6 Å wide |
|  | 2.8 Å deep | 7. Å deep |
| Diameter | 23 Å | 20 Å |

It is, however, important to realize that the structure of DNA is not static obeying the parameters typical for A- or B-type throughout the entire length. Depending on base sequence, a number of local variations occur [17]. Above all, changes in mutual orientation of the bases take place to maximize base-stacking and avoid repulsions caused by nonbonded interactions. B-DNA, for example, has sub types B', C, C' C", D, E, T. This is possible since the sugar-phosphate backbone is rather flexible. To fully describe its detailed conformation, six torsion angles defined in Figure 8.4A are needed. The detailed orientation of nucleobases is, in turn, described by parameters described in Figure 8.4B [17, 18].

At high salt concentrations, DNA may adopt an abnormal Z-DNA structure [2, 19]. This is an antiparallel left-handed double helix. One of the nucleosides in each base pair, usually the purine nucleoside, has N-type ring-puckering and a *syn*-conformation around the *N*-glycosidic bond, in contrast to normal *anti*. Consequently, the chain adopts a zigzag-like structure. There are 12 nucleosides per turn and the rise to base pair is 3.7 Å. The major groove is 8.8 Å wide and 3.7 Å deep. The minor groove is narrow and deep: 2 Å wide and 14 Å deep. Stabilization by high salt concentration may result from the fact that the distance between phosphate anions in opposite strands is smaller than with normal B-DNA, 8 Å vs. 12 Å.

A factor that enhances double-helix formation and influences on relative stability of various duplex types is hydration. B-DNA, that is the favored form of double helix, possesses a continuous three layers water network around it. When the water content is lowered to 10–20 molecules per base pair, conversion to A-DNA takes place [20]. The process is enforced by direct inner sphere binding of monovalent cations instead of water to phosphate oxygens. Normally B-DNA is in aqueous solution surrounded by

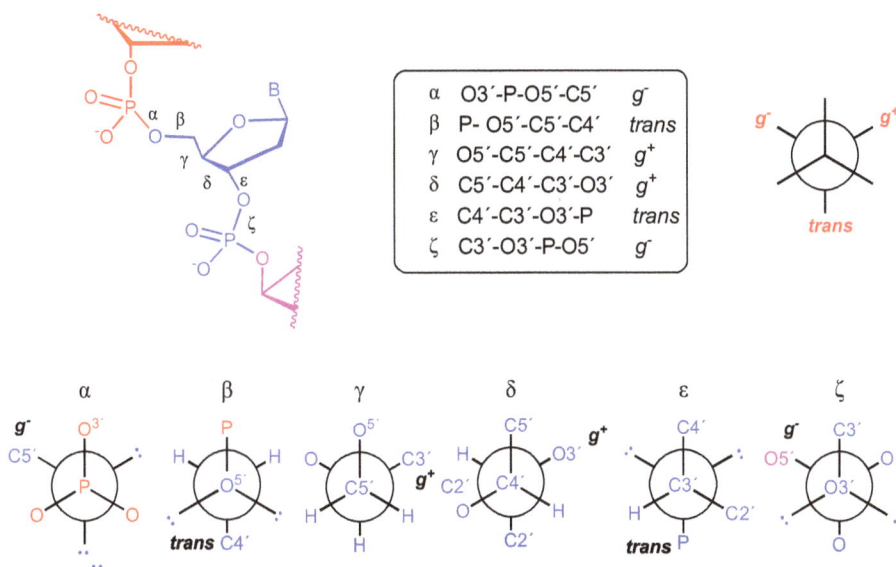

**Figure 8.4:** (A) Torsion angles defining the geometry of sugar-phosphate backbone in DNA and (B) parameters defining the position and orientation of nucleobases.

a cloud of monovalent metal ions that neutralize about 75% of the negative charge. Hydration is especially noticeable around the grooves, but the base moiety heteroatoms are also hydrated.

Figure 8.5 shows the overall shape of A-, B- and Z-DNA double helices. Each of the structures contains 22 base pairs. As shown, B-type DNA is more extended than A-DNA. One may wonder why so much attention is paid to A-DNA when B-DNA is the predominating structure in aqueous solution. The reason is that double-stranded RNA is of A-type, and hence what has been said about A-DNA largely applies to RNA double helices.

Eukaryotic DNA is linear, largely B-type double helix that evidently contains numerous local variations of helical parameters. It is packed into a folded structure by forming 1.7 turns loop around protein clusters consisting of eight individual histone proteins (Figure 8.6A) [21]. These basic units of packing are called nucleosomes. The double helical structure may, at least temporarily, contain bulges (slipped DNA, Figure 8.6B) [22] and hairpin loops (Figure 8.6C) [23], and it may form cruciform structures (Figure 8.6D) containing a so-called four-way junction [24]. The latter structure, called a Holliday junction, plays an important role as an intermediate of several key DNA processes such as insertion, recombination and repair.

Mitochondrial as well as bacterial DNA is circular. Such DNA can be wound around itself to a supercoiled form (Figure 8.6E).

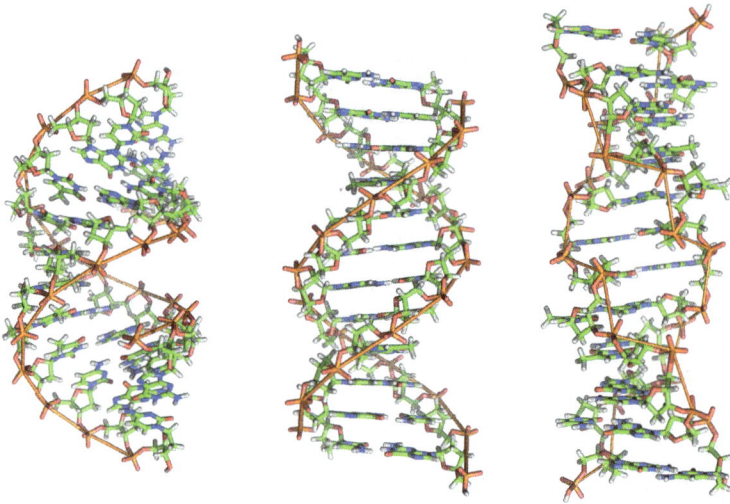

**Figure 8.5:** Comparison of the structure of A-, B- and Z-DNA (https://upload.wikimedia.org/wikipedia/com mons/b/b1/A-DNA%2C_B-DNA_and_Z-DNA.png).

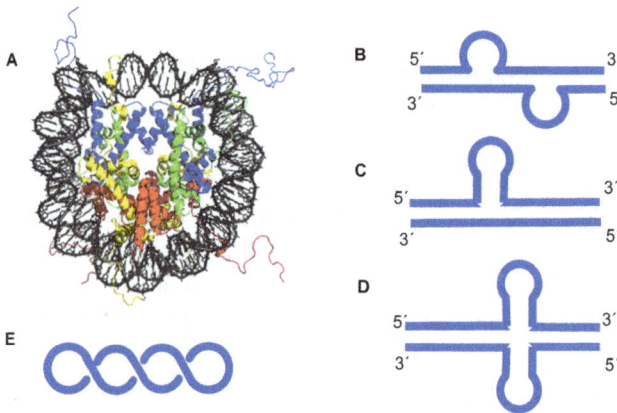

**Figure 8.6:** (A) Structure of nucleosome (https://en.wikipedia.org/wiki/ Nucleosome#/ media/File: Nucleosome_1KX5_colour_coded.png), (B) slipped DNA, (C) DNA containing a hairpin loop, (D) cruciform DNA and (E) supercoiled circular DNA.

### 8.1.3 Parallel DNA

While DNA usually occurs as a right-handed antiparallel duplex, parallel-oriented se-quences with noncanonical base pairing also exist [25]. Such sequences have been found in specific chromosome regions and in bacterial and insect genomes. Instead of WC base pairs, reverse WC (Figure 8.7) and Hoogsteen (Figure 8.1C) base pairs are formed. When engaged in a Hoogsteen GC⁺ base pair, cytosine base remains protonated even at pH 7 [26]. The overall shape of parallel DNA is rather similar to that of antiparallel B-DNA [27]. The base pairs are perpendicular to the helix axis and sugarring puckering is S. The two grooves are of similar size in contrast to normal B-DNA. Incorporation of 8-aminopurine bases [27, 28], isoguanine (6-amino-2-oxopurine) [29] and isocytosine (2-amino-4-oxopyrimidine) [29] markedly stabilize parallel nucleic acid structures. It is noteworthy that even the WC base pairs of canonical B-DNA exist in dynamic equilibrium with less than 1% population of Hoogsteen base pairs. The lifetime of such pairs is of the order of millisecond [30]. This dynamic structural change is known as Hoogsteen breathing since the distance of the anomeric carbons is in Hoogsteen base pair 2 Å shorter than in B-DNA.

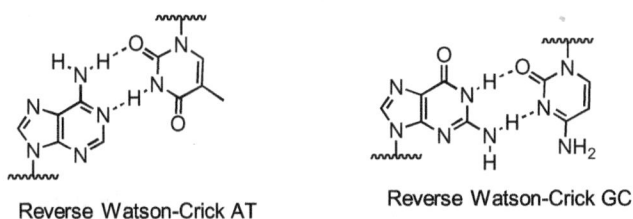

Reverse Watson-Crick AT

Reverse Watson-Crick GC

**Figure 8.7:** Reverse Watson–Crick base pairs.

## 8.1.4 Circular DNA

A minor proportion of DNA occurs in nature in a cyclized form either as a single stranded (ss) or a double stranded (ds) structure [31]. Circular DNAs exist in mitochondria, bacterial plasmids, chloroplasts, bacteriophages and viruses. The cyclic structure increases stability toward enzymatic degradation, above all against exonucleases. This has increased interest toward artificial ss circular DNAs as aptazymes, that is, catalytic DNAs the activity of which is regulated by aptamer-ligand interactions. A catalytic and a ligand-binding sequence is incorporated into a linear precursor of circular DNA and a chemical or enzymatic cyclization [32] is carried out.

## 8.1.5 Triple helical DNA

Polypurine/polypyrimidine sequences of B-DNA that are at least 10 base pairs long are able to bind an extra oligonucleotide into the major groove by Hoogteen or reversed Hoogsteen base pairing. In other words, a triple helical structure is formed. Two different binding modes exist. In the so-called H-DNA, a polypyrimidine is bound to the polypurine strand of B-DNA [33]. Base-triplets TAT or C⁺GC depicted in Figure 8.8A are formed. The additional Hoogsteen-paired pyrimidine strand is parallel with the purine strand and, hence, antiparallel to the other polypyrimidine strand. In H*-DNA, a polypurine strand is bound to the purine strand of B-DNA by reversed Hoogsteen base pairing. Base triplet AAT and GGC (Figure 8.8 B) are formed. The two purine strands are antiparallel. The sugar moieties retain *anti*-conformation around the *N*-glycosidic bond in both H- and H*-DNA triplexes. One should note that in C⁺GC triplet one of the cytosine bases is protonated, and hence the H-DNA-type triplexes are destabilized on approaching pH 7.

Formation of a clamp by chain invasion is also feasible. Polypyrimidine oligonucleotide may invade B-DNA by forming normal WC-paired duplex with the polypurine strand and then turn back forming Hoogsteen of reversed Hoogsteen binding to the same purine strand (Figure 8.8C) [34]. Triple helix formation in general and clamp formation in particular may well find applications as a means for gene recognition, as discussed in Section 11.4 in more detail.

### 8.1.6 G-quadruplexes

Guanine-rich DNA forms secondary structures containing planar guanine tetrads, where the guanines form Hoogsteen-type hydrogen bonds with each other and the resulting tetrads are vertically stacked (Figure 8.9A) [35]. The sugar ring puckering is either N or S. Interestingly, even guanosine 5′-monophosphate monomers aggregate in a similar manner in aqueous solution, as shown as early as in 1962 [36]. The distance between the tetrads is 3.25 Å. The structure, called G quadruplex, is stabilized by a monovalent ion, preferably $K^+$, situated in the center of the structure between two adjacent tetrads. Four separate strands may form a quadruplex [35], but it can also be formed from two [37] chains or even from one [38] chain. Figure 8.9B shows examples of known topologies [39]. $K^+$ favors parallel topology, whereas less tightly binding $Na^+$ enhances antiparallel topology [40]. Hydrophobic interactions with external species tend to destabilize the quadruplex structure regardless of topology [41].

The loops, that is, the oligonucleotide segments linking the participating guanine bases and *syn/anti*-equilibrium around the *N*-glycosidic bond largely determine the overall structure of the quadruplex. The structures fall in three major categories categories: parallel, antiparallel, hybrid [42]. Typical features of parallel structures are *anti*-conformation and three propeller-type loops running between the top and bottom tetrads. Antiparallel tetrads contain only lateral loops that link adjacent nucleosides within the top and/or bottom tetrads. Alternatively, they may contain, in addition to lateral loops, diagonal loops between opposite nucleosides within the top or bottom tetrad, and the *N*-glycosidic conformation may be *syn* or *anti*. Hybrid quarduplexes, in turn, may contain both *syn*- or *anti*-conformations and all kind of loops: propeller, lateral and diagonal. While DNA duplexes are in general subject to strand displacement of one prehybridized strand by a homologous single strand [43], related processes may also take place with quadruplex structures [44].

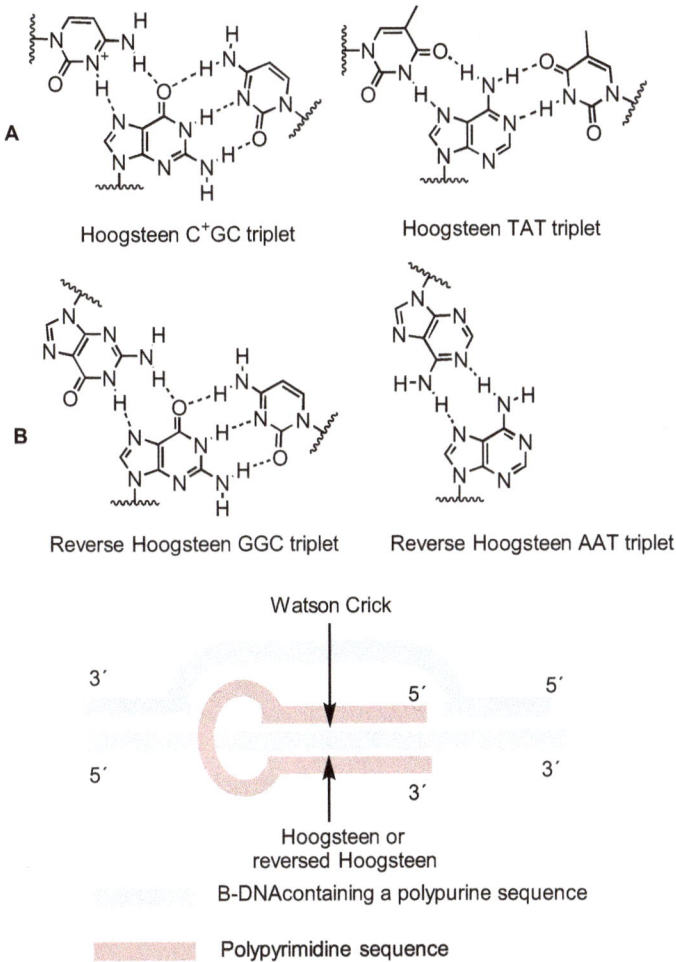

**Figure 8.8:** Base-triplets in triple helical H-DNA (A) and H*-DNA (B). Formation of a triple helical clamp by an invasion mechanism (C).

Human genome contains at least 370,000 sequences that could potentially form G quadruplexes [45], in particular, within regions that regulate transcription of genes [46]. Most extensively studied are those formed by four adjacent TTAGGG repeats at the terminal single-stranded DNA overhangs of chromosomes, called telomers. The quadruplex structures that these repeats form is, however, more versatile than the typical topologies described above. A tract of three or more consecutive guanosines is not an absolute prerequisite for quadruplex formation, but guanines rather distant from each other may participate in forming structures that resemble G-quadruplexs [47, 48].

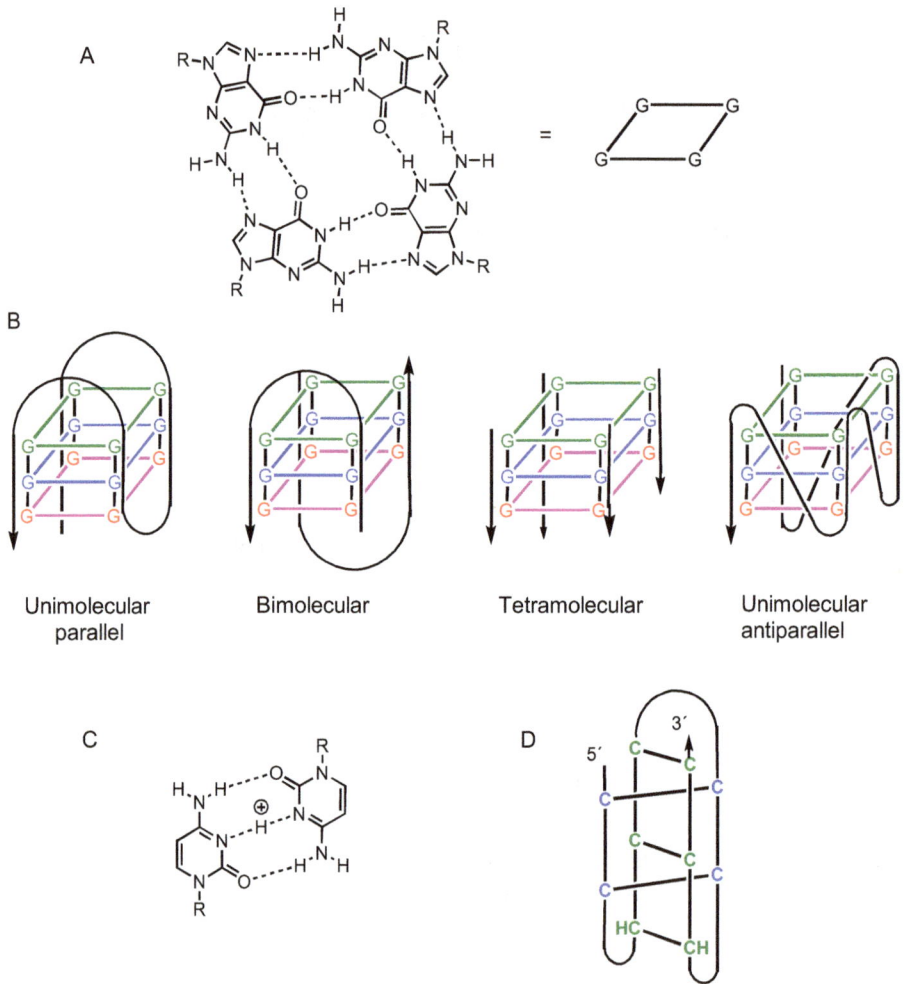

**Figure 8.9:** (A) Structure of a guanine tetrad, (B) topologies of G quadruplexes containing one (intramolecular), two (bimolecular) and four (tetramolecular) DNA strands [39], (C) structure of CC+ base pairs and (D) structure of i-motif [51].

### 8.1.7 i-Motifs

Another quadruplex structure of DNA is i-motif formed by stacking interactions of two parallel duplexes [49]. The duplexes involved consist of contiguous CC⁺ base pairs that according to theoretical calculations are more stable than the canonical CG pair [50]. Both duplexes are parallel but they form the quadruplex by antiparallel intercalation, as depicted in Figure 8.9C and D for sequence d[(5-MeC)CT₃CCT₃ACCT₃CC) [51, 52]. According to a recent observation [53], i-motif structures are present in regulatory regions of human genome. Their formation is cell-cycle and pH-dependent. The pH dependence is expected since protonation is required for the formation of $CC^+$ pairs. i-Motif is usually stable only at pH < 6. Nevertheless, the structure appears to be stable at physiological pH when the intercalating domains are longer than six CC pairs [54, 55]. Increasing concentration of $Li^+$, $Na^+$ or $K^+$ ions destabilizes i-motif [56]. Recently, stability of 271 sequences has been analyzed [57]. Each sequence contained four segments of 3–6 cytosine bases linked to each other by spacers of various lengths. The data allowed prediction of i-motif stability on the basis of primary structure. In addition, the occurrence of i-motif structure was verified in vivo by in-cell observation of imino proton NMR signals of $CC^+$ base pairs.

One should note that G-quadruplex and i-motif structures are both formed from a DNA sequence rich in GC base pairs. A recent study on double-stranded oncogenic c-Myc DNA promoter region sheds some light on the competition between the formations of these two quadruplex structures [58]. Under physiological pH, no i-motif is formed. Only random coil and G-quadruplex structures are detected. However, at lower pH both i-motif and G-quadruplex structures are formed.

## 8.2 Determination of DNA primary and secondary structure

### 8.2.1 Primary structure

Determination of the base sequence in DNA is called sequencing. The pioneering studies in this field date back to 1977. In that year, Maxam and Gilbert [59] published a method based on selective cleavage of nucleobases from DNA oligonucleotides and Sanger [60] a method based on enzymatic synthesis of a complementary oligonucleotide using the oligomer under sequencing as a template. The Maxam Gilbert approach is almost entirely nonenzymatic. Only the initial fragmentation of DNA to sequences of various lengths is enzymatic. The oligonucleotides obtained are first labeled with $^{32}P$ at the 5′-terminus and then each labeled oligomer is divided into four portions. Each portion is subjected to one of the four nucleobase selective reaction: (i) acid-catalyzed depurination with formic acid removes Ade and Gua, (ii) methylation of N7 of Gua in DMSO followed by alkaline imidazole ring opening removes Gua, (iii) treatment with hydazine cleaves Cyt as 3-aminopyrazole and Thy as 4-methyl-3-pyrazolone and (iv) the same reaction at high concentration of NaCl turns Cyt-selective. Treatment with piperidine at elevated temperature results in chain cleavage at the abasic sites. Aliquots of the product mixtures of the four reactions are arranged side by side in gel electrophoresis for size separation and the 5′-terminal $^{32}P$-labeled fragments are visualized by autoradiography. The sequence can then be read on the basis of the fragments produced by each of the four reactions (Figure 8.10).

The other first-generation sequencing, Sanger's sequencing, is based on a DNA polymerase catalyzed assembly of 5′-$^{32}P$-labeled oligonucleotide on a DNA template, not on p*-5'GCTACGTA3'

| | Cleavage at A&G | Cleavage at G | Cleavage at C&T | Cleavage at C |
|---|---|---|---|---|
| A | p*-GCTACGT | | | |
| T | | | p*-GCTACG | |
| G | p*-GCTAC | p*-GCTAC | | |
| C | | | p*-GCTA | p*-GCTA |
| A | p*-GCT | | | |
| T | | | p*-GC | |
| C | | | p*-G | p*-G |

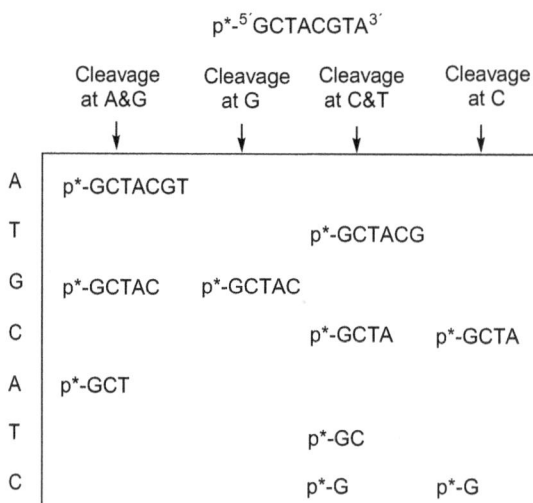

**Figure 8.10:** Principle of Maxam–Gilbert sequencing [59].

degradation of DNA. Again four parallel reactions, in this case four oligonucleotide syntheses are carried out as in Maxam–Gilbert method. In each of the four parallel synthesis, one 2',3'-dideoxy analog of the natural 2'-dNTPs is present in addition to all four natural ones. Incorporation of this dideoxy analog into the growing chain results in partial chain termination. In other words, a set of 5'-$^{32}$P-labeled ODNs of various lengths are produced by each synthetic reaction. By parallel polyacrylamide gel analysis, the lengths of the ODNs can be determined and the base sequence read (Figure 8.11).

Sanger's method may be regarded as the basis of several fully automated second generation sequencers [61]. One of the most widely used methods is the Solexa sequencing [62]. The purified DNA is first cut randomly by transposome enzymes to 200–300 base pair pieces. An adapter sequence and a primer sequence are then attached by ligases to the termini of these fragments. The adapter sequence is required for immobilization of the DNA fragments by hybridization on a solid support bearing hundreds of thousands of short catcher sequences. The primer sequence, in turn, allows enzymatic amplification on the solid-supported array. Each fragment is then sequenced separately but simultaneously. Fluorescently tagged dNTPs bearing an orthogonally removable protecting group at $O^{3'}$ are used as terminators instead of 2',3'-dideoxy-NTPs. The fluorescent dye is attached to the nucleobase via a bio-orthogonal linker to C5 of pyrimidines and N7 of purines. The essential feature of the linker is an α-azido substituted oligoethyleneglycol fragment. This linker is cleaved and the $O^{3'}$-CH$_2$N$_3$ group removed in a single step after DNA polymerase catalyzed incorporation to the 3'-terminus of the growing ODN chain. Staudinger reduction with triphenylphosphine converts the azido groups to amino groups generating a hydrolytically labile O-C-NH$_2$ linkage that undergoes C-O bond fission (Figure 8.12). The released dye can be identified and the $O^{3'}$ is free

Primer ⟶ p\*-5'TCAG3'
3'AGTCTAGCAACT5'

| | Terminated with ddA | Terminated with ddG | Terminated with ddC | Terminated with ddT |
|---|---|---|---|---|
| A | p\*-TCAGATCGTTGA<sup>dd</sup> | | | |
| G | | p\*-TCAGATCGTTG<sup>dd</sup> | | |
| T | | | | p\*-TCAGATCGTT<sup>dd</sup> |
| T | | | | p\*-TCAGATCGT<sup>dd</sup> |
| G | | p\*-TCAGATCG<sup>dd</sup> | | |
| C | | | p\*-TCAGATC<sup>dd</sup> | |
| T | | | | p\*-TCAGAT<sup>dd</sup> |
| A | p\*-TCAGA<sup>dd</sup> | | | |

Complementary sequence ³'AGTTGCTA⁵'
Sequenced sequence     ⁵'TCAACGAT³'

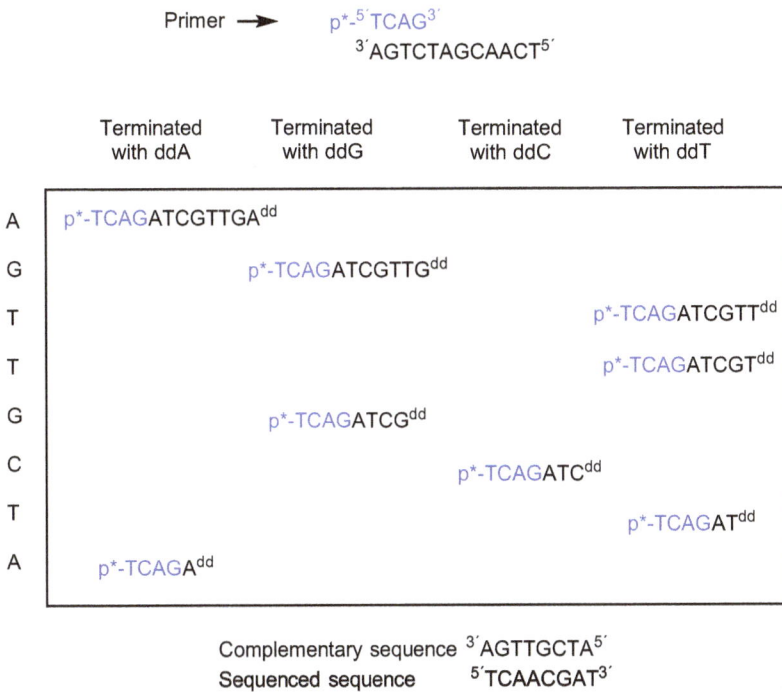

**Figure 8.11:** Principle of Sanger sequencing [60].

to react with the next fluorescently tagged NTP. Different nucleobases are naturally tagged with different dyes, and hence the identity of the incorporated nucleoside becomes identified. Owing to partial overlapping regions of the randomly generated ODNs, the entire base sequence of the DNA sample is obtained.

Solexa sequencing, as other second generation sequencing techniques, still depends on DNA amplification by polymerase chain reaction, although bright fluorophores together with laser excitation have been reported to allow sequencing of a single ODN chain [63].

Several third-generation sequencers allowing single-molecule sequencing are underdeveloped. The best known among these is the nanopore technology. DNA molecule is passed through a nanopore in a conductive material. Nucleotides, owing to their slightly different size, stretch the pore in a characteristic manner, allowing direct determination of the base sequence [64–66]. Nanopore sequencing is not limited to canonical bases only, but base modifications are also detected [67]. Even artificial duplexes containing four different base pairs (four canonical and four modified bases) have been sequenced by nanopore technology [68]. The epigenetic nucleic acid bases discussed in Section 1.1 cannot be detected by conventional sequencing methods but special techniques, often based on chromatographic separation and mass spectrometric analysis of the modified bases, are used [69].

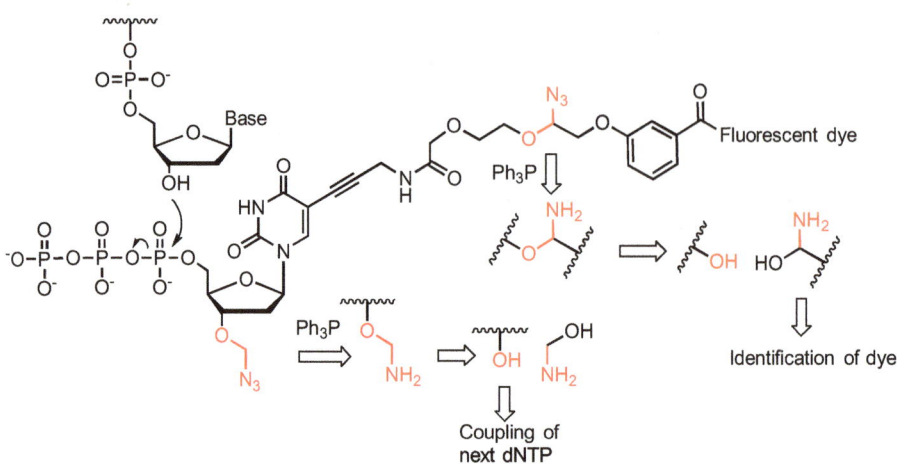

**Figure 8.12:** Principle of Solexa sequencing [62].

## 8.2.2 Secondary structure

Since the pioneering work of Watson and Crick in 1953 [70], thousands of X-ray structures of nucleic acids have been deposited in the Nucleic Acid Database [71]. The exact structural parameters of various families of nucleic acids are based on these studies. Additional techniques are, however, required to answer whether these data can be applied as such to aqueous solution and intracellular conditions. CD spectroscopy in the wavelength range 180–300 nm is the routinely used method [72]. The difference in molar absorptivity of the left- ($\varepsilon_L$) and right-handed ($\varepsilon_R$) circularly polarized light is usually reported. B-type of double helices typically show positive band ($\varepsilon_L - \varepsilon_R < 0$) in the 260–280 nm region and a negative band around 245 nm. With A-type helices, the positive band occurs at somewhat lower wavelengths, around 260 nm, and the negative band between 200 nm and 220 nm. The spectrum of the left-handed Z-DNA is very different: a negative band at 290 nm, a positive band at 260 nm and a very strong negative band at 190–200 nm. The exact shape of the spectrum, however, always depends on the base sequence. Figure 8.13 shows as an example the CD spectra of poly (dG:dC) in three different helical forms [73].

Another useful technique for the detection of conformational changes of nucleic acids in solution is FTIR spectroscopy. Several bands referring to a certain DNA conformational family are known. The sensitivity is so high that conformational changes may be followed even in functional cells [74]. IR spectroscopy also offers a straightforward method with which to assess whether a DNA sample has undergone dehydration and, hence, conversion to A-DNA during isolation and sample treatment. In case rehydration is accompanied by the appearance of bands typical for B-DNA, dehydration had taken place during isolation.

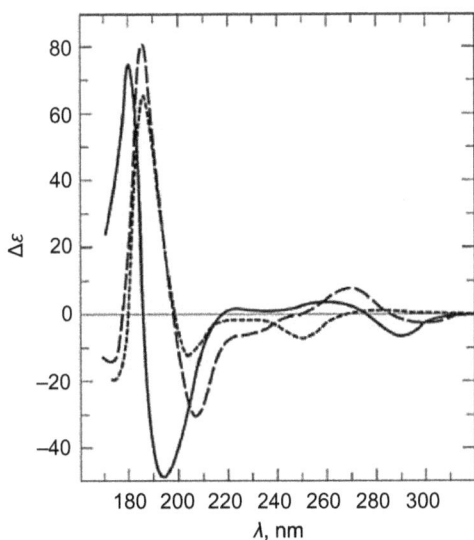

**Figure 8.13:** CD spectra of poly(dG:dC) in different double helical forms: B-DNA ($\cdots$), A-DNA (——) and Z-DNA ( $-$ ). (http://what-when-how.com/wp-content/uploads/2011/05/tmp2122_thumb12.jpg); http:// what-when-how.com/molecular-biology/circular-dichroism-part-2-molecular-biology/.

$^{31}$P NMR spectroscopy, although less sensitive than FTIR spectroscopy, also provides additional information about nucleic acid conformation in solution. It has been shown that $H^{3'}$-C3'-O$^{3'}$-P torsion angle and C4'-C3'-O$^{3'}$-P angle therefrom can be obtained from the three bond-coupling constants by Karplus relationship (cf. Section 1.3) [75]. This torsion angle has turned out to be fairly sequence independent for A-type helices, allowing distinguishing of A-DNA from B-DNA.

As far as oligonucleotide duplexes up to 100-mers are concerned, two-dimensional NMR spectroscopy can be used to determine the conformation in solution phase [76]. Total coherence transfer spectroscopy is applied to identify through-bond couplings and nuclear Overhauser effect (nOe) spectroscopy (NOESY) to detect couplings through space [77]. The nuclei used include $^1$H, $^{13}$C, $^{15}$N and $^{31}$P. A prerequisite for the detection of $^{13}$C and $^{15}$N nuclei is, however, incorporation of isotopically enriched nucleosides. nOe couplings of purine H8 and pyrimidine H6 with H1' protons play a decisive role in the conformational analysis. Each H6/H8 has a 2D cross-peak with H1' of the same nucleoside and its 5'-neighbor. Likewise, each H1' is cross-coupled to H6/H8 of the same molecule and its 3'-neighbor. This allows determination of the sequential connectivity along the chain as indicated in Figure 8.14. The imino proton resonances, in turn, indicate the base pairing. There are indications that even single-stranded oligonucleotides favor in solution a helical structure that closely resembles their structure in a double helix [78], but the structure is still too flexible to allow structure determination by NMR spectroscopy.

$^{19}$F substitution has received increasing interest as a means to detect conformational abnormalities of nucleic acids. Since the chemical shift of the $^{19}$F signal depends

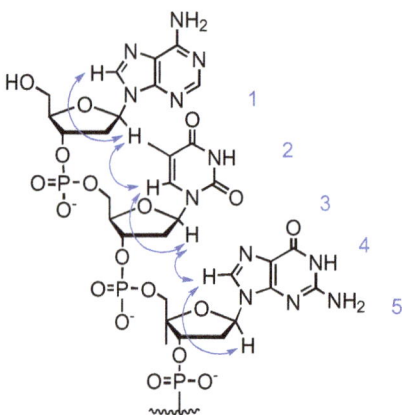

**Figure 8.14:** Determination of the conformation of the sugar-phosphate backbone with the aid of 2D nuclear Overhouse effect cross-couplings; the so-called nOe-walk.

on the molecular environment, formation of noncanonical structures, such as mismatched and bulged structures, triple helices and quadruplexes, is reflected to the $^{19}$F-shift. The $^{19}$F-probe may be present in the base [79, 80], sugar [81, 82] or phosphate [83] moiety. The $^{19}$F-shift is usually measured as a function of temperature. A similar technique may also be used to study interactions between nucleic acids and other biomolecules.

## 8.3 Stability of nucleic acids secondary structure

Determination of the stability of double helices and other secondary structures of nucleic acids is usually based on changes of UV absorption as a function of temperature. A single-stranded oligonucleotide has UV absorption maximum at ca. 260 nm, the molar absorptivity per nucleobase being of the order of $10^4$ $M^{-1}$ $cm^{-1}$. Stacking of base pairs upon formation of a double helix results in an up to 30% decrease in the absorptivity compared to a random single strand, a phenomenon known as a hypochromic effect. When a solution of double helix is slowly heated, the absorbance first remains almost unchanged, increases then by about 30% over a narrow temperature range and finally levels off to an almost constant value. A sigmoid melting curve is obtained, the inflection point of which is called the melting temperature ($T_m$) of the double helix in question. In addition to base composition, melting temperature $T_m$ depends on pH, ionic strength and even the identity of the salt used to adjust the ionic strength. Besides temperature, some H-bonding active compounds, such as urea or formamide, destabilize duplexes. 95% aqueous formamide is used to denature DNA completely at room temperature.

With triple helices, the melting curve is biphasic exhibiting two inflection points. The third strand bound to the major groove of B-DNA is dissociated at a temperature markedly below the melting temperature of the duplex. The formation and dissociation of triple helix is a slower process than duplex formation. That is why the temperature should be changed more slowly than at the rate 0.5 °C min$^{-1}$ routinely used in measurements of duplex melting temperature. In case the temperature has been changed too rapidly, a hysteresis occurs, i.e. the heating and cooling curves do not overlap.

The melting curve of G quadruplex is very different and helps to recognize the presence of a quadruplex structure. The quadruplex structure absorbs at 295 nm, at such a long wavelength where absorbtivity of nucleic acids otherwise is low. Upon breakdown of the quadruplex, this absorbance disappears.

$T_m$-value correlates with the stability of duplex. A decrease of 3–5 °C corresponds to weakening of the duplex by one order of magnitude. At the melting temperature, the concentration of the duplex is half of the initial concentration ($=\frac{1}{2}C_{init}$) and, hence, the equilibrium constant, $K_{eq}$, for the formation of the duplex from single strands is expressed as follows:

$$K_{eq} = (1/2\,C_{init}) / (1/2\,C_{init})^2 = 2/C_{init} \tag{8.1}$$

When $T_m$ is measured at various initial concentrations, $C_{init}$, and the first approximation of Van't Hoff equation ($\Delta H$ is independent of $T$) is applied, the enthalpy, entropy and free energy of the duplex formation may be calculated from the following equation:

$$\ln K_{eq} = -\Delta G^\theta/RT = \Delta H^\theta/RT + \Delta S^\theta/R \tag{8.2}$$

The melting curve, in fact, indicates the mole fraction of duplex and dissociated single strands at any point, and hence, a set of equilibrium constants at various temperatures referring to the ascending part of the melting curve is known. This allows calculation of the values of thermodynamic functions by eq. (8.2). The approach is extensively used, but it is more sensitive to errors than the one based on separate measurements at various initial concentrations.

Generally speaking, GC base pairs stabilize the duplex more than AT base pairs. The melting temperature of long polynucleotides usually depends on the percentage CG content according to eq. (8.3) at the ionic strength 0.3 M and pH 7:

$$T_m \cong 69.3 \; °C + 0.41(\%GC) \tag{8.3}$$

One should, however, bear in mind that vertical stacking of adjacent base pairs is more important than horizontal hydrogen bonding [1]. In other words, the identity of nearest neighbors influences on the contribution that an individual base pair has to the overall duplex stability. This can be taken into account by using $\Delta G^\ominus$ contributions for dimeric or trimeric fragments for prediction of the overall stability of a given oligomer. Table 8.2 records the so-called nearest-neighbor $\Delta G^\ominus$ parameters for dimeric DNA/DNA, RNA/RNA and RNA/DNA duplexes in 1M NaCl at 37 °C. The $\Delta G^\ominus$ value for a DNA/DNA duplex 5'-GCTAGC-3'/3'-CGATCG-5', for example, is obtained as a sum of the contributions of the five dimeric fragments, i.e., $2 \times \Delta G^\ominus$ (GC/CG), $\Delta G^\ominus$ (CT/GA), $\Delta G^\ominus$

**Table 8.2:** Nearest-neighbor parameters ($\Delta G^\ominus$) for formation of DNA/DNA, RNA/RNA and RNA/DNA duplexes in 1M aq NaCl at 37 °C.

| DNA/DNA[a] duplex | $\Delta G^\ominus$(37 °C) kcal mol⁻¹ | RNA/RNA[b] duplex | $\Delta G^\ominus$ (37 °C) kcal mol⁻¹ | RNA/DNA[c] duplex | $\Delta G^\ominus$ (37 °C) kcal mol⁻¹ |
|---|---|---|---|---|---|
| 5'AA3'/3'TT5' | -1.02 | 5'AA3'/3'UU5' | -0.93 | rAA/dTT | -1.0 |
| 5'TT3'/3'AA5' | | 5'UU3'/3'AA3' | | rUU/dAA | -0.2 |
| 5'AT3'/3'TA5' | -0.73 | 5'AU3'/3'UA5' | -1.10 | rAU/dTA | -0.9 |
| 5'TA3'/3'AT5' | -0.60 | 5'UA3'/3'AU5' | -1.33 | rUA/dAT | -0.6 |
| 5'CA3'/3'GT5' | -1.38 | 5'CA3'/3'GU5' | -2.11 | rCA/dGT | -0.9 |
| 5'TG3'/3'AC5' | | 5'UG3'/3'AC5' | | rUG/dAC | -1.6 |
| 5'GT3'/3'CA5' | -1.43 | 5'GU3'/3'CA5' | -2.24 | rGU/dCA | -1.1 |
| 5'AC3'/3'TG5' | | 5'AC3'/3'UG5' | | rAC/dTG | -2.1 |
| 5'CT3'/3'GA5' | -1.16 | 5'CU3'/3'GA5' | -2.08 | rCU/dGA | -0.9 |
| 5'AG3'/3'TC5' | | 5'AG3'/3'UC5' | | rAG/dTC | -1.8 |
| 5'GA3'/3'CT5' | -1.46 | 5'GA3'/3'CU5' | -2.35 | rGA/dCT | -1.3 |
| 5'TC3'/3'AG5' | | 5'UC3'/3'AG5' | | rUC/dAG | -1.5 |
| 5'CG3'/3'GC5' | -2.09 | 5'CG3'/3'GC5' | -2.36 | rCG/dGC | -1.7 |
| 5'GC3'/3'CG5' | -2.28 | 5'GC3'/3'CG5' | -3.42 | rGC/dCG | -2.7 |
| 5'GG3'/3'CC5' | -1.77 | 5'GG3'/3'CC5' | -3.26 | rGG/dCC | -2.9 |
| 5'CC3'/3'GG5' | | 5'CC3'/3'GG5' | | rCC/dGG | -2.1 |
| Initiation, at least one GC pair[d] | +1.82 | Initiation | +4.09 | Initiation | +3.1 |
| Initiation, only AT3 pairs[e] | +2.8 | Penalty for a terminal AU pair | +0.45 | | |
| Self-complementarity | +0.4 | | +0.43 | | |
| Penalty for terminal 5'TA3'pair[f] | +0.4 | | 0 | | |

[a]. Values taken from Ref. [84], [b]values taken from Ref. [85], [c]values taken from Ref. [86], [d]applied to duplexes that contain at least one GC pair, [e]applied to sequences containing only AT pairs, [f]does not concern 5'AT3'.

(TA/AT), $\Delta G^\ominus$ (AG/TC) and the contribution of initiation, $\Delta G^\ominus$(init) and symmetry correction $\Delta G^\ominus$(sym).

Duplex formation is an enthalpy-driven process that suffers from moderate entropy penalty. The $\Delta H^\ominus$ values of DNA/DNA duplexes range from −6.1 kcal mol$^{-1}$ (for 5′CT3′/3′ GA5′) to −11.1 kcal mol$^{-1}$ (for 5′GC3′/3′CG5′ the $-T\Delta S^\ominus$ values for these fragments being +5.0 kcal mol$^{-1}$ and +8.8 kcal mol$^{-1}$, respectively [84]. For RNA/RNA duplexes the enthalpy contributions vary from −6.8 kcal mol$^{-1}$ (for 5′AA3′/3′UU5′ to 14.9 kcal mol$^{-1}$ (for 5′GC3′/5′CG3′), referring to entropic contributions +5.9 kcal mol$^{-1}$ and +11.4 kcal mol$^{-1}$ [85].

## 8.4 Hybridization diagnostics

The occurrence of a given base sequence within genome is determined by hybridization test with an oligonucleotide probe having a complementary base sequence. Such studies are usually carried out for (i) verification of clinical symptoms of a genetic disorder, (ii) prenatal or postnatal evaluation of genetic risks without clinical symptom, (iii) population genetics and (iv) clarification of dynamic aspects of gene expression (functional genomics). For all these purposes, high-throughput diagnostics is needed. The most common approach is utilization of solid-supported oligonucleotide microarrays as a test platform. High-density microarrays may consist of thousands or even dozens of thousands oligonucleotides on a $2.5 \times 5$ cm glass plate. They are prepared either by synthesizing the oligonucleotides in situ on the glass plate or by immobilizing prefabricated oligonucleotides on a preactivated plate [87].

As regards in situ synthesis on a glass plate, a linker is first attached to glass by treatment with trimethoxysilane bearing a long chain ω-hydroxyalkyl group (Figure 8.15). This linker is then elongated with an appropriate spacer that allows oligonucleotide synthesis and warrants efficient hybridization [88]. The spacer may for instance be a short polyethylene glycol phosphoramidite having the terminal hydroxyl function protected with a photolabile protecting group [89]. Figure 8.29 shows examples of such groups. Alternatively, a 4,4′-dimethoxytrityl group may be used as in conventional solid-phase oligonucleotide synthesis. In this case, the deprotection is triggered with acid produced by photochemical decomposition of triphenylsulfonium hexafluorostilbate (Ph$_3$S$^+$SbF$_6^-$) [90].

A photolithographic approach is usually applied for the preparation of the microarray [91]. The glass plate is covered with a mask-containing holes on sites that are aimed at being photochemically deprotected and subsequently subjected to coupling with a given 5′-protected nucleoside 3′-phosphoramidite. The procedure is repeated until about 25-mer oligonucleotides are assembled. In a recent version of photolithographic microarray synthesis, the mask is replaced by a digitally controlled array of microscopic mirrors that project light very precisely on a desired point on the glass plate. The technique allows fabrication of very high density arrays [92]. Highly photolabile 2-[4-ethyl-2-nitro-5-(phenylthio)phenyl]propoxycarbonyl group [93] is used for

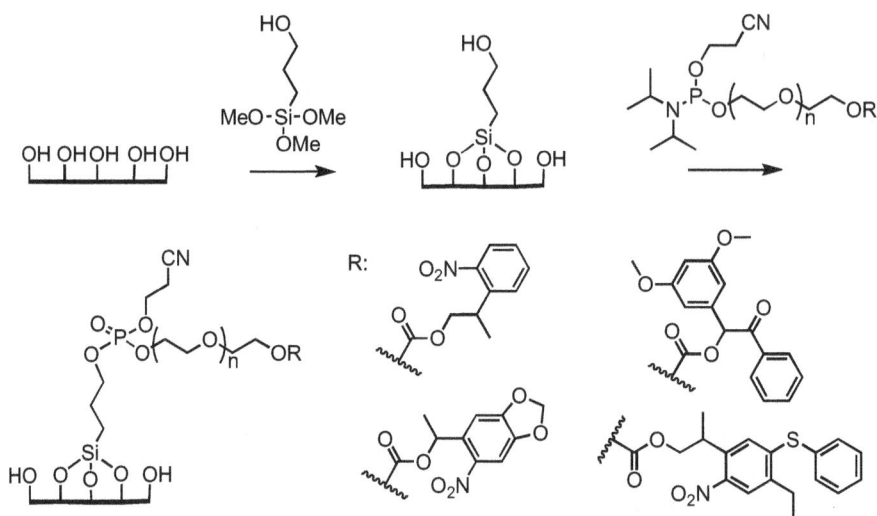

**Figure 8.15:** In situ synthesis of oligonucleotide microarrays using photochemically removable 5'-protecting groups.

5'-O-protection. On using 5-ethylthiotetrazole as an activator, coupling and deprotection times are 15 s and 9 s, respectively, and the whole coupling cycle takes only 50 s [94].

Microarrays are also prepared by spotting prefabricated oligonucleotides on glass slides [94], but the density of ODN probes in then lower than on slides obtained by in situ synthesis. The postsynthetic immobilization is based on a covalent bond formation between appropriately functionalized glass slide and ODN. For example, thiolated oligonucleotides can be attached to a maleimide-activated surface [95] or glyoxylyl oligonucleotides to semicarbazide bearing glass slides [96] (Figure 8.16). Depending on which terminus of the ODN has been used for functionalization, arrays having either 3'-OH or 5'-OH available for further reactions are obtained. Free 3'-terminus, for example, allows primer extension by DNA polymerases and, hence, enzymatic assays.

Microsphere-based suspension arrays offer an entirely different approach [97]. Polystyrene microspheres having a diameter of a few micrometers are loaded with a red and infrared dye at different concentration ratios to give a name for the particle. A specific oligonucleotide probe is then immobilized to each particle and a suspension of up to 100 particles is subjected to a hydridization assay with the biological sample containing fluorescently labeled oligonucleotides. The fluorescence emission is measured directly on the particle and the particle is recognized on the basis of VIS/IR absorption. Self-evidently, oligonucleotides on microarrays are not entirely homogeneous but most likely contain $n - 1$ and shorter truncated sequences. Nevertheless, the longmers undoubtedly predominate, and they form more stable duplexes than their shorter analogs. Accordingly, the results are reasonably reliable.

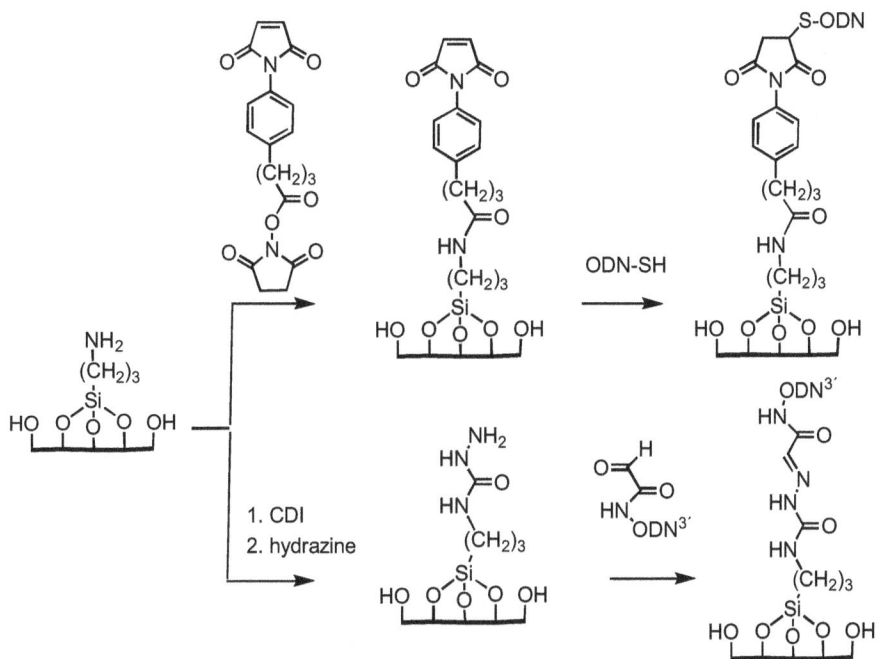

**Figure 8.16:** Fabrication of oligonucleotide microarrays by postsynthetic immobilization on an amino functionalized glass plate [94–96].

On applying microarray technique, oligonucleotides subjected to analysis must somehow be visualized for the detection of their binding to the microarray. This can be done by introducing a fluorescently labeled nucleoside 5′-monophosphate enzymatically to the 3′-terminus of the oligonucleotide with the aid of terminal deoxynucleotidyl transferase [98]. This kind of labeling does not markedly disturb hybridization with the solid-supported oligonucleotide probes, providing that the probes are anchored through the 3′-terminus. Upon antiparallel duplex formation, the fluorophore, hence, points to solution phase, not toward the glass support. Some commonly used fluorophores are depicted in Figure 8.17. BODIPY and Cy-dyes, in particular, are extensively used. They are highly absorbing, have high quantum yield and the difference between the excitation and emission wavelength, the so-called Stokes shift, is large. Fluorescein suffers from photobleaching and rhodamine from somewhat low quantum yield. The fluorescence emission of dansyl and NBD is environment-sensitive. All these fluorophores are usually tethered to the nucleoside base moiety via an aminolinker.

Although the sensitivity of fluorescence emission to environment is usually regarded as a disadvantage rather than advantage, this phenomenon can also be utilized in hydrization diagnostics. Molecular beacons offer an example for the utilization of environment-sensitive fluorophores. Molecular beacons are ODN hairpins enabling hybridization assays in solution phase. Originally, beacons bore a fluorescent dye at one

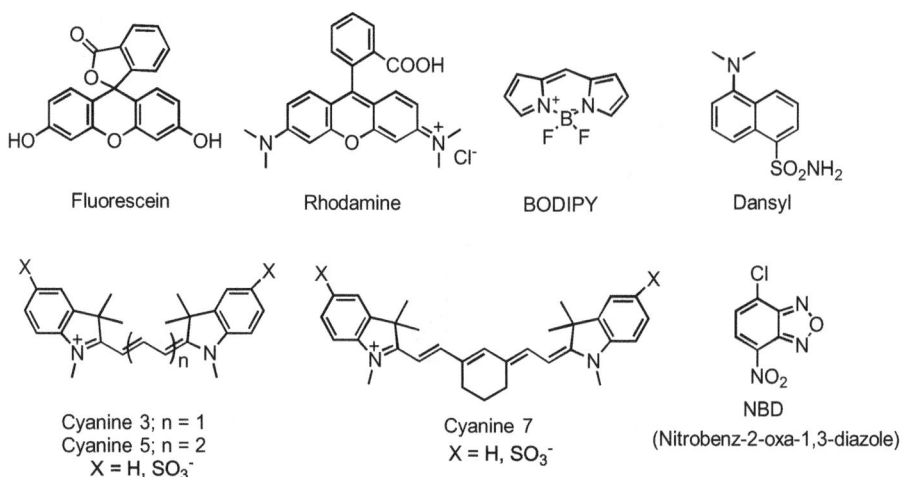

| Fluorescein | Rhodamine | BODIPY | Dansyl |

Cyanine 3; n = 1
Cyanine 5; n = 2
X = H, SO₃⁻

Cyanine 7
X = H, SO₃⁻

NBD
(Nitrobenz-2-oxa-1,3-diazole)

**Figure 8.17:** Fluorophores frequently used in hybridization diagnostics.

terminus of the oligomer and a quencher, e.g., 4-(dimethylaminoazo)benzene-4-carbox-ylic acid (dabcyl), at the other end (Figure 8.18A). Owing to the hairpin structure, the fluorophore and quencher were so close to each other that the fluorescent emission was entirely quenched. Upon hydridization with a fully complementary sequence, the distance between the fluorophore and quencher increased and fluorescence emission was detected [99]. Quencher-free molecular beacons have more recently been developed [100]. Pyrene or fluorene is attached to C5 of a pyrimidine or C8 or a purine nucleoside via a rigid alkyne linker, and this nucleoside is placed in the middle of the hairpin loop (Figure 8.18B). The neighboring nucleobases, above all guanine bases, markedly quench the fluorescence emission. Upon double helix formation, the fluorophore becomes oriented outside the double helix and the emission intensity increases. Environment sensitive nucleosides, 2-aminopurine and pyrrolopyrimidine ribosides (cf. Section 1.5), have been used for the same purpose [101]. The fluorescence spectrum of these nucleosides is markedly altered upon single strand/double strand transition, and hence by proper beacon design a reliable test is been obtained.

A factor that limits the sensitivity of assays based on organic fluorophores is a short-lived background fluorescence of biologic material. The fluorescence life time of organic fluorophores also is short and hence the background fluorescence cannot be entirely eliminated. By contrast, the fluorescence life time of some lanthanide chelates is long, allowing time-resolved (time-gated) measurement [102]. Measurement can be initiated after decay of the background fluorescence. $Eu^{3+}$, $Tb^{3+}$, $Dy^{3+}$ and $Sm^{3+}$ chelates can be used, the $Eu^{3+}$ chelates being most popular. The underlying idea is to bind the metal ion to a UV-absorbing ligand that becomes excited by UV irradiation to the first electronic single state. This is spontaneously converted to a more stable excited triplet state, and in case the chelate structure is optimal, the triplet state energy is through

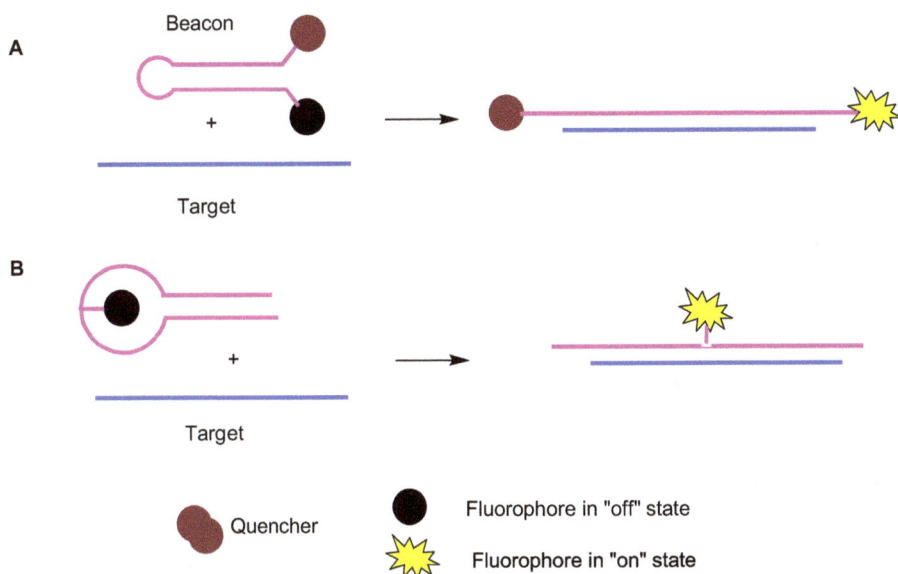

**A** Beacon

+

Target

**B**

+

Target

Quencher

Fluorophore in "off" state

Fluorophore in "on" state

**Figure 8.18:** (A) Hybridization assay in solution with a conventional molecular beacon. (B) Hybridization assay with a quencher-free beacon.

coordinative bonds transmitted to the central ion that becomes excited. Return of the metal ion to the ground state is accompanied with lanthanide emission at 540–650 nm. Since the wavelength of emission is much longer than that of excitation, an emitted quantum is not absorbed and hence concentration quenching is negligible. The dynamic range remains linear over many orders of magnitude, in striking contrast to the situation with organic fluorophores. The fluorescence life time is hundreds of microseconds. Since the background fluorescence is usually quenched in less than 10 µs, the chelates well allow time-resolved measurement: the data can be collected, e.g., 20–150 µs after excitation pulse. Figure 8.19 shows a couple of examples of lanthanide chelates useful for labeling of oligonucleotides and a phosphoramidite building block for the synthesis of a $Eu^{3+}$ chelate-labeled oligodeoxyribonucleotides [103].

For sandwich-type assays on a solid support, gold nanoparticles coated with oligonucleotides have been introduced for visualization instead of fluorophores. This approach consists of two subsequent hybridizations of oligonucleotides obtained by enzymatic amplification of DNA samples. These oligomers contain, in addition to the sequence under interest, a catcher sequence that is identical in the entire pool. After hybridization with support bound catcher probes, the hybridized oligonucleotides are visualized by a second hybridization. Instead of fluorescently tagged detection probe, gold nanoparticles coated with the detection probe are used (Figure 8.20) [104]. Spots are then visualized by silver staining, that is, by reducing $Ag^+$ ions to $Ag^0$ atom on the surface of gold nanoparticle. This approach is more sensitive than the methods utiliz-

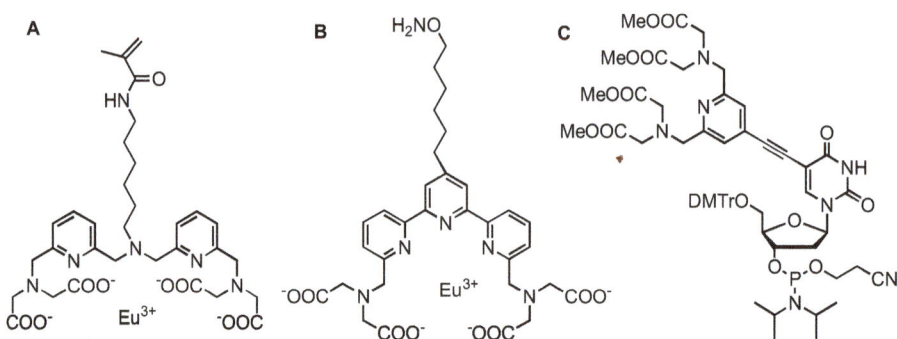

**Figure 8.19:** Fluorescent lanthanide chelates used for postsynthetic labeling (A and B) and for solid-supported synthesis (C) of oligonucleotide conjugates.

ing fluorescently tagged detection probe. The assay may even be converted to a multi-plexed version when detection probes containing a Raman-active dye are used for coating the gold nanoparticles. After silver staining, the nanoparticle is recognized on the basis of its Raman spectrum [105].

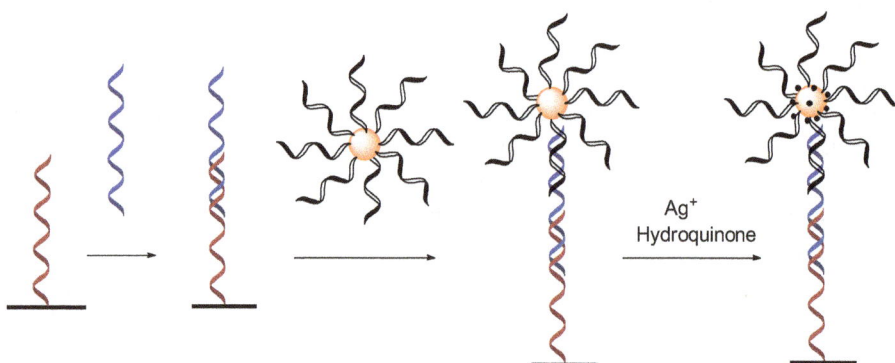

**Figure 8.20:** Hybridization assay on a solid support based on utilization of gold nanoparticles coated with the detection probes for visualization [104].

## 8.5 In vivo visualization of DNA

Covalent labeling of DNA in live cells enables elucidation of the dynamics of DNA's functions in native environment. The procedure makes use of modified nucleosides or nucleobases that are accepted by enzymes of the salvage pathways and by DNA polymerases. In addition, the nucleoside must bear a functionality that enables postsynthetic attachment of a fluorophore. Figure 8.21 shows nucleosides used for this purpose [106]. The most common among them is 5-ethynyl-2′-deoxyuridine (Figure 8.21A). Reaction with azido-functionalized fluorescent dye then enables visualization. 2′-Deoxy-2′-fluoro-*arabino* counterpart (Figure 8.21B) of 5-ethynyl-2′-deoxyuridine has turned out to be less toxic, which is a noticceable advantage in studies where long-time survival of cells is essential [107]. 5-Ethynyl-2′-deoxycytidine (Figure 8.21D) works as efficiently as nucleoside A [108]. The advantage of azido-functionalized nucleosides (Figure 8.21F and G) is that $Cu^{2+}$-free strain-promoted click reaction can be used for the attachment of the fluorophore [109]. 5-Vinyl-2′-deoxyuridine (Figure 8.21H) can be labeled with fluorescent tetrazine by inverse electron-demand Diels–Alder cycloaddition (IEDDA) [110].

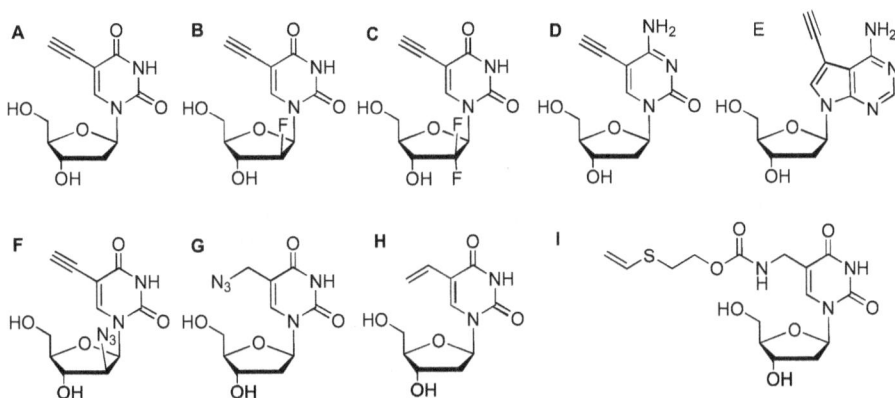

**Figure 8.21:** Nucleosides used for metabolic labeling of DNA [106].

Instead of a modified nucleoside, a prodrug of appropriately modified nucleoside 5′-triphosphate may be introduced into a living cell and labeled there with a cell permeable dye [111]. Figure 8.22 describes a recently published procedure. Protected dCTP tethered with cyclooctene at C5 is internalized into cells. Intracellular removal of the phosphate protecting groups takes place and the released triphosphate becomes incorporated into DNA. Treatment with cell permeable tetrazine conjugate of coumarin then leads to IEDDA reaction between the tetrazine conjugate and the cyclooctene tether.

## 8.6 Interactions with groove binders, intercalators and metal ions

### 8.6.1 Groove binding

As discussed in Section 8.1, polypurine/polypyrimidine sequences in B-DNA can be recognized by selective binding of an oligonucleotide into the major groove. Sequence-specific DNA-binding proteins likewise read the base sequence through the major

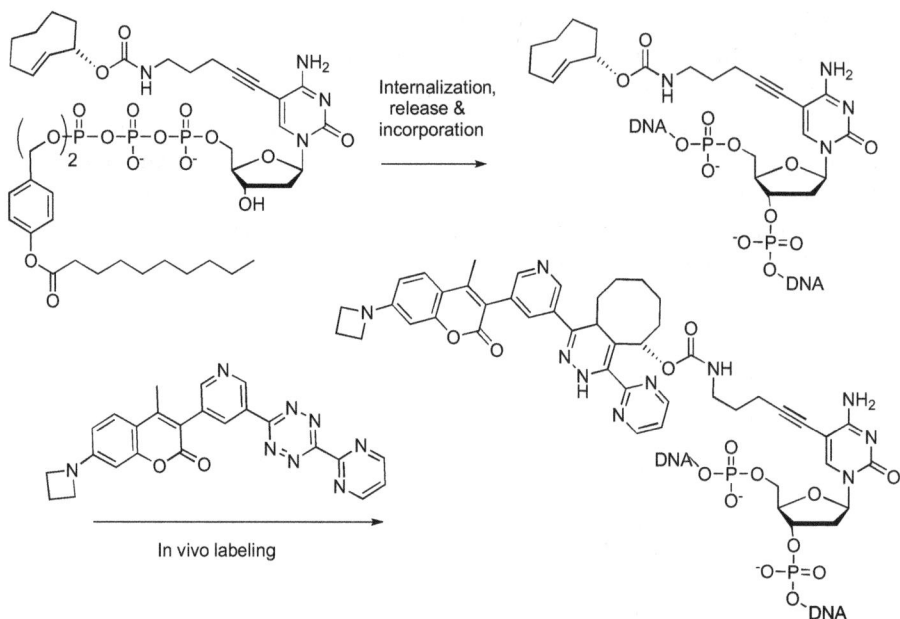

**Figure 8.22:** Metabolic labeling by internalization of a protected modified nucleoside 5′-triphosphate [111].

groove. Two bacterial oligoamide antibiotics, distamycin [112] and netropsin [113], in turn, bind tightly into the minor groove (Figure 8.23A and B). Mimicking their structures, numerous oligoamides have been constructed as candidates for sequence recognition of B-DNA through the minor groove [114]. 4-Amino-2-carboxy-1-methylpyrrole, its 3-hydroxy derivative and 4-amino-2-carboxy-1-methylimidazole have been used as building blocks. Figure 8.23C shows how these heteroaromatic amino acids allow recognition of canonical base pairs through the minor groove. The 4-amido groups form H-bonds to purine N3 or pyrimidine $O^2$ atoms. van der Waals interaction between H2 of adenine and H3 of pyrrole ring serves as an additional stabilizing factor for recognition of adenine. N3 of the imidazole ring, in turn, serves as an H-bond acceptor for guanine 2-amino function, providing G selectivity, and 4-amino-2-carboxy-3-hydroxy-1-methylpyrrole exhibits enhanced pyrimidine selectivity by additional H-bonding of the hydroxyl function to $O^2$.

Figure 8.24 shows an illustrative example of an oligoamide construct recognizing a 9-mer sequence of B-DNA [115]. Short oligoamides consisting of two or three pyrrole/imidazole units are linked together via β-alanine or γ-aminobutyric acid units giving a longer oligomer. γ-Aminobutyric acid allows formation of the so-called γ-turn that allows the oligoamide to adopt a hairpin structure [116]. Accordingly, the amino terminal and carboxy terminal moieties of the oligopeptide are able to bind to the antiparallel strands of B-DNA and hence selectively recognize base pairs. The γ-amido proton is additionally H-bonded to thymine. The β-alanine linkers between the oligopyrrole/imidazole amides H-bond like a pyrrole amide. This kind of groove binders have found several applications as research

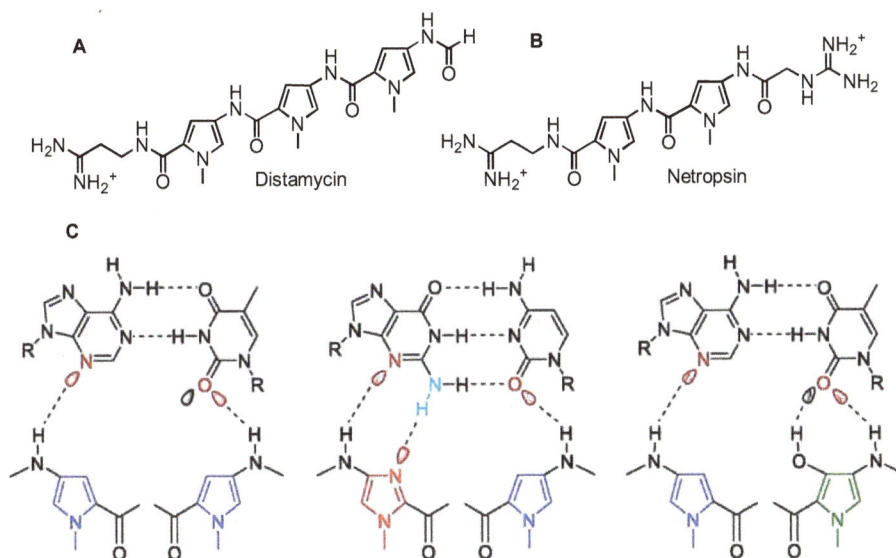

**Figure 8.23:** Structures of distamycin (A) and netropsin (B) and recognition of canonical base pairs by their constituents through the minor groove [114].

tools in cell biology, especially as regulators or repressors of gene expression, in cellular imaging and in the development of sequence selective cleaving agents for DNA [114]. Conjugates of pyrrole/imidazole oligoamides with known DNA alkylating cancer drugs, above all with chlorambucil and seco-CBI, have been studied as a means to convert unspecific cancer drugs sequence selective [117, 118].

## 8.6.2 Intercalation

Polycyclic aromatic compounds are able to penetrate between the base pairs of double-stranded DNA [119]. This results in local unwinding of the double helix by 20°–30° and increases the bending persistence [120]. The intercalated aromatic molecule serves as an extra base pair and hence the distance between the intercalated base pairs is approximately doubled. This disrupts replication and transcription leading to cell death. In fact, several anticancer drugs and drug candidates are intercalators [121]. Another common application is covalent conjugation of an oligonucleotide probe for the enhancement of oligonucleotide binding to both single and double-stranded nucleic acid targets. Some commonly used intercalators are depicted in Figure 8.25A. Proflavine is a disinfectant, daunomycin and doxorubicin are cancer drugs, and ethidium bromide and berberine are fluorescent compounds used for the staining of biological samples.

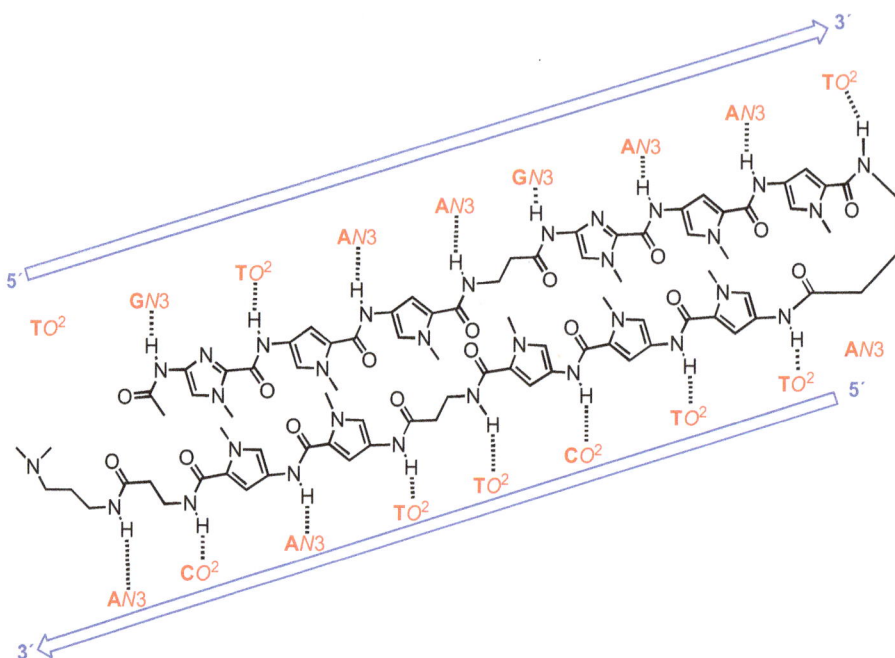

**Figure 8.24:** Recognition of a 9-mer sequence within B-DNA by a pyrrole/imidazole oligoamide construct [115].

As regards structural studies of DNA, metallointercalators are more interesting than their purely organic counterparts. In particular, interaction of DNA with substitution stable hexadentate complexes of $Ru^{2+}$ has received wide interest. Among such complexes, $[Ru(bpy)_3]^{2+}$ and $[Ru(bpy)_2(dppz)]^{2+}$ (bpy = bipyridyl, dppz = dipyridophenazine) have been extensively studied [122]. These complexes are chiral, as evident from Figure 8.25B. The right-handed Δ-complexes bind to right-handed B-DNA more tightly than their left-handed Λ-counterparts [123]. Interestingly, Δ-$[Ru(bpy)_2(dppz)]^{2+}$ itself is not markedly fluorescent, but upon binding to B-DNA the emission intensity is increased $10^4$-fold [124]. The binding usually takes place through the major groove, one of the bidentate ligands serving as the actual intercalator. In case of Δ-$[Ru(bpy)_2(dppz)]^{2+}$, dipyridophenazine is the intercalating ligand.

A closely related $Rh^{3+}$ complex $[Rh(bpy)_2(chrysi)]^{3+}$ (chrysi = 5,6-chrysenediimine) exhibits selectivity toward mismatch base pairs [125]. The reason for this selectivity evidently is reduced stability of the mismatch pair compared to a canonical base pair. Among the eight possible mismatches, five (TT, TC, CC, CA, AA) become recognized, resulting in photochemical cleavage of the chain at the 3'-side of the mismatch. The binding mode of $[Rh(bpy)_2(chrysi)]^{3+}$ differs from that of the $Ru^{2+}$-complexes discussed above. The binding takes place from the minor groove. No unwinding/lengthening of the double helix takes place, but the chrysenediimine ligand displaces the mispaired

**A**

Proflavine

Daunomycin

Doxorubicin

Ethidium bromide

Berberine

**B**

$\Delta$-[Ru(bpy)$_3$]$^{2+}$    $\Lambda$-[Ru(bpy)$_3$]$^{2+}$    $\Delta$-[Ru(bpy)$_2$(dppz)]$^{2+}$    $\Delta$-[Rh(bpy)$_2$(chrysi)]$^{3+}$

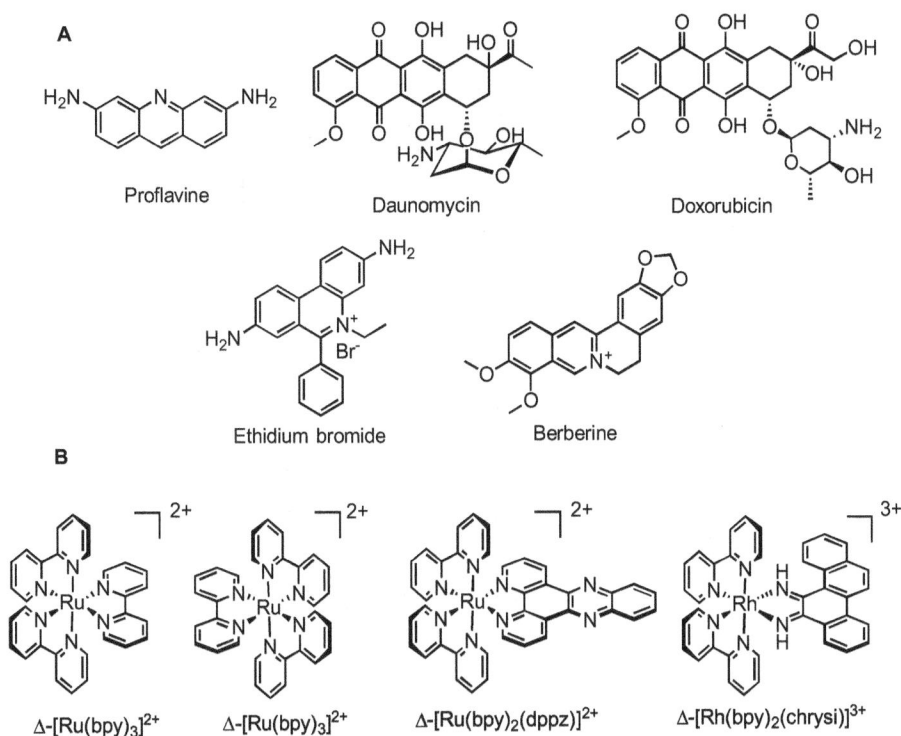

**Figure 8.25:** Examples of nonmetallic (A) and metallic (B) intercalators of B-DNA.

bases [126]. Accordingly, [Rh(bpy)$_2$(chrysi)]$^{3+}$ and its congeners are often called metalloinsertors to make a difference to metallointercalators.

### 8.6.3 Metal-ion-mediated base pairs

Metal ions may participate in the formation of base pairs although the biological relevance of this kind of interactions is unknown. Instead of H-bonding, the bases on opposite strands are coordinated to a single metal ion. As far as canonical WC base pairs are concerned, metal-ion-mediated binding has been observed with divalent metal ions, above all with Zn$^{2+}$, but only at such a high pH where the metal ion is able to displace the N3 proton of thymine or N1 proton of guanine [127]. The mode of interaction is, however, still under debate; partial unwinding of long DNA upon Zn$^{2+}$ appears possible [128], in particular, since recent studies have shown that 40–45 nucleotides long DNA oligomers can bind Zn$^{2+}$ and Cd$^{2+}$ very tightly [129]. A DNA duplex containing only Ag$^+$-mediated canonical base pairs has been crystallized and an electric conducting up to 0.1 mm long nanowire is obtained [130].

The Hg$^{2+}$-mediated TT-mispair, originally reported already in 1963 [131], has been studied much more thoroughly [132, 133]. Both thymine bases are deprotonated and coordinated to the same Hg$^{2+}$ ion (Figure 8.26A). Several contiguous Hg$^{2+}$-mediated base pairs may occur. When the binding is an alternating TA sequence, a T-Hg$^{2+}$-T base pair can be formed by local slippage of the double helix. Another mispair stabilized by metal ion binding is Ag$^+$ stabilized CC pair [134]. The mutual orientation of the cytosine bases can be either *cis* or *trans* [135] as depicted in Figure 8.26B and C.

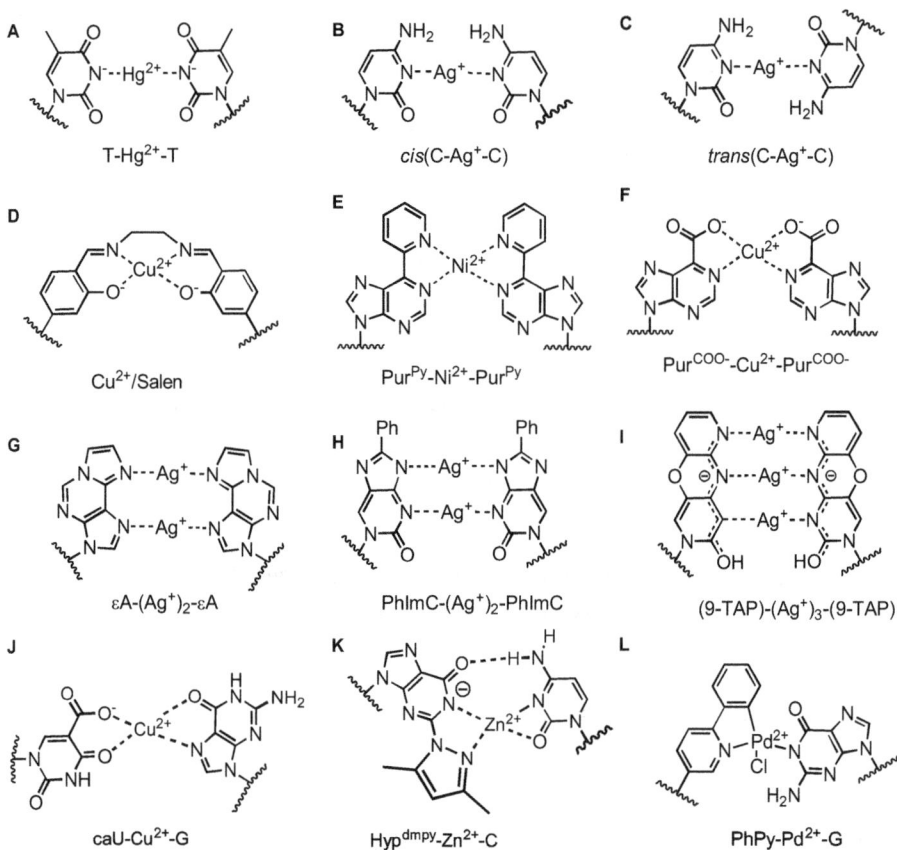

**Figure 8.26:** Metal-ion-mediated base pairs: A [132, 133], B [135], C [135], D [138], E [139], F [140], G [141], H [142], I [143], J [144], K [145] and L [151].

Numerous metal-ion-mediated DNA-duplexes containing a metal-ion-binding surrogate of nucleobase on both strands have been prepared and characterized [136, 137]. Many of such surrogate bases are entirely artificial structures derived from well-known aromatic complexing agents, while others are derivatives of canonical nucleobases. This kind of metal-ion-base-paired duplexes expectedly find applications as components of various nanoscale devices, diagnostics and possibly also in biology [137]. Illustrative examples are given in Figure 8.26. Some purely artificial base surrogates allow very large duplex stabilizations by metal ion coordination. A classic example of a particularly stable artificial base pair is the interchain salen complex formed by salicylaldehyde surrogate bases in the presence of ethylenediamine and $Cu^{2+}$ (Figure 8.26D) [134]. Introduction of one such pair into a 15-mer duplex increases the melting temperature by 40 °C. Other highly stabilizing base pairs include a $Ni^{2+}$ complex of 6-(pyridine-2-yl)purine bases (Figure 8.26E) [139] and a $Cu^{2+}$ complex of 6-carboxypurine bases (Figure 8.26F) [140]. Dinuclear $Ag^+$ complexes of 1,$N^6$-ethenoadenine (Figure 8.26G) [141] and 8-phenylimidazolo-cytosine (Figure 8.26H) [142] are examples of base pairs formed by moderately modified canonical nucleobases. The latter increased the melting temperature of a 12-mer duplex by almost 40 °C. Comparable stabilization has been achieved by trinuclear $Ag^+$ complexes of 1,3,9-triaza-2-oxophenoxazine base pairs (Figure 8.26I) [143]. Metal-ion-mediated base pairs may also be formed between modified and canonical nucleobases. Illustrative examples are offered by $Cu^{2+}$-mediated binding of 5-carboxyuracil to guanine (Figure 8.26J) [144] and $Cu^{2+}$- and $Zn^{2+}$-mediated binding of 2-(3,5-dimethylpyrazol-1-yl)hypoxanthine to any of the natural nucleobases (Figure 8.26K) [145]. Formation of a metal-ion-mediated base pair may also enhance DNA-strand displacement. It has been shown that an ODN double helix containing several consecutive 5-hydroxyuracil (5-OH-U) –adenine base pairs undergoes with another ODN duplex Gd(III)-induced strand switching driven by the formation of consecutive 5-OH-U – Gd(III) – 5-OH-U pairs [146].

As far as biological applications are concerned, substitution inert organometallic base surrogates offer one clear advantage compared to kinetically labile metal ion complexes, that is, stability under intracellular conditions where concentration of free metal ions is very low. The data on organometallic oligonucleotide probes is still scanty [147]. 5-Mercuricytosine and 5-mercuriuracil at the 3′-terminus of oligothymidine have been observed to enhance triple helix formation by interaction with a 5′-AT -3′ or 5′-TA-3′ pair [148]. A 2,6-dimercuriphenol-derived C-nucleoside, in turn, has been shown to form stable dinuclear $Hg^{II}$-mediated base triplets with Ade, Cyt and Thy within a triple helix [149] and 1,8-dimercury-6-phenyl-1H-carbazole as a surrogate base in double-stranded ODN forms a $Hg^{2+}$-mediated bond to both $O^2$ and $O^4$ of a thymine base on the opposite strand [150]. Palladacyclic surrogate base is another organometallic modification studied (Figure 8.26L). It shows upon hybridization selectivity to G over T, C and, in particular, A [151]. Cyclopalladated oligonucleotide probes turned out to be more efficient splice-correcting agents than their unmodified counterpart in the HeLa Luc/705 cell line [152].

In addition to metal-mediated base pairing, formation of metal-based triples and teterads with nucleic acids has been extensively studied [136]. The main topics include stabilization of triple helices [149, 153] three- [154] and four-way [155] junctions, i-motifs [156] and G-quadruplexes [157] with metal ions. As with base pairing, future will show whether some of these promising studies lead to biological applications.

## 8.7 DNA base modifications

### 8.7.1 Depurination and deamination as a source of mutagenesis

The sugar phosphate backbone of DNA is very stable under physiological conditions, as discussed in Chapter 6. As long as the base moieties remain intact, the phosphodiester linkages can be cleaved only by a few tetravalent metal ions ($Ce^{4+}$, $Zr^{4+}$, $Th^{4+}$) and $Co^{3+}$. Base moieties in DNA are much more susceptible to nonenzymatic reactions. In the absence of any environmental contaminant, the fastest polar reactions are depurination and deamination. These reactions constitute the source of spontaneous mutagenesis.

Depurination most likely proceeds by departure of a protonated base. Even under physiological conditions, this reaction has been estimated to generate daily 2,000–10,000 abasic sites in human genome [158]. Although majority of them most likely become repurinated by repair enzymes, the remaining ones still constitute a risk for misreading during replication. Deamination of cytosine to uracil is another source of spontaneous mutagenesis [159]. The predominant mechanism presumably is direct displacement of the 4-amino group by a water molecule that simultaneously protonates N3 (Figure 8.27), although kinetically equivalent attack of hydroxide ion on N3-protonated base or pre-equilibrium hydration of the 5,6-double bond cannot be strictly excluded.

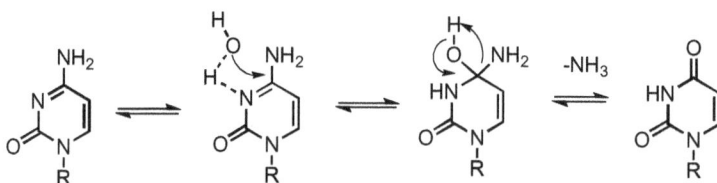

**Figure 8.27:** Plausible mechanism for the deamination of DNA cytosines to uracils.

Modification of nucleic acid bases by environmental contaminants may greatly increase the frequency of mutations [160]. Alkylation of N3 or N7 position of purines, for instance, generates a positive charge on the nucleobase which accelerate the loss of purines under physiological conditions by several orders of magnitude [161]. Likewise, $O_2$ of pyrimidines is prone to alkylation. These modifications does not necessarily lead to the formation of abasic sites, but DNA glycosylases that are responsible for base excision repair, remove and replace the alkylated bases and hence the probability of mispairing is increased [162].

## 8.7.2 Oxidation of nucleobases

Another source of increased mutagenesis is oxidation of nucleobases. Even normal cellular metabolism generates as by-products reactive oxidative species such as hydroxyl and superoxide radicals (HO · and $O_2$ ·) and hydrogen peroxide. External oxidative conditions and ionizing γ-radiation induce the formation of similar oxidizing agents. Among the reactive oxidative species formed by various mechanisms, hydroxyl radical deserves the main attention [163]. The preferred site of attack of hydroxyl radical is guanine [164]. Carbon atoms 4, 5 and 8 are attacked (Figure 8.28A). Usually the C4- and C5-adducts are, however, reduced back to guanine, whereas the C8-adduct is under oxidative conditions converted to 7,8-dihydro-8-oxoguanine and under reductive conditions to 2,6-diamino-5-formamido-4-hydroxypyrimidine [165]. In addition, 2,5-diamino-4 *H*-imidazol-4-one and 2,2,4-triaminooxazol-5(2 *H*)-one are formed [166]. It has been suggested that all the hydroxyl adducts may possibly be dehydrated to guanine radical that is then oxidized by singlet oxygen to 2,5-diamino-4*H*-imidazol-4-one and finally hydrolyzed to 2,2,4-triaminooxazol-5(2*H*)-one. Adenine is modified rather similarly to guanine (Figure 8.28B) [167].

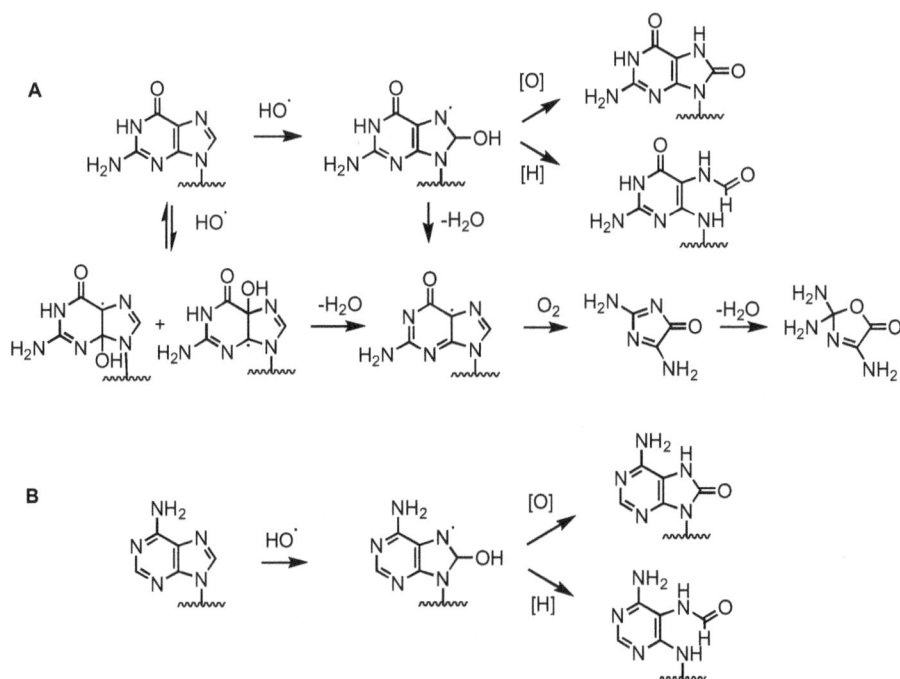

**Figure 8.28:** Reactions of hydroxyl radical with guanine (A) [164–166] and adenine (B) [167] bases.

Pyrimidine bases yield a variety of products depicted in Figure 8.29 [168–170]. With thymine, hydroxyl radical initially forms C5- and C6-adducts and abstracts hydrogen atom from the methyl group [171]. With cytosine, C5- and C6-adducts are likewise formed [172]. Since partial reduction of the 5,6-double bond accelerates deamination, a variety of products are obtained depending under oxidative or reductive conditions.

In addition to hydroxyl radicals, ionizing radiation forms alkoxy (RO ·), alkylperoxyl (ROO ·) and superoxide ($O_2 \cdot^-$) radicals. Among these, alkylperoxyl radicals are of special importance since lipid peroxides are products of exposure to oxidative stress. All these radicals generate base modifications that closely resemble those produced by hydroxyl radical.

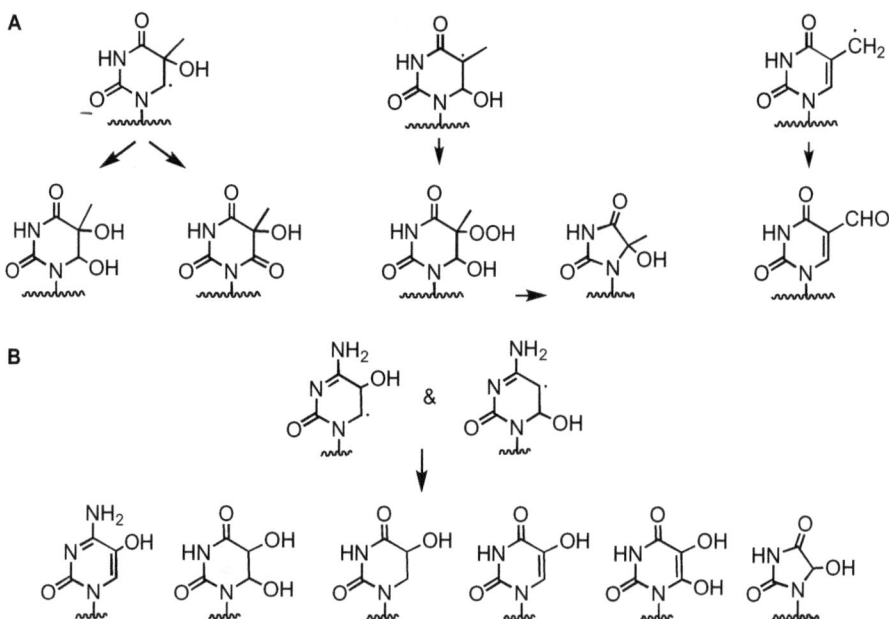

**Figure 8.29:** Reactions of hydroxyl radical with thymine [171] and cytosine [172] nucleosides, and products [168–170] that the initially formed radical intermediates yield.

Ozone deserves special attention as a major component of air pollution. In vitro experiments with calf thymus DNA have revealed that thymidine is oxidized most readily, followed by 2′-deoxyguanosine and 2′-deoxycytidine. 2′-Deoxyadenosine was considerably more tolerant toward ozone [173]. The main isolated oxidation products of pyrimidine nucleosides were 5-hydroxyhydantoin and 5,6-dihydroxy-5,6-dihydropyrimidine derivatives. 2′-Deoxyguanosine evidently degraded through initial formation of 8-oxo-7,8-dihydroguanine intermediate.

Ultraviolet radiation also modifies DNA bases. Short wavelength radiation below 250 nm leads to the formation of 7,8-dihydro-8-oxoguanine [174]. At 254 nm, mimicking the effect of sunlight, pyrimidine [2 + 2] and [4–6] photodimers are formed (Figure 8.30). Among these, the [2 + 2] cyclobutane dimer is twice as abundant as the [6–4] dimer.

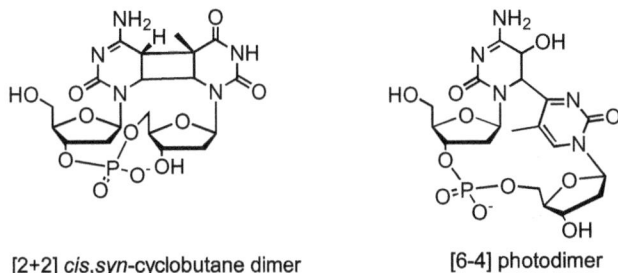

[2+2] *cis,syn*-cyclobutane dimer

[6-4] photodimer

**Figure 8.30:** Photodimers obtained by irradiation at 254 nm [174].

Base modifications can also lead to scission of the sugar-phosphate backbone since several repair enzymes exhibit lyase activity [175]. In some cases, a nucleobase radical may abstract hydrogen radical from C1′ or C2′ site of the 5′-neighboring nucleoside and this leads to nonenzymatic strand scission [170]. In addition, abstraction of hydrogen atom from the sugar moiety by ionizing radiation, a metal ion or a photoactive metal ion complex is possible [176, 177].

### 8.7.3 Base modifications by nucleophilic substitution

Strong nucleophiles, including hydrazines (H$_2$N–NHR), alkoxyamines (H$_2$NOR) and bisulfite ion (HSO$_3^-$), attack on C6 of pyrimidines leading to multistep transformation reactions of Cyt, Thy and Ura, as discussed in more detail in Chapter 2. Bases engaged in double helix are, however, much less vulnerable to nucleophiles than monomeric nucleosides or even the base moieties of single stranded sequences. For example, the reactivity of hydroxylamine (H$_2$NOH) toward a cytosine base decreases in the order: 5′-CMP > polyC > ds DNA, the relative rates being 1,500, 300 and 1, respectively [178].

### 8.7.4 Base modifications by electrophilic substitution

Electrophiles tend to alkylate nucleic acid bases. Those preferring S$_N$1 susbtitution alkylate N7 of Gua, N3 of Ade, O$^6$ of Gua and O$^4$ of Thy [179]. Such alkylating agents are toxic at lower concentrations than those reacting by S$_N$2 mechanism. The latter tends to alkylate N1 of Ade and N3 of pyrimidines [178]. The monofunctional alkylating agents consists of alkyl alkanesulfonates (R$^1$-SO$_2$-OR$^2$) and dialkyl sulfates (R$^1$O-SO$_2$-OR$^2$) and nitrosoamines, viz. alkyl nitrosoureas, alkyl N′-nitro-N-nitrosoguanidine and N,N-dialkylnitrosoamines. All these nitrosoamines are decomposed to alkyldiatsonium ions

(R–N$^+$ ≡ N) that serve as the actual alkylating agent [180]. Compared to dialkyl sulfates and alkyl alkanesulfonates, alkyldiatsonium ions exhibit higher tendency to alkylate oxygen atoms. Ura, Thy and Gua are alkylated more readily than Cyt or Ade. 2-Haloethylamines and 2-haloethyl thioethers are produced by intramolecular nucleophilic displacement highly alkylating iminium and sulfonium ions, respectively. The predominant sites of alkylation are the purine ring-nitrogens.

1,2- and 1,3-dicarbonyl compounds and their congeners, such as 2-haloaldehydes or α,β-unsaturated aldehydes form cyclic adducts with nucleic acid bases. Some of these compounds, above all malonaldehyde are formed in vivo as a catabolic product of lipid peroxidation caused by the so-called oxidative stress. Chloroacetaldehyde forms ethenoadducts by reacting with the amidine fragments, N-C = N, of Ade, Gua and Cyt [181]. Gua, however, is considerably less susceptible than Ade and Cyt to etheno adduct formation. The carbonyl carbon becomes bound to the primary amino group and the α-carbon to ring nitrogen. Numerous 1,2-bifunctional electrophiles undergo similar reactions giving etheno or substituted etheno adducts. These include N-substituted 2-bromoacetamides, bromomalonaldehyde, epoxy carbonyl compounds, 1-haloepoxides and 1,2-dicarbonyl compounds [182]. Since the primary amino groups of nucleobases are less nucleophilic than ring nitrogens, it appears likely that the reaction is initiated by attack of ring-nitrogen on the α-carbon. The halogen substituent is displaced and alkylation of the ring nitrogen leads to conversion of the amino tautomer to an imino tautomer, and attack of the imino nitrogen on the carbonyl carbon completes the cyclization (Figure 8.31) [183]. An alternative mechanistic interpretation is that a carbinolamine intermediate is formed in a preequilibrium step and rate-limiting attack of ring nitrogen on the α-carbon results in the formation of a cyclic intermediate that finally undergoes dehydration. Anyway, both mechanisms lead to the same regioselectivity. 1,$N^6$-Ethenoadenosine, 1,$N^4$-ethenocytidine and their 2′-deoxyribo counterparts are fluorescent and hence used as probes in mechanistic studies of biological reactions. Etheno nucleosides also offer access to structural transformations that are otherwise difficult to obtain. The best known example is pyrimidine ring opening of 1,$N^6$-ethenoadenosine by aqueous alkali giving 3-(β-D-ribofuranosyl)-4-amino-5-(imidazole-2-yl)-imidazole [184], a compound that allows recyclization to several 2-substituted adenosine derivatives [185].

α,β-Unsaturated aldehydes, enals, form propano adducts with Gua, Ade and Cyt bases. With acrolein, two sets of regioisomers are formed: the carbonyl carbon is bound either to the primary amino group or to ring nitrogen [186]. Similar adducts are obtained with Cyt [187] and Ade [188]. With crotonaldehyde and its congeners having the β carbon alkylated, only the isomer having the carbonyl carbon bound to ring nitrogen is formed for steric reasons [189]. *trans*-4-Hydroxy-2-nonenal (HNE) and malonaldehyde are catabolic products of lipid peroxides. *trans*-4-Hydroxy-2-nonenal forms adducts similar to those of crotonaldehyde [190]. The 4-hydroxy group markedly enhances adduct formation by increasing the susceptibility of β carbon to the attack of $N^2$ of Gua. Malonaldehyde forms with Gua a cyclic propeno adduct [191] and with Ade and Cyt acyclic oxopropenyl adducts [192].

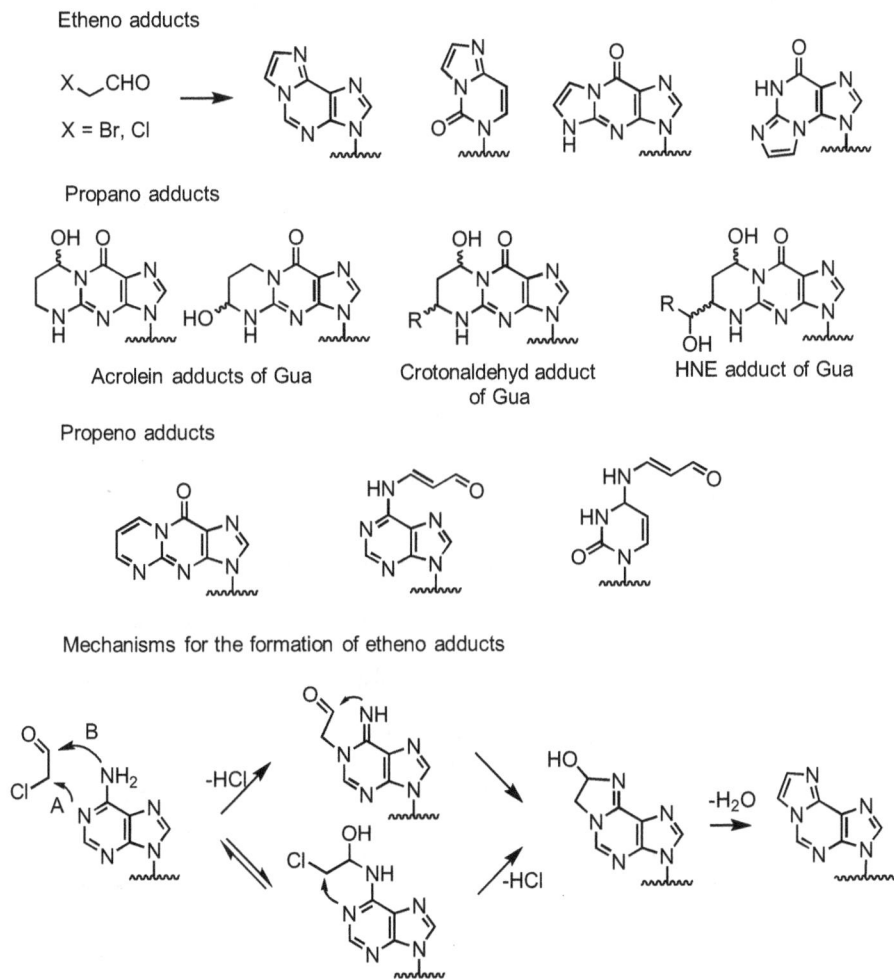

Etheno adducts

X = Br, Cl

Propano adducts

Acrolein adducts of Gua　　　Crotonaldehyd adduct　　HNE adduct of Gua
　　　　　　　　　　　　　　　of Gua

Propeno adducts

Mechanisms for the formation of etheno adducts

Figure 8.31: Structures of adducts of purine bases with carbonyl compounds.

## Further reading

Benabou S, Avino A, Eritja R, Gonzalez C, Gargallo R. Fundamental aspects of the nucleic acid i-motif structures. RSC Adv 2014, 4, 26956–26980.

Blackburn GM, Gait MJ, Loakes D, Williams DM. Nucleic Acids in Chemistry and Biology 3rd ed. Cambridge, UK: RSC Publishing, 2006, Chapter 2.

Burrows CJ, Muller JG. Oxidative nucleobase modifications leading to strand scission. Chem Rev 1998, 98, 1109–1151.

Bünzli J-CG. Lanthanide luminescence for biomedical analyses and imaging. Chem Rev 2010, 110, 2729–2755.

Friedberg EC. DNA damage and repair. Nature 2003, 421, 436–440.

Gao X, Gulari E, Zhou X. In situ synthesis of oligonucleotide microarrays. Biopolymers 2004, 73, 579–596.

Kawamoto Y, Bando T, Sugiyama H. Sequence-specific DNA binding pyrrole–imidazole polyamides and their applications. Bioorg Med Chem 2018, 26, 1393–1411.

Kypr J, Kejnovska I, Bednarova K, Vorlickova M. Circular Dichroism Spectroscopy of Nucleic Acids. In Comprehensive Chiroptical Spectroscopy, Volume 2: Applications in Stereochemical Analysis of Synthetic Compounds, Natural Products, and Biomolecules 1st ed. Berova N, Polavarapu PL, Nakanishi K, Woody RW, eds. John Wiley & Sons, Inc, 2012, 573–584.

Ma D-L, Zhang Z, Wang M, Lu L, Zhong H-J, Leung C-H. Recent developments in G-quadruplex probes. Chem Biol 2015, 22, 812–828.

Miller MB, Tang YW. Basic concepts of microarrays and potential applications in clinical microbiology. Clin Microbiol Rev 2009, 22, 611–633.

Monsen RC, Trent JO, Chaires JB. G-quadruplex DNA: A longer story. Acc Chem Res 2022, 55, 3242–3252.

Mukherjee A, Sasikala WD. Drug–DNA intercalation: From discovery to the molecular mechanism. Adv Prot Chem Struct Biol 2013, 92, 1–62.

Naskar S, Guha R, Müller J. Metal-modified nucleic acids: Metal-mediated base pairs, triples, and tetrads. Angew Chem Int Ed 2020, 59, 1397–1406.

Neidle S. Beyond the double helix: DNA structural diversity and the PDB. J Biol Chem 2021, 296, 100553 (11 pages).

Pirrung MC. How to make a DNA chip. Angew Chem Int Ed 2002, 41, 1276–1289.

Rombouts K, Braeckmans K, Remaut K. Fluorescent labeling of plasmid DNA and mRNA: Gains and losses of current labeling strategies. Bioconjugate Chem 2016, 27, 280–297.

Schatzschneider U. Metallointercalators and Metalloinsertors: Structural Requirements for DNA Recognition and Anticancer Activity in Metallo-drugs. In Development and Action of Anticancer Agents. Metal Ions in Life Sciences. vol. 18, Sigel A, Sigel H, Freisinger E, Sigel RKO, eds. Berlin, Germany: Walter de Gruyter GmbH, 2018, 388–422.

Takezawa Y, Shionoya M, Müller J. Artificial DNA base pairing mediated by diverse metal ions. Chem Lett 2017, 46, 622–633.

Tian B, Bevilacqua PC, Diegelman-Parente A, Mathews MB. The double-stranded-RNA binding motif: Interference and much more. Nat Rev Mol Cell Biol 2004, 5, 1013–1023.

Varshney D, Spiegel J, Zyner K, Tannahill D, Balasubramanian S. The regulation and functions of DNA and RNA G-quadruplexes. Nat Rev Mol Cell Biol 2020, 21, 459–474.

Venkatesan N, Seo YJ, Kim BH. Quencher-free molecular beacons: A new strategy in fluorescence based nucleic acid analysis. Chem Soc Rev 2008, 37, 648–663.

Wang Y, Zhao Y, Bollas A, Wang Y, Au KF. Nanopore sequencing technology, bioinformatics and applications. Nat Biotechnol 2021, 39, 1348–1365.

# References

[1]     Li R, Mak CH. A deep dive into DNA base pairing interactions under water. J Phys Chem B 2020, 124, 5559–5570.

[2]     Yakovchuk P, Protozanova E, Frank-Kamenetskii MD. Base-stacking and base-pairing contributions into thermal stability of the DNA double helix. Nucleic Acids Res 2006, 34, 564–574.

[3]     Zacharias M. Base-pairing and base-stacking contributions to double-stranded DNA formation. J Phys Chem B 2020, 124, 10345–10352.

[4]     Sklenar V, Feigon J. Formation of a stable triplex from a single DNA strand. Nature 1990, 345, 836–838.

[5]     Varshney D, Spiegel J, Zyner K, Tannahill D, Balasubramanian S. The regulation and functions of DNA and RNA G-quadruplexes. Nat Rev Mol Cell Biol 2020, 21, 459–474.

[6]     Rice PA, Yang SW, Mizuuchi K, Nash HA. Crystal structure of an IHF-DNA complex: A protein-induced DNA U-turn. Cell 1996, 87, 1295–1306.

[7]     Wang AHJ, Ughetto G, Quigley GJ, Hakoshima T, Vandermarel GA, Van Boom JH, Rich A. The Molecular-structure of a DNA triostin-a complex. Science 1984, 225, 1115–1121.

[8]     Manandhar M, Chun E, Romesberg FE. Genetic code expansion: Inception, development, commercialization. J Am Chem Soc 2021, 143, 4859–4878.

[9]     Kimoto M, Hirao I. Genetic alphabet expansion technology by creating unnatural base pairs. Chem Soc Rev 2020, 49, 7602–7626.

[10]    Wang L. Engineering the genetic code in cells and animals: Biological considerations and impacts. Acc Chem Res 2017, 50, 2767–2775.

[11]    Yang Z, Hutter D, Sheng P, Sismour AM, Benner SA. Artificially expanded genetic information system: A new base pair with an alternative hydrogen bonding pattern. Nucleic Acids Res 2006, 34, 6095–6101.

[12]    Schweitzer BA, Kool ET. Hydrophobic, non-hydrogen-bonding bases and base pairs in DNA. J Am Chem Soc 1995, 117, 186318–186372.

[13]    Kimoto M, Yamashige R, Matsunaga K, Yokoyama S, Hirao I. Generation of high-affinity DNA aptamers using an extended genetic alphabet. Nat Biotechnol 2013, 31, 453–457.

[14]    Gonzalez-Olvera JC, Durec M, Marek R, Fiala R, Morales-Garcia MDRJ, Gonzalez-Jasso E, Pless RC. Protonation of nucleobases in single- and double-stranded DNA. ChemBioChem 2018, 19, 2088–2098.

[15]    Kennard O, Hunter WN. Single-crystal X-ray diffraction studies of oligonucleotides and oligonucleotide-drug complexes. Angew Chem Int Ed Engl 1991, 30, 1254–1277.

[16]    Sinden RR. 1994-01-15. DNA Structure and Function 1st ed. Academic Press, 1994, 398. ISBN 0–12–645750-6.

[17]    Hartmann B, Lavery R. DNA structural forms. Quart Rev Biophys 1996, 29, 309–368.

[18]    Olson WK, Bansal M, Burley SK, Dickerson RE, Gerstein M, Harvey SC, Heinemann U, Lu X-J, Neidle S, Shakked Z, Sklenar H, Suzuki M, Tung C-S, Westhof E, Wolberger C, Berman HM. A standard reference frame for the description of nucleic acid base-pair geometry. J Mol Biol 2001, 313, 229–237.

[19]    Rich A, Norheim A, Wang AH. The chemistry and biology of left-handed Z-DNA. Annu Rev Biochem 1984, 53, 791–846.

[20]    Brovchenko I, Krukau A, Oleinikova A, Mazur A. Water percolation governs polymorphic transition and conductivity of DNA, from computational biophysics to systems biology (CBSB07), Proceedings of the NIC Workshop 2007, eds. Hansmann UHE, Meinke J, Mohanty S, Zimmermann O John von Neumann Institute for Computing, Jülich, NIC Series, 2007, 36, 195–197.

[21]    Richmond T, Finch JT, Rushton B, Rhodes D, Klug A. The structure of the nucleosome core particle at 7Å resolution. Nature 1984, 311, 532–537.

[22] Sinden RR, Pytlos-Sinden MJ, Potaman VN. Slipped strand DNA structures. Front Biosci 2007, 12, 4788–4799.

[23] Xodo LE, Manzini G, Quadrifoglio F, van der Marel G, van Boom J. DNA hairpin loops in solution. Correlation between primary structure, thermostability and reactivity with single-strand-specific nuclease from mung bean. Nucleic Acids Res 1991, 19, 1505–1511.

[24] Timsit Y, Moras D. Cruciform structures and functions. Quart Rev Biophys 1996, 29, 279–307.

[25] Szabat M, Kierzek R. Parallel-stranded DNA and RNA duplexes – Structural features and potential applications. FEBS J 2017, 284, 3986–3998.

[26] Bhaumik SR, Chary KV, Govil G, Liu K, Miles HT. A novel palindromic triple-stranded structure formed by homopyrimidine dodecamer d-CTTCTCCTCTTC and homopurine hexamer d-GAAGAG. Nucleic Acids Res 1998, 26, 2981–2988.

[27] Cubero E, Avino A, de la Torre BG, Frieden M, Eritja R, Luque FJ, Gonzalez C, Orozco M. Hoogsteen-based parallel-stranded duplexes of DNA. Effect of 8-amino-purine derivatives. J Am Chem Soc 2002, 124, 3133–3142.

[28] Garcia RG, Ferrer E, Macias MJ, Eritja R, Orozco M. Theoretical calculations, synthesis and base pairing properties of oligonucleotides containing 8-amino-2′-deoxyadenosine. Nucleic Acids Res 1999, 27, 1991–1998.

[29] Yang XL, Sugiyama H, Ikeda S, Saito I, Wang AH. Structural studies of a stable parallel-stranded DNA duplex incorporating isoguanine: Cytosine and isocytosine:guanine basepairs by nuclear magnetic resonance spectroscopy. Biophys J 1998, 75, 1163–1171.

[30] Nikolova EN, Kim E, Wise AA, O'Brien PJ, Andricioaei I, Al-Hashimi HM. Transient Hoogsteen base pairs in canonical duplex DNA. Nature 2011, 470, 498–502.

[31] Li J, Mohammed-Elsabagh M, Paczkowski F, Li Y. Circular nucleic acids: discovery, functions and applications. ChemBioChem 2020, 21, 1547–1566.

[32] Lietard J, Meyer A, Vasseur -J-J, Morvan F. New strategies for cyclization and bicyclization of oligonucleotides by click chemistry assisted by microwaves. J Org Chem 2008, 73, 191–200, and references therein.

[33] Lyamichev VI, Mirkin SM, Frank-Kamenetskii MD. Structures of homopurine-homopyrimidine tract in superhelical DNA. Biomol Struct Dyn 1986, 3, 667–669.

[34] Giovannangeli C, Thuong NT, Helene C. Oligonucleotide clamps arrest DNA synthesis on a single-stranded DNA target. Proc Natl Acad Sci USA 1993, 90, 10013–10017.

[35] Sen D, Gilbert W. Formation of parallel four-stranded complexes by guanine-rich motifs in DNA and its implications for meiosis. Nature 1988, 334, 364–366.

[36] Gellert M, Lipsett MN, Davies DR. Helix formation by guanylic acid. Proc Natl Acad Sci USA 1962, 48, 2013–2018.

[37] Sundquist W, Klug A. Telomeric DNA dimerizes by formation of guanine tetrads between hairpin loops. Nature 1989, 342, 825–829.

[38] Largy E, Mergny J-L, Gabelica V. Role of alkali metal ions in G-quadruplex nucleic acid structure and stability. In Sigel A, Sigel H, Sigel RKO. The Alkali Metal Ions: Their Role in Life. Metal Ions in Life Sciences. 16. Springer, 2016, Chapter 7, 203–258.

[39] Mendoza O, Bourdoncle A, Boule J-B, Brosh RM, Jr, Mergny JL. G-quadruplex and helicases. Nucleic Acids Res 2016, 44, 1989–2006.

[40] David DA, Nesbitt J. Kinetic and thermodynamic control of G-quadruplex polymorphism by Na⁺ and K⁺ Cations. J Phys Chem B 2023, 127, 6842–6855.

[41] Bisoi A, Sarkar S, Singh PC. Hydrophobic interaction-induced topology-independent destabilization of G-quadruplex. Biochemistry 2023, 62, 3430–3439.

[42] Esposito V, Galeone A, Mayol L, Oliviero G, Virgilio A, Randazzo L. A topological classification of G-quadruplex structures. Nucleosides Nucleotides Nucleic Acids 2007, 26, 1155–1159.

[43] Simmel FC, Yurke B, Singh HR. Principles and applications of nucleic acid strand displacement reactions. Chem Rev 2019, 119, 6326–6369.

[44] Hu W, Jing H, Fu W, Wang Z, Zhou J, Zhang N. Conversion to trimolecular G-quadruplex by spontaneous Hoogsteen pairing-based strand displacement reaction between bimolecular G-quadruplex and double G-rich probes. J Am Chem Soc 2023, 145, 18578–18590.

[45] Huppert JL, Balasubramanian S. Prevalence of quadruplexes in the human genome. Nucleic Acids Res 2005, 33, 2908–2916.

[46] Rhodes D, Lipps HJ. G-quadruplexes and their regulatory roles in biology. Nucleic Acids Res 2015, 43, 8627–8637.

[47] Monsen RC, Trent JO, Chaires JB. G-quadruplex DNA: A Longer Story. Acc Chem Res 2022, 55, 3242–3252.

[48] Jana J, Mohr S, Vianney YM, Weisz K. Structural motifs and intramolecular interactions in non-canonical G-quadruplexes. RSC Chem Biol 2021, 2, 338–353.

[49] Benabou S, Avino A, Eritja R, Gonzalez C, Gargallo R. Fundamental aspects of the nucleic acid i-motif structures. RSC Adv 2014, 4, 26956–26980.

[50] Yang B, Rodgers MT. Base-pairing energies of proton-bound heterodimers of cytosine and modified cytosines: Implications for the stability of DNA i-motif conformations. J Am Chem Soc 2014, 136, 282–290.

[51] Han X, Leroy J-L, Gueron M. An intramolecular i-motif: The solution structure and base-pair opening kinetics of d(5mCCT3CCT3ACCT3CC). J Mol Biol 1998, 278, 949–965.

[52] Day HA, Pavlou P, Waller ZAE. i-Motif DNA: Structure, stability and targeting with ligands. Bioorg Med Chem 2014, 22, 4407–4418.

[53] Zeraati M, Langley DB, Schofield P, Moye AL, Roue R, Hughes WE, Bryan TM, Dinger ME, Christ D. i-Motif DNA structures are formed in the nuclei of human cells. Nat Chem 2018, 10, 631–637.

[54] Wright EP, Huppert JL, Waller ZAE. Identification of multiple genomic DNA sequences which form i-motif structures at neutral pH. Nucleic Acids Res 2017, 45, 2951–2959.

[55] Li T, Famulok M. i-Motif-programmed functionalization of DNA nanocircles. J Am Chem Soc 2013, 135, 1593–1599.

[56] Kim SE, Hong S-C. Two opposing effects of monovalent cations on the stability of i-motif structure. J Phys Chem B 2023, 127, 1932–1939.

[57] Cheng M, Qiu D, Tamon L, Istvankova E, Viskova P, Amrane S, Guedin A, Chen J, Lacroix L, Ju H, Trantirek L, Sahakyan AB, Zhou J, Mergny J-L. Thermal and pH stabilities of i-DNA: Confronting in vitro experiments with models and in-cell NMR data. Angew Chem Int Ed 2021, 60, 10286–10294.

[58] Pandey A, Roy S, Srivatsan SG. Probing the competition between duplex, G-quadruplex and i-motif structures of the oncogenic c-Myc DNA promoter region. Chem Asian J 2023, 18, e202300510 (9 pages).

[59] Maxam AM, Gilbert W. A new method for sequencing DNA. Proc Natl Acad Sci USA 1977, 74, 560–564.

[60] Sanger F, Nicklen F, Coulson AR. DNA sequencing with chain-terminating inhibitors. Proc Natl Acad Sci USA 1977, 74, 5463–5467.

[61] Heather JM, Chain B. The sequence of sequencers: The history of sequencing DNA. Genomics 2016, 107, 1–8.

[62] Balasubramanian S. Sequencing nucleic acids: From chemistry to medicine. Chem Commun 2011, 47, 7281–7286.

[63] Braslavsky I, Herbert B, Kartalov E, Quake SR. Sequence information can be obtained from single DNA molecules. Proc Natl Acad Sci USA 2003, 100, 3960–3964.

[64] Liu Z, Wang Y, Deng T, Chen Q. Solid-state nanopore-based DNA sequencing technology. J Nanomaterials 2016, 0, Article ID 5284786, 1–13.

[65] Wang Y, Zhao Y, Bollas A, Wang Y, Au KF. Nanopore sequencing technology, bioinformatics and applications. Nat Biotechnol 2021, 39, 1348–1365.

[66]  Ding T, Yang J, Pan V, Zhao N, Lu Z, Ke Y, Zhang C. DNA nanotechnology assisted nanopore-based analysis. Nucleic Acids Res 2020, 48, 2791–2806.

[67]  Rand AC, Jain M, Eizenga JM, Musselman-Brown A, Olsen HE, Akeson M, Paten B. Mapping DNA methylation with high-throughput nanopore sequencing. Nat Methods 2017, 14, 411–413.

[68]  Thomas CA, Craig JM, Hoshika S, Brinkerhoff H, Huang JR, Abell SJ, Kim HC, Franzi MC, Carrasco JD, Kim -J-J, Smith DC, Gundlach JH, Benner SA, Laszlo AH. Assessing readability of an 8-letter expanded deoxyribonucleic acid alphabet with nanopores. J Am Chem Soc 2023, 145, 8560–8568.

[69]  Yuan B-F. Assessment of DNA epigenetic modifications. Chem Res Toxicol 2020, 33, 695–708.

[70]  Watson JD, Crick FH. Molecular structure of nucleic acids: A structure for deoxyribose nucleic acid. Nature 1953, 171, 737–738.

[71]  http://ndbserver.rutgers.edu.

[72]  Kypr J, Kejnovska I, Renciuk D, Vorlıckova M. Circular dichroism and conformational polymorphism of DNA. Nucleic Acids Res 2009, 37, 1713–1725.

[73]  http://what-when-how.com/molecular-biology/circular-dichroism-part-2-molecular-biology/

[74]  Wood BR. The importance of hydration and DNA conformation in interpreting infrared spectra of cells and tissues. Chem Soc Rev 2016, 45, 1980–1998.

[75]  Gorenstein DG. $^{31}$P NMR of DNA. Methods Enzymol 1992, 211, 254–286.

[76]  Wemmer D. Structure and Dynamics by NMR. In Bloomfield VA, Crothers DM, Tinoco I. Nucleic Acids: Structures, Properties, and Functions. Sausalito, California: University Science Books, 1996, Chapter 6, 281–312 ISBN 0-935702-49-0.

[77]  James TL. NMR determination of oligonucleotide structure. Curr Protoc Nucleic Acid Chem 2001, Unit 7.2.8.

[78]  Capobianco A, Velardo A, Peluso A. Single-stranded DNA oligonucleotides retain rise coordinates characteristic of double helices. J Phys Chem B 2018, 122, 7978–7989.

[79]  Puffer B, Kreutz C, Rieder U, Ebert MO, Konrat R, Micura R. 5-Fluoro pyrimidines: Labels to probe DNA and RNA secondary structures by 1D $^{19}$F NMR spectroscopy. Nucleic Acids Res 2009, 37, 7728–7740.

[80]  Tanabe K, Tsuda T, Ito T, Nishimoto S-I. Probing DNA mismatched and bulged structures by using $^{19}$F NMR spectroscopy and oligodeoxynucleotides with a $^{19}$F-labeled nucleobase. Chem Eur J 2013, 19, 15133–15140.

[81]  Granqvist L, Virta P. 4′-C-[(4-Trifluoromethyl-1H-1,2,3-triazol-1-yl)methyl]thymidine as a sensitive $^{19}$F NMR sensor for the detection of oligonucleotide secondary structures. J Org Chem 2014, 79, 3529–3536.

[82]  El-Khoury R, Damha MJ. 2′-Fluoro-arabinonucleic acid (FANA): A versatile tool for probing biomolecular interactions. Acc Chem Res 2021, 54, 2287–2297.

[83]  Baranowski MR, Warminski M, Jemielity J, Kowalska J. 5′-Fluoro(di)phosphate-labeled oligonucleotides are versatile molecular probes for studying nucleic acid secondary structure and interactions by $^{19}$F NMR. Nucleic Acids Res 2020, 48, 8209–8224.

[84]  SantaLucia J Jr, Allawi HT, Seneviratne PA. Improved nearest-neighbor parameters for predicting DNA duplex stability. Biochemistry 1996, 35, 3555–3562.

[85]  Xia T, SantaLucia J Jr, Burkard ME, Kierzek R, Schroeder SJ, Jiao X, Cox C, Turner DH. Thermodynamic parameters for an expanded nearest-neighbor model for formation of RNA duplexes with Watson-Crick base pairs. Biochemistry 1998, 37, 14719–14735.

[86]  Sugimoto N, Nakano S-I, Katoh M, Matsumura A, Nakamuta H, Ohmichi T, Yoneyam M, Sasaki M. Thermodynamic parameters to predict stability of RNA/DNA hybrid duplexes. Biochemistry 1995, 34, 11211–11216.

[87]  Pirrung MC. How to Make a DNA Chip. Angew Chem Int Ed 2002, 41, 1276–1289.

[88]  Shchepinov MS, Case-Green SG, Southern EM. Steric factors influencing hybridisation of nucleic acids to oligonucleotide arrays. Nucleic Acids Res 1997, 25, 1155–1161.

[89] Southern EM, Maskos U, Elder JK. Analyzing and comparing nucleic acid sequences by hybridization to arrays of oligonucleotides: Evaluation using experimental models. Genomics 1992, 13, 1008–1017.

[90] Gao X, Yu P, LeProust E, Sonigo L, Pellois JP, Zhang H. Oligonucleotide synthesis using solution photogenerated acids. J Am Chem Soc 1998, 120, 12698–12699.

[91] Dalma-Weiszhausz DD, Warrington J, Tanimoto EY, Miyada CG. The Affymetrix GeneChip platform: An overview. Methods Enzymol 2006, 410, 3–28, adopted by Miller MB, Tang YW. Basic concepts of microarrays and potential applications in clinical microbiology. Clin Microbiol Rev 2009, 22, 611–633.

[92] Sack M, Hölz K, Holik A-K, Kretschy N, Somoza V, Stengele KP, Somoza MM. Express photolithographic DNA microarray synthesis with optimized chemistry and high-efficiency photolabile groups. J Nanobiotechnol 2016, 14, 1–13.

[93] Kretschy N, Holik A-K, Somoza V, Stengele K-P, Somoza MM. Next-generation o-nitrobenzyl photolabile groups for light-directed chemistry and microarray synthesis. Angew Chem Int Ed 2015, 54, 8555–8559.

[94] Okamoto T, Suzuki T, Yamamoto N. Microarray fabrication with covalent attachment of DNA using bubble jet technology. Nat Biotechnol 2000, 18, 438–441.

[95] Chrisey LA, Lee GU, O'Ferrall CE. Covalent attachment of synthetic DNA to self-assembled monolayer films. Nucleic Acids Res 1996, 24, 3031–3039.

[96] Olivier C, Hot D, Huot L, Ollivier N, El-Mahdi O, Gouyette C, Huynh-Dinh T, Gras-Masse H, Lemoine Y, Melnyk O. α-Oxo semicarbazone peptide or oligodeoxynucleotide microarrays. Bioconjugate Chem 2003, 14, 430–439.

[97] Dunbar SA. Applications of LuminexR xMAPi technology for rapid, high-throughput multiplexed nucleic acid detection. Clin Chim Acta 2006, 363, 71–82.

[98] Kumar A, Tchen P, Poullet F, Cohen J. Nonradioactive labeling of synthetic oligonucleotide probes with terminal deoxynucleotidyl transferase. Anal Biochem 1988, 169, 376–382.

[99] Tyagi S, Kramer FR. Molecular beacons: Probes that fluoresce upon hybridization. Nat Biotechnol 1996, 14, 303–308.

[100] Venkatesan N, Seo YJ, Kim BH. Quencher-free molecular beacons: A new strategy in fluorescence based nucleic acid analysis. Chem Soc Rev 2008, 37, 648–663.

[101] Marti AA, Jockusch S, Li Z, Ju J, Turro NJ. Molecular beacons with intrinsically fluorescent nucleotides. Nucleic Acids Res 2006, 34, e50 (7 pages).

[102] Bünzli J-CG. Lanthanide luminescence for biomedical analyses and imaging. Chem Rev 2010, 110, 2729–2755.

[103] Hovinen J, Guy PM. Bioconjugation with stable luminescent lanthanide(III) chelates comprising pyridine subunits. Bioconjugate Chem 2009, 20, 404–421.

[104] Taton TA, Mirkin CA, Letsinger RL. Scanometric DNA array detection with nanoparticle probes. Science 2000, 289, 1757–1759.

[105] Cao Y-WC, Jin R, Mirkin CA. Nanoparticles with Raman spectroscopic fingerprints for DNA and RNA detection. Science 2002, 297, 1536–1539.

[106] Klöcker N, Weissenboeck FP, Rentmeister A. Covalent labeling of nucleic acids. Chem Soc Rev 2020, 49, 8749–8773.

[107] Neef AB, Luedtke NW. Dynamic metabolic labeling of DNA in vivo with arabinosyl nucleosides. Proc Natl Acad Sci USA 2011, 108, 20404–20409.

[108] Qu D, Wang G, Wang Z, Zhou L, Chi W, Cong S, Ren X, Liang P, Zhang B. 5-Ethynyl-2'-deoxycytidine as a new agent for DNA labeling: Detection of proliferating cells. Anal Biochem 2011, 417, 112–121.

[109] Neef AB, Luedtke NW. An azide-modified nucleoside for metabolic labeling of DNA. ChemBioChem 2014, 15, 789–793.

[110] Rieder U, Luedtke NW. Alkene–tetrazine ligation for imaging cellular DNA. Angew Chem Int Ed 2014, 53, 9168–9172.

[111] Sterrenberg VT, Stalling D, Knaack JIH, Soh TK, Bosse JB, Meier C. A TriPPPro-nucleotide reporter with optimized cell-permeable dyes for metabolic labeling of cellular and viral DNA in living cells. Angew Chem Int Ed 2023, 62, e202308271 (9 pages).

[112] Arcamone F, Penco P, Orezzi P, Nicolella V, Pirelli A. Structure and synthesis of distamycin A. Nature 1964, 203, 1064–1065.

[113] Finlay AC, Hochstein FA, Sobin BA, Murphy FX. Netropsin, a new antibiotic produced by a Streptomyces. J Am Chem Soc 1951, 73, 341–343.

[114] Kawamoto Y, Bando T, Sugiyama H. Sequence-specific DNA binding pyrrole–imidazole polyamides and their applications. Bioorg Med Chem 2018, 26, 1393–1411.

[115] Lai Y-M, Fukuda N, Ueno T, Matsuda H, Saito S, Matsumoto K, Ayame H, Bando T, Sugiyama H, Mugishima H, Serie K. Synthetic pyrrole-imidazole polyamide inhibits expression of the human transforming growth factor-β1 gene. J Pharm Exp Therapeutics 2005, 315, 571–575.

[116] Mrksich M, Parks ME, Dervan PB. Hairpin peptide motif. A new class of oligopeptides for sequence-specific recognition in the minor groove of double-helical DNA. J Am Chem Soc 1994, 116, 7983–7988.

[117] Maeda R, Bando T, Sugiyama H. Application of DNA-alkylating pyrrole-imidazole polyamides for cancer treatment. ChemBioChem 2021, 22, 1538–1545.

[118] Hirose Y, Hashiya K, Bando T, Sugiyama H. Evaluation of the DNA alkylation properties of a chlorambucil-conjugated cyclic pyrrole-imidazole polyamide. Chem Eur J 2021, 27, 2782–2788.

[119] Mukherjee A, Sasikala WD. Drug–DNA intercalation: From discovery to the molecular mechanism. Adv Prot Chem Struct Biol 2013, 92, 1–62.

[120] Tibbs J, Ali Tabei SM, Kidd TE, Peters JP. Effects of intercalating molecules on the polymer properties of DNA. J Phys Chem B 2020, 124, 8572–8582.

[121] El-Adl K, Ibrahim M-K, Alesawy MSI, Eissa IH. [1,2,4]Triazolo[4,3-c]quinazoline and bis([1,2,4]triazolo) [4,3-a:4′,3′-c] quinazoline derived DNA intercalators: design, synthesis, in silico ADMET profile, molecular docking and anti-proliferative evaluation studies. Bioorg Med Chem 2021, 30, 115958 (16 pages).

[122] Schatzschneider U. Metallointercalators and Metalloinsertors: Structural Requirements for DNA Recognition and Anticancer Activity in Metallo-drugs. In Development and Action of Anticancer Agents. Metal Ions in Life Sciences. vol. 18, Sigel A, Sigel H, Freisinger E, Sigel RKO, eds. Berlin, Germany: Walter de Gruyter GmbH, 2018, 388–422.

[123] Barton JK, Danishhefsky AT, Goldberg JM. Tris(phenanthroline)ruthenium(II): Stereoselectivity in binding to DNA. J Am Chem Soc 1984, 106, 2172–2176.

[124] Friedman AE, Chambron J-C, Sauvage J-P, Turro NJ, Barton JK. A molecular light switch for DNA: Ru (bpy)2(dppz)2+. J Am Chem Soc 1990, 112, 4960–4962.

[125] Jackson BA, Barton JK. Recognition of DNA base mismatches by a rhodium intercalator. J Am Chem Soc 1997, 119, 12986–12987.

[126] Boyle KM, Barton JK. Targeting DNA mismatches with rhodium metalloinsertors. Inorg Chim Acta 2016, 452, 3–11.

[127] Wettig SD, Wood DO, Lee JS. Thermodynamic investigation of M-DNA: A novel metal ion–DNA complex. J Inorg Biochem 2003, 94, 94–99.

[128] Spring BQ, Clegg RM. Fluorescence measurements of duplex DNA oligomers under conditions conducive for forming M−DNA (a Metal−DNA Complex). J Phys Chem B 2007, 111, 10040–10052.

[129] Zhou M, Xu T, Xia K, Gao H, Li W, Zhai T, Gu H. Small DNAs that specifically and tightly bind transition metal ions. J Am Chem Soc 2023, 145, 8776–8780.

[130] Kondo J, Tada Y, Dairaku T, Hattori Y, Saneyoshi H, Ono A, Tanaka Y. A metallo-DNA nanowire with uninterrupted one-dimensional silver array. Nat Chem 2017, 9, 956–960.

[131] Katz S. The reversible reaction of Hg(II) and double-stranded polynucleotides a step-function theory and its significance. Biochim Biophys Acta, Spec Sect Nucleic Acids Related Subjects 1963, 68, 240–253.

[132] Kondo J, Yamada T, Hirose C, Okamoto I, Tanaka Y, Ono A. Crystal structure of metallo DNA duplex containing consecutive Watson-Crick-like T-Hg(II)-T base pairs. Angew Chem Int Ed 2014, 63, 2385–2388.

[133] Yamaguchi H, Sebera J, Kondo J, Oda S, Komuro T, Kawamura T, Dairaku T, Kondo Y, Okamoto I, Oni A, Burda JV, Kojima C, Sychrovsky V, Tanaka Y. The structure of metallo-DNA with consecutive thymine–HgII–thymine base pairs explains positive entropy for the metallo base pair formation. Nucleic Acids Res 2014, 42, 4094–4099.

[134] Torigoe H, Okamoto I, Dairaku T, Tanaka Y, Ono A, Kozasa T. Thermodynamic and structural properties of the specific binding between Ag+ ion and C:C mismatched base pair in duplex DNA to form C-Ag-C metal-mediated base pair. Biochimie 2012, 94, 2431–2440.

[135] Fortino M, Marino T, Russo N. Theoretical study of silver-ion-mediated base pairs: The case of C−Ag−C and C−Ag−A systems. J Phys Chem A 2015, 119, 5153–5157.

[136] Naskar S, Guha R, Müller J. Metal-modified nucleic acids: metal-mediated base pairs, triples, and tetrads. Angew Chem Int Ed 2020, 59, 1397–1406.

[137] Jash B, Müller J. Metal-mediated base pairs: From characterization to application. Chem Eur J 2017, 23, 17166–17178.

[138] Clever GH, Polborn K, Carell T. A highly DNA-duplex-stabilizing metal-salen base pair. Angew Chem Int Ed Engl 2005, 44, 7204–7208.

[139] Switzer C, Sinha S, Kim PH, Heuberger BD. A purine-like nickel(II) base pair for DNA. Angew Chem Int Ed 2005, 44, 1529–1532.

[140] Kim E-K, Switzer C. Bis(6-carboxypurine)-Cu$^{2+}$: A possibly primitive metal-mediated nucleobase pair. Org Lett 2014, 16, 4059–4061.

[141] Mandal S, Hepp A, Müller J. Unprecedented dinuclear silver(I)-mediated base pair involving the DNA lesion 1,$N^6$-ethenoadenine. Dalton Trans 2015, 44, 3540–3543.

[142] Mei H, Ingale SA, Seela F. Imidazolo-dC metal-mediated base pairs: Purine nucleosides capture two Ag+ ions and form a duplex with the stability of a covalent DNA cross-link. Chem Eur J 2014, 20, 16248–16257.

[143] Fujii A, Nakagawa O, Kishimoto Y, Okuda T, Nakatsuji Y, Nozaki N, Kasahara Y, Obika S. 1,3,9-Triaza-2-oxophenoxazine: An artificial nucleobase forming highly stable self-base pairs with three AgI ions in a duplex. Chem Eur J 2019, 25, 7443–7448.

[144] Takezawa Y, Suzuki A, Nakaya M, Nishiyama K, Shionoya M. Metal-dependent DNA base pairing of 5-carboxyuracil with itself and all four canonical nucleobases. J Am Chem Soc 2020, 142, 21640–21644.

[145] Taherpour S, Golubev O, Lonnberg T. Metal-ion-mediated base pairing between natural nucleobases and bidentate 3,5-dimethylpyrazolyl-substituted purine ligands. J Org Chem 2014, 79, 8990–8999.

[146] Takezawa Y, Mori K, Huang W-E, Nishiyama K, Xing T, Nakama T, Shionoya M. Metal-mediated DNA strand displacementand molecular device operations based on base-pair switching of 5-hydroxyuracil nucleobases. Nat Commun 2023, 14, 4759 (10 pages).

[147] Lönnberg TA, Hande MA, Ukale DU. Oligonucleotide complexes in bioorganometallic chemistry. In Parkin G, Meyer K, O'Hare D, eds. Comprehensive Organometallic Chemistry IV. 2022, Section 15, 146–182.

[148] Ukale DU, Lönnberg T. Triplex formation by oligonucleotides containing organomercurated base moieties. ChemBioChem 2018, 19, 1096–1101.

[149] Ukale DU, Lönnberg T. 2,6-Dimercuriphenol as a bifacial dinuclear organometallic nucleobase. Angew Chem Int Ed 2018, 57, 16171–16175.

[150] Ukale DU, Tahtinen P, Lonnberg T. 1,8-Dimercuri-6-phenyl-1*H*-carbazole as a monofacial dinuclear organometallic nucleobase. Chem Eur J 2020, 26, 2164–2168.

[151] Maity SK, Lönnberg T. Oligonucleotides incorporating palladocyclic nucleobase surrogates. Chem Eur J 2018, 24, 1274–1277.

[152] Hande M, Saher O, Lundin KE, Smith CIE, Zain R, Lönnberg T. Oligonucleotide–palladacycle conjugates as splice-correcting agents. Molecules 2019, 24, 1180 (13 pages).

[153] Tanaka K, Yamada Y, Shionoya M. Formation of silver(I)-mediated DNA duplex and triplex through an alternative base pair of pyridine nucleobases. J Am Chem Soc 2002, 124, 8802–8803.

[154] Phongtongpasuk S, Paulus S, Schnabl J, Sigel RKO, Spingler B, Hannon MJ, Freisinge Er. Binding of a designed anti-cancer drug to the central cavity of an RNA three-way junction. Angew Chem Int Ed 2013, 52, 11513–11516.

[155] Craig JS, Melidis L, Williams HD, Dettmer SJ, Heidecker AA, Altmann PJ, Guan S, Campbell C, Browning DF, Sigel RKO, Johannsen S, Egan RT, Aikman B, Casini A, Pöthig A, Hannon MJ. Organometallic pillarplexes that bind DNA 4-Way Holliday junctions and forks. J Am Chem Soc 2023, 145, 13570–13580.

[156] Abdelhamid MAS, Fabian L, MacDonald CJ, Cheesman MR, Gates AJ, Waller ZAJ. Redox-dependent control of i-motif DNA structure using copper cations. Nucleic Acids Res 2018, 46, 5886–5893.

[157] Engelhard DM, Nowack J, Clever GH. Copper-induced topology switching and thrombin inhibition with telomeric DNA G-quadruplexes. Angew Chem Int Ed 2017, 56, 11640–11644.

[158] Lindahl T. Instability and decay of the primary structure of DNA. Nature 1993, 362, 709–715.

[159] Frederico LA, Kunkel TA, Ramsay Shaw B. Cytosine deamination in mismatched base pairs. Biochemistry 1993, 32, 6523–6530.

[160] Lindahl T, Nyberg B. Rate of depurination of native deoxyribonucleic acid. Biochemistry 1972, 11, 3610–3618.

[161] Wilson DM III, Barsky D. The major human abasic endonuclease: Formation, consequences and repair of abasic lesions in DNA. Mutat Res DNA Repair 2001, 485, 283–307.

[162] Lindahl T. DNA repair enzymes. Annu Rev Biochem 1982, 51, 61–87.

[163] Cadet J, Delatour T, Douki T, Gasparutto D, Pouget J-P, Ravanat J-L, Sauvaigo S. Hydroxyl radicals and DNA base damage. Mutat Res 1999, 424, 9–21.

[164] Cadet J, Douki T, Gasparutto D, Ravanat JL. Oxidative damage to DNA: Formation, measurement and biochemical features. Mutat Res 2003, 531, 5–23.

[165] Dizdaroglu M. Formation of an 8-hydroxyguanine moiety in deoxyribonucleic acid on gamma-irradiation in aqueous solution. Biochemistry 1985, 24, 4476–4481.

[166] Cadet J, Berger M, Buchko GW, Joshi PC, Raoul S, Ravanat JL. 2,2-Diamino-4-[(3,5-di-O-acetyl-2-deoxy-beta-D-erythro-pentofuranosyl)amino]-5-(2H)-oxazolone: A novel and predominant radical oxidation product of 3′,5′-di-O-acetyl-2′-deoxyguanosine. J Am Chem Soc 1994, 116, 7403–7404.

[167] Vieira AJSC, Steenken S. Pattern of hydroxy radical reaction with adenine and its nucleosides and nucleotides. Characterization of two types of isomeric hydroxy adduct and their unimolecular transformation reactions. J Am Chem Soc 1990, 112, 6986–6994.

[168] Wagner JR, van Lier JE, Decarroz C, Berger M, Cadet J. Photodynarnic methods for oxy radical-induced DNA damage. Meth Enzymol 1990, 186, 502–511.

[169] Wagner JR, van Lier JW, Berger M, Cadet J. Thymidine hydroperoxides: Structural assignment, conformational features, and thermal decomposition in water. J Am Chem Soc 1994, 116, 2235–2242.

[170] Burrows CJ, Muller JG. Oxidative nucleobase modifications leading to strand scission. Chem Rev 1998, 98, 1109–1151.

[171] Jovanovic SV, Simic MG. Mechanism of OH radical reactions with thymine and uracil derivatives. J Am Chem Soc 1986, 108, 5968–5972.

[172] Hazra DK, Steenken S. Pattern of hydroxyl radical addition to cytosine and 1-, 3-, 5-, and 6-substituted cytosines. Electron transfer and dehydration reactions of the hydroxyl adducts. J Am Chem Soc 1983, 105, 4380–4386.

[173] Wagner JR, Madugundu GS, Cadet J. Ozone-induced DNA damage: A pandora's box of oxidatively modified DNA bases. Chem Res Toxicol 2021, 34, 80–90.

[174] Candeias LP, Steenken S. Ionization of purine nucleosides and nucleotides and their components by 193-nm laser photolysis in aqueous solution: Model studies for oxidative damage of DNA. J Am Chem Soc 1992, 114, 699–704.

[175] Krokan HE, Standal R, Slupphaug G. DNA glycosylases in the base excision repair of DNA. Biochem J 1997, 325, 1–16.

[176] Cadet J. DNA Adducts: Identification and Biological Significance. In Hemminki A, Dipple A, Shuker DEG, Kadlubar FF, Segerbäck D, Bartsch H, eds. IARC Scientific Publication No. 125. Lyon, 1994.

[177] Pogozelski WK, Tullius TD. Oxidative strand scission of nucleic acids: Routes Initiated by hydrogen abstraction from the sugar moiety. Chem Rev 1998, 98, 1089–1107.

[178] Brown DM. Basic Principles in Nucleic Acid Chemistry, Ts'o POP, ed. NY and London: Academic Press Inc. Vol II, 1974, Chapter 1.

[179] Beranek DT. Distribution of methyl and ethyl adducts following alkylation with monofunctional alkylating agents. Mutat Res 1990, 231, 11–30.

[180] Rokita SE, Yang J, Pande P, Greenberg WA. Quinone methide alkylation of deoxycytidine. J Org Chem 1997, 62, 3010–3012.

[181] Jahnz-Wechmann Z, Framski GR, Januszczyk PA, Boryski J. Base-modified nucleosides: Etheno derivatives. Front Chem 2016, 4, Article 19.

[182] Guengerich FP, Raney VM. Formation of etheno adducts of adenosine and cytidine from 1-halooxiranes. Evidence for a mechanism involving initial reaction with the endocyclic nitrogen atoms. J Am Chem Soc 1992, 114, 1074–1080.

[183] Krzyzosiak WJ, Biernat J, Ciesiolka J, Gornicki P, Wiewiorowski M. Comparative studies on reactions o fadenosine and cytidine with chloroacetaldehyde, α-bromopropionaldehyde and chloroacetone synthesis of 1,$N^6$-ethenoadenosine and 3,$N^4$-ethenocytidine derivatives methylated at the ethenobridge. Polish J Chem 1983, 57, 779–787.

[184] Yip KF, Tsou KC. Synthesis of fluorescent adenosine derivatives. Tetrahedron Lett 1973, 14, 3087–3090.

[185] Yamaji N, Yuasa Y, Kato M. The synthesis of 2-substituted 1,$N^6$-ethenoadenosine-3′, 5′-cyclicphosphate by ring reclosure of alkali-hydrolyzate of 1,$N^6$-ethenoadenosine-3′,5′-cyclic phosphate. Chem Pharm Bull 1976, 24, 1561–1567.

[186] Eder E, Hoffman C, Bastian H, Deininger C, Scheckenbach S. Molecular mechanisms of DNA damage initiated by alpha, beta-unsaturated carbonyl compounds as criteria for genotoxicity and mutagenicity. Environ Health Perspect 1990, 88, 99–106.

[187] Chenna A, Iden RC. Characterization of 2′-deoxycytidine and 2′-deoxyuridine adducts formed in reactions with acrolein and 2-bromoacrolein. Chem Res Toxicol 1993, 6, 261–268.

[188] Kawai Y, Furuhata A, Toyokuni S, Aratani Y, Uchida K. Formation of acrolein-derived 2′-deoxyadenosine adduct in an iron-induced carcinogenesis model. J Biol Chem 2003, 278, 50346–50354.

[189] Chung F-L, Chen H-JC, Nath RG. Lipid peroxidation as a potential endogenous source for the formation of exocyclic DNA adducts. Carcinogenesis 1996, 17, 2105–2111.

[190] Douki T, Ames BN. An HPLC-EC assay for 1,$N^2$-propano adducts of 2'-deoxyguanosine with 4-hydroxynonenal and other alpha,beta-unsaturated aldehydes. Chem Res Toxicol 1994, 7, 511–518.

[191] Marnett LJ. Lipid peroxidation-DNA damage by malondialdehyde. Mutat Res 1999, 424, 83–95.

[192] Dedon PC, Plastaras JP, Rouzer CA, Marnett LJ. Indirect mutagenesis by oxidative DNA damage: Formation of the pyrimidopurinone adduct of deoxyguanosine by base propenal. Proc Natl Acad Sci USA 1998, 95, 11113–11116.

# 9 Nucleic acids: RNA

## 9.1 Structure of RNA

### 9.1.1 Global structure

RNA forms an A-type double helix very similar to A-DNA but even more stable (see Table 8.1). RNA also forms a duplex with DNA. The conformation of hybrid DNA/RNA duplex differs from the conformation of both A- and B-type duplexes, but is biased toward A-type duplex [1]. In biological systems, RNA usually occurs as a single strand, although its ability to hybridize with DNA and other RNA molecules plays a crucial role in RNA's biology. The global structure of single-stranded RNA is complicated. Several secondary structural elements are recognized that by mutual interactions give RNA a protein like tertiary structure [2]. H-bonding interactions play a crucial role in addition to Watson–Crick, reversed Watson–Crick, Hoogsteen and reversed Hoogsteen base pairs. The chain folding takes place in a hierarchical manner, proceeding by formation of small structural elements, the mutual interaction of which then gives the overall shape of chain folding. Figure 9.1 depicts the typical structural motifs [3]. Figure 9.2, in turn, gives an example of how interaction of various motifs of secondary structure

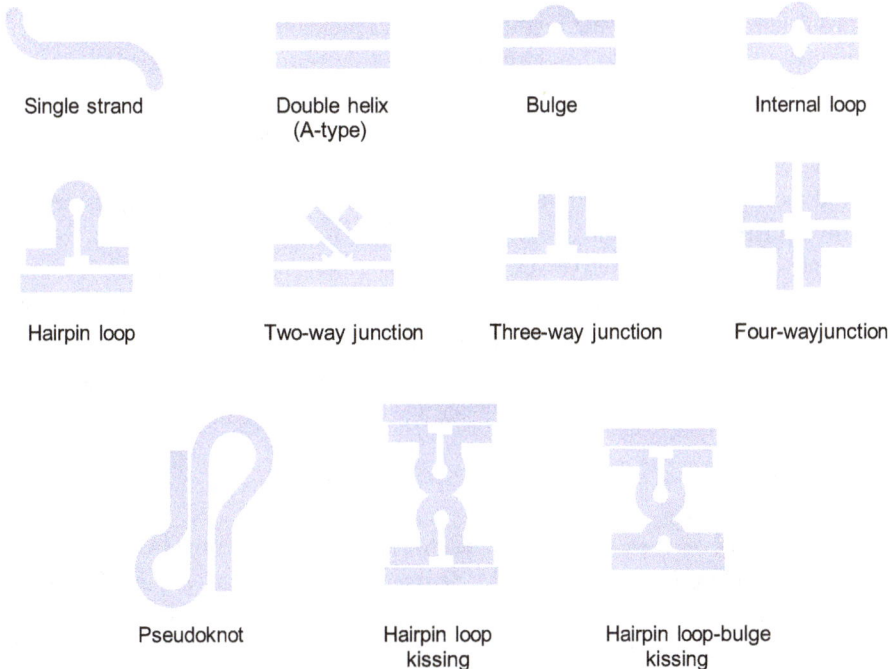

| Single strand | Double helix (A-type) | Bulge | Internal loop |

| Hairpin loop | Two-way junction | Three-way junction | Four-wayjunction |

| Pseudoknot | Hairpin loop kissing | Hairpin loop-bulge kissing |

**Figure 9.1:** RNA secondary and tertiary structure motifs [3].

https://doi.org/10.1515/9783111325637-009

results in chain folding to protein-like tertiary structure. Figure 9.3 records a rather comprehensive collection of possible noncanonical base–base interactions [4]. Besides these, RNA contains more than 100 rare nucleosides, above all in transfer and ribosomal RNA, but also in messenger RNA (cf. Section 9.2) [5]. The actual role of most nucleoside modifications remains obscure. Several of them occur in functional regions of ribosomal RNA (rRNA), suggesting that a functional role is possible. Finally, one should note that in single-stranded RNA, the amount of unpaired single strand regions is low, usually of the order of a few percent. RNA is also able to form G-quadruplex structures similar to those formed by DNA [6] and, interesting enough, hetero G-quadruplexes with DNA [7]. The latter type of quadruplexes is assumed to play a role in regulation of gene transcription.

### 9.1.2 Determination of primary structure

Sequencing of RNA customarily consists of enzymatic reverse transcription to complementary DNA and its high-throughput sequencing. The benefit of this approach is high efficiency that allows insight to the whole transcriptome of a cell [8]. The shortcoming is that base modifications are not retained during transcription, but base modification becomes replaced by a canonical base, the efficiency of transcription differing from that of the canonical counterpart [9]. This situation may be improved by development of third-generation sequencing methods. Direct sequencing of RNA by the nanopore technology (see Section 8.2) has already allowed detection of $N^6$-methyladenosine and inosine modifications [10]. For specific quantification of a certain modification several enzymatic and chemical techniques have been developed [11]. For example, a reverse transcriptase selected by directed evolution recognizes 1-methyladenosine [10] and a catalytic A3A protein (known as eA3A-v10) deaminates both cytosine and 5-methylcytosine, but not 5-hydroxymethyl cytosine [12]. During sequencing, deaminated cytosine and 5-methylcytosine are, hence, converted to uracil and thymine, whereas 5-hydroxymethylcytosine becomes read as cytosine. Glyoxal/nitrite deamination, in turn, has been used to make a difference between $N^6$-methyladenosine and adenosine [13] and bisulfite promoted cleavage for quantification of pseudouridine [14].

### 9.1.3 Determination of secondary and tertiary structures

As mentioned above, even single-stranded RNA is largely base-paired. A conventional chemical method for determination of single-stranded regions is alkylation of nitrogen atoms to find out which nucleosides undergo alkylation and which remain shielded from alkylation due to base-pairing. Methylation of N1 atom of both purines and pyrimidines by dimethylsulfate is a traditional method [15]. The other related approaches consist of $N^1,N^2$-cyclization of guanine with glyoxals [16] and N1 alkylation of uracil

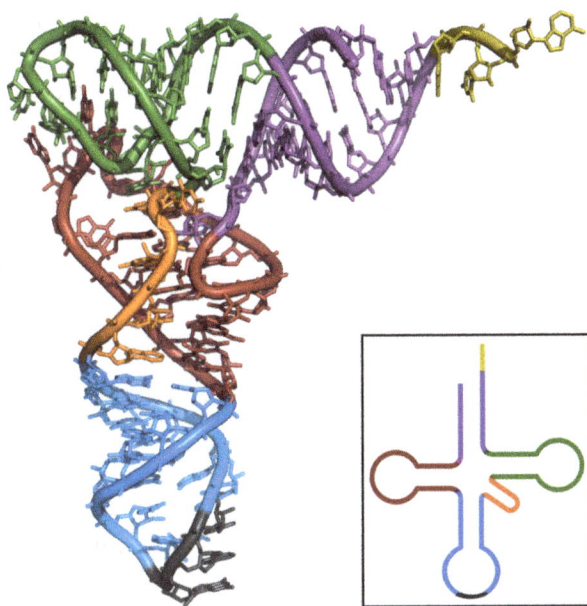

**Figure 9.2:** An example of folded secondary structure of tRNA; https://upload.wikime dia.org/wikipedia/commons/b/ba/TRNA-Phe_yeast_1ehz.png.

with carbodiimides [17]. Carbodiimides are sufficiently basic to deprotonate N1H of uracil, which enhances the attack of N1 on the carbodiimide center. Guanine is also subject to N1-alkylation by a similar mechanism, but the reaction takes place less readily. Besides base moiety alkylation, 2′-OH acylation is increasingly used for identification of flexible regions in RNA. The acronym of the method is SHAPE (selective 2′-hydroxyl acylation analyzed by primer extension) [18].

Accessibility of various regions of RNA strand to solvent offers another approach to probe the chain folding. Hydroxyl radicals generated by Fenton reagent (a solution of hydrogen peroxide and $FeSO_4$) are traditionally used for the purpose. When radicals reach the sugar ring of an unpaired nucleoside, they abstract H3′ or H4′ resulting in chain cleavage [19]. N-Aroyl azides, such as pyridine-3-carbonylazide, are currently used for the same purpose. Upon irradiation with long-wavelength UV light, aroyl azides decompose to aroyl nitrenium ions (Ar-C(O)-N) that react with purine nucleosides yielding C8-amidated adenosine and guanosine [20]. The latter method is even applicable for probing interactions of RNA with proteins in living cells.

As regards studies on hybridization equilibria of RNA or the interaction with other biomolecules, fluorine labelling plays an important role since the $^{19}F$-shifts are sensitive to environment. The focus of pioneering studies that date back to late 1980s was on conformation of 5-$^{19}F$-uracil substituted transfer RNAs (tRNA) [21]. 2′-$^{19}F$-ribonucleosides, in turn, have enabled quantification of alternative hairpin forms within a single-stranded

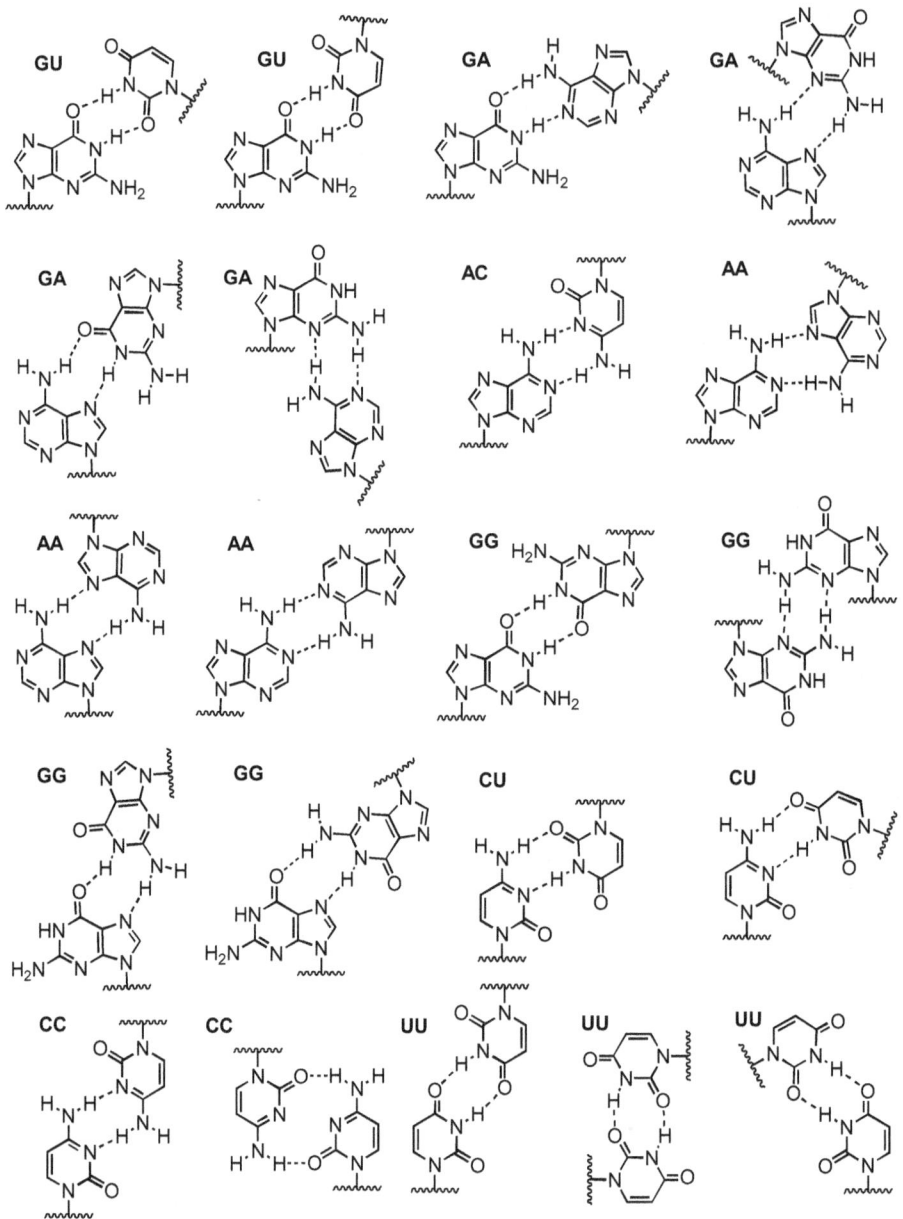

**Figure 9.3:** Noncanonical base pairs occurring in RNA [4].

oligoribonucleotide [22]. Dependence of $^{19}$F-shift on temperature offers an extensively used method for determination of melting temperature for RNA duplexes [23]. Labeling of the stem of an RNA hairpin with a highly sensitive tris(trifluoromethyl) group has enabled following of oligoribonucleotide invasion into the hairpin structure [24]. 2'-

Fluoroarabinonucleoside [25] and 2'-*O*-trifluoromethylribonucleosides [26, 27] have received interest as enzymatically stable oligonucleotides that still hybridize with native nucleic acids and may find pharmaceutical applications.

## 9.2 Types and functions of RNA

The most abundant type of cellular RNA is **ribosomal RNA (rRNA)**, the main component of ribosomes, that is, cytoplasmic nucleoprotein complexes within which protein synthesis takes place. The length of human rRNA typically falls in the range 4,000–5,000 nucleotides. rRNA is non-coding; it does not mediate the code of protein synthesis from DNA. Instead, it is ribozyme, a catalytic nucleic acid that promotes formation of peptide bonds (cf. Section 10.3). Another non-coding RNA, **transfer RNA (tRNA)** participates in protein synthesis by carrying amino acids to ribosomes. They are 76- to 90-nucleotide-long single-stranded molecules that have a cloverleaf type secondary structure (see Figure 9.2). Each tRNA carries a given amino acid esterified to the 3'-terminus. The middle one of the three stem loops contains a three-letter anticodon region that recognizes the complementary codon within **messenger RNA (mRNA)**, the coding RNA that mediates the genetic code from DNA to ribosomes.

The RNA transcript of DNA (pre-RNA) formed in nucleus contains, in addition to coding regions (exons), intervening non-coding regions (introns). These are removed by splicing, a process catalyzed by spliceosome ribozymes (see Section 10.3). The 5'-terminus of mRNA is then capped with 7-methylguanosine that becomes linked to the 5'-terminal nucleoside via a 5',5'-triphosphate bridge [28]. This penultimate nucleoside of the capped-mRNA often is a 2'-*O*-methyl nucleoside. Abbreviations cap, cap-1 and cap-2 refer to state of methylation: none, one or two terminal sugars methylated. The role of cap structure is to stabilize RNA against 5'-exonucleolytic degradation and to serve as a recognition site for numerous proteins participating in buildup of the splicing machinery, polyadenylation of the 3'-terminus, export from nucleus and initiation of translation [29].

As regards chemical properties of mRNA, N7-methyl acidifies N1H by two orders magnitude, reducing the p$K_a$ value of cap structure to 7.35 ($T$ = 25 °C, $I$ = 0.15 M) [30]. As long as the penultimate 5'-nucleoside is a 2'-*O*-methyl purine ribonucleoside, the cap structure exhibits strong intramolecular stacking [31, 32]. As regards chemical stability, 7-methylguanine base is susceptible to nucleophilic attack on C8. However, the half-life for imidazole ring opening still is around 1 week at pH 9 and 25 °C [33]. For comparison, the cellular half-life of mRNA is only minutes. Macrocyclic oligoamine complexes of $Cu^{2+}$ and $Eu^{3+}$ have been shown to cleave the triphosphate bridge of 5'-capped oligoribonucleotides hybridized with complementary DNA sequences [34, 35]. Although the cap structure is stable under biological conditions, its lability hampers chemical synthesis of capped ORNs required for biological studies. Reasonably large amounts of capped ORNs has been obtained by assembling the ORN on solid support from 2'-*O*-

pivaloylmethyl protected phosphoramidites, coupling guanosine5'-diphosphate to the 5'-terminal hydroxyl function by imidazole activation, purifying the released ORN and carrying out the N7 methylation with human (guanine-N7)-methyl transferase [36]. Recently, an entirely non-enzymatic solid-supported synthesis for capped RNAs has been reported [37]. The sequence is first assembled on solid support from 2'-O-propionyloxymethyl protected phosphoramidite building blocks, converted to 5'-phosphoroimidazolide and capped with 7-methyl-GDP in DMF in the presence of zinc chloride. The 2'-O-propionyloxymethyl protection is sufficiently labile to be removed under conditions that the cap moiety withstands.

According to very recent studies, diadenosine α,δ-tetraphosphate capped RNA is present in human and rat cell lines [38]. It does not cause immune response, which imply to occurrence as a natural component of transcriptome. It becomes enzymatically de-capped, but not translated.

The other naturally occurring RNAs consist of small nuclear RNA, micro-RNA, long noncoding RNA and circular RNA. **Small nuclear RNAs (snRNA)** are around 150 nucleotides long constituents of spliceosomes, that is, nucleoprotein complexes that catalyze splicing. The cap of snRNA differs from that of mRNA: the 2-amino group of 7-methylguanosine is di-methylated. **Micro-RNAs (miRNA)** are around 22 nucleotides long non-coding RNAs that mediate post-transcriptional gene silencing by a so-called RNA interference mechanism (cf. Chapter 12). The biological functions of **long non-coding RNAs (lncRNA) and circular RNAs (circRNA)** are not fully known. lncRNAs are more than 200 nucleotides long RNA transcripts [39]. The knowledge of their biological functions is still rather limited. CircRNAs are single-stranded RNAs covalently closed to a cyclic form with a phosphodiester linkage between 3'- and 5'-terminal hydroxyl functions [40]. In eukaryotes, they most likely are formed from introns during splicing of pre-RNA (cf. Chapter 10). The functions of circRNAs are not fully known. At least they serve as scaffolds for proteins and encode synthesis of short peptides [41]. Owing to cyclic structure, the biological half-life of circRNAs is longer than that of their linear counterparts. This has increased interest in cyclic RNAs as therapeutic of diagnostic agents [42]. Solid-phase synthesis of cyclic ORNs of moderate ring-size has been developed [43]. However, the ring closure inevitably becomes more difficult with increasing the ring size. Finally, it is noteworthy that all types of RNA may contain modified nucleosides. Altogether 170 modifications, both base and sugar modifications, have been identified [44].

## 9.3 Imaging of RNA in cells

Metabolic labeling is the technique most extensively used to follow transcriptional dynamics of RNA in living cells. The procedure is very similar to that discussed in Section 8.5 for in vivo imaging of DNA. Ribonucleosides bearing alkyne, alkene or azido substituents (Figure 9.4) are taken up by cells and converted via salvage pathways to 5'-

triphosphates, that is, substrates of RNA polymerases [45]. The most widely used among them is 5-ethynyluridine (Figure 9.4A). A potential side reaction is ribonucleoside reductase-catalyzed deoxygenation to the corresponding 2′-deoxyribonucleoside. The ethynyl- or azido-substituted nucleosides (Figure 9.4A–D) incorporated into RNA by transcription are visualized in fixed cells by Cu-catalyzed azide–alkyne cycloaddition of a fluorophore [46]. However, the approach is not fully bio-orthogonal but upon long treatment, proliferation tends to slow down [47]. The cytotoxicity of copper may be reduced by careful choice of conditions and addition of a water-soluble multidentate triazole-based ligand, such as tris(3-hydroxypropyltriazolylmethyl)amine (THPTA), that reduces reactive oxygen species [48]. Still, vinyl substituted ribonucleosides (Figure 9.4E–G) visualized by IEDDA reaction (see Section 8.5) may well offer better bioorthogonality [47].

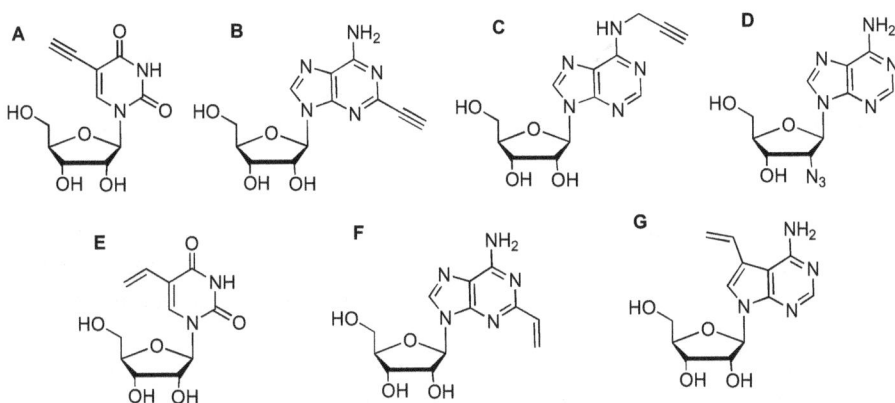

**Figure 9.4:** Modified nucleosides for imaging of RNA in cells [45].

## 9.4 Riboswitches

Riboswitches are noncoding elements within untranslated regions of bacterial mRNAs. Their role is to regulate the expression of genes of biosynthetic pathways. The end product of a metabolic pathway can, for example, halt its own production by binding to the riboswitch domain. Change in the 3D structure of this domain triggers a change in the secondary or tertiary structure of the translated region of mRNA. Consequently, translation of mRNA to an enzyme essential for production of the end metabolite becomes prevented, as schematically depicted in Figure 9.5 [49, 50]. In addition to bacteria, riboswitches occur in archaea, plants, fungi and algae [51]. The spectrum of ligands recognized by riboswitches is wide. Identification of the first family of riboswitches that consisted of vitamins thiamine pyrophosphate (TPP) [52, 53], flavin mononucleotide (FMN) [53], and adenosylcobalamin (AdoCbl) [54], dates back to the first years of this century. Since then, more than 50 different types of riboswitches have been identified [55], including riboswitches recognizing nucleobases, amino acids, sugar phosphates, secondary messengers [56] and some ions [52, 57, 58]. Many of these ligands are those that assumedly were relevant for primitive life, that is, during the era of "RNA world," that is, prior to "protein world" [55].

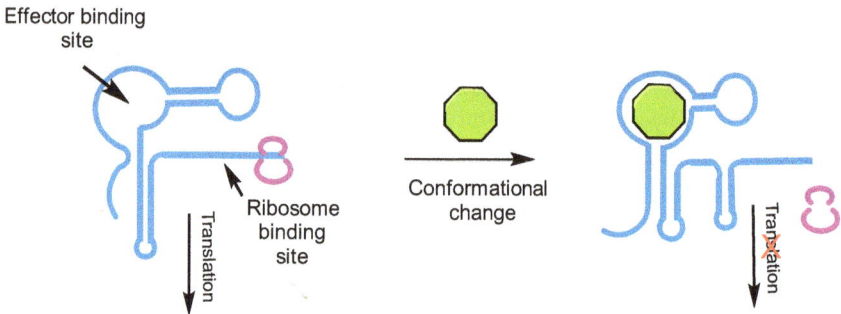

**Figure 9.5:** Schematic presentation of ligand-induced termination of translation by binding to a riboswitch.

Representative examples of the secondary structures of riboswitches are given in Figure 9.6 [59]. A common feature is presence of several double helical stems that are linked to each other by single-stranded connectors. The conserved nucleotides are shown in red. They most likely play the major role in chain folding and ligand binding. An appreciable proportion of these nucleotides are situated in internal or terminal loops of the helical stems and in the single-stranded connectors. Interaction between the conserved loop structures results in clustering of the stems, whereas the connectors often form a binding pocket for ligand. The family of purine riboswitches that bind adenine [60], hypoxanthine [61] and guanine [62], serves as an illustrative example of the ligand binding modes that riboswitches use. Their regulatory domain consists of three double helical segments centered upon a three way junction (Figure 9.7). Two of the helical stem loops clustered coaxially through stacking of the conserved nucleotides in their terminal loops (shaded in orange) [63]. The nucleotides in the connector region are, in turn, responsible for selectivity of the purine binding.

The ligand recognition mode of several other riboswitches is closely related to that of purine riboswitches. With a larger lysine riboswitch (Figure 9.6E), for example, a co-axial three helix bundle is formed. Among the double helical stems, the one having an internal loop is clustered with the two other stem loops: with the longest one through stacking of the terminal loops and with a shorter one by interaction of the internal loop with the terminal loop [64]. As with purine riboswitches, the ligand is bound in the junction region of the three helices. The riboswitch for cyclic diguanylate, a bacterial secondary messenger, likewise consists of three helices, two of which adopt a parallel orientation by internal loop/terminal loop interaction [65]. The third one, a shorter heli-

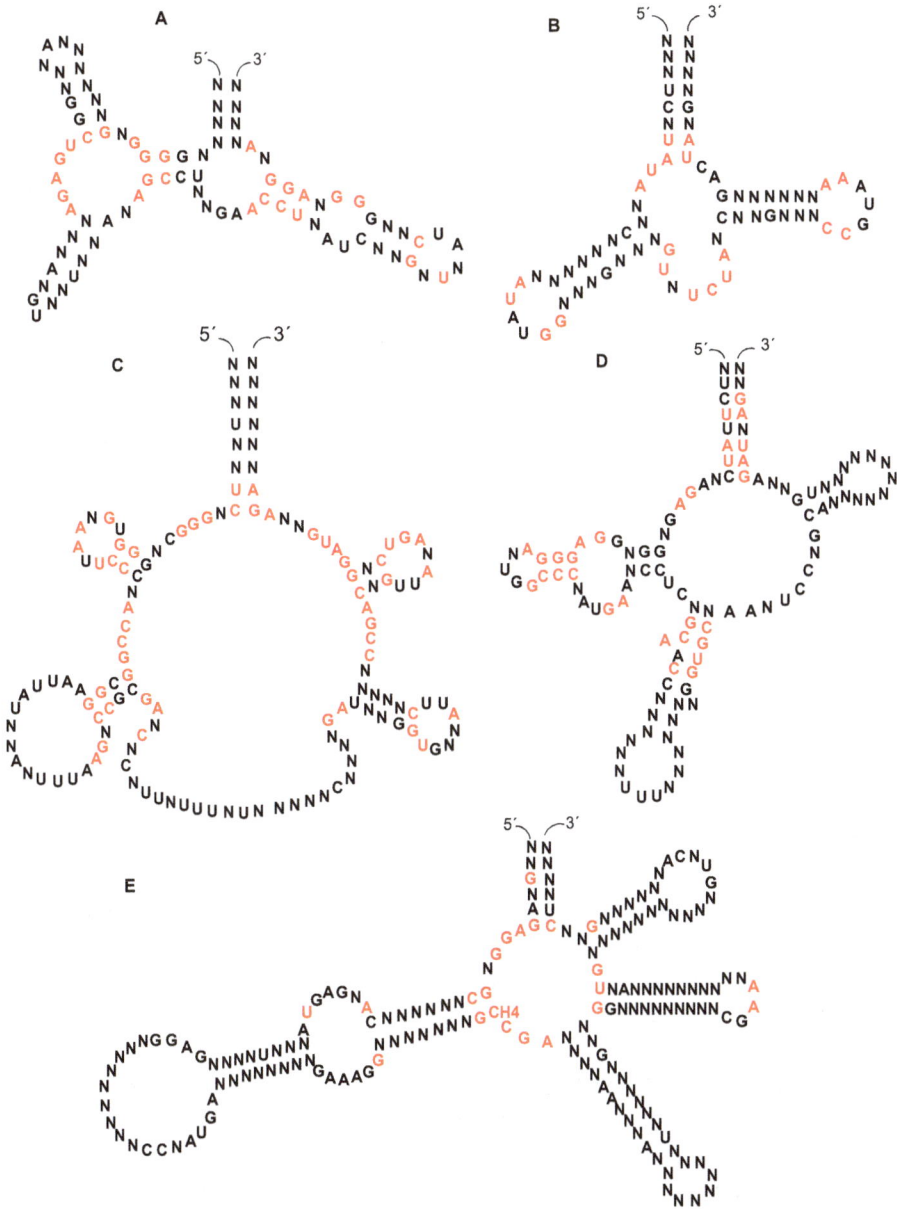

**Figure 9.6:** Schematic presentation of secondary structures of the regulatory domains of riboswitches for: thiamine pyrophosphate (A), guanine (B), flavin adenine moninucleotide (C), S-adenosylmethionine (D) and lysine (E). For more detailed information, see Ref. [59].

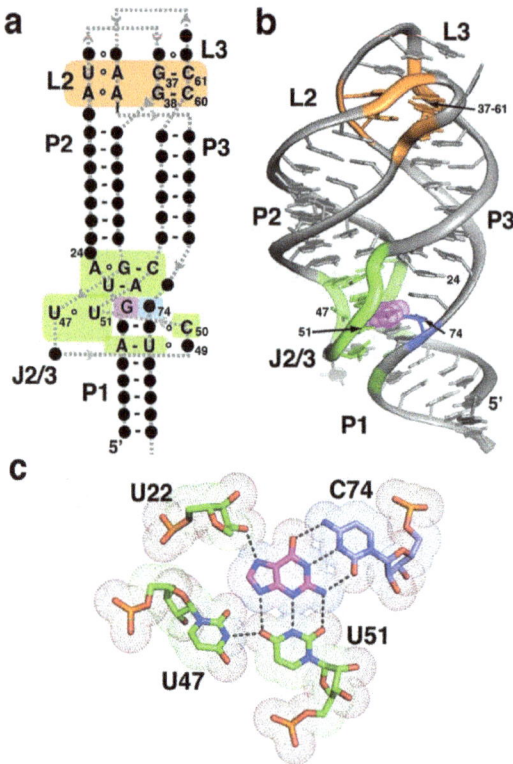

**Figure 9.7:** Structure of guanine riboswitch taken from Ref. [63] with permission from Cambridge University Press.

cal segment, is oriented perpendicular to those two. The binding pocket is again located in the joining region of the helices. 7-Aminomethyl-7-deazaguanine (PreQ₁) riboswitch that consists of only 34 nucleotides serves as an example of small riboswitches [66]. Two stems that are separated by short-loop structures are folded into a pseudoknot binding-pocket.

Tiamine pyrophosphate riboswitch (Figure 9.6A), in turn, exhibits a basically different binding mode: the ligand is bound to two internal loops and, hence, bridges two helical stems [67, 68]. It is also worth bearing in mind that riboswitches for a single molecule can show marked structural diversity. The best known example is S-adenosylmethionine riboswitch (SAM) that occurs in six different forms [69, 70]. All of them have their own structural features. SAM-I (or SAM, Figure 9.6D) contains a four way junction, while SAM-II has a rather small pseudoknot structure and SAM-III contains a three-way junction. The binding site of SAM-IV resembles that of SAM-I, but the scaffolds are different. SAM-V is, in turn, a variant of SAM-II.

Among known riboswitches, glucosamine-6-phosphate (glmS) riboswitch is the only one that is catalytically active, that is, a ribozyme [71]. Binding of the glucoamine-6-phosphate promotes self-cleavage of the riboswitch (cf. Section 10.2), which down-regulates glucosamine-6-phosphate production. The 3D structure is rather complex comprising several more or less coaxial helical stems and three pseudoknots that bury glmS inside the structure [72].

Interestingly enough, riboswitches do not only bind metabolites, but they also sense ion concentrations. A riboswitch in the 5′-untranslated region of mRNA encoded by $Mg^{2+}$ transport gene of *Salmonella enterica* has been shown to undergo marked change in tertiary structure upon $Mg^{2+}$ binding [73, 74]. These structural changes then determine whether the transcription reading stops or not within the 5′-untranslated region of mRNA. Even more unexpectedly, riboswitches that selectively sense fluoride ions have been identified in several bacterial and archaeal species [75]. They activate expression of genes that encode enzymes required to lessen the toxic effects of fluoride ion.

Naturally occurring riboswitches are attractive targets for drug discovery [49]. Riboswitches occur predominantly in bacteria being involved in vital metabolic pathways. Their function could in principle be interfered with small drug-like molecules. From this point of view, they seem to be ideal targets for discovery of novel antibiotics that actually are desperately needed.

## Further reading

Breaker RR The biochemical landscape of riboswitch ligands. Biochemistry 2022, 61, 137–149.

England WE, Garfio CM, Spitale RC Chemical approaches to analyzing RNA structure transcriptome-wide. ChemBioChem 2021, 22, 1114–1121.

Hendrix DK, Brenner SE, Holbrook SR RNA structural motifs: Building blocks of a modular biomolecule. Quart. Rev Biophys 2005, 38, 221–243.

Klöcker N, Weissenboeck FP, Rentmeister A Covalent labeling of nucleic acids. Chem Soc Rev 2020, 49, 8749–8773.

Mandal M, Boese B, Barrick JE, Winkler WC, Breaker RR Riboswitches control fundamental biochemical pathways in Bacillus subtilis and other bacteria. Cell 2003, 113, 577–586.

Mattick JS, Amaral PP, Carninci P, Carpenter S, Chang HY, Chen -L-L, Chen R, Dean C, Dinger ME, Fitzgerald KA, Gingeras TR, Guttman M, Hirose T, Huarte M, Johnson R, Kanduri C, Kapranov P, Lawrence JB, Lee JT, Mendell JT, Mercer TR, Moore KJ, Nakagawa S, Rinn JL, Spector DL, Ulitsky I, Wan Y, Wilusz JE, Wu M Long non-coding RNAs: Definitions, functions, challenges and recommendations. Nat Rev Mol Cell Biol 2023, 24, 430–447.

Motorin Y, Helm M General principles and limitations for detection of RNA modifications by sequencing. Acc Chem Res 2024, 57, 275–288.

Xiao MS, Ai Y, Wilusz JE Biogenesis and functions of circular RNAs come into focus. Trends Cell Biol 2020, 30, 226–240.

Zhang L-S, Dai Q, He C Base-resolution sequencing methods for whole-transcriptome quantification of mRNA modifications. Acc Chem Res 2024, 57, 1, 47–58.

# References

[1]    Liu JH, Xi K, Zhang X, Bao L, Zhang X, Tan ZJ. Structural flexibility of DNA-RNA hybrid complexes: Stretching and twist-stretch coupling. Biophys J 2019, 117, 74–86.

[2]    Hendrix DK, Brenner SE, Holbrook SR. RNA structural motifs: Building blocks of a modular biomolecule. Quart Rev Biophys 2005, 38, 221–243.

[3]    Tian B, Bevilacqua PC, Diegelman-Parente A, Mathews MB. The double-stranded-RNA binding motif: Interference and much more. Nat Rev Mol Cell Biol 2004, 5, 1013–1023.

[4]    Tinoco I Jr. Appendix 1 of the RNA World. Gesteland RF, Atkins JF, Eds, Cold Spring Harbor Laboratory Press, 1993, 603–607.

[5]    Cantara WA, Crain PF, Rozenski J, McCloskey JA, Harris KA, Zhang X, Vendeix FA, Fabris D, Agris PF. The RNA modification database, RNAMDB: 2011 update. Nucleic Acids Res 2011, 39, Database issue, D195–201.

[6]    Banco MT, Ferre-D'Amare AR. The emerging structural complexity of G-quadruplex RNAs. RNA 2021, 27, 390–402.

[7]    Berlyoung AS, Armitage BA. Assembly and characterization of RNA/DNA hetero-G-quadruplexes. Biochemistry 2020, 59, 4072–4080.

[8]    Kukurba KR, Montgomery SB. Sequencing and analysis. Cold Spring Harb Protoc 2015, 11, 951–969.

[9]    Motorin Y, Helm M. General principles and limitations for detection of RNA modifications by sequencing. Acc Chem Res 2024, 57, 275–288.

[10]   Workman RE, Tang AD, Tang PS, Jain M, Tyson JR, Razaghi R, Zuzarte PC, Gilpatrick T, Payne A, Quick J, Sadowski N, Holmes N, de Jesus JG, Jones KL, Soulette CM, Snutch TP, Loman N, Paten B, Loose M, Simpson JT, Olsen HE, Brooks AN, Akeson M, Timp W. Nanopore native RNA sequencing of a human poly(A) transcriptome. Nature Methods 2019, 16, 1297–1305.

[11]   Zhang L-S, Dai Q, He C. Base-resolution sequencing methods for whole-transcriptome quantification of mRNA modifications. Acc Chem Res 2024, 57, 1, 47–58.

[12]   Xie N-B, Wang M, Chen W, T-T J, Guo X, Gang F-Y, Wang Y-F, Feng Y-Q, Liang Y, Ci W, Yuan B-F. Resolution by bisulfite-free single-step deamination with engineered cytosine deaminase. ACS Cent Sci 2023, 9, 2315–2325.

[13]   Zhang M, Xiao Y, Jiang Z, Yi C. Quantifying m6A and Ψ modifications in the transcriptome via chemical-assisted approaches. Acc Chem Res 2023, 56, 2980–2991.

[14]   Burrows CJ, Fleming AM. Bisulfite and nanopore sequencing for pseudouridine in RNA. Acc Chem Res 2023, 56, 2740–2751.

[15]   Tijerina P, Mohr S, Russell R. DMS footprinting of structured RNAs and RNA–protein complexes. Nat Protoc 2007, 2, 2608–2623.

[16]   Mitchell D 3rd, Ritchey LE, Park H, Babitzke P, Assmann SM, Bevilacqua PC. Glyoxals as in vivo RNA structural probes of guanine base-pairing. RNA 2018, 24, 114–124.

[17]   Mitchell D 3rd, Renda AJ, Douds CA, Babitzke Assmann SM, Bevilacqua PC. vivo RNA structural probing of uracil and guanine base-pairing by 1-ethyl-3-(3-dimethylaminopropyl)carbodiimide (EDC). RNA 2019, 25, 147–157.

[18]   Wilkinson KA, Merino EJ, Weeks KM. Selective 2′-hydroxyl acylation analyzed by primer extension (SHAPE): Quantitative RNA structure analysis at single nucleotide resolution. Nat Protoc 2006, 1, 1610–1616.

[19]   Ingle S, Azad RN, Jain SS, Tullius TD. Chemical probing of RNA with the hydroxyl radical at single-atom resolution. Nucleic Acids Res 2014, 42, 12758–12767.

[20]   Feng C, Chan D, Joseph J, Muuronen M, Coldren WH, Dai N, Correa IR, Jr, Furche F, Hadad CM 2, Spitale RC. Light-activated chemical probing of nucleobase solvent accessibility inside cells. Nat Chem Biol 2018, 14, 276–283.

[21] Chu W-C, Kintanar A, Horowitz J. Correlations between fluorine-19 nuclear magnetic resonance chemical shift and the secondary and tertiary structure of 5-fluorouracisl substituted tRNA. J Mol Biol 1992, 227, 1173–1181.

[22] Kreutz C, Kählig H, Konrat R, Micura R. Ribose 2′-F Labeling: A simple tool for the characterization of RNA secondary structure equilibria by $^{19}$F NMR Spectroscopy. J Am Chem Soc 2005, 127, 11558–11559.

[23] Graber D, Moroder H, Micura R. $^{19}$F NMR spectroscopy for the analysis of RNA secondary structure populations. J Am Chem Soc 2008, 130, 17230–17231.

[24] Kivimäki A, Virta P. Characterization of RNA Invasion by $^{19}$F NMR Spectroscopy. J Am Chem Soc 2010, 132, 8560–8562.

[25] El-Khoury R, Damha MJ. 2′-Fluoro-arabinonucleic acid (FANA): A versatile tool for probing biomolecular Interactions. Acc Chem Res 2021, 54, 2287–2297.

[26] Himmelstoss M, Erharter K, Renard E, Ennifar E, Kreutz C, Micura R. 2′-O-Trifluoromethylated RNA – A powerful modification for RNA chemistry and NMR spectroscopy. Chem Sci 2020, 11, 11322–11330.

[27] Eichler C, Himmelstoss M, Plangger R, Weber LI, Hartl M, Kreutz C, Micura R. Advances in RNA labeling with trifluoromethyl . Chem Eur J 2023, e202302220 (1 of 10).

[28] Shatkin AJ, Manley JL. The ends of the affair: Capping and polyadenylation. Nature Struct Mol Biol 2000, 7, 838–842.

[29] Lewis JD, Izaurralde E. The role of the cap structure in RNA processing and nuclear export. Eur J Biochem 1997, 247, 461–469.

[30] Wieczorek Z, Stepinski J, Jankowska M, Lönnberg H. Fluorescence and absorption spectroscopic properties of RNA 5′-cap analogues derived from 7-methyl-, N2,7-dimethyl- and N2,N2,7-trimethyl-guanosines. J Photochem Photobiol B 1995, 28, 57–63.

[31] Kim CH, Sarma RH. Spatial configuration of the bizarre 5′ terminus of mammalian mRNA. J Am Chem Soc 1978, 100, 1571–1590.

[32] Wieczorek Z, Zdanowski K, Chlebicka L, Stepinski J, Jankowska M, Kierdazuk B, Temeriusz A, Darzynkiewicz E, Stolarski R. Fluorescence and NMR studies of intramolecular stacking of mRNA cap-analogues. Biochim Biophys Acta 1997, 1354, 145–152.

[33] Darzynkiewicz E, Stepinski J, Tahara SM, Stolarski R, Ekiel I, Haber D, Neuvonen K, Lehikoinen P, Labadi I, Lönnberg H. Synthesis, conformation and hydrolytic stability of P1,P3–dinucleoside triphosphates related to mRNA 5′-cap, and comparative kinetic studies on their nucleoside and nucleoside monophosphate analog.s. Nucleosides & Nucleotides 1990, 9, 599–618.

[34] Baker BF. Decapitation of a 5′-capped oligoribonucleotide by o-phenanthroline:copper(II). J Am Chem Soc 1993, 115, 3378–3379.

[35] Baker BF, Khalili H, Wei N, Morrow JR. Cleavage of the 5′ cap structure of mRNA by a europium(III) macrocyclic complex with pendant alcohol groups. J Am Chem Soc 1997, 119, 8749–8755.

[36] Thillier Y, Decroly E, Morvan F, Canard B, Vasseur -J-J, Debart F. Synthesis of 5′ cap-0 and cap-1 RNAs using solid-phase chemistry coupled with enzymatic methylation by human (guanine-N7)-methyl transferase. RNA 2012, 18, 856–868.

[37] Noel M, Guez T, Thillier Y, Vasseur -J-J, Debart F. Access to high-purity 7mG-cap RNA in substantial quantities by a convenient all-chemical solid-phase method. ChemBioChem 2023, 24, e202300544 (7 pages).

[38] Potuzník JF, Nesuta O, Skríba A, Voleníkova B, Mititelu M-B, Mancini F, Serianni V, Fernandez H, Spustova K, Trylcova J, Vopalensky P, Cahova H. Diadenosine tetraphosphate (Ap4A) serves as a 5′ RNA cap in mammalian cells. Angew Chem Int Ed 2024, 63, e202314951 (7 pages).

[39] Mattick JS, Amaral PP, Carninci P, Carpenter S, Chang HY, Chen -L-L, Chen R, Dean C, Dinger ME, Fitzgerald KA, Gingeras TR, Guttman M, Hirose T, Huarte M, Johnson R, Kanduri C, Kapranov P, Lawrence JB, Lee JT, Mendell JT, Mercer TR, Moore KJ, Nakagawa S, Rinn JL, Spector DL, Ulitsky I,

Wan Y, Wilusz JE, Wu M. Long non-coding RNAs: Definitions, functions, challenges and recommendations. Nat Rev Mol Cell Biol 2023, 24, 430–447.

[40] Zhou W-Y, Cai Z-R, Liu J, Wang D-S, Ju H-Q, Xu R-H. Circular RNA: Metabolism, functions and interactions with proteins. Mol Cancer 2020, 19, 172 (19 pages).

[41] Xiao MS, Ai Y, Wilusz JE. Biogenesis and functions of circular RNAs come into focus. Trends Cell Biol 2020, 30, 226–240.

[42] Liang X, Chen H, Li L, An R, Komiyama M. Ring-structured DNA and RNA as key players in vivo and in vitro. Bull Chem Soc Jpn 2021, 94, 141–157.

[43] Micura R. Cyclic oligoribonucleotides (RNA) by solid-phase synthesis. Chem Eur J 1999, 5, 2077–2082.

[44] Boccaletto P, Machnicka A, Purta M, Piatkowski E, Baginski P, Wirecki TK, De Crecy-Lagard V, Ross R, Limbach P, Kotter A, et al.. MODOMICS: A database of RNA modification pathways. 2017 update. Nucleic Acids Res 2018, 46, D303–D307.

[45] Klöcker N, Weissenboeck FP, Rentmeister A. Covalent labeling of nucleic acids. Chem Soc Rev 2020, 49, 8749–8773.

[46] Li L, Zhang Z. Development and applications of the copper-catalyzed azide-alkyne cycloaddition (CuAAC) as a bioorthogonal reaction. Molecules 2016, 21, 1393 (22 pages).

[47] Kubota M, Nainar, Parker SM, England W, Furche F, Spitale RC. Expanding the scope of RNA metabolic labeling with vinyl nucleosides and inverse electron-demand Diels–Alder chemistry. ACS Chem Biol 2019, 14, 1698–1707.

[48] Hong V, Presolski SI, Ma C, Finn MG. Analysis and optimization of copper-catalyzed azide–alkyne cycloaddition for bioconjugation. Angew Chem Int Ed 2009, 48, 9879–9883.

[49] Deigan KE, Ferre-D'Amare AR. Riboswitches: Discovery of drugs that target bacterial gene-regulatory RNAs. Acc Chem Res 2011, 44, 1329–1338.

[50] Garst AD, Edwards AL, Batey RT. Riboswitches: Structures and mechanisms. Cold Spring Harb Perspect Biol 2011, 3, a003533 (13 pages).

[51] Serganov A, Nudler EA. Decade of riboswitches. Cell 2013, 152, 17–24.

[52] Mironov AS, Gusarov I, Rafikov R, Lopez LE, Shatalin K, Kreneva RA, Perumov DA, Nudler E. Sensing small molecules by nascent RNA: A mechanism to control transcription in bacteria. Cell 2002, 111, 747–756.

[53] Winkler W, Nahvi A, Breaker RR. Thiamine derivatives bind messenger RNAs directly to regulate bacterial gene expression. Nature 2002, 419, 952–956.

[54] Nahvi A, Sudarsan N, Ebert MS, Zou X, Brown KL, Breaker RR. Genetic control by a metabolite binding mRNA. Chem Biol 2002, 9, 1043–1049.

[55] Breaker RR. The biochemical landscape of riboswitch ligands. Biochemistry 2022, 61, 137–149.

[56] Sherlock ME, Sudarsan N, Breaker RR. Riboswitches for the alarmone ppGpp expand the collection of RNA-based signaling systems. Proc Natl Acad Sci USA 2018, 115, 6052–6057.

[57] Hallberg ZH, Su Y, Kitto RZ, Hammond MC. Engineering and in vivo applications of riboswitches. Annu Rev Biochem 2017, 86, 515–539.

[58] Mandal M, Boese B, Barrick JE, Winkler WC, Breaker RR. Riboswitches control fundamental biochemical pathways in Bacillus subtilis and other bacteria. Cell 2003, 113, 577–586.

[59] https://en.wikipedia.org/wiki/Riboswitch (Aug 17, 2019)

[60] Mandal M, Breaker RR. Adenine riboswitches and gene activation by disruption of a transcription terminator. Nat Struct Mol Biol 2004, 11, 29–35.

[61] Batey RT, Gilbert SD, Montange RK. Structure of a natural guanine-responsive riboswitch complexed with the metabolite hypoxanthine. Nature 2004, 432, 411–415.

[62] Kim JN, Roth A, Breaker RR. Guanine riboswitch variants from Mesoplasma florum selectively recognize 2′-deoxyguanosine. Proc Natl Acad Sci USA 2007, 104, 16092–16097.

[63] Batey RT. Structure and mechanism of purine binding riboswitches. Q Rev Biophys 2012, 45, 345–381.

[64] Garst AD, Heroux A, Rambo RP, Batey RT. Crystal structure of the lysine riboswitch regulatory mRNA element. J Biol Chem 2008, 283, 22347–22351.

[65] Kulshina N, Baird NJ, Ferre-D'Amare AR. Recognition of the bacterial second messenger cyclic diguanylate by its cognate riboswitch. Nat Struct Mol Biol 2009, 16, 1212–1217.

[66] Klein DJ, Edwards TE, Ferre-D'Amare AR. Cocrystal structure of a class-I preQ1 riboswitch reveals a pseudoknot recognizing an essential hypermodified nucleobase. Nat Struct Mol Biol 2009, 16, 343–344.

[67] Serganov A, Polonskaia A, Phan AT, Breaker RR, Patel DJ. Structural basis for gene regulation by a thiamine pyrophosphate-sensing riboswitch. Nature 2006, 441, 1167–1171.

[68] Thore S, Leibundgut M, Ban N. Structure of the eukaryotic thiamine pyrophosphate riboswitch with its regulatory ligand. Science 2006, 312, 1208–1211.

[69] Wang JX, Breaker RR. Riboswitches that sense. Biochem Cell Biol 2008, 86, 157–168.

[70] Sun A, Gasser C, Li F, Chen H, Mair S, Krasheninina O, Micura R, Ren A. SAM-VI riboswitch structure and signature for ligand discrimination. Nat Commun 2019, 10, 5728 (13 pages).

[71] Roth A, Nahvi A, Lee M, Jona I, Breaker RR. Characteristics of the glmS ribozyme suggest only structural roles for divalent metal ions. RNA 2006, 12, 607–619.

[72] Klein DJ, Ferre-D'Amare AR. Structural basis of glmS ribozyme activation by glucosamine-6-phosphate. Science 2006, 313, 1752–1756.

[73] Cromie MJ, Shi Y, Latifi T, Groisman EA. An RNA sensor for intracellular Mg(2+). Cell 2006, 125, 71–84.

[74] Dann CE III, Wakeman CA, Sieling CL, Baker SC, Irnov I, Winkler WC. Structure and mechanism of a metal-sensing regulatory RNA. Cell 2007, 130, 878–892.

[75] Baker JL, Sudarsan N, Weinberg Z, Roth A, Stockbridge RB, Breaker RR. Widespread genetic switches and toxicity resistance proteins for fluoride. Science 2012, 335, 233–235.

# 10 Catalytic nucleic acids

## 10.1 Introduction

Researchers interested in the origin of life started in late 1960s to speculate with the idea that polynucleotides could have served as primitive catalysts of replication, although displaced later by more efficient peptide catalysts [1, 2]. The first piece of experimental evidence for catalytic nucleic acids was reported 13 years later. It was shown that an intervening sequence within ribosomal RNA of *Tetrahymena thermophila* strain B VII nuclei catalyzed its own splicing. The entire process consisted of excision of the intervening sequence, its cyclization and religation of the RNA sequences neighboring the excised intervening sequence [3]. Two years later, the RNA moiety was shown to be the catalytically active component of ribonuclease P, a ribonucleoprotein that tailors the precursor form of transfer RNA to the functional form [4]. Since then, a number of catalytic RNA sequences of biological origin have been identified [5], and engineered by in vitro selection [6].

The naturally occurring ribozymes are usually divided into small and large ribozymes. Small ribozymes are typically 50- to 150-nucleotide-long RNA sequences that catalyze the cleavage of their own phosphodiester bond by an attack of the 2'-OH on phosphorus atom of the neighboring phosphodiester linkage. In other words, the mechanism resembles that of nonenzymatic cleavage of RNA [7, 8]. Large ribozymes, in turn, consist of hundreds of nucleotides, often associated with proteins [5, 8]. The cleavage of phosphodiester bond usually proceeds by participation of a nucleoside situated at some distance from the scissile bond [9]. Among naturally occurring ribozymes, some are present in only a few organisms, while others play an essential role in all forms of life, as discussed below in more detail. Ribozymes may require metal ions for full catalytic activity, but, with one exception, they do not use other cofactors. The catalytic activity is generally somewhat lower than that of protein enzymes.

The third group of catalytic nucleic acids consists of artificially made nucleic acid catalysts. These may be either RNA- or DNA sequences obtained by the so-called chemical evolution, that is, by stepwise enrichment from a pool of random nucleic acid sequences.

## 10.2 Small ribozymes

The number of identified small ribozymes has steadily increased during the past 30 years. At present the list contains 9 nucleolytic ribozymes: hammerhead, hairpin, Varkud satellite (VS), hepatitis delta virus (HDV), twister, glucosamine-6-phosphate riboswitch (glmS), pistol, twister-sister (TS) and hatchet ribozyme [10]. As mentioned above, all small ribozymes catalyze the cleavage of their own phosphodiester linkage.

https://doi.org/10.1515/9783111325637-010

The ribozyme can, however, be converted to a "trans-acting" version by separating the strand fragment containing the cleavage site from the rest of the ribozyme. Accordingly, one strand serves as the substrate and the other strand as the enzyme.

### 10.2.1 Hammerhead ribozyme

Hammerhead ribozyme was discovered more than 30 years ago [11, 12] and it undoubtedly is the most thoroughly studied nucleolytic ribozyme [13, 14]. It is widely distributed among both prokaryotes and eukaryotes [15], but the biological function still is obscure. The ribozyme consists of three short helices connected to a junction of 15 largely conserved nucleotides. Figure 10.1A represents a typical trans-acting ribozyme [14]. Guanosines G8 and G12 are assumed to play a key role in catalysis. As discussed in Chapter 6.1, a prerequisite to cleavage of an RNA phosphodiester linkage is (i) deprotonation of the attacking 2'-OH, (ii) protonation of the departing 5'-O and (iii) stabilization of the oxyphosphorane intermediate/transition state. In addition, the angle O2'–P–O3' should be close to 180°, since both the attacking and departing nucleophile must take an apical position within the phosphorane intermediate/transition state. With spontaneous pH-independent cleavage of RNA, this is achieved by water-mediated proton transfer from 2'-OH to nonbridging phosphoryl oxygen concerted with P–O$^{2'}$ bond formation, followed by water-mediated proton transfer from hydroxyl ligand of the phosphorane intermediate to the departing 5'-O. In ribozyme catalysis, the proton transfers may take place more or less simultaneously. It has been proposed that G12 serves as a general base that accepts proton form the attacking 2'-OH of C17 (Figure 10.1B) [13]. To be able to serve as a base, the N1H site of G12 must first become deprotonated. A specifically bound water molecule is assumed to accept the N1 proton and shuttle it to 2'-OH of G8 that is H-bonded to the departing 5'-O. This proton shuttle is believed to increase the acidity of the 2'-OH of G8 sufficiently to enable its function as a general acid. The mechanistic suggestion is based on the X-ray structure of full length hammerhead ribozyme [16]. The N1H of G12 is at H-bonding distance from the attacking 2'-OH. The latter group is in position appropriate for in-line attack and the 2'-OH of G8 is at H-bonding distance from the in-line departing 5'-O. It has, however, been argued that Mg$^{2+}$ participates in the catalysis [17]. A Mg$^{2+}$ ion that bridges the scissile phosphate (C17) to another phosphodiester bond (A9) simultaneously interacts with 2'-OH of G8 increasing its acidity and, hence, ability to serve as a general acid. According to another mechanistic model [18], a conformational rearrangement taking place before the actual bond cleavage allows the Mg$^{2+}$ ion in the vicinity of G12 to become directly coordinated to O$^6$ and, hence, the deprotonation of N1 that serves as a general base, is facilitated. Another Mg$^{2+}$ ion is believed to bridge the scissile phosphate to N7 of the departing G. A hammerhead ribozyme having a vanadate diester linkage in place of the scissile phosphodiester linkage has been prepared as a transition state analog and shown to exhibit a crystal structure that is consistent with the two Mg$^{2+}$ ion mechanism [19].

Finally, it is worth noting that Hammerhead is the most efficient among the nucleo-lytic ribozyme so far known. The rate-enhancement compared to uncatalyzed rate of RNA cleavage is $10^9$-fold, while RNase A still is two orders of magnitude more efficient as a catalyst [20]. Bulky organic cations, above all pentyl or benzyl substituted tetra-substituted ammonium ions, still accelerate the turnover, assumedly by promoting dis-sociation of the products and refolding to the optimal catalytic conformation [21].

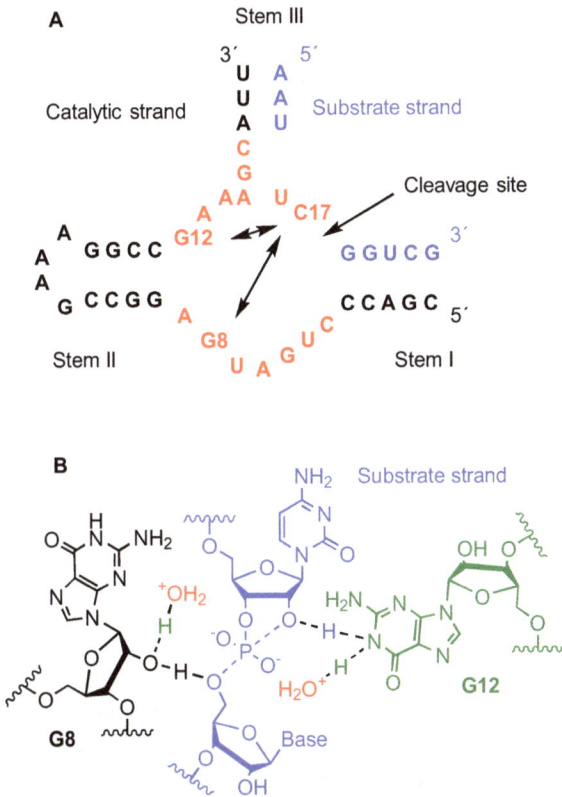

**Figure 10.1:** The catalytic center of a typical trans-acting version of a hammerhead ribozyme (A) [14] and a description of the role of the two catalytic guanine residues, G8 and G12, in the cleavage reaction at C17 of the substrate strand (B). The two appropriately situated water molecules that play a crucial role are highlighted in red [13].

## 10.2.2 Hairpin ribozyme

Hairpin ribozyme found in RNA satellites of plant viruses [22], consists of two double helical segments, both of which contain a rather large internal loop. The cleavage site is within one of these [23]. The naturally occurring ribozyme comprises four helical stems joined by a 4-way junction (Figure 10.2A). Only two of them, however, are essential for the catalytic activity [24] and, hence, the ribozyme may be converted to a trans-acting enzyme depicted in Figure 10.2B. In fact, mixing of the two pre-fabricated stems in solutions in the presence of divalent cations, gives the catalytically active ribozyme [25]. The cleavage mechanism resembles that of hammerhead ribozyme. G8 situated in the A-loop opposite to the 5′-ApG-3′ site, serves as a general base that deprotonates the attacking 2′-OH (Figure 10.2C), similarly to G12 in hammerhead ribozyme. N1-Protonated A38 in the B-loop is, in turn, the general acid that protonates the departing 5′-O. The proton transfer chain, hence, resembles the one operating in hammerhead ribozymes: the proton of the attacking 2′-OH becomes transferred through G8, water molecules and A38 to the departing oxygen [26]. X-ray structures of both the natural [27, 28] and trans-acting hairpin ribozyme [29] lend support for this mechanism. Simulations by QM/MM metadynamics suggest the mechanism to be associative. In other words, the pentacoordinated phosphorane obtained by the attack of 2′-OH of 3′-linked adenosine on phosphorus is rather an intermediate than transition state [30]. The rate-limiting step, hence, is departure of the 5′-linked guanosine.

In striking contrast to hammerhead ribozyme, the internal equilibrium of hairpin ribozyme favors ligation over cleavage [31]. This makes, together with the relatively small size, hairpin ribozyme as an attractive candidate for design of variants that catalyze either ligation or phosphodiester cleavage. By extending the structure of minimal trans-acting ribozyme, artificial ribozymes have been engineered that bind small molecular effectors and catalyze RNA recombination, circularization or oligomerization [6, 32].

### 10.2.3 Varkud satellite ribozyme (VS)

VS ribozyme is the largest among nucleolytic ribozymes. It is a 154-long insert in long noncoding satellite RNA found first in mitochondria of Varkud-1C Neurospora strain [33]. The ribozyme consists of seven helices linked together through three-way junctions, as indicated in Figure 10.3 [34]. The cleavage site is situated within an internal loop of helix I. This helix may be disconnected from the rest of the ribozyme and the self-cleavage catalysis is restored by mixing the two separated components [35]. Helix I becomes anchored to the binding pocket formed by helices II–VII and the 5′-GpA-3′ linkage within the internal loop of helix I is cleaved. In spite of very different overall structure of VS and hairpin ribozymes, their catalytic mechanisms are very similar. Kinetic [36] and crystallographic [37] studies strongly suggest that G638 serves as a general base (as G8 in hairpin ribozyme) and A756 as a general acid (as A38 in hairpin ribozyme). Proton abstraction by the general base (G638) is almost complete, while the proton donation by general acid (A756) is only partial [38, 39]. Evidently, the latter step is rate-limiting. Additionally, a $Mg^{2+}$ ion plays an important role by creating with

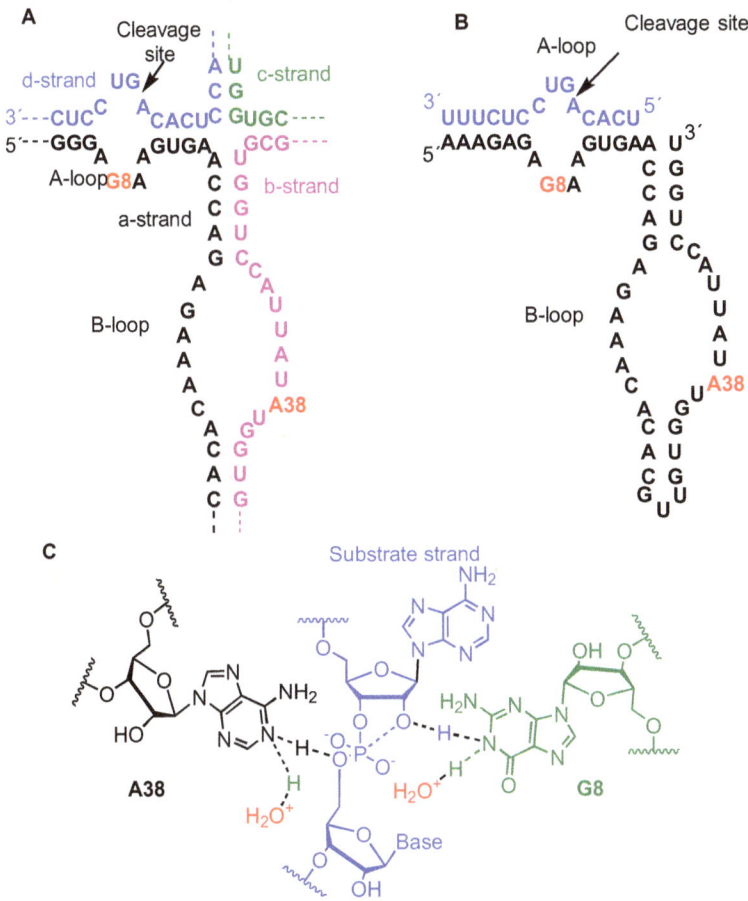

Figure 10.2: Naturally occurring hairpin ribozyme (A) [24], its trans-acting version (B) [24] and description of the role of the catalytic guanine residues, A38 and G8, in the cleavage reaction [26].

the aid of coordinative interactions favorable conditions for co-linear formation of O2′–P–O5′ bonds upon formation of the pentacoordinated intermediate.

## 10.2.4 Hepatitis delta virus ribozyme (HDV)

HDV ribozyme is an 85-nucleotide-long sequence found originally embedded in non-coding RNA of hepatitis delta virus [40]. Similar ribozymes are, however, present in numerous organisms including viruses, bacteria, nematodes, plants, fungi and marine organisms [41]. It consists of five helical segments forming a double pseudoknot structure [42, 43]. Figure 10.4A shows the secondary structure for a trans-acting version. In contrast to the nucleolytic ribozymes discussed above, HDV ribozyme most likely is a

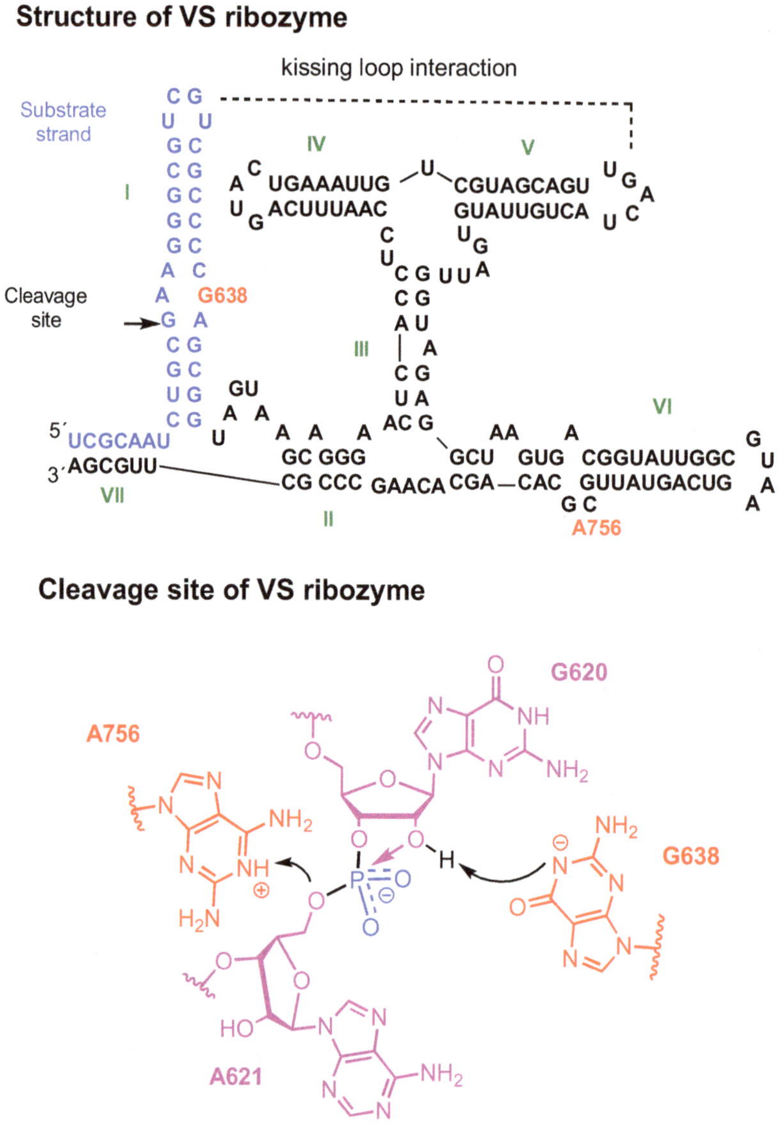

## Structure of VS ribozyme

## Cleavage site of VS ribozyme

**Figure 10.3:** Secondary structural elements of VS ribozyme [34] and mechanism of the chain cleavage [38, 39].

metalloenzyme. $Mg^{2+}$ undergoes bidentate coordination to 2′-OH of U(−1) and a non-bridging oxygen of neighboring phosphodiester linkage, resulting in deprotonation of 2′-OH [44]. The nucleophilicity of 2′-O is increased and simultaneously the electron density at phosphorus is reduced facilitating the nucleophilic attack. N3 of C75 is exceptionally basic, the $pK_a$ being $6.3 \pm 0.2$ [45]. Accordingly, it is partly protonated under physiological conditions and serves as a general acid protonating the departing 5'-O (Figure 10.4B) [46]. The naturally occurring cis-acting ribozyme processes the RNA transcripts by this mechanism to unit lengths during replication. According to $^{18}O$ kinetic isotope effects, the bidentate binding of $Mg^{2+}$ to the attacking O2' and to one of the nonbridging phosphoryl oxygen atoms stabilizes the developing phosphorane intermediate to the extent that cleavage of the P–O5' bond by proton transfer from C75 becomes rate-limiting [44].

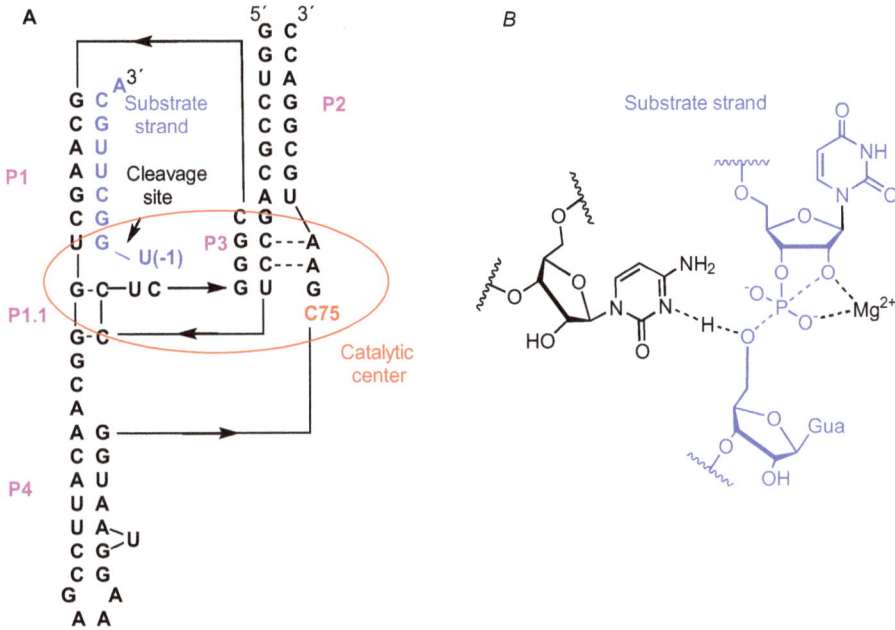

**Figure 10.4:** Trans-acting version of HDV ribozyme (A) [42, 43] and the mechanism of chain cleavage [46, 47].

### 10.2.5 Glucosamine 6-phosphate riboswitch (glmS)

GlmS ribozyme occurs within the 5'-untranslated region of the mRNA that encodes glucosamine 6-phosphate synthase in Gram positive bacteria [48]. It is the only known small ribozyme the activity of which is regulated by an external compound. Binding of glucosamine 6-phosphate accelerates the nucleolytic self-cleavage by a factor of $10^5$, constituting a feedback inhibition for its biosynthesis [49]. The ribozyme contains 150 nucleotides in three helices packed coaxially side-by-side (Figure 10.5) [50]. The structure is rigid. Binding of glucosamine 6-phosphate does not induce any marked structural change [51]. Still the role of this external effector is crucial in catalysis. It has been suggested that the glucosamine 2-amino group deprotonates N1 of G40 enabling

function as a general base that deprotonates the attacking 2'-OH of A(−1). The protonated 2-amino group of glucosamine 6-phosphate, in turn, serves as a general acid protonating the departing 5'-O. It has been suggested that in the absence of glmS, the 2'-
OH of A(−1) is H-bonded to *pro-R*$_P$ oxygen which prevents its action as a nucleophile.
Deprotonation of this hydroxy function by glmS, hence, activate the 2'-OH as a catalytic nucleophile [52].

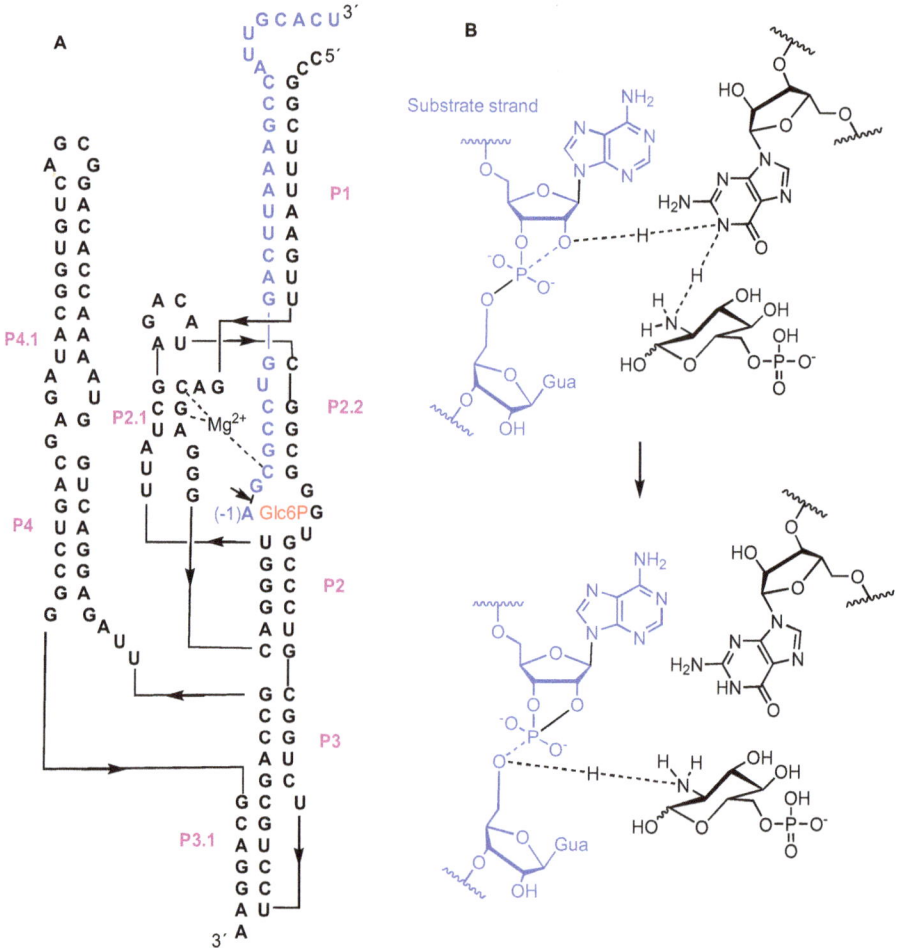

**Figure 10.5:** Secondary structure of glucosamine 6-phosphate riboswitch (glmS) (A) [50] and the
mechanism of chain cleavage (B).

## 10.2.6 Twister, pistol, TS and hatchet ribozymes

Several new nucleolytic ribozymes have been identified by bioinformatics analysis of genomics [10, 53]. The oldest and most extensively studied among these is the twister ribozyme, found to be widely distributed in prokaryotic genomes but additionally present in some eukaryotic genomes [53]. The secondary structure is rather simple: a long hairpin with two internal loops in the double helical stem. The two pseudoknot interactions indicated in Figure 10.6A are crucial for the tertiary structure and catalytic activity [54, 55]. Crystal structure [55, 56] together with mutation studies and pH-rate profiles [57] suggest that N1 of G33 deprotonates the attacking 2′-OH of U(−1) and N3H$^+$ of A(+1) intramolecularly protonates the departing 5′-O (Figure 10.6B) [7]. In addition, H-bonding of the 2-amino group of G33 to *pro-R$_P$* oxygen of the scissile phosphodiester linkage stabilizes the pentacoordinated intermediate/transition state and, hence, facilitates its formation.

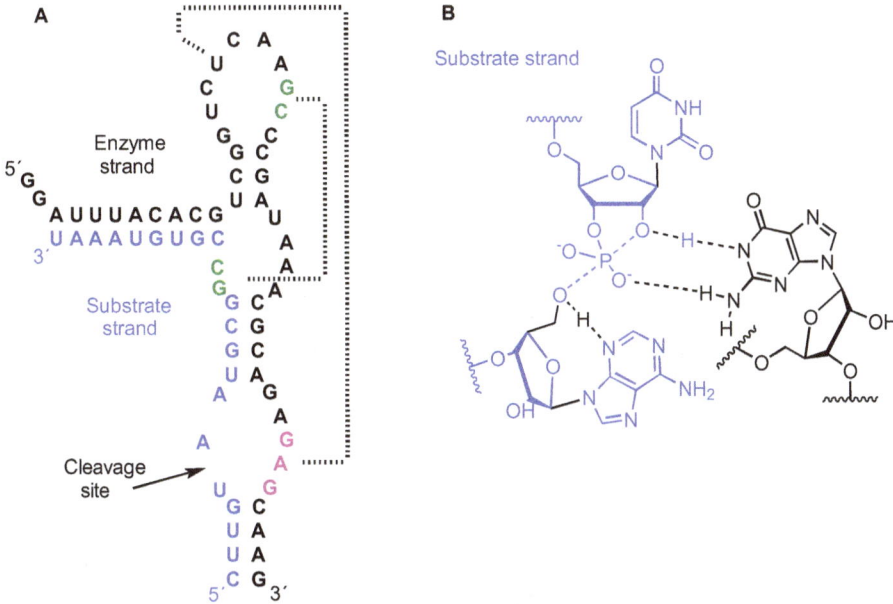

**Figure 10.6:** Secondary structure of a trans-acting twister ribozyme (A) and a probable mechanism (B) [57].

The mechanisms of nucleolytic cleavage by the pistol, TS and hatchet ribozymes are still poorly known. TS ribozyme resembles structurally the twister ribozyme but is believed to be a metalloenzyme. The detailed mechanism still remains open to various interpretations [58, 59]. Pistol ribozyme also seems to be a metalloenzyme [60]. Most likely, hydrated Mg$^{2+}$ ion bound to N7 of adenosine in the neighborhood of the scissile phosphodiester linkage serves as a general acid protonating the departing O5′ [60–63].

A second $Mg^{2+}$ is assumed to stabilize the developing 2′,3′-cyclic phosphate by outer-sphere coordination [60]. However, alternative mechanisms for the latter step have been presented [64, 65]. Theoretical studies by classical molecular dynamics has led to suggestion that the catalytically relevant $Mg^{2+}$ ion does not stay in the inner sphere of N7 but is shifted to inner sphere of *pro-R*$_P$ oxygen in the course of the catalytic process [66].

Crystal structure of hatchet ribozyme together with cleavage assays on its mutants has led to the conclusion that the catalytic mechanism of hatchet ribozyme resembles that of HDV ribozyme, although no clear evidence for the direct involvement of $Mg^{2+}$ ion has been obtained [67].

Another recently identified self-cleaving small ribozyme is Hovlinc that occurs in long noncoding RNAs of humans and chimpanzees [68]. It consists of three stem loops joined in a central loop and large pseudoknots that rigidify the structure, altogether 168 nucleotides. Previously, only three small ribozymes have earlier been reported to occur in human RNA: HDV-like mammalian CPEB3 ribozyme [69], hammerhead-like HH9 and HH10 motifs [70], and 2 SINE retrotransposons [71].

## 10.3 Large ribozymes

### 10.3.1 Introduction

Large ribozymes were the first nucleic acid sequences recognized to be catalytic. They include the RNA subunit of RNase P [4], group I [3] and group II [72] introns and peptidyl transferase 23S rRNA [73]. In addition, the catalytic center of eukaryotic spliceosome that catalyzes the removal of introns from pre-mRNA consists of oligoribonucleotides, small nuclear RNAs, as main components. These undoubtedly play an essential role in the process, the mechanism of which largely resembles the action of group II introns [74]. The role of protein components, however, appears to be so important for shaping the catalytic center that spliceosomes may as also be regarded as ribonucleoprotein enzymes rather than ribozymes [75, 76].

Large ribozymes differ from small ones, besides the size, by the mechanism of catalysis. The nucleophilic attack on phosphorus is intermolecular, not intramolecular as with small ribozymes. The external nucleophile may be either a 2′- or 3′-OH of a distant nucleoside, or a water molecule [77]. All large ribozymes are metalloenzymes, more than one $Mg^{2+}$ ions being directly involved in the catalysis [78]. Another common feature is that all transesterification and hydrolysis reactions proceed by 100% inversion at phosphorus. In other words, the pentacoordinate intermediate, even if having a finite lifetime, does not pseudorotate. It is also noteworthy that although large ribozymes do not rely on the intramolecular nucleophilic attack of the vicinal 2′-OH group to initiate transesterification, they still generally exhibit a preference for RNA over DNA substrates.

### 10.3.2 Ribonuclease P

RNase P was the first large ribozyme [4]. It is widely distributed in nature occurring in bacteria, archaea and eukaryotes as a protein complex. The function of RNase P is to convert pre- $^t$RNA to its active form by site-specific hydrolytic removal of the so-called leader sequence. Human nuclear RNase P additionally participates in transcription of several small noncoding RNAs [79]. The RNA component of the nucleoprotein complex consists of around 340 nucleotides and the protein component around 120 amino acids. RNA is the catalytic subunit in the complex. It incorporates a catalytic domain that is directly involved in the phosphodiester cleavage and a specificity domain that participates in substrate binding and orientation. The protein subunit participates in substrate binding and product release but not in the actual cleavage step [80]. Unlike small ribozymes, RNase P cleaves the P–O3′ bond, not the P–O5′ bond.

Interestingly, the substrates of RNase P do not have a conserved base sequence in proximity of the cleavage site [81]. In other words, complementarity of the enzyme and substrate sequences around the cleavage site is not a prerequisite for catalysis by RNase P [82]. Only a double helical stem on the 3′-side of the cleavage′ site and a 5′-CCA-3′ sequence attached to the 3′-terminal nucleoside of this stem seem to be required for recognition (Figure 10.7A). The co-called 5′-leader sequence of pre-tRNA that becomes cut off by RNase P, mainly interacts with the protein subunit.

Three $Mg^{2+}$ ions play a catalytic role, as indicated in Figure 10.7B. Pro-$R_P$ oxygen is the central coordination site. Thiosubstitution at this site retards the cleavage by three orders of magnitude, most likely owing to low affinity of hard $Mg^{2+}$ ion to sulfur. Replacement of $Mg^{2+}$ with more thiophilic $Cd^{2+}$ or $Mn^{2+}$ ion largely restores the cleavage rate [83, 84]. The rescue effect is of second-order in concentration of the metal ion, suggesting that two metal ions bind to pro-$R_P$ oxygen. One of these is the $Mg^{2+}$ ion that additionally is coordinated to PNase P through $O^4$ of U52 and one of the nonbridging oxygen atoms of A50. A hydroxide ligand of this $Mg^{2+}$ ion serves as an intracomplex nucleophile attacking the phosphorus atom [80]. The second $Mg^{2+}$ ion that is coordinated to a nonbridging oxygen atom of G51, is bound to the pro-$R_P$ oxygen via a water molecule. Bidentate binding of $Mg^{2+}$ to the nonbridging oxygen atoms expectedly stabilizes the phosphorane intermediate obtained by the attack of a ribozyme bound hydroxyl group on phosphorus. Bidentate binding to nonbridging and departing 3′ oxygen, in turn, stabilizes the leaving group. The third $Mg^{2+}$ ion is possibly coordinated, either directly or through a water molecule, to 2′-O of the leaving nucleoside [85]. In summary, the role of the catalytic domain of PNase P is to serve as a template for the metal ions that by inner and outer sphere coordination create an environment allowing rupture of the P–O3′ bond.

**A**

**B**

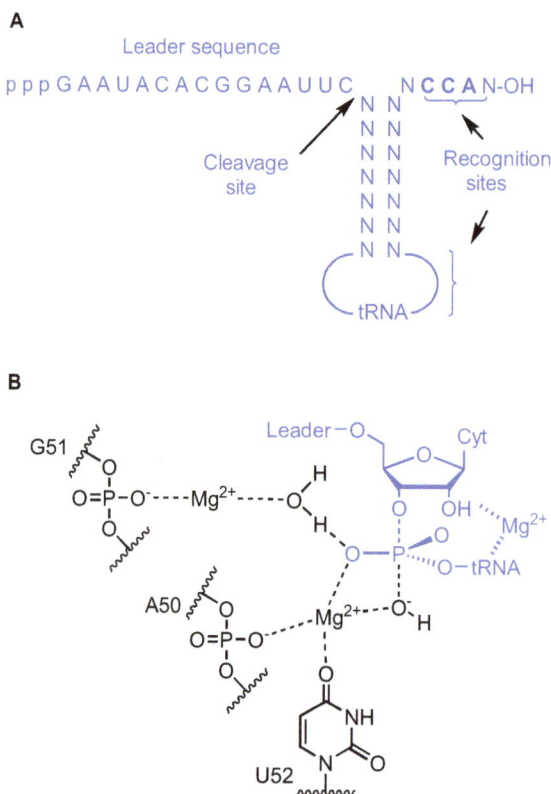

**Figure 10.7:** Recognition of pre-tRNA by RNase P (A) and the structure of transition state (B) [80].

### 10.3.3 Group I introns

Group I introns are 250- to 500-nucleotide-long autocatalytic sequences occurring in precursor forms of mRNA, tRNA and rRNA in eukaryotic microorganisms, bacteria and plants [86]. They catalyze their own cleavage from adjacent exons with concomitant ligation of the exons with each other, a process called splicing (Figure 10.8A) [87]. The ribozyme first hybridizes with the 5′-terminal exon (exon 1) through the internal guiding sequence (IGS) [88]. Additionally, several tertiary interactions stabilize the intron/exon interaction creating a binding pocket for a guanosine monomer. The 3′-OH of this guanosine attacks on the 5′-terminal phosphorus atom of the intron, resulting in departure of exon 1 by transesterification. Exon 1 still remains hybridized to the IGS of the intron. The 3′-terminal guanosine of the intron then displaces the 5′-linked guanosine from the guanosine binding pocket. This brings the 3′-terminal exon (exon 2) in proximity of exon 1. The 3′-OH of exon 1 attacks on the phosphodiester linkage between intron and exon 2. The exons become ligated with concomitant release of intron.

Based on crystallographic data [88, 89], rescue effects of thio [90] and amino [91] substitutions and functional studies [92], three metal ions are required for full catalytic activity of group I intron. Both steps of the self-splicing proceed through a very similar transition state, although to opposite directions. The catalytic mechanism of both steps, hence, is rather similar. As indicated in Figure 10.8B, two $Mg^{2+}$ ions are coordinated to the pro-$S_P$ oxygen of the scissile phosphodiester linkage. One of them is additionally coordinated to three other internucleosidic phosphate groups within the intron and to the 3'-O of the departing (step 1)/attacking (step 2) nucleoside. The other one is coordinated to one additional phosphodiester linkage and to 2'- and 3'-OH of the attacking (step 1)/departing (step 2) nucleoside. The third $Mg^{2+}$ is bound to 3'-O of the attacking (step 1)/departing (step 2) nucleoside. Evidently, this kind of metal ion network markedly stabilizes the pentacoordinated intermediate accelerating the reactions to both directions.

The 2'-OH of the departing (step 1)/attacking (step 2) nucleoside is not a binding site for $Mg^{2+}$, but it still is essential for catalysis. Replacement of the departing nucleoside with its 2'-deoxy analogue (step 1) retards the transesterification by three orders of magnitude [93] and 2'-deoxyguanosine is a competitive inhibitor for the attack of guanosine monomer [94]. As regards the departing nucleoside in the first stage, the 2'-OH evidently enhances the departure of the neighboring 3'-O by intramolecular H-bonding [95]. The 2'-OH of the attacking guanosine in the second stage also serves as an H-bond donor, but the acceptor remains uncertain [96]. Anyway, the acceptor seems not to be the neighboring 3'-O, as in the first stage [97].

### 10.3.4 Group II introns

Group II introns are present in organelles of fungi, bacteria, eukaryotic microorganisms and plants, incorporated in precursors of rRNA, tRNA and mRNA [98]. They contain more than 400 nucleotides in six stem-loop structures. The number of conserved nucleotides is very limited. The tertiary structure is complicated, the catalytically important nucleotides being distributed over the complete structure.

Group II introns, like group I introns, catalyze their own excision from pre-RNAs. The reaction starts, as with group I introns, by binding of the 5'-terminal exon (exon 1) to the IGS. In this case, the IGS is not contiguous but consists of two hexameric segments [99]. The 3'-terminal exon (exon 2) forms, in turn, only one base pair with the intron. The main mechanistic difference compared to group I introns is that instead of external guanosine, an intrachain nucleoside, usually but not necessarily a bulged adenosine, serves as a nucleophile in the first step of the splicing [100, 101]. At least two $Mg^{2+}$ ions, one coordinated to the departing 3'-O and the other to the pro-$R_P$ oxygen of the scissile linkage, are essential for the catalysis (Figure 10.9) [102]. In both steps of splicing, the metal ion enhances the transesterification by coordination to the departing 3'-O. Displacement of proton from the attacking 3'-O by the pro-$R_P$ oxygen

**A**

**B**

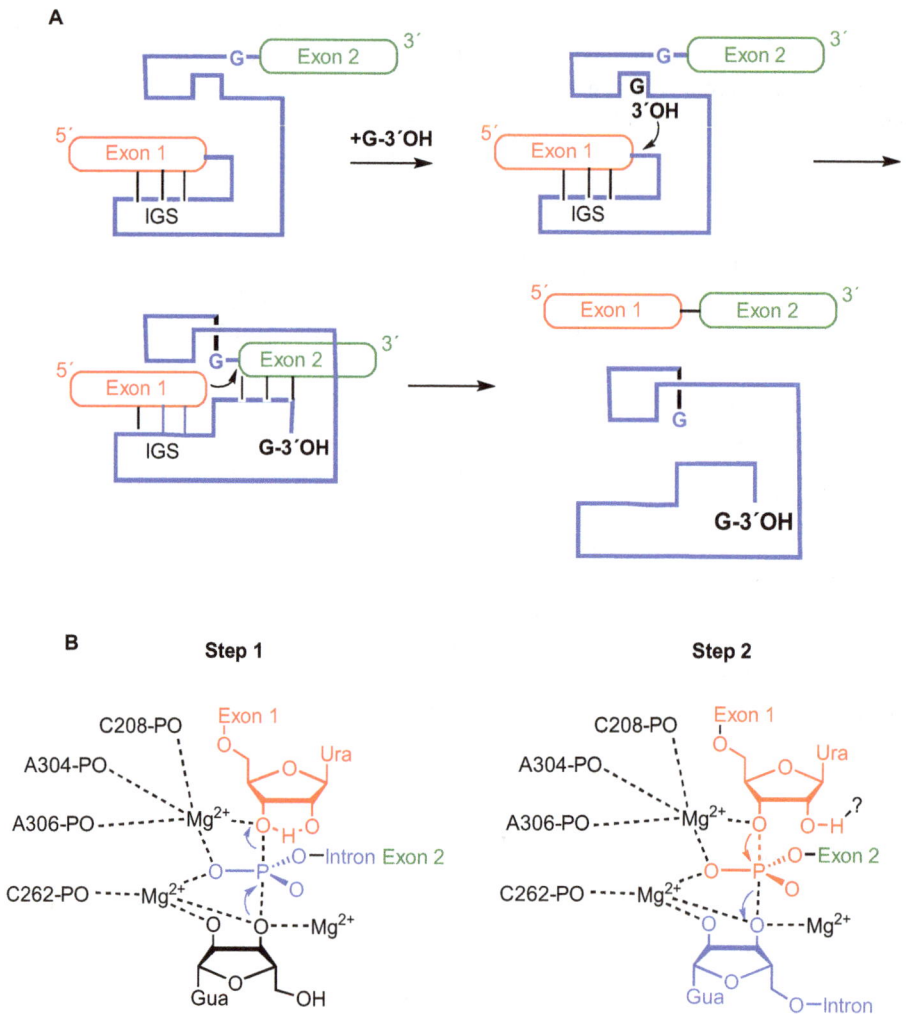

**Figure 10.8:** Pre-mRNA splicing catalyzed by group I intron (A) [87] and the transition states for bond cleavage (step 1) and religation (step 2) (B) [88–92].

coordinated $Mg^{2+}$ additionally enhances the second step. The role of metal ions, hence, closely resembles that in group I intron. The 2′-OH of the departing nucleoside is not as essential as with PNase P or group I introns. Replacing the departing ribonucleoside with a 2′-deoxyribonucleoside retards the first step of splicing by only one order of magnitude [103]. In the second step, the influence is more marked but still less deleterious than with group I introns [104].

**A**

**B**

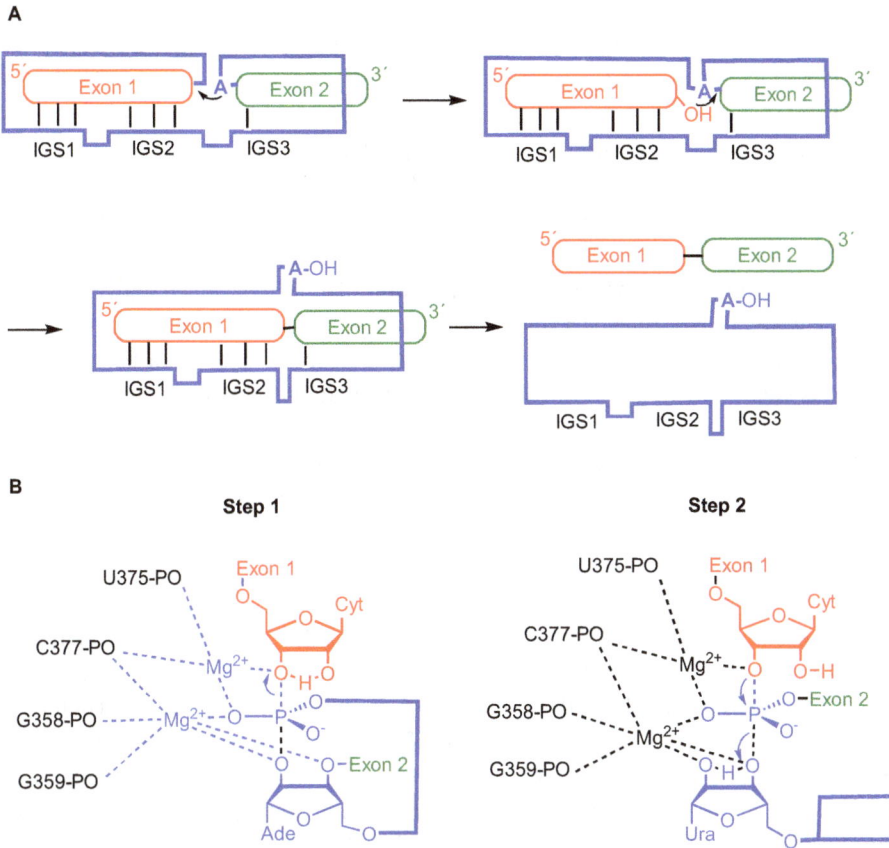

**Figure 10.9:** Pre-mRNA splicing by group II intron (A) [99–101] and the transition states for the bond cleavage (step 1) and relegation (step 2) (B) [102].

## 10.3.5 Spliceosome

Spliceosome is a large complex of five 100- to 300-nucleotide-long oligoribonucleotides rich in uridine, called small nuclear RNAs (U1, U2, U4, U5 and U6 snRNAs), and around 80 protein units. snRNAs are associated with proteins to small nuclear ribonucleoprotein particles, U1, U2, U4/U6 and U5 snRNPs. Spliceosome complex catalyzes splicing of pre-mRNA in nucleus of eukaryotic cells [76]. The catalytic complex is a highly dynamic structure assembled across the intron removed. The base sequences at the termini of pre-mRNA introns are conserved, the 5′- and 3′-splicing sites being GU and AG, respectively [105]. U1 snRNP recognizes the 5′-spicing site and U1 snRNA hybridizes with the GU site. U2 snRNP recognizes an intrachain adenosine, the 2′-OH of which later attacks the 5′-GU site. Binding of U4/U6 and U5 snRNPs then completes assembly of the functional spliceosome. The actual splicing process involves the cleavage of 5'-intron/(exon 1) linkage by attack adenosine 2'-OH on the 5'-GU site and subsequent ligation of exon 1 to exon 2 by attack of exon 1 5'-OH on the 3'-AG spicing site. In the course of splicing, the snRNAs that form the catalytic center of the complex undergo continuous remodeling by protein units. The products and intermediates closely resemble those of group II introns, lending support for similarity of mechanism [106]. According to thiosubstitution studies [107–109], the role of metal ions seems to be very similar in both cases. The role of protein subunits to shaping the environment optimal for the catalysis, however, is so important that, as mentioned above, spliceosome may well be regarded as a ribonucleoprotein enzyme instead of ribozyme.

Interestingly, a model for a primitive splicing process based on a hairpin ribozyme has been recently established [110]. This bipartite hairpin ribozyme RNA cut itself at two sites and underwent ligation of the resulted ends. Moreover, the cut-out segment was shown to regulate the activity of the ribozyme activity.

### 10.3.6 Ribosome

Ribosome is a very large complex of RNA and proteins that catalyzes the protein synthesis in all living organisms [111]. The complex consists of two subunits, the molecular weights of which are 1.5 and 0.8 million Daltons, respectively. The RNA content is as high as two thirds. The smaller subunit offers a platform for recognition of the three-nucleoside codes of mRNA by the anticodons of tRNAs. The larger subunit, in turns, contains the machinery for the transfer of amino acid residues from tRNAs to the end of the carboxy end of the growing peptide chain. The principle of the process is depicted in Figure 10.10A. Ribosome contains three binding sites for tRNAs: A-site for tRNA carrying the incoming amino acid residue, P-site for tRNA bearing the growing peptide chain, and E-site for tRNA departing after deacylation. Ribosome moves along the mRNA. The aminoacylated $tRNA(aa_{1+1})$ in A-site attacks on peptidyl bearing $tRNA(aa_1)$ in P-site, as shown in Figure 10.10B. The peptidyl chain is transferred to $tRNA(aa_{1+1})$ and $tRNA(aa_1)$ is released through the E-site. This process shifts the ribosome stepwise toward the 3'-terminus of mRNA. The next aminoacylated $tRNA(aa_{1+2})$ occupies the A-site and the same process is repeated. Crystallographic studies have shown that binding of aminoacylated tRNA to the A-site and peptidyl tRNA to the P-site is mediated only by rRNA. Only rRNA is within ribosome in such a position that enables catalysis of the peptide bond formation [112–114]. Figure 10.10B shows the mechanism. The amino group of the aminoacylated tRNA in A-site attacks on the carbonyl carbon of peptidyl tRNA in the P-site. The attack is accompanied by proton shuttle from the amino group to the departing 3'-O group [115, 116]. The reaction, however, is rather stepwise than fully concerted. Breakdown of the tetrahedral intermediate obtained takes place in a separate step.

**A**

**B**

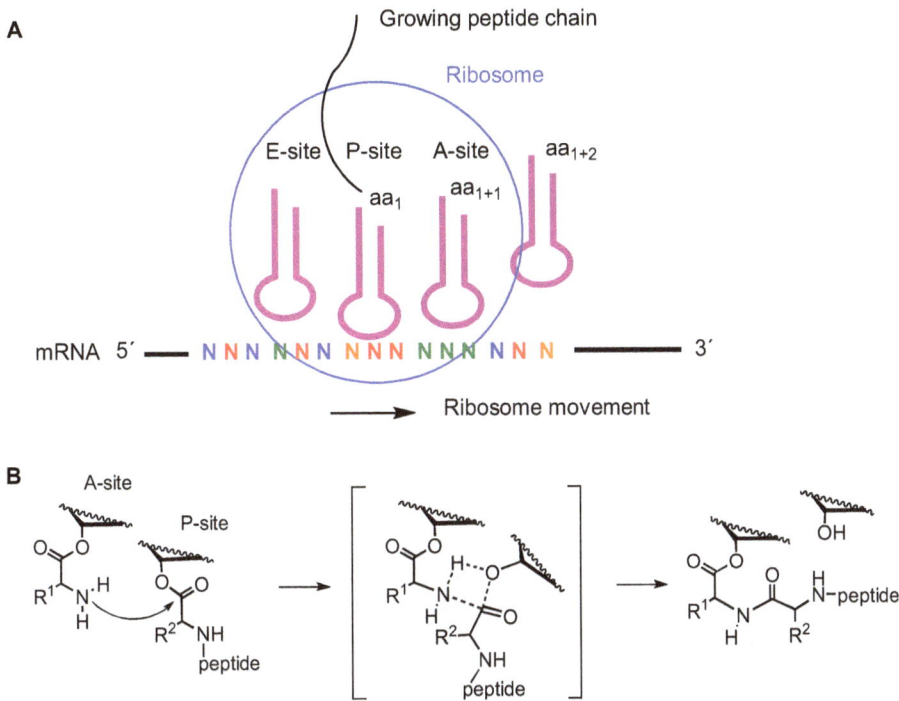

**Figure 10.10:** Schematic presentation of ribosome-catalyzed peptide synthesis (A) and the mechanism of acyl transfer between A- and P-sites (B) [115, 116].

## 10.4 Artificial ribozymes

### 10.4.1 Deoxyribozymes

Oligodeoxyribonucleotides that show catalytic activity are called deoxyribozymes, DNA-zymes or catalytic DNA. Unlike ribozymes that are of biological origin, deoxyribozymes are artificial catalysts discovered by in vitro selection [117, 118]. Various experimental procedures are used for selection [119, 120], but the underlying principle is always the same. The structure aimed at serving as the substrate, often an RNA sequence, is treated under desired reaction conditions with a random pool of oligodeoxyribonucleotides. The sequences that are able to result in the desired reaction are separated, enzymatically amplified and subjected again to the reaction conditions with the substrate. In this manner, the catalytically most active sequence is gradually enriched to such an extent that it can be recognized. The procedure used for discovery of the first DNAzyme [121] is presented in Figure 10.11 as an example. As in this example, the selection is virtually always carried out in the presence of some metal ion. Besides $Pb^{2+}$, $Zn^{2+}$, $Cu^{2+}$, $UO_2^{2+}$, $Cd^{2+}$, $Ca^{2+}$, $Hg^{2+}$, $Ag^+$

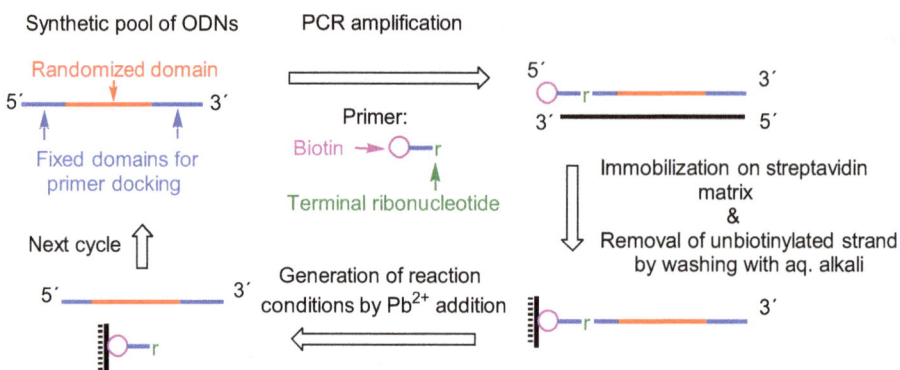

**Figure 10.11:** The in vitro selection of the first DNAzyme [122].

and trivalent lanthanide ions have been used, and even Na$^+$ selective DNAzymes have been obtained by using Co(NH$_3$)$_6$$^{3+}$ in the selection [122].

The by far most extensively studied DNAzyme-catalyzed reaction is cleavage of RNA. The best known catalysts are 10–23 and 8–17 deoxyribozymes (Figure 10.12A). The 10–23 DNAzymes consist of a 15-mer loop as the catalytic core and two 7- to 8-nucleotide-long wings that recognize the RNA substrate [123]. The cleavage site is 5′-N$^1$pN$^2$-3′, where N$^1$ is an unpaired purine and N$^2$ a paired pyrimidine nucleoside. A millimolar concentration of Mg$^{2+}$ is required for full catalytic activity. As with small ribozymes, the reaction proceeds by the attack of the 2′-OH of N$^1$ on the phosphorus atom with concomitant rupture of the P–ON$^2$-3′ bond. The catalytic efficiency is high, the $k_{cat}/K_M$ value being of the order of 10$^9$ M$^{-1}$ min$^{-1}$.

8–17 DNAzyme has a 14-nucleotide catalytic group that consists of a 9-nucleotide hairpin and a 5-nucleotide bulge [124]. The catalytically active nucleoside in the substrate strand is G wobble paired with T in DNAzyme. The tolerance to structural variations within the substrate strand is not quite as high as with 10–23 ribozymes. The cleavage reaction evidently proceeds by a general acid–base catalysis mechanism [125, 126], resembling the mechanism of self-cleaving ribozymes. The base that deprotonates the 2′-OH of the attacking guanine residue is a guanine residue in the DNA-zyme bulge. The N1 site of this guanine residue is deprotonated, and hence basic, owing to involvement in a noncanonical GA base pair. The general acid that protonates the departing O5′ evidently is a hydrated metal ion, Pb$^{2+}$ or Zn$^{2+}$.

DNAzymes also catalyze ligation of both DNA and RNA. As regards DNA ligation, the substrate is, however, unnatural, viz. ODN 3′-phosphorimidazolide attacked by the 5′-OH of another ODN, both oligonucleotides being hybridized with the DNAzyme (E47 deoxyribozyme). The catalytic core consists of one hairpin loop and two small bulges and is Zn$^{2+}$- or Cu$^{2+}$-dependent [127]. More interestingly, the DNAzyme-catalyzed native 3′,5′-ligation of RNA proceeds by an attack of the terminal 3′-OH on a natural substrate, viz. ORN 5′-triphosphate (Figure 10.12B) [128]. Both Mg$^{2+}$- and Zn$^{2+}$-dependent

DNAzymes are known. The DNAzyme-catalyzed formation of branched and lariat ORNs likewise utilizes 5'-triphosphate as a substrate (Figure 10.12C) [129]. The latter approach enables introduction of conjugate groups into an intrachain position in RNA [130]. A short RNA sequence incorporating 5-(3-aminoprop-1-en-1-yl)cytidine as the penultimate 5'-terminal nucleoside (next to $N^2$ in 8.12C) is first prepared by in vitro transcription and the desired conjugate group is attached to this amino side arm. DNAzyme (10DM24)-catalyzed reaction similar to that in Figure 10.12C then allows attachment of the short labeled RNA sequence to an internal 2'-OH ($N^1$ in Figure 10.12C) of a longer RNA.

DNAzymes do not catalyze only cleavage and formation of phosphodiester linkages but also a variety of biologically relevant reactions. Replacing thymidine in the random-sequence DNA pool with a modified nucleoside broadens the catalytic scope. For example, DNA catalysts that are able to cleave a peptide bond, have been obtained by using 5-(3-aminoprop-2-en-1-yl), 5-(2-carboxyvinyl) or 5-hydroxymethyl substituted uracil instead of thymine [131]. Figure 10.13A shows the experimental setup.

DNAzymes exhibiting tyrosine kinase [132] or phosphatase [133] activity have been obtained even without introduction of any modified nucleosides. The kinase deoxyribozyme (8EA101, Figure 10.13B) and the phosphatase deoxyribozyme (14WM9, Figure 10.13C) are both $Zn^{2+}$-dependent metalloenzymes. The phosphatase DNAzyme catalyzes the hydrolysis of serinyl phosphates in addition to tyrosinylphosphates, whereas the kinase DNAzyme is selective for tyrosine. It is noteworthy that the kinase DNAzyme can use free GTP instead of ODN 5'-triphosphate as the source of phosphate group. An efficient artificial metal-ion-independent RNAase has been obtained by using 8-histaminyl-dATP, 5-guanidinoallyl-dUTP and 5-aminoallyl-dCTP in the selection process [134]. The groups attached to the nucleobases mimic the amino acid residues playing a central role in the catalysis of RNase A.

Photoreversion of thymidine dimer [135] and peroxidation by hemin-quadruplex complexes [136] are other DNA-catalyzed biologically relevant reactions that deserve attention. The common feature of both DNA catalysts is a G-quadruplex structure. The DNA serving as a photolyase of thymidine photodimer is a 42-mer guanine rich trans-acting ODN (UV1C). The highest activity is achieved when irradiated with 305 nm light (Figure 10.13D) [136]. Evidently, the quadruplex structure harvests the light and mediates the energy through a guanine base by electron donation to the thymidine dimer anchored by hybridization opposite to the quadruplex. Peroxidation by hemin-quadruplex complexes is an exceptional DNA-catalyzed reaction in the sense that the substrate is not anchored to the DNAzyme by ODN hybridization (Figure 10.13E). Among several quadruplexes studied, the quadruplex cmyc-2345, $TGAG_3TG_4AG_3TG_4AA$, present in human c-myc gene, has turned out to be the most efficient peroxidating agent [136].

Since DNAzymes are able to recognize and cleave a given RNA-sequence, they in principle show marked potential as chemotherapeutic agents. One of the thresholds that have to be passed is avoidance of biodegradation by nucleases. Incorporation of structurally modified nucleosides that retard enzymatic cleavage without impairing

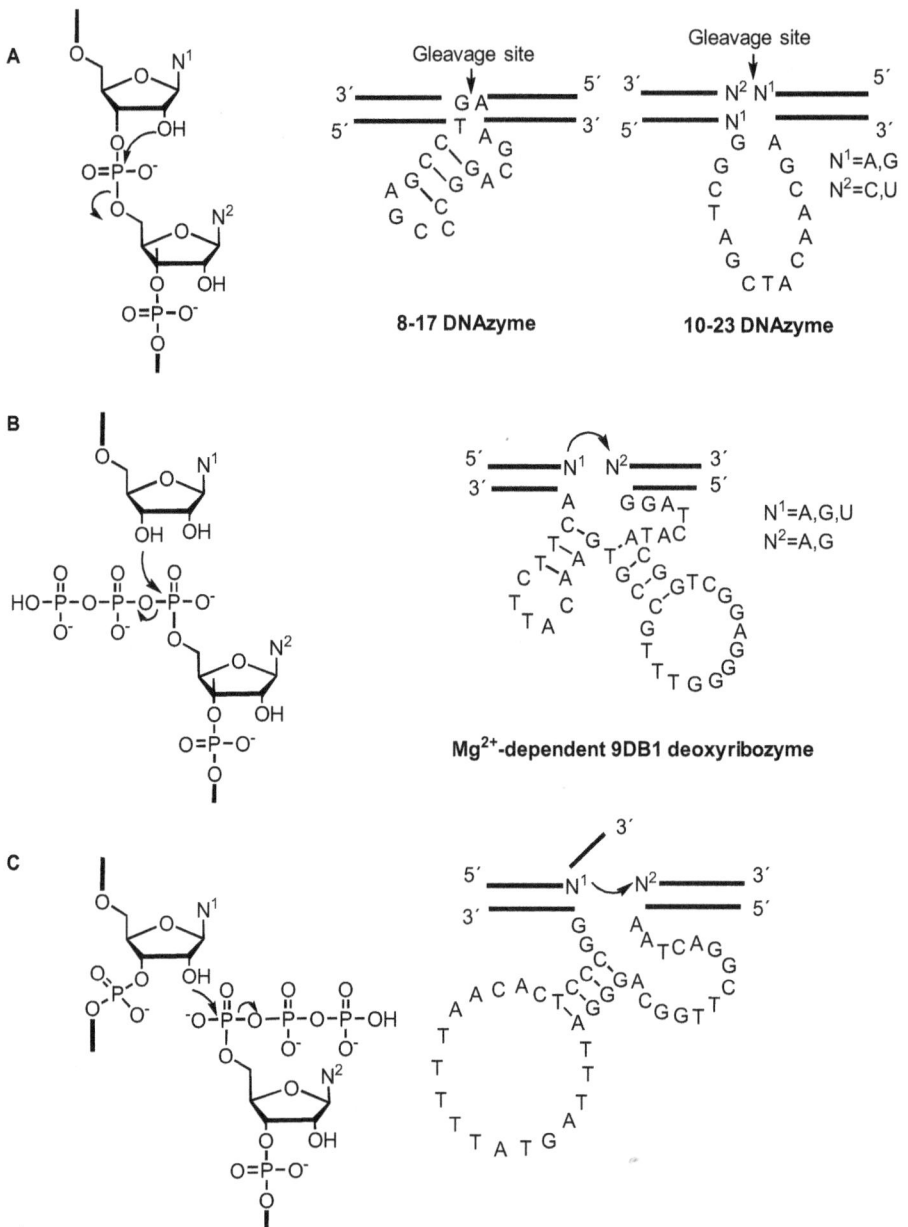

**Figure 10.12:** (A) Cleavage of RNA [125, 126], (B) ligation of RNA [128] and (C) branching of RNA [129].

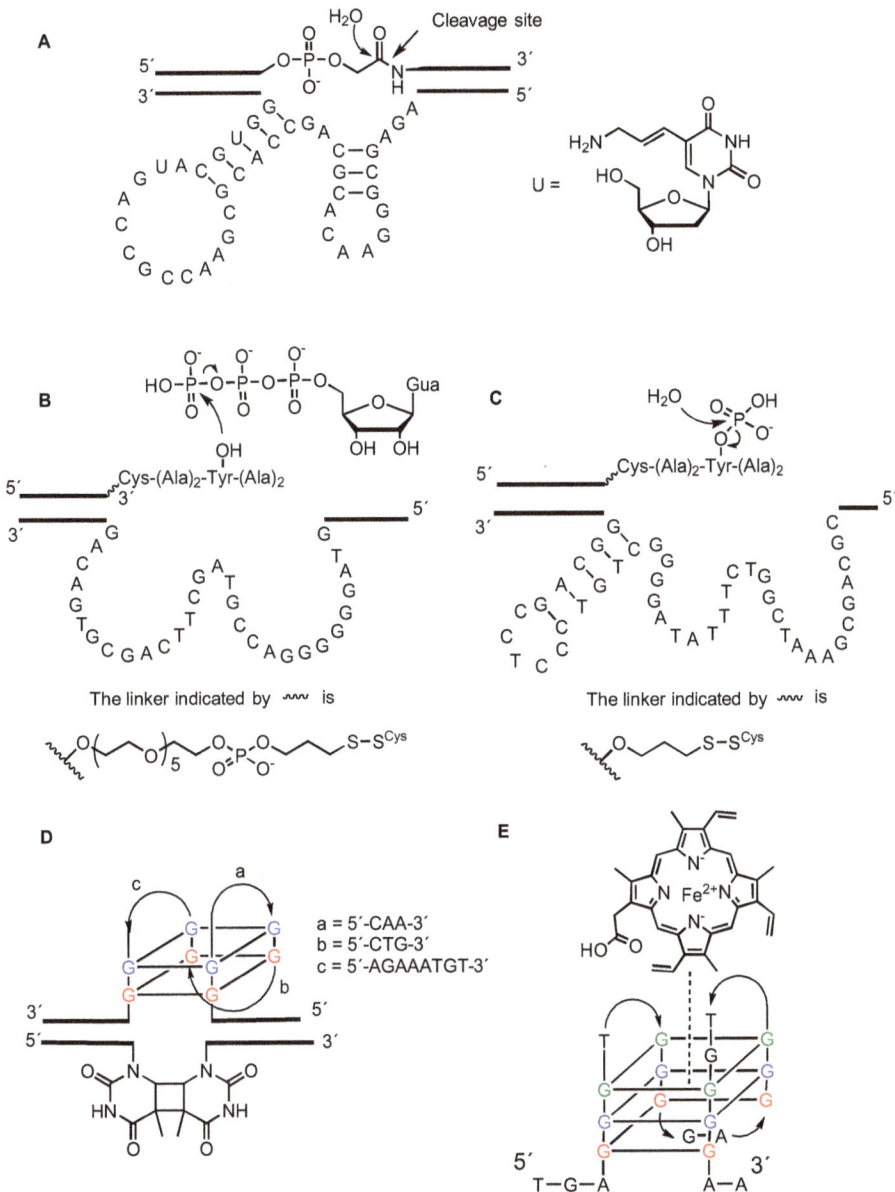

**Figure 10.13:** (A) Peptide bond cleavage [131], (B) peptide phosphorylation [132], (C) phosphopeptide hydrolysis [133], (D) photoreversion of thymidine dimer [135] and (E) peroxidating hemin-quadruplex complex [136].

catalytic activity self-evidently is an approach worth trying. The progress achieved is discussed in discussed in Chapter 12.7.

### 10.4.2 Artificial RNA ribozymes

Selection of artificial ribozymes from pools of oligoribonucleotides has received much less attention than selection of DNAzymes. A few observations, however, seems worth mentioning. An RNA polymerase ribozyme has been reported to recognize the template sequence, primer sequence and entering nucleoside 5'-triphosphate by tertiary interactions, not making use of canonical base-pairing [137]. Another interesting observation is that $Cu^{2+}$ complex of 3',5'-cyclic di(adenosine 5'-monophosphate) can be regarded as a primitive ribozyme. It catalyzes Friedel–Crafts reaction enantioselectively in aqueous solution [138]. Compared to $Cu^{2+}$, the acceleration is 20-fold. Studies on artificial RNA ribozymes are often motivated by desire to develop models for catalysts that possible enabled development of primitive life. For example, an $Yb^{3+}$-dependent ribozyme has been selected that converts nucleosides to their 5'-triphosphates using cyclic trimetaphosphate as the source [139], and 3-nucleotide sequences have been recognized that accelerate RNA-ligation on template [140].

## Further reading

Jimenez RM, Polanco JA, Lupták A. Chemistry and biology of self-cleaving ribozymes. Trends Biochem Sci 2015, 40, 648–661.

Hollenstein M. DNA catalysis: The chemical repertoire of DNAzymes. Molecules 2015, 20, 20777–20804.

Lilley DM. How RNA acts as a nuclease: Some mechanistic comparisons in the nucleolytic ribozymes. Biochem Soc Trans 2017, 45, 683–691.

Lönnberg T. Understanding catalysis of phosphate-transfer reactions by the large ribozymes. Chem Eur J 2011, 17, 7140–7153.

Morrison D, Rothenbroker M, Li Y. DNAzymes: Selected for applications. Small Methods 2018, 2, 1700319 (12 pages).

Müller U. Design and experimental evolution of trans-splicing group I Intron Ribozymes. Molecules 2017, 22, 75–83.

Peng H, Latifi B, Müller S, Luptak A, Chen IA. Self-cleaving ribozymes: Substrate specificity and synthetic biology applications. RSC Chem Biol 2021, 2, 1370–1383.

Ren A, Micura R, Patel DJ. Structure-based mechanistic insights into catalysis by small self-cleaving ribozymes. Curr Opin Chem Biol 2017, 41, 71–83.

Silverman SK. Catalytic DNA: Scope, applications, and biochemistry of deoxyribozymes. Trends Biochem Sci 2016, 41, 595–609.

Steitz TA, Moore PB. RNA, the first macromolecular catalyst: The ribosome is a ribozyme. Trends Biochem Sci 2002, 28, 411–418.

Will CL, Lührmann R. Spliceosome structure and function. Cold Spring Harbor Perspectives in Biology 2011, 3 a003707.

Zimmerly S, Semper C. Evolution of group II introns. Mobile DNA 2015, 6, 1–19.

# References

[1]  Orgel LE. Evolution of the genetic apparatus. J Mol Biol 1968, 88, 367–379.

[2]  Crick FHC. The origin of the genetic code. J Mol Biol 1968, 88, 381–393.

[3]  Cech TR, Zaug AJ, Grabowski PJ. In vitro splicing of the ribosomal RNA precursor of Tetrahymena: Involvement of a guanosine nucleotide in the excision of the intervening sequence. Cell 1981, 27, 487–496.

[4]  Guerrier-Takada C, Gardiner K, Marsh T, Pace N, Altman S. The RNA moiety of ribonuclease P is the catalytic subunit of the enzyme. Cell 1983, 35, 849–857.

[5]  Ward L, Plakos K, DeRose VJ. Nucleic acid catalysis: Metals, nucleobases, and other cofactors. Chem Rev 2014, 114, 4318–4342.

[6]  Müller S. Engineering of ribozymes with useful activities in the ancient RNA world. Ann NY Acad Sci 2015 1341, 54–60.

[7]  Lilley DM. How RNA acts as a nuclease: Some mechanistic comparisons in the nucleolytic ribozymes. Biochem Soc Trans 2017, 45, 683–691.

[8]  Fedor MJ, Williamson JR. The catalytic diversity of RNA. Nat Rev Mol Cell Biol 2005, 6, 399–412.

[9]  Lönnberg T. Understanding catalysis of phosphate-transfer reactions by the large ribozymes. Chem Eur J 2011, 17, 7140–7153.

[10]  Weinberg Z, Kim PB, Chen TH, Li S, Harris KA, Lünse CE, Breaker RR. New classes of self-cleaving ribozymes revealed by comparative genomics analysis. Nat Chem Biol 2015, 11, 606–610.

[11]  Branch AD, Robertson HD. A replication cycle for viroids and small infectious RNAs. Science 1984, 223, 450–455.

[12]  Uhlenbeck OC. A small catalytic oligoribonucleotide. Nature 1987, 328, 596–600.

[13]  Scott WG, Horan LH, Martick M. The hammerhead ribozyme: Structure, catalysis and gene regulation. Prog Mol Biol Transl Sci 2013, 120, 1–23.

[14]  de la Pena M, García-Robles I, Cervera A. The hammerhead ribozyme: A long history for a short RNA. Molecules 2017, 22, 78–90.

[15]  Hammann C, Luptak A, Perreault J, de la Pena M. The ubiquitous hammerhead ribozyme. RNA 2012, 18, 871–885.

[16]  Martick M, Scott WG. Tertiary contacts distant from the active site prime a ribozyme for catalysis. Cell 2006, 126, 309–320.

[17]  Lee T-S, López CS, Giambaşu GM, Martick M, Scott WG, York DM. Role of Mg2+ in hammerhead ribozyme catalysis from molecular simulation. J Am Chem Soc 2008, 130, 3053–3064.

[18]  Mir A, Chen J, Robinson K, Lendy E, Goodman J, Neau D, Golden BL. Two divalent metal ions and conformational changes play roles in the hammerhead ribozyme cleavage reaction. Biochemistry 2015, 54, 6369–6381.

[19]  Mir A, Golden BL. Two active site divalent ions in the crystal structure of the hammerhead ribozyme bound to a transition state analogue. Biochemistry 2016, 55, 633–636.

[20]  Thompson JE, Kuteladze TG, Schuster MC, Venegas FD, Messmore JM, Raines RT. Limits to catalysis by ribonuclease A. Bioorg Chem 1995, 23, 471–481.

[21]  Nakano S-I, Yamashita H, Sugimoto N. Enhancement of the catalytic activity of hammerhead ribozymes by organic cations. ChemBioChem 2021, 22, 2721–2728.

[22]  Symons RH. Plant pathogenic RNAs and RNA catalysis. Nucleic Acids Res 1997, 25, 2683–2689.

[23]  Ferre -D'Amare AR. The hairpin ribozyme. Biopolymers 2004, 73, 71–78.

[24]  Butcher SE, Heckman JE, Burke JM. Reconstitution of hairpin ribozyme activity following separation of functional domains. J Biol Chem 1995, 270, 29648–29651.

[25]  Murchie AI, Thomson JB, Walter F, Lilley DMJ. Folding of the hairpin ribozyme in its natural conformation achieves close physical proximity of the loops. Mol Cell 1998, 1, 873–881.

[26] Kath-Schorr S, Wilson TJ, Li N-S, Lu J, Piccirilli JA, Lilley DMJ. General acid–base catalysis mediated by nucleobases in the hairpin ribozyme. J Am Chem Soc 2012, 134, 16717–16724.

[27] Rupert PB, Ferre-D'Amare AR. Crystal structure of a hairpin ribozyme inhibitor complex with implications for catalysis. Nature 2001, 410, 780–786.

[28] Rupert PB, Massey AP, Sigurdsson ST, Ferre-D'Amare AR. Transition state stabilization by a catalytic RNA. Science 2002, 298, 1421–1424.

[29] MacElrevey C, Salter JD, Krucinska J, Wedekind JE. Structural effects of nucleobase variations at key active site residue Ade38 in the hairpin ribozyme. RNA 2008, 14, 1600–1616.

[30] Kumar N, Marx D. Deciphering the self-cleavage reaction mechanism of hairpin ribozyme. J Phys Chem B 2020, 124, 4906–4918.

[31] Nesbitt SM, Erlacher HA, Fedor MJ. The internal equilibrium of the hairpin ribozyme: Temperature, ion and pH effects. J Mol Biol 1999, 286, 1009–1024.

[32] Hieronymus R, Müller S. Engineering of hairpin ribozyme variants for RNA recombination and splicing. Ann NY Acad Sci 2019, 1447, 135–143.

[33] Saville BJ, Collins RA. A site-specific self-cleavage reaction performed by a novel RNA in Neurospora mitochondria. Cell 1990, 61, 685–696.

[34] Jones FD, Ryder SP, Strobel SA. An efficient ligation reaction promoted by a varkud satellite ribozyme with extended 5′- and 3′- termini. Nucleic Acids Res 2001, 29, 5115–5120.

[35] Guo HC, Collins RA. Efficient trans-cleavage of a stem-loop RNA substrate by a ribozyme derived from neurospora vs RNA. EMBO J 1995, 14, 368–376.

[36] Wilson TJ, Lilley DMJ. Do the hairpin and VS ribozymes share a common catalytic mechanism based on general acid–base catalysis? A critical assessment of available experimental data. RNA 2011, 17, 213–221.

[37] Suslov NB, DasGupta S, Huang H, Fuller JR, Lilley DMJ, Rice PA, Piccirilli JA. Crystal structure of the VS ribozyme. Nat Chem Biol 2015, 11, 840–846.

[38] Ganguly A, Weissman BP, Giese TJ, Li N-S, Hoshika S, Rao S, Benner SA, Piccirilli JA, York DM. Confluence of theory and experiment reveals the catalytic mechanism of the Varkud satellite ribozyme. Nat Chem 2020, 12, 193–201.

[39] DasGupta S, Piccirilli JA. The varkud satellite ribozyme: A thirty-year journey through biochemistry, crystallography, and computation. Acc Chem Res 2021, 54, 2591–2602.

[40] Kuo MY, Sharmeen L, Dinter-Gottlieb G, Taylor J. Characterization of self-cleaving RNA sequences on the genome and antigenome of human hepatitis delta virus. J Virol 1988, 62, 4439–4444.

[41] Webb C-HT, Riccitelli NJ, Ruminski DJ, Luptak A. Widespread occurrence of self-cleaving ribozymes. Science 2009, 326, 953.

[42] Ferre-D'Amare AR, Zhou K, Doudna JA. Crystal structure of a hepatitis delta virus ribozyme. Nature 1998, 395, 567–574.

[43] Ke A, Zhou K, Ding F, Cate JH, Doudna JA. A conformational switch controls hepatitis delta virus ribozyme catalysis. Nature 2004, 429, 201–205.

[44] Chen J-H, Yajima R, Chadalavada DM, Chase E, Bevilacqua PC, Golden BL. A 1.9 Å crystal structure of the HDV ribozyme precleavage suggests both Lewis acid and general acid mechanisms contribute to phosphodiester cleavage. Biochemistry 2010, 49, 6508–6518.

[45] Gong B, Chen JH, Chase E, Chadalavada DM, Yajima R, Golden BL, Bevilacqua PC, Carey PR. Direct measurement of a pK(a) near neutrality for the catalytic cytosine in the genomic HDV ribozyme using Raman crystallography. J Am Chem Soc 2007, 129, 13335–13342.

[46] Golden B. Two distinct catalytic strategies in the HDV ribozyme cleavage reaction. Biochemistry 2011, 50, 9424–9433.

[47] Weissman B, Ekesan S, Lin H-C, Gardezi S, Li N-S, Giese TJ, McCarthy E, Harris ME, York DM, Piccirilli JA. Dissociative transition state in Hepatitis Delta Virus ribozyme catalysis. J Am Chem Soc 2023, 145, 2830–2839.

[48] Winkler WC, Nahvi A, Roth A, Collins JA, Breaker RR. Control of gene expression by a natural metabolite-responsive ribozyme. Nature 2004, 428, 281–286.

[49] Roth A, Nahvi A, Lee M, Jona I, Breaker RR. Characteristics of the glmS ribozyme suggest only structural roles for divalent metal ions. RNA 2006, 12, 607–619.

[50] Klein DJ, Ferre-D'Amare AR. Structural basis of glmS ribozyme activation by glucosamine-6-phosphate. Science 2006, 313, 1752–1756.

[51] Cochrane JC, Lipchock SV, Strobel SA. Structural investigation of the GlmS ribozyme bound to its catalytic cofactor. Chem Biol 2007, 14, 97–105.

[52] Bingaman JL, Zhang S, Stevens DR, Yennawar NH, Hammes-Schiffer, Bevilacqua PC. The GlcN6P cofactor serves multiple catalytic roles in the glmS ribozyme. Nat Chem Biol 2017, 13, 439–445.

[53] Roth A, Weinberg Z, Chen AGY, Kim PB, Ames TD, Breaker RR. A widespread self-cleaving ribozyme class is revealed by bioinformatics. Nat Chem Biol 2014, 10, 56–60.

[54] Gasser C, Gebetsberger J, Gebetsberger M, Micura R. SHAPE probing pictures Mg2+-dependent folding of small self-cleaving ribozymes. Nucleic Acids Res 2018, 46, 6983–6995.

[55] Liu Y, Wilson TJ, McPhee SA, Lilley DMJ. Crystal structure and mechanistic investigation of the twister ribozyme. Nat Chem Biol 2014, 10, 739–744.

[56] Eiler D, Wang J, Steitz TA. Structural basis for the fast self-cleavage reaction catalyzed by the twister ribozyme. Proc Natl Acad Sci USA 2014, 111, 13028–13033.

[57] Ren A, Kosutic M, Rajashankar KR, Frener M, Santner T, Westhof E, Micura R, Patel DJ. In-line alignment and Mg2+ coordination at the cleavage site of the env22 twister ribozyme. Nat Commun 2014, 5, 5534 (10 pages).

[58] Liu Y, Wilson TJ, Lilley DMJ. The structure of a nucleolytic ribozyme that employs a catalytic metal ion. Nat Chem Biol 2017, 13, 508–513.

[59] Zheng L, Mairhofer E, Teplova M, Zhang Y, Ma J, Patel DJ, Micura R, Ren A. Structure-based insights into self-cleavage by a four-way junctional twister-sister ribozyme. Nat Commun 2017, 8, 1180 (12 pages).

[60] Teplova M, Falschlunger C, Krasheninina O, Egger M, Ren A, Patel DJ, Micura R. Crucial roles of two hydrated $Mg^{2+}$ ions in reaction catalysis of the pistol ribozyme. Angew Chem Int Ed 2020, 59, 2837–2843.

[61] Harris KA, Lünse CE, Li S, Brewer KI. Breaker RR Biochemical analysis of pistol self-cleaving ribozymes. RNA 2015, 21, 1852–1858.

[62] Wilson TJ, Liu Y, Li NS, Dai Q, Piccirilli JA, Lilley DMJ. Comparison of the Structures and Mechanisms of the Pistol and Hammerhead Ribozymes. J Am Chem Soc 2019, 141, 7865–7875.

[63] Neuner S, Falschlunger C, Fuchs E, Himmelstoss M, Ren A, Patel DJ. Micura R Atom-Specific Mutagenesis Reveals Structural and Catalytic Roles for an Active-Site Adenosine and Hydrated Mg(2+) in Pistol Ribozymes. Angew Chem Int Ed Engl 2017, 56, 15954–15958.

[64] Ekesan S, York DM. Who stole the proton? Suspect general base guanine found with a smoking gun in the pistol ribozyme. Org Biomol Chem 2022, 20, 6219–6230.

[65] Yoon S, Ollie E, York DM, Piccirilli JA, Harris ME. Rapid kinetics of pistol ribozyme: Insights into limits to RNA catalysis. Biochemistry 2023, 62, 13, 2079–2092.

[66] Serrano-Aparicio N, Swiderek K, Tunon I, Moliner V, Bertran J. Theoretical studies of the self-cleavage Pistol ribozyme mechanism. Topics in Catalysis 2022, 65, 505–516.

[67] Zheng L, Falschlunger C, Huang K, Mairhofer E, Yuan S, Wang J, Patel DJ, Micura R, Ren A. Hatchet ribozyme structure and implications for cleavage mechanism. Proc Natl Acad Sci USA 2019, 116, 10783–10791.

[68] Chen Y, Qi F, Gao F, Cao H, Xu D, Salehi-Ashtiani K, Kapranov P. Hovlinc is a recently evolved class of ribozyme found in human lncRNA. Nat Chem Biol 2021, 17, 601–607.

[69] Chadalavada DM, Gratton EA, Bevilacqua PC. The Human HDV-like CPEB3 Ribozyme Is Intrinsically Fast-Reacting. Biochemistry 2010, 49, 5321–5330.

[70]  De la Pena M, García-Robles I. Intronic hammerhead ribozymes are ultraconserved in the human genome. EMBO Reports 2010, 11, 711–716.

[71]  Hernandez AJ, Zovoilis A, Cifuentes-Rojas C, Han L, Bujisic B, Lee JT. B2 and ALU retrotransposons are self-cleaving ribozymes whose activity is enhanced by EZH2. Proc Nat Acad Sci 2020, 117, 415–425.

[72]  Michel F, Jacquier A, Dujon B. Comparison of fungal mitochondrial introns reveals extensive homologies in RNA secondary structure. Biochimie 1982, 64, 867–881.

[73]  Noller HF, Hoffarth V, Zimniak L. Unusual resistance of peptidyl transferase to protein extraction procedures. Science 1992, 256, 1416–1419.

[74]  Toor N, Keating KS, Taylor SD, Pyle AM. Crystal structure of a self-spliced Group II Intron. Science 2008, 320, 77–82.

[75]  Butcher SE. The spliceosome as ribozyme hypothesis takes a second step. Proc Natl Acad Sci USA 2009, 106, 12211–12212.

[76]  Will CL, Lührmann R. Spliceosome structure and function. Cold Spring Harbor Perspectives in Biology 2011, 3, a003707.

[77]  Lönnberg T. Understanding catalysis of phosphate-transfer reactions by the large ribozymes. Chem Eur J 2011, 17, 7140–7153.

[78]  Pyle AM. Metal ions in the structure and function of RNA. J Biol Inorg Chem 2002, 7, 679–690.

[79]  Reiner R, Ben-Asouli Y, Krilovetzky I, Jarrous N. A role for the catalytic ribonucleoprotein RNase P in RNA polymerase III transcription. Genes Dev 2006, 20, 1621–1635.

[80]  Reiter NJ, Osterman A, Torres-Larios A, Swinger KK, Pan T, Mondragón A. Structure of a bacterial ribonuclease P holoenzyme in complex with tRNA. Nature 2010, 468, 784–789.

[81]  Fournier MJ, Ozeki H. Structure and organization of the transfer ribonucleic acid genes of Escherichia coli K-12. Microbiol Rev 1985, 49, 379–397.

[82]  Forster AC, Altman S. External guide sequences for an RNA enzyme. Science 1990, 249, 783–786.

[83]  Warnecke JM, Fürste JP, Hardt WD, Erdmann VA, Hartmann RK. Ribonuclease P (RNase P) RNA is converted to a Cd(2+)-ribozyme by a single Rp-phosphorothioate modification in the precursor tRNA at the RNase P cleavage site. Proc Natl Acad Sci USA 1996, 93, 8924–8928.

[84]  Pfeiffer T, Tekos A, Warnecke JM, Drainas D, Engelke DR, Seraphin B, Hartmann RK. Effects of phosphorothioate modifications on precursor tRNA processing by eukaryotic RNase P enzymes. J Mol Biol 2000, 298, 559–565.

[85]  Brännvall M, Kikovska E, Kirsebom LA. Cross talk between the 173/294 interaction and the cleavage site in RNase P RNA mediated cleavage. Nucleic Acids Res 2004, 32, 5418–5429.

[86]  Nielsen H, Johansen SD. Group I introns: Moving in new directions. RNA Biology 2009, 6, 375–383.

[87]  Cech TR. Self-splicing of group I introns. Annu Rev Biochem 1990, 59, 543–568.

[88]  Vicens Q, Cech TR. Atomic level architecture of group I introns revealed. Trends Biochem Sci 2006, 31, 41–51.

[89]  Stahley MR, Adams PL, Wang JM, Strobel SA. Structural metals in the group I intron: A ribozyme with a multiple metal ion core. J Mol Biol 2007, 372, 89–102.

[90]  Shan S-o, Kravchuk AV, Piccirilli JA, Herschlag D. Defining the catalytic metal ion interactions in the Tetrahymena ribozyme reaction. Biochemistry 2001, 40, 5161–5171.

[91]  Shan S-o, Herschlag D. Probing the role of metal ions in RNA catalysis: Kinetic and thermodynamic characterization of a metal ion interaction with the 2′-moiety of the guanosine nucleophile in the Tetrahymena group I ribozyme. Biochemistry 1999, 38, 10958–10975.

[92]  Forconi M, Lee J, Lee JK, Piccirilli JA, Herschlag D. Functional identification of ligands for a catalytic metal ion in group I introns. Biochemistry 2008, 47, 6883–6894.

[93]  Herschlag D, Eckstein F, Cech TR. Catalysis by the Tetrahymena ribozyme. An energetic picture of an active site composed of RNA. Biochemistry 1993, 32, 8299–8311.

[94] Bass BL, Cech TR. Ribozyme inhibitors: Deoxyguanosine and dideoxyguanosine are competitive inhibitors of self-splicing of the Tetrahymena ribosomal ribonucleic acid precursor. Biochemistry 1986, 25, 4473–4477.

[95] Yoshida A, Shan S, Herschlag D, Piccirilli JA. The role of the cleavage site 2′-hydroxyl in the Tetrahymena group I ribozyme reaction. Chem Biol 2000, 7, 85–96.

[96] Forconi M, Sengupta RN, Liu M-C, Sartorelli AC, Piccirilli JA, Herschlag D. Structure and function converge to identify a hydrogen bond in a group I ribozyme active site. Angew Chem Int Ed 2009, 48, 7171–7175.

[97] Lu J, Li NS, Sengupta RN, Piccirilli JA. Synthesis and biochemical application of 2′-O-methyl-3′-thioguanosine as a probe to explore group I intron catalysis. Bioorg Med Chem 2008, 16, 5754–5760.

[98] Lambowitz AM, Zimmerly S. Group II introns: Mobile ribozymes that invade DNA. Cold Spring Harb Perspect Biol 2011, 3, a003616 (19 pages).

[99] Jacquier A, Michel F. Multiple exon-binding sites in class II self-splicing introns. Cell 1987, 50, 17–29.

[100] Peebles CL, Perlman PS, Mecklenburg KL, Petrillo ML, Tabor JH, Jarrell KA, Cheng HL. A self-splicing RNA excises an intron lariat. Cell 1986, 44, 213–223.

[101] Chu VT, Adamidi C, Liu QL, Perlman PS, Pyle AM. Control of branch-site choice by a group II intron. EMBO J 2001, 20, 6866–6876.

[102] Gordon PM, Fong R, Piccirilli JA. A second divalent metal ion in the group II intron reaction center. Chem Biol 2007, 14, 607–612.

[103] Griffiin EA, Qin ZF, Michels WJ, Pyle AM. Group II intron ribozymes that cleave DNA and RNA linkages with similar efficiency, and lack contacts with substrate 2′-hydroxyl groups. Chem Biol 1995, 2, 761–770.

[104] Gordon PM, Sontheimer EJ, Piccirilli JA. Kinetic characterization of the second step of group II intron splicing: Role of metal ions and the cleavage Site 2′-OH in catalysis. Biochemistry 2000, 39, 12939–12952.

[105] Hardison RC Splicing of introns in pre-mRNA. LibreTextsTM Biology 12.6. https://bio.libretexts.org/Bookshelves/Genetics/Book%3A_Working_with_Molecular_Genetics_(Hardison)/Unit_III%3A_The_Pathway_of_Gene_Expression/12%3A_RNA_processing/12.6%3A_Splicing_of_introns_in_pre%E2%80%91mRNAs. June 5, 2019.

[106] Jacquier A. Self-splicing group II and nuclear pre-mRNA introns: How similar are they?. Trends Biochem Sci 1990, 15, 351–354.

[107] Sontheimer EJ, Gordon PM, Piccirilli JA. Metal ion catalysis during group II intron self-splicing: Parallels with the spliceosome. Gene Dev 1999, 13, 1729–1741.

[108] Sontheimer EJ, Sun SG, Piccirilli JA. Metal ion catalysis during splicing of premessenger RNA. Nature 1997, 388, 801–805.

[109] Gordon PM, Sontheimer EJ, Piccirilli JA. Metal ion catalysis during the exon-ligation step of nuclear pre-mRNA splicing: Extending the parallels between the spliceosome and group II introns. RNA 2000, 6, 199–205.

[110] Zhu J, Hieronymus R, Müller S. A Hairpin Ribozyme Derived Spliceozyme. ChemBioChem 2023, 24, e202300204 (7 pages).

[111] Steitz TA, Moore PB. RNA, the first macromolecular catalyst: The ribosome is a ribozyme. Trends Biochem Sci 2002, 28, 411–418.

[112] Nissen P, Hansen J, Ban N, Moore PB, Steitz TA. The structural basis of ribosome activity in peptide bond synthesis. Science 2000, 289, 920–930.

[113] Schmeing TM, Seila AC, Hansen JL, Freeborn B, Soukup JK, Scaringe SA, Strobel SA, Moore PB, Steitz TA. A pre-translocational intermediate in protein synthesis observed in crystals of enzymatically active 50S subunits. Nat Struct Biol 2002, 9, 225–230.

[114] Hansen JL, Schmeing TM, Moore PB, Steitz TA. Structural insights into peptide bond formation. Proc Natl Acad Sci USA 2002, 99, 11670–11675.

[115] Hiller DA, Singh V, Zhong M, Strobel SA. A two-step chemical mechanism for ribosome-catalyzed peptide bond formation. Nature 2011, 476, 236–239.

[116] Kazemi M, Socan J, Himo F, Åqvist J. Mechanistic alternatives for peptide bond formation on the ribosome. Nucleic Acids Res 2018, 46, 5345–5354.

[117] Silverman SK. Catalytic DNA: Scope, applications, and biochemistry of deoxyribozymes. Trends Biochem Sci 2016, 41, 595–609.

[118] Hollenstein M. DNA catalysis: The chemical repertoire of DNAzymes. Molecules 2015, 20, 20777–20804.

[119] Silverman SK. Catalytic DNA (deoxyribozymes) for synthetic applications – Current abilities and future prospects. Chem Commun 2008, 0, 3467–3485.

[120] Diafa S, Hollenstein M. Generation of aptamers with an expanded chemical repertoire. Molecules 2015, 20, 16643–16671.

[121] Breaker RR, Joyce GF. A DNA enzyme that cleaves RNA. Chem Biol 1994, 1, 223–229.

[122] Ma L, Liu J. An in vitro–selected DNAzyme mutant highly specific for Na+ under slightly acidic sonditions. ChemBioChem 2019, 20, 537–542, and references therein.

[123] Santoro SW, Joyce GF. A general purpose RNA-cleaving DNA enzyme. Proc Natl Acad Sci USA 1997, 94, 4262–4266.

[124] Li J, Zheng W, Kwon AH. Lu Y In vitro selection and characterization of a highly efficient Zn(II)-dependent RNA-cleaving deoxyribozyme. Nucleic Acids Res 2000, 28, 481–488.

[125] Cepeda-Plaza M, McGhee CE, Lu Y. Evidence of a general acid−base catalysis mechanism in the 8−17 DNAzyme. Biochemistry 2018, 57, 1517–1522.

[126] Cepeda-Plaza M, Peracchi A. Insights into DNA catalysis from structural and functional studies of the 8−17 DNAzyme. Org Biomol Chem 2020, 18, 1697–1709.

[127] Cuenoud B, Szostak JW. A DNA metalloenzyme with DNA ligase activity. Nature 1995, 375, 611–614.

[128] Purtha WE, Coppins RL, Smalley MK, Silverman SK. General deoxyribozyme-catalyzed synthesis of native 3′-5′ RNA Linkages. J Am Chem Soc 2005, 127, 13124–13125.

[129] Wang Y, Silverman SK. Deoxyribozymes that synthesize branched and lariat RNA. J Am Chem Soc 2003, 125, 6880–6888.

[130] Baum DA, Silverman SK. Deoxyribozyme-catalyzed labeling of RNA. Angew Chem Int Ed 2007, 46, 3502–3504.

[131] Zhou C, Avins JL, Klauser PC, Brandsen BM, Lee Y, Silverman SK. DNA-catalyzed amide hydrolysis. J Am Chem Soc 2016, 138, 2106–2109.

[132] Walsh SM, Sachdeva A, Silverman SK. DNA catalysts with tyrosine kinase activity. J Am Chem Soc 2013, 135, 14928–14931.

[133] Chandrasekar J, Silverman SK. Catalytic DNA with phosphatase activity. Proc Natl Acad Sci USA 2013, 110, 5315–5320.

[134] Wang Y, Liu E, Lam CH, Perrin DM. A densely modified M2+-independent DNAzyme that cleaves RNA efficiently with multiple catalytic turnover. Chem Sci 2018, 9, 1813–1821.

[135] Chinnapen DJ-F, Sen DA. Deoxyribozyme that harnesses light to repair thymine dimers in DNA. Proc Natl Acad Sci USA 2004, 101, 65–69.

[136] Cheng X, Liu X, Bing T, Cao Z, Shangguan D. General peroxidase activity of G-quadruplex-hemin complexes and Its application in ligand screening. Biochemistry 2009, 48, 7817–7823.

[137] Kakoti A, Joyce GF. RNA polymerase ribozyme that recognizes the template−primer complex through tertiary interactions. Biochemistry 2023, 62, 1916–1928.

[138] Wang C, Hao M, Qi Q, Dang J, Dong X, Lv S, Xiong L, Gao H, Jia G, Chen Y, Hartig JS, Li C. Highly efficient cyclic dinucleotide based artificial metalloribozymes for enantioselective Friedel–Crafts reactions in water. Angew Chem Int Ed 2020, 59, 3444–3449.

[139] Sweeney KJ, Han, Müller UF. A ribozyme that uses lanthanides as cofactor. Nucleic Acids Res 2023, 51, 7163–7173.

[140] Nomura Y, Yokobayashi Y. RNA ligase ribozymes with a small catalytic core. Nat Sci Reports 2023, 13, 8584 (8–pages.

# 11 Nucleic acids as drugs and drug targets 1: oligodeoxyribonucleotides in chemotherapy

## 11.1 Introduction

The common aim of nucleic acid drugs is to specifically interfere the expression of a given gene and, hence, the production of a given protein. While the final goal is always the same, the targets and techniques of interference are numerous (Figure 11.1) [1]. In most cases the key step is highly selective recognition of the base sequence of the target nucleic acid, either DNA or RNA. The genomic information is translated into protein structures via transcription to pre-RNA, splicing of this to mRNA and finally binding of mRNA to ribosome where protein synthesis takes place. Any of these steps is a potential target for nucleic acid drugs. Transcription of DNA to pre-RNA may be prevented by sequence-selective blocking [2, 3] or cleavage [4, 5] of double-stranded DNA. The splicing sites within pre-RNA can be altered by oligonucleotide hybridization [6], and mature mRNA can be arrested or cleaved by oligonucleotides, either chemically [7] or by triggering an enzymatic process [8, 9]. Besides mRNA, DNA encodes synthesis of numerous noncoding RNAs [10]. These play an important role in regulation of gene expression and they, hence, are potential targets of oligonucleotide drugs. In addition, oligonucleotides are able to stimulate immunosystem [11] and oli-

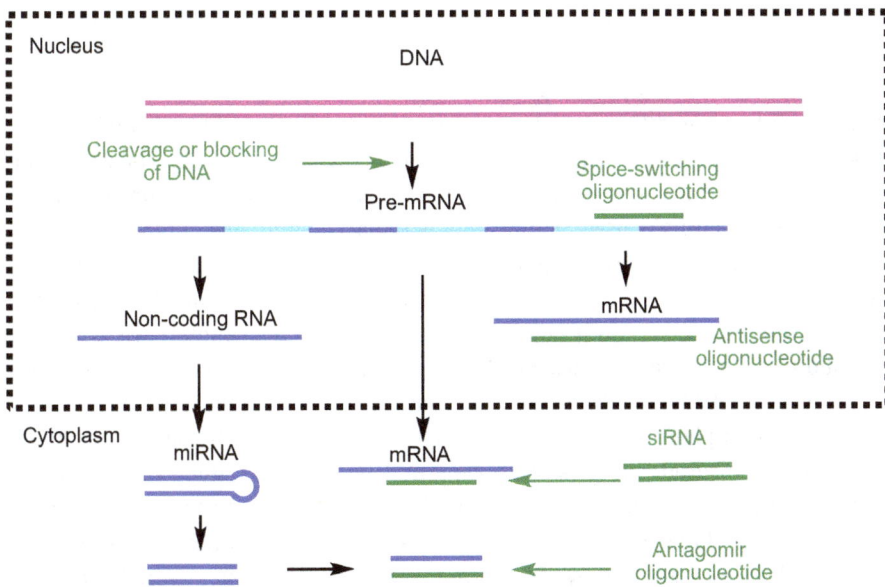

**Figure 11.1:** Alternative mechanisms for interference in transfer of genetic information with oligonucleotides.

https://doi.org/10.1515/9783111325637-011

gonucleotide aptamers, nucleic acid counterparts of antibodies, enable recognition of biological targets, such as cell membrane structures [12]. All these subjects are discussed in this chapter and also two subsequent chapters. The common feature of all the approaches is that the therapeutic oligonucleotides are structurally modified, and they often bear covalently linked conjugate groups.

## 11.2 Antisense oligonucleotides

### 11.2.1 Structural modifications

The origin of oligonucleotide drugs was the observation in 1978, according to which the replication of Rous sarcoma virus could be specifically inhibited by a tridecamer d(AATGGTAAAATG), which was complementary to the 3'- and 5'-terminal sequences of its 35S RNA [13]. Statistically, a 17-mer oligonucleotide is sufficiently long to warrant fully specific binding within the sequence of human genome [7]. The problems that have to be solved on the way to therapeutic oligonucleotides include prevention of rapid excretion, stabilization against nucleases, targeting to desired organs of cell types, internalization through cell membrane to cytoplasm and availability of the cognate sequence within mRNA [8, 14]. Structural modifications and conjugate groups are obviously necessary to fulfill these demands. These modifications have to be non-toxic and technically feasible, and they should not appreciably reduce the efficiency or selectivity of hybridization. Finally, the drug action should rather be catalytic than stoichiometric and the target sequence within RNA should be accessible.

On the way to therapeutic oligonucleotides, the base, sugar and phosphate moieties have all been modified [14]. Among these approaches, modification of internucleosidic phosphodiester linkages offers an obvious way to prolong the biological half-life that with natural oligodeoxyribonucleotides (ODNs) is around 15 min. The oldest and still most extensively used modification is replacement of one of the nonbridging oxygen atoms with sulfur [15]. These phosphorothioate ODNs, called first-generation antisense oligonucleotides (AONs), are sufficiently stable toward nucleases [16], and their distribution into tissues is better than with their oxygen counterparts [17]. They also bind to cell surface proteins, which expectedly facilitate internalization into cell [8]. The major advantage over many other backbone modifications, however, is that the action of phosphorothioate ODNs is catalytic. Mammalian cells contain two RNase H enzymes that degrade the RNA component in RNA/DNA heteroduplexes [18]. One of these, RNase H1, is present in nucleus, cytoplasm and mitochondria, while the other RNase H2 is bound to chromatin. Like natural ODNs, phosphorothioate ODNs are able to activate RNase H1 enzyme by binding to mRNA. mRNA becomes degraded and the phosphorothioate AON is available for hybridization with another target mRNA. More than half of oligonucleotide drugs accepted for clinical use are phosphorothioates. Phosphorothioate ASOs are often gap-mers. This means that the central part of AON is a 10-mer ODN phosphorothioate that serves as the primary target for RNase H. The 3'- and 5'-terminal wings, in turn, consist of sugar-modified nucleotides aimed at enhancing hybridization with the target.

Unfortunately, phosphorothioate oligonucleotides also suffer from some shortcomings. Hybridization with RNA is less efficient compared to unmodified ODNs. The relatively high affinity to proteins has dual consequences. On one hand, circulation in blood is prolonged by binding to serum albumins, and interaction with cell surface proteins enhances cellular uptake by endocytosis and subsequent release from endosomes [19]. On the other hand, the risk for toxic effects on using high doses is obvious [20]. One reason for toxicity is high affinity to paraspeckle proteins that result in delocalization in nucleolus [21]. However, replacement of even one deoxynucleoside phosphorothioate unit in the gap-region with some other modification may reduce toxicity. For example, C5'-substitution [22], insertion of a 2',5'-linkage [23], and replacement of a single phosphorothioate linkage with an alkylphosphonate [24] or a methanesulfonyl phosphoramidate linkage [25] are all modifications that lower the toxicity and improve the therapeutic index [26].

Replacement of nonbridging oxygen in phosphodiester linkages with sulfur creates a new stereogenic center, making the linkage chiral (cf. Section 6.4). Although it is known that stereopure $R_P$-isomer hybridizes with RNA and activates RNase H more efficiently than the $S_P$-isomer [27], $R_P/S_P$-racemates are usually utilized in biological studies. The $S_P$-isomer is more stable toward nucleases [28], and it has been argued that a mixed $R_P/S_P$-sequence offers a good compromise between activity and stability [29]. In addition, a mixed $3'$-$S_P S_P R_P$ triplet within a stereopure oligomer has been reported to promote the target cleavage [30]. So far no clear evidence for importance of stereochemistry in phosphorothioate AONs has been obtained [31, 32], but in cell-free assays, interruption of a $S_P S_P S_P S_P S_P$ sequence in the gap region with a single $R_P$ stereomer has influence on RNase H1 selectivity and activity [32].

The other phosphate modified ODNs that are stable toward nucleases include borano phosphate ODNs [33], N3'-phosphoramidate ODNs [34], thio-N3'-phosphoramidate [35], phosphoryl guanidine [36] and methanesulfonylphosphoramidate ODNs [37] (Figure 11.2). The last one has been argued to exhibit lower toxicity and more efficient activation of RNase H than phosphorothioate ODNs. Borano phosphate ODNs, like phosphorothioates, activate RNase H. As rather hydrophobic oligomers, they penetrate through the cell membrane better than unmodified ODNs [38]. This modification has not, however, proceeded into drug development pipeline. N3'-phosphoramidate ODNs [39] and their thioate analogs [35] hybridize with RNA more efficiently than phosphorothioates, but they do not activate RNase H. Their antisense influence is, hence, based on passive arresting. The silencing effect of stereopure phosphoryl guanidine oligonucleotides has turned out to be in cultured neurons superior to their stereopure phosphorothioate analogs [31, 32]. Protocols for synthesis of several other phosphorus modified oligonucleotides are available, but data on their behavior in biological systems is scarce [40].

Two heavily modified ODN analogs, phosphorodiamidate morpholino oligomers (PMO) [41, 42] and peptide nucleic acids (PNA) [43], have recently received increasing interest as candidates of antisense drugs. The monomeric unit in PMO is 6-hydroxymethylmorpholine that bears a nucleobase at C2 on the β-face (*cis* to the hydroxymethyl group) (see Section 5.4). These units are linked to each other by N4′→O$^{6'}$-dimethylaminophosphorodiamidate linkages. The resulting oligomers are resistant to nucleases and hybridize as efficiently as unmodified DNA. They interact only weakly with proteins and their cellular uptake, hence, is very limited. They do not support RNase H cleavage. PNA, in turn, has a peptide-like backbone composed of neutral *N*-(2-aminoethyl)glycin units, the nucleobases being attached to the glycine amino function via an acetyl bridge. The affinity of PNA to RNA, in particular to polypyrimidine sequences by clam-type triple helix formation, is exceptionally high [44], and replacement of some of the *N*-(2-aminoethyl)glycin units with their 2-aminocyclopentyl counterparts still strengthens the binding [45]. Both PMO and PNA oligomers show excellent nuclease stability [43, 46], but neither activates RNase H. Poor cellular uptake is another serious shortcoming.

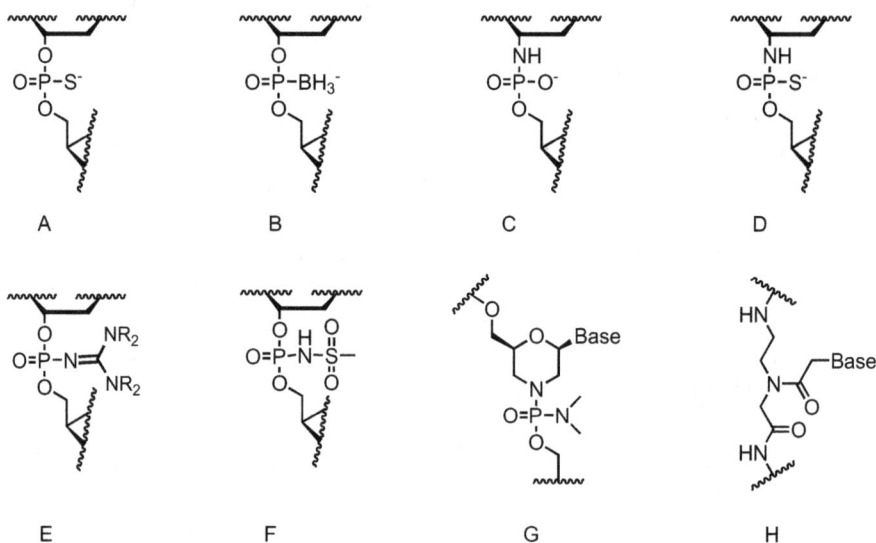

**Figure 11.2:** Phosphodiester modifications that prolong the biological half-life of ODNs:
(A) phosphorothioate ODN [15], (B) boranophosphate ODN [33], (C) *N*3′-phosphoroamidate ODN [34],
(D) *N*3′-phosphorothioamidate ODN [35], (E) phosphoryl guanidine ODN [36],
(F) methanesulfonylphosphoramidate ODN [37], (G) phosphorodiamidate morpholino oligomer [41, 42]
and (H) peptide nucleic acid [43].

The sugar moiety modifications introduced in AON drug candidates are numerous. They are aimed at increasing the affinity to RNA and hopefully nuclease stability as well [47]. Among them, 2′-*O*-(2-methoxyethyl)ribonucleoside (2′-*O*-MOE, Figure 11.3A)

[48] stands out. It is a constituent of several oligonucleotide drugs accepted for clinical use or passed to late phase clinical trials. When used in a phosphorothioate AON, the melting temperature of the RNA hybrid is increased by 2 °C per modification. 2'-*O*-MOE group increases hydration by trapping a water molecule and forms a steric hindrance for the attack on the neighboring 3'-phosphodiester, hence giving a better protection 2'-*O*-Me against nucleases than 2'-*O*-Me. The plasma clearance of 2'-*O*-MOE phosphorothioates mainly takes place to tissues, only a minor proportion being excreted to urine. Unfortunately, it does not activate RNase H.

Another sugar modification that increases affinity to RNA complements, even more than 2'-*O*-MOE, is 2'-*O*,4'-*C*-methylene bridge (Figure 11.3B) [49, 50]. The methylene bridge locks the sugar ring in N-type (C3'-endo) conformation, that is, to a conformation present in A-form RNA duplexes. Abbreviation LNA (locked nucleic acid) or BN (bicyclic nucleoside) is used for this modification. Introduction of three LNA nucleosides into a 9-mer sequence increases the melting temperature of the duplex by more than 20 °C compared to unmodified RNA [50]. A drawback is that a moderate risk for hepatotoxicity exists [51]. A diastereomeric form of LNA (α-L-LNA, Figure 11.3C) also shows high affinity to RNA and high stability against nucleases [52, 53].

In addition to 2'-*O*-MOE and LNA/BN, numerous other sugar modifications have shown encouraging behavior in vitro studies and some of these may well end up into AONs. 2'-Deoxy-2'-fluoro-β-D-arabinonucleoside (2'-F-ANA, Figure 11.3D) [54], cyclohexene (CeNA, Figure 11.3E) [55], oxepane (OXE, Figure 11.3F) [56], and (S)-5'-C-aminopropyl-2'-*O*-methyl ribonucleosides (5'-Pr^NH2-2'-OMe, Figure 11.3G) [57], for example, enhance binding to RNA and, more importantly, activate RNase H. DiF-tricyclo-DNA (11.3 H) hybridizes somewhat less efficiently than DNA, but is able to activate RNaseH [58]. 4'-Thio-ORNs (4'-Thio-RNA, Figure 11.3I) show good nuclease resistance and affinity to RNA is comparable to that of 2'-*O*-methyl ORNs [59]. In addition, several analogs of LNA/BNA (Figure 11.3J–R) [61–68, 70, 71] exhibit high affinity to RNA. Future will show whether some of these will end up to constituents of AON drugs. The ethyl bridged analog, cEtBNA (Figure 11.3L) [62], for example, has shown promise as a surrogate of LNA exhibiting a lower risk of hepatotoxicity [69].

Base moiety modifications have played a less important role in development of AONs. The most common base modification is 5-methyl substitution of cytosine (Figure 11.4A) [72]. It stabilizes the duplex with RNA by half a degree centigrade per modification and diminishes the risk for immune reaction. 5-Alkynyl substitutions (Figure 11.4B) are more stabilizing, owing to higher polarizability of the more extensive π-electron delocalization, but they have turned out to be toxic [73]. 5-Thiazolyl modification (Figure 11.4C) has been used as a surrogate of alkynyl group [74]. Phenoxazine base modification [75] also allows normal Watson-Crick base pairing and improves stacking. Its 9-(2-aminoethoxy) derivative, called G-clamp (Figure 11.4D, has been reported to enhance antisense activity, evidently due to stabilization of GC-base pair by H-bonding of the protonated amino group to guanine $O^6$ [76]. Replacement of adenine with 2,6-diamino purine allows an additional H-bond forma-

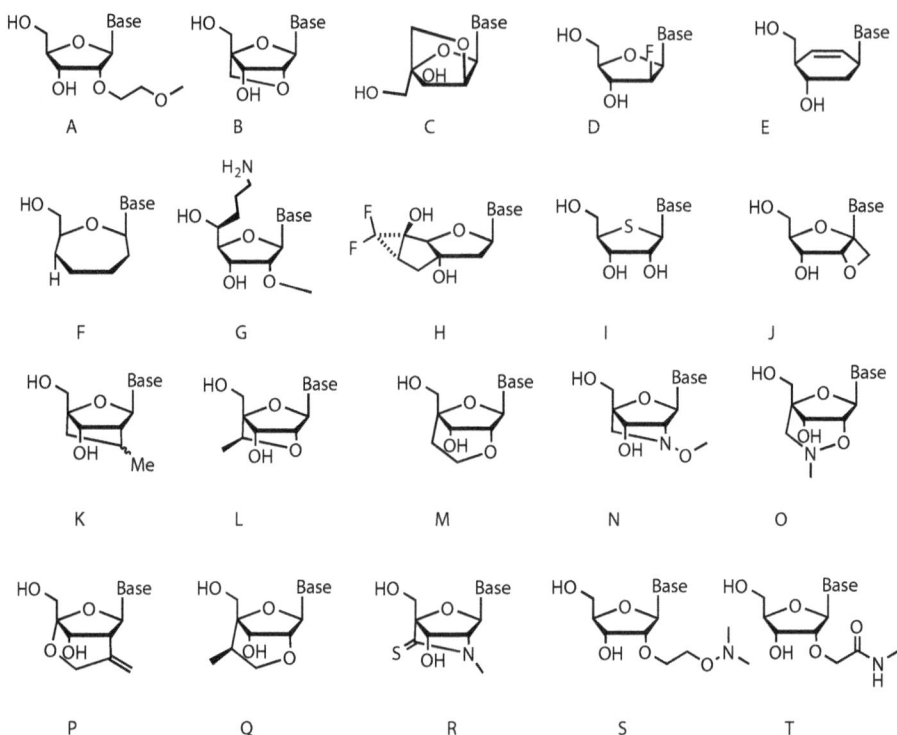

**Figure 11.3:** Promising sugar moiety modifications for antisense oligonucleotide chemotherapy: (A) 2′-O-MOE [48], (B) LNA [49, 50], (C) α-L-LNA [52,53], (D) 2′F-ANA [54], (E) CeNA [55], (F) OXE [56], (G) 5′-Pr^NH2-2′-OMe, 10.3 G [57], (H) diF-tricyclo-DNA [58], (I) 4′-Thio-RNA [59], (J) Oxetane-DNA [60], (K) Carbocyclic LNA [61], (L) cEt-BNA [62], (M) ENA [63], (N) N-MeO-amino-BNA [64], (O) 2′,4′-BNA^NC[NMe] [65], (P) Methylene-EoDNA [66], (Q) 6′R-Me-ENA [67], (R) thioAmNA [68], (S) 2′-O-DMAOE RNA [70] and (T) 2′-O-NMA [71].

tion to uracil (Figure 11.4E) and 3-amino- or 3-imidazolyl-propyl group at guanine $N^2$ stabilizes the duplex by electrostatic interaction with a phosphodiester bond within the opposite strand [72]. One should, however, bear in mind that enhanced in vitro hybridization does not necessarily mean enhanced antisense effect in vivo. Pharmacokinetics, cellular uptake and intracellular distribution also play a role.

## 11.2.2 Antisense oligonucleotide drugs

The first antisense oligonucleotide drug, approved by FDA in 1998, was a 21-mer phosphorothioate ODN, Fomivirsen (brand name Vitravene, Table 11.1). The drug was targeted against mRNA encoded by UL123 gene of cytomegalovirus. This infection caused blindness for AIDS patients. The drug was administered by intraocular injection. The action was based on formation of a heteroduplex with mRNA in cytoplasm which trig-

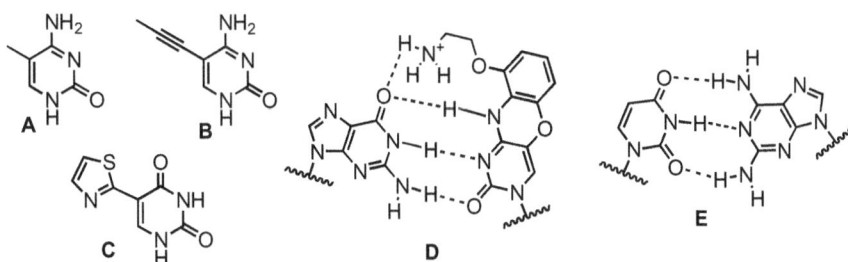

**Figure 11.4:** Base moiety modifications used in ASOs and their candidates. (A) 5-Methylcytosine [72], (B) 5-(propyn-1-yl)cytosine [73], (C) 5-thiazolyluracil [74], (D) G-clamp [76] and (E) 2,6-diaminopurine [72], base pairing with uracil indicated.

gered RNase H catalyzed mRNA degradation [77]. Fomivirsen was, however, withdrawn from market quite soon, in Europe in 2002 and in the USA 4 years later. Advanced anti-HIV chemotherapy had dramatically reduced the number of patients.

The other approvals of antisense oligonucleotides for therapeutic use are much more recent. Mipomersen (brand name Kynamro), designed for treatment of familial hypercholesterolemia, a genetic disorder leading to high cholesterol levels, was approved in 2013. The action is based on retardation of synthesis of ApoB protein that plays a central role in production of low-density lipoprotein (LDL) [78, 79]. Mipomersen is the first systemically administered AON. It is a 20-mer phosphorothioate ODN having a so-called gap-mer structure. This means that the central part is a 10-mer ODN phosphorothioate, known as a thioate window, and the terminal sequences consists of five 2'-O-MOE nucleosides (Table 10.1). As discussed above, the underlying idea is that the central part triggers RNase H activity upon hybridization with the target mRNA, whereas the terminal sequences warrant more efficient hybridization with the target.

Inotersen and Volanesorsen also are gap-mer phosphorothioates with 2'-O-MOE wings. Inotersen inhibits production of transthyretin in liver and is used for treatment of hereditary transthyretin amyloidosis that leads to heart dysfunction [80]. The drug is administered by subcutaneous injection. Volanesorsen, in turn, is used for reduction of triglycerides by targeting apolipoprotein $C_3$ [81].

Imetelstat is a 5'-plamitoyl conjugate of a 13-mer N3'-phosphorothioamidate oligonucleotide (cf. Figure 11.2D) that is aimed at inhibiting the telomerase ribonucleoprotein in cancer cells [82]. The target evidently is the RNA component of this enzyme rather than mRNA. The terminus of chromosomes, telomere, consists of a $(TTdAdGdGdG)_n$ repeat. When DNA is replicated, telomer is shortened since replication does not start from the very end of telomer. Finally, continuous shortening of telomere leads to situation that DNA cannot any more replicate and cell dies. In contrast to somatic cells, most cancer cells upregulate telomerase. A reverse transcriptase that binds the telomere 3'-terminus to its own RNA and synthesizes new TTdAdGdGdG repeats, avoiding, hence, cell death.

In cell line studies, AON complementary to the telomerase RNA component has blocked telomerase and decreased cell proliferation [83]. Accordingly, telomerase inhibition represents an exceptionally broad spectrum approach in combat against cancer. Imetelstat, however, still waits final approval.

**Table 11.1:** Sequences of therapeutic antisense oligonucleotides.

| | |
|---|---|
| Fomivirsen (Vitravene) | 5′-dG-dC-dG-T-T-T-dG-dC-T-dC-T-T-dC-T-T-dC-T-T-dG-dC-dG-3′ phosphorothioate |
| Mipomersen (Kynamro) | 5′-G$^{MOE}$-C$^{MOE}$-C$^{MOE}$-U$^{MOE}$-C$^{MOE}$-dA-dG-T-dC-T-dG-dC$^m$-T-T-dC$^m$-G$^{MOE}$-C$^{MOE}$-A$^{MOE}$-C$^{MOE}$-C$^{MOE}$-3′-phosphorothioate |
| Inotersen Volanesorsen | 5′-mU$^{MOE}$-mC$^{MOE}$-mU$^{MOE}$-mU$^{MOE}$-G$^{MOE}$-dG-T-T-dA-dC$^m$-dA-T-dG-dA-dA-A$^{MOE}$-mU$^{MOE}$-mC$^{MOE}$-mC$^{MOE}$-mC$^{MOE}$-3′ phosphorothioate 5′mU$^{MOE}$A$^{MOE}$mU$^{MOE}$mU$^{MOE}$mU$^{MOE}$dC$^m$dGdA-dC$^m$ dC$^m$TdGTTdC$^m$mU$^{MOE}$mU$^{MOE}$mC$^{MOE}$G$^{MOE}$A$^{MOE}$-3′-phosphorothioate |
| Imetelstat | Me(CH$_2$)$_{14}$C(O)NHCH$_2$CH(OH)CH$_2$OP(O)(S⁻)-T-dA-dG-dG-dG-T-T-dA-dG-dA-dC-dA-dA-3′-*N*3′-thiophosphoramidate |

T refers to thymidine, dA, dG and dC to 2′-deoxy-adenosine, -guanosine and –cytidine, respectively. dCm stands for 5-methyl-2′-deoxycytidine, mCMOE for 2′-*O*-(2-methoxyethyl)-5-methylcytidine, mUMOE for 2′-*O*-(2-methoxyethyl)-5-methyluridine and AMOE, GMOE, CMOE and UMOE to 2′-*O*-(2-methoxyethyl) derivatives of adenosine, guanosine, cytidine and uridine, respectively.

## 11.3 Splice-switching oligonucleotides

### 11.3.1 Mechanism of action

Therapeutic oligonucleotides may also be targeted toward pre-mRNA instead of mRNA. The aim is to modulate the splicing of pre-mRNA to mRNA. Pre-mRNA consists of, in addition to 5′- and 3′-terminal untranslated regions (UTR), alternating exons and introns. Exons contain the genetic material that is ultimately used to encode the protein synthesis, whereas introns (with some exceptions) become eliminated during the maturation of pre-mRNA to mRNA. The excision of intron and ligation of the flanking exons is catalyzed by spliceosome, as discussed in Section 10.3. This splicing can, however, take place in different manners. In fact, 95% of human genes have been argued to encode splice-variant proteins [84]. In other words, a single pre-mRNA can be transformed to several different mRNAs, depending on which ones of the exons become ligated. Oligonucleotides can modulate the splicing process in several different manners, as depicted in Figure 11.5 [85]. Several promising oligonucleotide drug candidates fall into this category [86]. Firstly, exon skipping may be achieved. When one of the splicing sites is blocked by hybridization, the neighboring exon is removed together with the flanking introns (Figure 11.5A).

Alternatively, a special exonic splicing enhancer site (ESE) within the exon may be blocked with an oligonucleotide. This prevents assembly of the spliceosome on the splicing site and, hence, splicing is prevented or redirected to take place along another pathway. Secondly, exon incorporation into the mature mRNA may also be enhanced. Some exons contain an exonic splicing silencer (ESS) element that retards splicing and, hence, their inclusion into mRNA. Hybridization with ESS restores the inclusion of the exon into mRNA (Figure 11.5B). Finally, defective pre-RNAs can be corrected with the aid of oligonucleotides. In some cases, an intron may be folded in such a way that it forms a false splicing site within the intron instead of intron/exon junction. Accordingly, part of the intron is incorporated into mRNA (Figure 11.5C). Hybridization with an external oligonucleotide prevents this aberrant splicing and restores formation of correct mRNA (Figure 11.5D).

The effect of splice-switching ODNs is based on steric blocking by hybridization with the target sequence. Unlike with AONs, no activation of any intracellular enzyme is required. The therapeutic effect is only based on efficiency of hybridization, cellular uptake and release in cytoplasm. Morpholino oligomers (PMOs) and fully modified 2'-MOE-RNA phosphorothioates have turned out to be viable candidates for therapeutic purposes. In addition, α-L-LNA [87] and 7',5'-α-bicyclo-DNA [88] modification have shown promise as constituents of exon-skipping oligonucleotides.

### 11.3.2 Splice-switching oligonucleotides in chemotherapy

Five splice-switching oligomers have so far been approved for clinical use. Eteplirsen, a 30-mer exon-skipping morpholino oligomer (Table 11.2), was approved in 2016 for treatment of Duchenne muscular dystrophy (DMD). DMD is a progressive neuromuscular disease that leads to loss of mobility and death at the age of 30, due to failure of respiratory and cardiac functions. The reason for this genetic disorder is DNA deletions within the gene encoding dystrophin protein. These deletions disrupt the reading frame of base sequence and prevent synthesis of dystrophin. Dystrophin gene is very large, encoding 79 exons. Eteplirsen alters the splicing by skipping exon 51 of dystrophin pre-RNA [89]. The truncated protein still retains part of the dystrophin functionality. Unfortunately, eteplirsen appears to help only a minor proportion of patients [90]. Golodirsen [91], Viltolarsen [92] and Casimersen [93] are structural analogs of Eteplirsen that result in skipping of exon 53, 53 and 45, respectively.

Nusinersen, in turn, is an 18-mer 2'-O-MOE RNA phosphorothioate approved in 2016 for treatment of spinal muscular atrophy (SMA). This disease most commonly originates from mutations in the survival motor neuron 1 (SMN1) gene [94]. It results in progressive paralysis and may lead to infant death. A small amount of SMN protein is, however, produced by SMN2 gene, which allows survival of patients. The activity of SMN2 is low, owing to the presence of an exonic splicing silencer (ESS) within one of the exons [95]. Nusinersen is able to modulate the splicing by blocking this ESS and, hence, allowing splicing to SNM1 mRNA [96]. In other words, exon retention takes

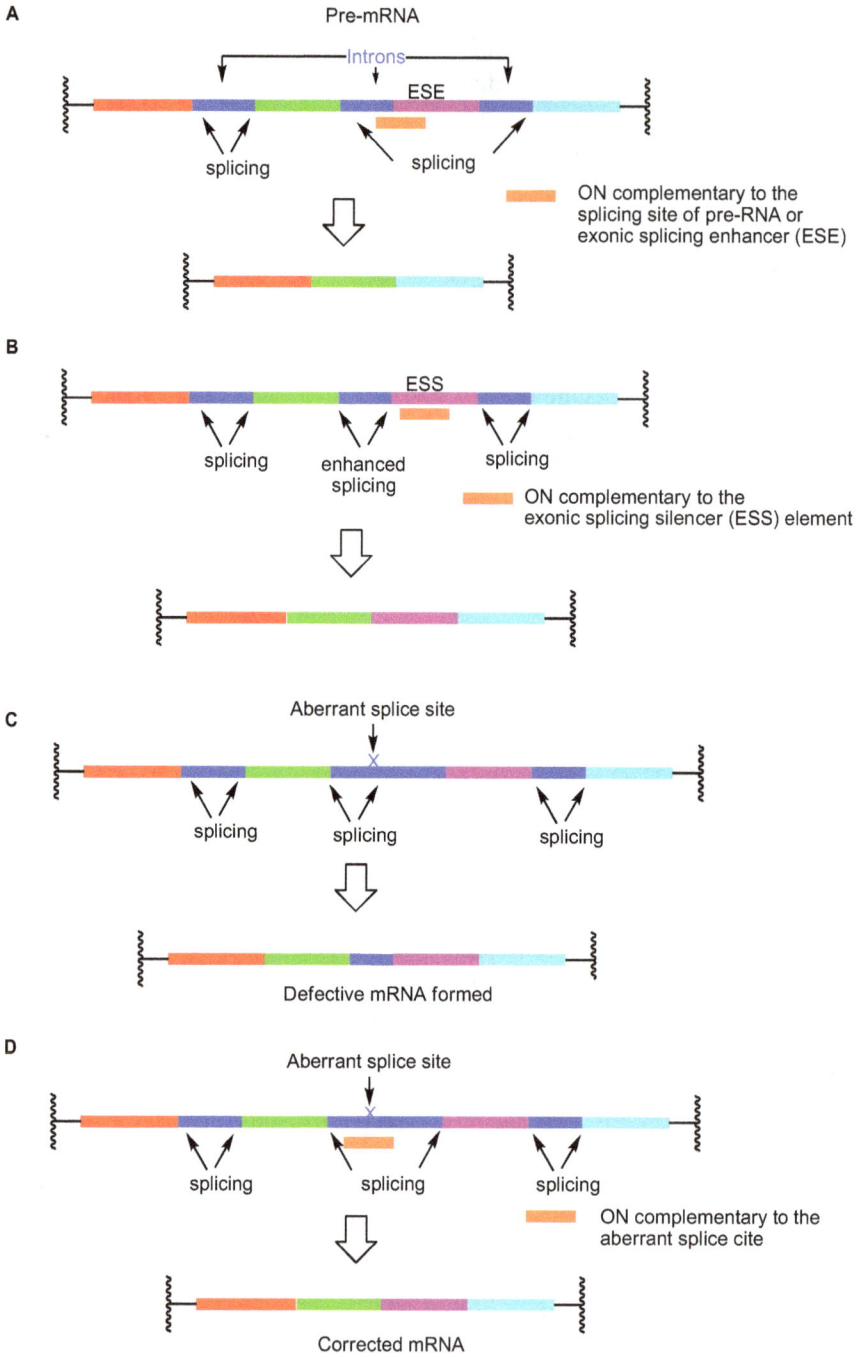

**Figure 11.5:** (A) Exon skipping, (B) exon retention, (C) aberrant splicing and (D) correction of aberrant splicing.

place (cf. Figure 11.5B). The drug is administered by intrathecal injection into the central nervous system.

**Table 11.2:** Sequences of therapeutic splice-switching oligonucleotides.

| | |
|---|---|
| Eteplirsen (Exondys51) | HO(CH$_2$CH$_2$O)$_3$C(O)piperazinyl-P(O)(NMe$_2$)O-C-T-C-C-A-A-C-A-T-C-A-A-G-G-A-A-G-A-T-G-G-C-A-T-T-T-C-T-A-G-3′ <br> morpholino phosphorodiamidate |
| Nusinersen (Spinraza) | 5′-mU$^{MOE}$- mC$^{MOE}$- A$^{MOE}$- mC$^{MOE}$- mU$^{MOE}$- mU$^{MOE}$- mC$^{MOE}$- mU$^{MOE}$- mC$^{MOE}$- A$^{MOE}$- mU$^{MOE}$- A$^{MOE}$- A$^{MOE}$- mU$^{MOE}$- G$^{MOE}$- mC$^{MOE}$- mU$^{MOE}$- G$^{MOE}$- G$^{MOE}$-3′ <br> phosphorothioate |

mC$^{MOE}$, mU$^{MOE}$, A$^{MOE}$ and G$^{MOE}$ stand for 2′-O-(2-methoxyethyl) substituted 5-methylcytidine, 5-methyluridine, adenosine and guanosine, respectively. C, T, and A refer to morpholino phosphorodiamidates derived from cytosine, thymine and adenine.

No splice-switching oligonucleotide drugs based on correction of aberrant splicing (cf. Figure 11.5D) are in therapeutic use or in late phase clinical trials. Beta thalassemias are inherited blood disorders that cause reduced synthesis of the hemoglobin β-chains. The most frequently occurring mutations result in aberrant splicing of β-globin pre-mRNA. It has been shown [97] that the correct splicing can really be restored in vitro with splice-switching oligonucleotide, but the results of in vivo studies have not yet lead to a sufficiently promising drug candidate. The problems related to bio-stability and pharmacokinetics has to be solved on the way to an approved drug.

In addition to the antisense and splice-switching oligonucleotides discussed above, dozens of candidates are in various phases of clinical trials [98]. The target diseases include cancer, cardiovascular, metabolic, neurological, inflammatory and infectious diseases, just to mention the most common ones.

## 11.4 Triplex-forming oligonucleotides

Triple helix formation by binding an oligonucleotide to the major groove of double-stranded DNA in principle offers an approach to selective recognition of genes and, hence, inhibition of their expression. As discussed in Section 8.3 and depicted in Figure 8.9, a double-helical polypurine/polypyrimidine sequence may be recognized by a polypyrimidine or a polypurine triple-helix-forming oligonucleotide (TFO). In the so-called parallel recognition mode, a polypyrimidine TFO sequence is bound in parallel manner to the polypurine strand of the duplex by formation of Hoogsteen C$^+$GC and TAT triplets (Figure 11.6). In antiparallel recognition, a polypurine TFO binds in an antiparallel manner to the polypurine strand forming reverse Hoogsteen GGC and AAT triplets. The polypurine/polypyrimidine sequence should, however, be rather long, preferably more than 10 base pairs, to warrant sufficient stability for the triplex.

**Figure 11.6:** Alternative modes for recognition of polypurine/polypyrimidine sequences in double-stranded DNA.

An obvious way to increase the triplex stability is to use base modifications. Replacement of cytosine with 5-methylcytosine (Figure 11.7A) and thymine with 3-(1-propynyl) uracil (Figure 11.7B) enhances stacking [99], while 2-aminopyridine (Figure 11.7C) and 2-amino-3-methylpyridine (Figure 11.7D) that are more basic than cytosine (p$K_a$ 6.9 vs. 4.2), allow stable H-bonding upon formation of a C$^+$GC triplet under physiological conditions [100]. Conjugation to intercalators also increases stability [101]. On using polypurine TFOs, G-quartet formation may compete with triplex formation. To prevent this, guanosine analogs have been used: 7-deazaxanthine (Figure 11.7E) [102], 6-thioguanine (Figure 11.7F) [93], and 7-chloro-7-deazaguanine (Figure 11.7G) [104].

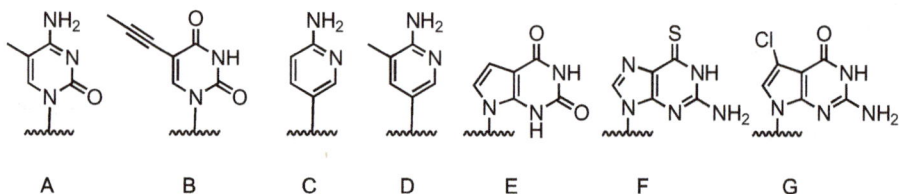

**Figure 11.7:** Base modifications used to stabilize triple helices: A [99], B [99], C [100], D [100], E [102], F [103] and G [104].

An attractive approach is recognition of base pairs instead of individual bases [105]. The underlying idea is to eliminate the destabilizing effect of a single base pair interruption within a polypurine/polypyrimidine sequence with the aid of a modified surrogate base that binds efficiently to the interrupting base pair. Numerous modified nucleobases have been introduced, but none of them has clearly stood out. Some noteworthy examples are depicted in Figure 11.8. As regards the parallel binding of polypyrimidine TFOs to a polypurine sequence, $N^4$-(6-aminopyridine-2-yl)cytosine (Figure 11.8A) forms with CG interruption an triplet that is as stable as the native C$^+$GC triad [106]. Binding is not, however, selective since AT pair is recognized almost as efficiently. Evidently, the cytosine base participates as a minor tautomer. The TA interruption, in turn, is recognized, although not selectively, with a thiazolylaniline nucleoside surrogate (Figure 11.8B) [107]. The triplet formed is approximately as stable as the TAT triad. In case of antiparallel binding to polypurine sequence, W-Shaped nucleoside analogs (Figure 11.8C [108] and 11.8D [109]) have been used to alleviate the destabilizing influence of CG and TA interruptions, respectively. While the 4-methylpyrazol-3-one and thymine base evidently interacts with the base pair, the phenyl group enforces the binding by stacking. The efficiency of binding is somewhat sequence-dependent. 2'-Deoxyformycin (Figure 11.8E) [110] may also be used to lessen the destabilizing effect of CG.

The ultimate goal, however, is development of a set of unnatural nucleobases with which all natural base pairs (AT, TA, GC, CG) could be recognized. Figure 11.8F–I shows structures recently suggested for GC, CG, AT and TA base pairs [111, 112], but firm experimental evidence has not yet published.

Sugar and phosphate modifications also enhance triple helix formation. Most common sugar modifications are 2'-methyl- and 2'-$O$-(2-aminoethyl)- ribonucleosides [113] and LNA [114] that all exhibit N-type sugar-ring-puckering. Although the sugar moieties of double-stranded DNA adopt an S-type conformation, the third strand bound to the major groove prefers N-puckering [115]. For the same reason, phosphoramidate ODNs that exhibit RNA-like N-conformation [116] form stable parallel but unstable antiparallel triplexes [117]. Particularly stabilizing are 2'-amino-LNAs that bear an aliphatic triamine at 2'-$N$. The amino groups undergo protonation under neutral conditions and, hence, weaken electrostatic repulsion between negatively charged phosphate groups [118]. The stabilizing effects of base and sugar moiety modifications may in some cases be additive. 2'-$O$-(2-Aminoethyl)-5-(3-aminoprop-1-yn-1-yl)uridine serves as an example. The two amino groups presumably interact with different phosphate groups, resulting in an additive increase in triplex melting point [119].

Compared to TFOs, PNA as a neutral biopolymer exhibits higher affinity to double-stranded DNA. The binding behavior, however, differs from that of TFOs. When the concentration of polypyrimidine PNA is low and ionic strength is high, the behavior still very much resembles parallel binding TFOs, the amino terminus of PNA facing the 5'-terminus of the polypurine strand in the duplex [120]. However, at low ionic strength and high concentration of polypyrimidine PNA, the situation is changed. The polypyrimidine strand of the target duplex is displaced by the invading polypyrimidine PNA and a stable PNA-DNA-PNA triplex is formed by TAT and C$^+$GC triplets as shown in Figure 11.9 [121]. Either, two separate PNA strands bind to the polypurine sequence, or a single PNA forms a bimolecular triple-helical clamp. Consequently, the overall shape of DNA is altered more markedly than by binding of TFO to the major groove. Accordingly, effects on DNA transcription, replication and repair are expectedly more drastic. C5-linked 2-aminopyridine (isoorotamde) as a base surrogate has been shown to improve recognition of AU base pairs [122].

Besides low stability of triple helices, and hence, demand for a reasonably long polypurine/polypyrimidine target sequence, intracellular localization to nucleus and access to tightly packed chromosomal DNA form extra hurdles on the way to medical use of TFOs. In case these thresholds are passed, two mechanisms of action appear possible. Firstly, PNA has shown some promise in preclinical gene editing studies with mouse models [123]. Formation of a PNA-DNA-PNA heterotriplex clamp results in displacement of the opposite strand that may then recruit endogenous repair. Secondly, photochemi-

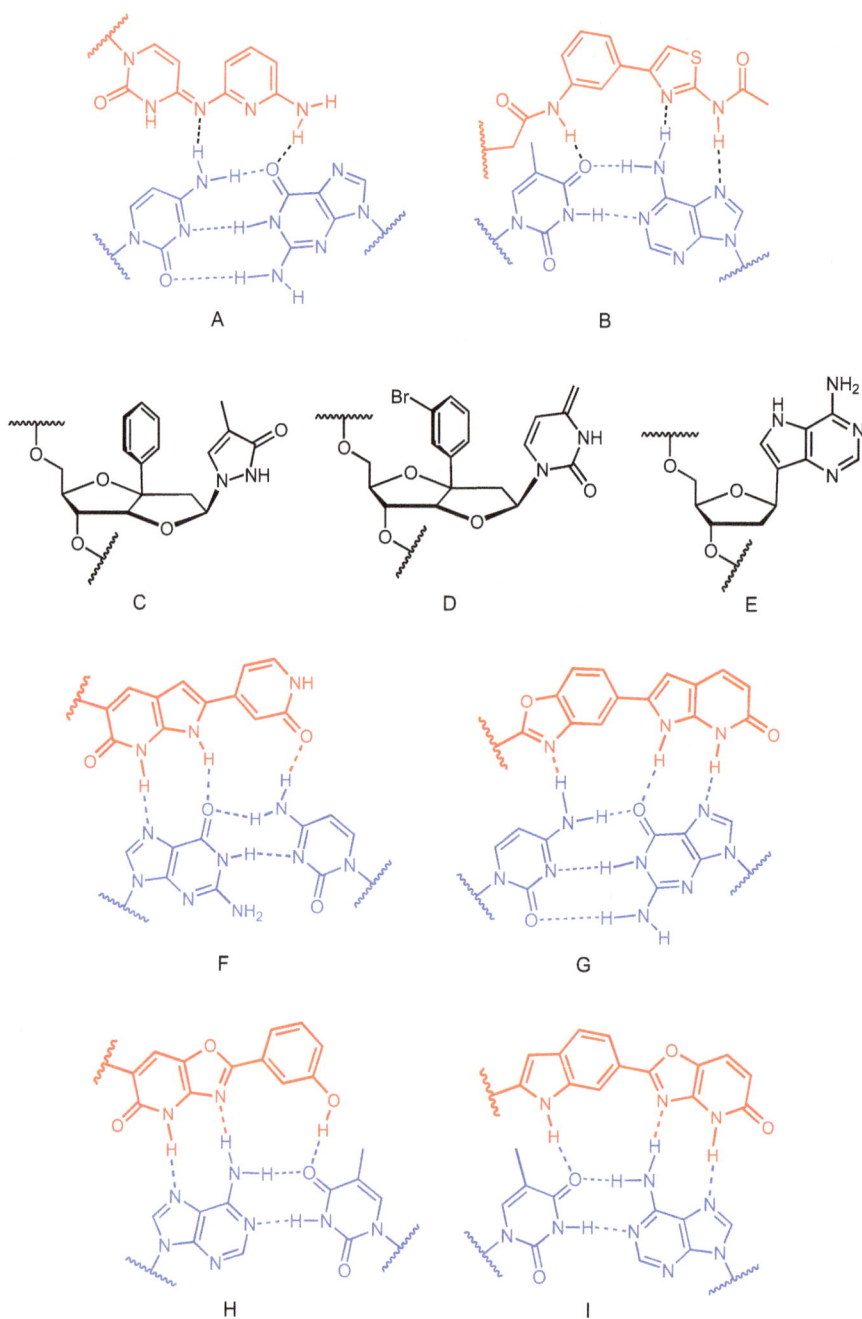

**Figure 11.8:** Artificial nucleoside analogs used to recognize CG and TA interruptions within a polypurine/polypyrimidine: A [106], B [107], C [108], D [109] and E [110]. Unnatural nucleosides suggested to recognize to all canonical base pairs (GC; CG; AT; TA) F-I [111, 112].

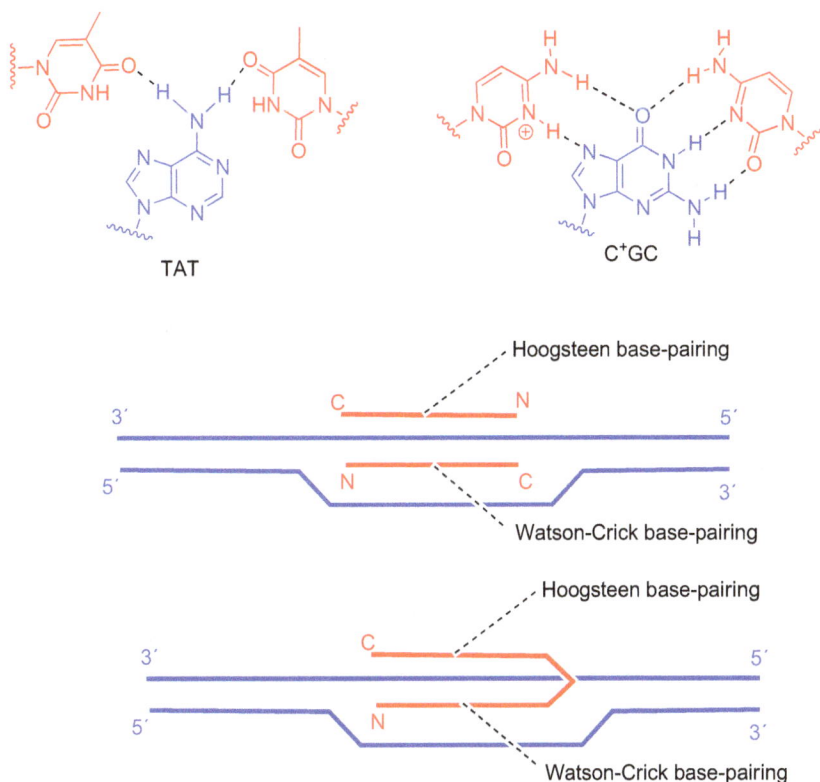

**Figure 11.9:** Base triplets and structures of PNA-DNA-PNA triplexes.

cally active TFO conjugate can induce a site specific mutation either via covalent cross-linking or by cleavage and repair of DNA double strand [124, 125]. Therapeutically relevant chromophores are those absorbing light at wavelengths of visible light (>380 nm). Such radiation penetrates tissues rather well without modifying biomolecules. At the site of chromophore excitation, reactive oxygen derived species are formed that, in turn, result in cross-linking and other structural changes. The most extensively studied chromophore is psoralen that intercalates DNA and undergoes upon photoactivation at 400 nm [2 + 2] cycloaddition with pyrimidine bases [126]. TFO conjugates of metal ion complexes that oxidative cleave DNA have also received some interest. The feasibility of this approach has been proven by two different $Cu^{2+}$-bearing conjugates in vitro [127, 128], but no in vivo data exist.

Compared to antisense oligonucleotides, the progress of TFOs toward drugs has been slow. Despite quite extensive in vitro and in vivo studies, results on animal models are rather limited and no FTOs have passed to major clinical trials. An observation of long noncoding RNA interaction with genomic DNA by RNA/DNA/DNA triplex formation and its role in gene regulation [129], possibly activates research on medical applications of TFOs.

## 11.5 Immune stimulatory oligonucleotides

Some oligonucleotides stimulate exceptionally efficiently the innate immune system, that is, the evolutionary developed defense systems of vertebrates that are based on recognition of certain molecular structures by a set of receptors [130]. As regards stimulation by oligonucleotides, the key structural unit is 5′-CpG-3′ that is recognized by toll like receptor 9 (TLR9) [131]. The CpG dimer occurs frequently in DNA of viruses and bacteria, less frequently in vertebrates where the cytosine base is usually methylated. Two other receptors, TLR7 and TLR8, in turn, recognize single-stranded RNA and TLR3 double-stranded RNA [132]. All these receptors are located in the endosomal compartment of cell, that is, in a discontinuous closed membrane system through which the content of vesicles taken up by endocytosis is released. Oligonucleotides that usually are taken into cells by endocytosis may, hence, trigger the immune response without release to cytoplasm [133]. In addition to the endosomal receptors TLR3, TLR7, TLR8, and TLR9, one RNA recognizing receptor, RIF-I, is present in cytosol. This receptor recognizes viral RNAs on the basis of their 5′-triphosphate [134]. The 5′-triphosphate of vertebrates′ RNA is either capped with 7-methylguanosine or the 5′-terminus is not phosphorylated at all.

Although unmethylated CpG dimer plays a central role in binding to TLR9, its stimulatory effect still is sequence dependent. The dimer becomes more efficiently recognized when incorporated in 6-mer sequence GT**CG**TT, and the immune response is even stronger when several of such hexanucleotides occur close to each other in the sequence [131]. Three distinct classes of strongly immune stimulatory ODNs, known as CpG ODNs, have been recognized (Table 11.3). The immune stimulatory profile, that is, transmission of the immune response via interferon production or B lymphocyte proliferation, is different in every class. The location of the CpG dimers also affects the potency of TLR9 activation. A CpG motif close to the 5′-terminus appears to be important [135]. Evidently the 5′-terminus plays a dominant role in receptor binding since CpG ODNs that are 3′-3′-linked to each other, are better agonists than their 5′-5′-counterparts [136]. Besides CpG dimer, an $N^{Py}$TTTTGT-motif is highly stimulatory ODN [137].

Some antisense oligonucleotides that are phosphorothioate gap-mers may induce TLR9-dependent innate immune response. The exact mechanism of interaction with TLR9 still is unknown, but replacement of two phosphorothioate linkages in the gap with mesyl phosphoroamidate linkages mitigates the immune response [138].

As with antisense oligonucleotides, nuclease resistance and cellular uptake of immune stimulatory ONs are enhanced by structural modifications. In addition, binding

**Table 11.3:** Distinct classes of immune stimulatory CpG ODNs [139].

| | |
|---|---|
| A-Class | **G$^s$G$^s$GGACGATCGTCG$^s$G$^s$G$^s$C$^s$G$^s$G**; forms multimers via G-tetrads. Strong interferon induction & moderate B lymphocyte proliferation |
| B-Class | **T$^s$C$^s$G$^s$T$^s$C$^s$G$^s$T$^s$T$^s$T$^s$T$^s$G$^s$T$^s$C$^s$G$^s$T$^s$T$^s$T$^s$T$^s$G$^s$T$^s$C$^s$G$^s$T$^s$T** Moderate interferon induction & strong B lymphocyte proliferation |
| C-Class | **T$^s$C$^s$G$^s$T$^s$C$^s$G$^s$T$^s$T$^s$T$^s$T$^s$C$^s$G$^s$G$^s$C$^s$G$^s$C$^s$G$^s$C$^s$G$^s$C$^s$C$^s$G** **G$^s$C$^s$C$^s$G$^s$C$^s$G$^s$C$^s$G$^s$C$^s$G$^s$G$^s$C$^s$T$^s$T$^s$T$^s$T$^s$G$^s$C$^s$T$^s$G$^s$C$^s$**; mixture of two oligonucleotides that form a duplex with the purple sequences. Strong interferon induction & strong B lymphocyte proliferation |

to TLR9 and efficiency of activation may be influenced by base modifications. Phosphorothioate linkages are usually used to warrant stability against degradation, especially with B- and C-class CpG ODNs [140]. With A-class ODNs that form more complicated secondary and tertiary structures, utilization of phosphorothioate-phosphodiester copolymers may be advantageous. The $R_P$ linkage is more immune stimulatory than the $S_P$ linkage [141], in spite of the fact that the latter is degraded more slowly by endonucleases. It is worth noting that phosphorothioate linkage is not inherently more immune stimulatory than unmodified phosphodiester linkage, but its utilization is based on longer biological half-lives. Immune response to neutral methylphosphonate and N3'-phosphoramidate linkages is weaker [142]. A single methylphosphonate CpG linkage in an otherwise all-phosphorothioate ODN is, however, well tolerated [143].

Several variations in the structure of the CpG motif markedly reduce, but do not entirely abolish, the immune response. These include switching of the positions of C and G, C5-substitution or N$^4$-alkylation of the cytosine base, and replacing of either C or G with a universal base (3-nitropyrrole and 5-nitroindole, respectively) or an abasic sugar moiety analog [144–146]. In summary, either cytosine or guanine base in the correct position plays an important role for immune stimulation, but both of them are not absolutely required. Evidently, C5 and N$^4$ are the crucial sites in cytosine base. The guanine base most likely becomes recognized by the N7 and O$^6$ sites, since hypoxanthine and 6-thioguanine, but not 7-deazaguanine, in place of guanine result in immune response. With 2-aminopurine, 2,6-diaminopurine and 8-oxoguanine analogs, the response is half of that produced by guanine. Evidently, the guanine base is recognized by a Hoogsteen type interaction. As regards sugar moiety modifications at the CpG site, the *S*-type ring puckering of 2'-deoxysugar is clearly favored. 2'-OMe substitution decreases the immune stimulation by 50% and LNA modification abolishes it [147, 148]. This concerns only the vicinity of the CpG site. Outside this region, these modifications enforce the immune response. Figure 11.10 summarizes the structural features essential for binding to human TLR9.

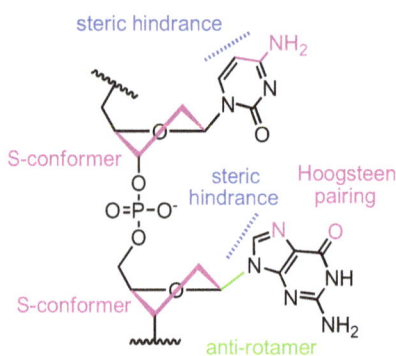

**Figure 11.10:** Structural features essential for binding of CpG motif to TLR9.

The most extensively studied immunostimulatory ODN drug candidates are B-class phosphorothioates injected subcutaneously [149]. They have shown to enhance anticancer chemotherapy in mouse models and entered to clinical trials in several cancer indications, including breast cancer, renal cell cancer, melanoma, nonsmall-cell lung cancer and non-Hodgkin's lymphoma [150, 151]. However, no CpG ODN drug has so far been approved.

## 11.6 G4-quadruplex structures as drug targets

As discussed in Section 11.2, telomeres, the single-stranded overhangs of chromosomes, consist of thousands of nucleotides long TTdAdGdGdG repeats that fold to G4-quadruplex structures. Upon DNA replication in somatic cells, telomere is shortened, which finally leads to cell death. Cancer cells, however, upregulate a telomerase enzyme that catalyzes the synthesis of new TTdAdGdGdG avoiding cell death. Stabilization of the G-quadruplex structure by small molecular compounds inhibits the action of telomerase [152]. These inhibitors usually are planar aromatic molecules that stack with the G-tetrads and by positively charged side chains interact electrostatically with the phosphodiester linkages [153].

Telomeres are not, however, the only G-rich sites in genome. The promoter region of proto-oncogenes of carcinogenic pathways is often G-rich, too [154]. A G4 structure may block interaction with RNA polymerase preventing transcription. But how to affect selectively only to a desired transcription process? Conjugation of the G4-stabilizing compound to an oligonucleotide offers, at least in principle, a way to improve selectivity by base pairing of the oligonucleotide moiety with the loops and flanking sequences of the G-quadruplex [155]. In other words, oligonucleotide may serve as a carrier that bring the G-stabilizing compound close enough to the target quadruplex, but a G-rich oligonucleotide may additionally invade into the target quadruplex forming a new more stable quadruplex. Alternatively, oligonucleotides might be used as carriers of G-quadruplex destabilizing agents. It has been recently shown that amphiphilic molecules containing a polar head and a long hydrocarbon tail destabilize G-quadruplex [156].

Some approaches have quite recently passed to clinical studies. The most widely known example is a G4-forming oligonucleotide targeted to nucleolin, a phosphoprotein highly expressed both intracellularly and on the cell surface in several cancers [157]. Another interesting feature for G4-forming oligonucleotides is their good cellular internalization and good nuclease resistance. Accordingly, they have received interest as carriers of small molecules into cancer cells [158].

## Further reading

Asanuma H, Kamiya Y, Kashida H, Murayama K. Xeno nucleic acids (XNAs) having non-ribose scaffolds with unique supramolecular properties. Chem Commun 2022, 58, 3993–4004.

Bekkouche I, Shishonin AY, Vetcher AV. Recent development in biomedical applications of oligonucleotides with triplex-forming ability. Polymers 2023, 15, 858 (17 pages).

Bodera P. Immunostimulatory oligonucleotides. Recent Pat Inflamm Allergy Drug Discov 2011, 5, 87–93.

Cadoni E, De Paepe L, Manicardi A, Madder A. Beyond small molecules: Targeting G-quadruplex structures with oligonucleotides and their analogues. Nucleic Acids Res 2021, 49, 6638–6659.

Crooke ST, Baker BF, Crooke RM, Liang X-H. Antisense technology: An overview and prospectus. Nat Rev Drug Discov 2021, 20, 427–453.

Hari Y, Obika S, Imanishi T. Towards the sequence-selective recognition of double-stranded DNA containing pyrimidine-purine Interruptions by triplex-forming oligonucleotides. Eur J Org Chem 2012, 0, 2875–2887.

Havens MA, Hastings ML. Splice-switching antisense oligonucleotides as therapeutic drugs. Nucleic Acids Res 2016, 44, 6549–6563.

Kole R, Krainer AR, Altman S. RNA therapeutics: Beyond RNA interference and antisense oligonucleotides. Nat Rev Drug Discov 2012, 11, 125–140.

Kolevzon N, Yavin E. Site-specific DNA photocleavage and photomodulation by oligonucleotide conjugates. Oligonucleotides 2010, 20, 263–275.

McKenzie LK, El-Khoury R, Thorpe JD, Damha MJ, Hollenstein M. Recent progress in non-native nucleic acid modifications. Chem Soc Rev 2021, 50, 5126–5164.

Nan Y, Zhang Y-J. Antisense phosphordiamidate morpholino oligomers as novel antiviral compounds. Front Microbiol 2018, 9, 750 (15 pages).

Rinaldi C, Wood MJA. Antisense oligonucleotides: The next frontier for treatment of neurological disorders. Nat Rev Neurology 2018, 14, 9–21.

Rusling DA, Broughton-Head VJ, Brown T, Fox KR. Towards the targeted modulation of gene expression by modified triplex-forming oligonucleotides. Curr Chem Biol 2008, 2, 1–10.

Shen X, Corey DR. Chemistry, mechanism and clinical status of antisense oligonucleotides and duplex RNAs. Nucleic Acids Res 2018, 46, 1584–1600.

Smith CIE, Zain R. Therapeutic oligonucleotides: State of the art. Annu Rev Pharmacol Toxicol 2019, 59, 605–630.

Uhlmann E. Immune Stimulatory Oligonucleotides. In RSC Biomolecular Sciences. Therapeutic Oligonucleotides. Kurreck J. ed. RSC 2008 Ch 6, 142–162.

Wan WB, Seth PP. The medicinal chemistry of therapeutic oligonucleotides. J Med Chem 2016, 59, 9645–9667.

Zhang JJ. RNA-cleaving DNAzymes: Old catalysts with new tricks for intracellular and in vivo applications. Catalysts 2018, 8, 550–570.

# References

[1] Smith CIE, Zain R. Therapeutic oligonucleotides: State of the art. Annu Rev Pharmacol Toxicol 2019, 59, 605–630.

[2] Moreno PMD, Geny S, Pabon YV, Bergquist H, Zaghloul EM, Rocha CSJ, Oprea II, Bestas B, EL Andaloussi S, Jørgensen PT, Pedersen EB, Lundin KE, Zain R, Wengel J, Smith CIE. Development of bis-locked nucleic acid (bisLNA) oligonucleotides for efficient invasion of supercoiled duplex DNA. Nucleic Acids Res 2013, 41, 3257–3273.

[3] Engholm M, Christensen L, Dueholm KL, Buchardt O, Coull J, Nielsen PE. Efficient pH-independent sequence-specific DNA binding by pseudoisocytosine-containing bis-PNA. Nucleic Acids Res 1995, 23, 217–222.

[4] Kelley ML, Strezoska Z, He K, Vermeulen A, van Brabant Smith A. Versatility of chemically synthesized guide RNAs for CRISPR-Cas9 genome editing. J Biotechnol 2016, 233, 74–83.

[5] Shigi N, Sumaoka J, Komiyama M. Applications of PNA-based artificial restriction DNA cutters. Molecules 2017, 22, 1586 (14 pages).

[6] Kole R, Krainer AR, Altman S. RNA therapeutics: Beyond RNA interference and antisense oligonucleotides. Nat Rev Drug Discov 2012, 11, 125–140.

[7] Uhlmann E, Peyman A. Antisense oligonucleotides: A new therapeutic principle. Chem Rev 1990, 90, 544–579.

[8] Crooke ST. Molecular mechanisms of antisense oligonucleotides. Nucleic Acid Ther 2017, 27, 70–77.

[9] Nakanishi K. Anatomy of RISC: How do small RNAs and chaperones activate Argonaute proteins. Wiley Interdisip Rev RNA 2016, 7, 637–660.

[10] Bartel DP. MicroRNAs: Target recognition and regulatory functions. Cell 2009, 136, 215–233.

[11] Uhlmann E. Immune stimulatory oligonucleotides. In Kurreck J, eds. RSC Biomolecular Sciences. Therapeutic Oligonucleotides. RSC (Royal Society of Chemistry) 2008, Ch 6, 142–162.

[12] Famulok M, Hartig JS, Mayer G. Functional aptamers and aptazymes in biotechnology, diagnostics and therapy. Chem Rev 2007, 107, 3715–3743.

[13] Zamecnik PC, Stephenson ML. Inhibition of Rous sarcoma virus replication and cell transformation by a specific oligodeoxynucleotide. Proc Natl Acad Sci USA 1978, 75, 280–284.

[14] Wan WB, Seth PP. The medicinal chemistry of therapeutic oligonucleotides. J Med Chem 2016, 59, 9645–9667.

[15] Eckstein F. Phosphorothioate oligodeoxynucleotides: What is their origin and what is unique about them?. Antisense Nucleic Acid Drug Dev 2000, 10, 117–121.

[16] Agrawal S, Temsamani J, Tang JY. Pharmacokinetics, biodistribution, and stability of oligodeoxynucleotide phosphorothioates in mice. Proc Natl Acad Sci 1991, 88, 7595–7599.

[17] Geary RS, Yu RZ, Levin AA. Pharmacokinetics of phosphorothioate antisense oligodeoxynucleotides. Curr Opin Investig Drugs 2001, 2, 562–573.

[18] Cerritelli SM, Crouch RJ. Ribonuclease H: The enzymes in eukaryotes. FEBS J 2009, 276, 1494–1505.

[19] Seth PP, Tanowitz M, Bennett CF. Selective tissue targeting of synthetic nucleic acid drugs. J Clin Invest 2019, 129, 915–925.

[20] Iannitti T, Morales-Medina JC, Palmieri B. Phosphorothioate oligonucleotides: Effectiveness and toxicity. Curr Drug Targets 2014, 15, 663–673.

[21] Shen W, De Hoyos CL, Migawa MT, Vickers TA, Sun H, Low A, Bell TA, 3rd, Rahdar M, Mukhopadhyay S, Hart CE, Bell M, Riney S, Murray SF, Greenlee S, Crooke RM, Liang X-H, Seth PP, Crooke ST. Chemical modification of PS-ASO therapeutics reduces cellular protein-binding and improves the therapeutic index. Nat Biotechnol 2019, 37, 640–650.

[22] Vasquez G, Freestone GC, Wan WB, Low A, De Hoyos CL, Yu J, Prakash TP, Ostergaard ME, Liang X-H, Crooke ST, Swayze EE, Migawa MT, Seth PP. Site-specific incorporation of 5′-methyl DNA enhances the therapeutic profile of gapmer ASOs. Nucleic Acids Res 2021, 49, 1828–1839.

[23] Prakash TP, Yu J, Shen W, De Hoyos CL, Berdeja A, Gaus H, Liang X-H, Crooke ST, Seth PP. Sites pecific incorporation of 2',5'-linked nucleic acids enhances therapeutic profile of antisense oligonucleotides. ACS Med Chem Lett 2021, 12, 922–927.

[24] Migawa MT, Shen W, Wan WB, Vasquez G, Oestergaard ME, Low A, De Hoyos CL, Gupta R, Murray S, Tanowitz M, Bell M, Nichols JG, Gaus H, Liang X-H, Swayze EE, Crooke ST, Seth PP. Site-specific replacement of phosphorothioate with alkyl phosphonate linkages enhances the therapeutic profile of gapmer ASOs by modulating interactions with cellular proteins. Nucleic Acids Res 2019, 47, 5465–5479.

[25] Anderson BA, Freestone GC, Low A, De-Hoyos CL, Drury WJ, III, Østergaard ME, Migawa MT, Fazio M, Wan WB, Berdeja A, Scandalis E, Burel SA, Vickers TA, Crooke ST, Swayze EE, Liang X, Seth PP. Towards next generation antisense oligonucleotides: Mesylphosphoramidate modification improves therapeutic index and duration of effect of gapmer antisense oligonucleotides. Nucleic Acids Res 2021, 49, 9026–9041.

[26] Vasquez G, Migawa MT, Wan WB, Low A, Tanowitz M, Swayze EE, Seth PP. Evaluation of phosphorus and non-phosphorus neutral oligonucleotide backbones for enhancing therapeutic index of gapmer antisense oligonucleotides. Nucleic Acid Ther 2022, 32, 40–50.

[27] Koziolkiewicz M, Krakowlak A, Kwinkowski M, Boczkowska M, Stec WJ. Stereodifferentiation––the effect of P chirality of oligo(nucleoside phosphorothioates) on the activity of bacterial RNase H. Nucleic Acids Res 1995, 23, 5000–5005.

[28] Koziolkiewicz M, Wojcik M, Kobylanska A, Karwowski B, Rebowska B, Guga P, Stec WJ. Stability of stereoregular oligo(nucleoside phosphorothioate)s in human plasma: Diastereoselectivity of plasma 3'-exonuclease. Antisense Nucleic Acid Drug Dev 1997, 7, 43–48.

[29] Wan WB, Migawa MT, Vasquez G, Murray HM, Nichols JG, Gaus H, Berdeja A, Lee S, Hart CE, Lima WF, Swayze EE, Seth PP. Synthesis, biophysical properties and biological activity of second generation antisense oligonucleotides containing chiral phosphorothioate linkages. Nucleic Acids Res 2014, 42, 13456–13468.

[30] Iwamoto N, Butler DCD, Svrzikapa N, Mohapatra S, Zlatev I, Sah DWY, Meena, Standley SM, Lu G, Apponi LH, Frank-Kamenetsky M, Zhang JJ, Vargeese C, Verdine GL. Control of phosphorothioate stereochemistry substantially increases the efficacy of antisense oligonucleotides. Nat Biotechnol 2017, 35, 845–851.

[31] Østergaard ME, De Hoyos CL, Wan WB, Shen W, Low A, Berdeja A, Vasquez G, Murray S, Migawa MT, Liang X-H, Swayze EE, Crooke ST, Seth PP. Understanding the effect of controlling phosphorothioate chirality in the DNA gap on the potency and safety of gapmer antisense oligonucleotides. Nucleic Acids Res 2020, 48, 1691–1700.

[32] Meena, Lemaitre MM. Stereocontrolled oligonucleotides for nucleic acid therapeutics: A perspective. Nucleic Acid Ther 2021, 31, 1–6.

[33] Rait VK, Ramsey Shaw B. Boranophosphates support the RNase H cleavage of polyribonucleotides. Antisense Nucleic Acid Drug Dev 1999, 9, 53–60.

[34] Gryaznov SM, Lloyd DH, Chen JK, Schultz RG, DeDionisio LA, Ratmeye L, Wilson WD. Oligonucleotide N3'–>P5' phosphoramidates. Proc Natl Acad Sci USA 1995, 92, 5798–5802.

[35] Pongracz K, Gryaznov S. Oligonucleotide N3'→P5' thiophosphoramidates: Synthesis and properties. Tetrahedron Lett 1999, 40, 7661–7664.

[36] Kandasamy P, Liu Y, Aduda V, Akare S, Alam R, Andreucci A, Boulay D, Bowman K, Byrne M, Cannon M, Chivatakarn O, Shelke JD, Iwamoto N, Kawamoto T, Kumarasamy J, Lamore S, Lemaitre M, Lin X, Longo K, Looby R, Marappan S, Metterville J, Mohapatra S, Newman B, Paik I-H, Patil S, Purcell-Estabrook E, Shimizu M, Shum P, Standley S, Taborn K, Tripathi S, Yang H, Yin Y, Zhao X, Dale E, Vargeese C. Impact of guanidine-containing backbone linkages on stereopure antisense oligonucleotides in the CNS. Nucleic Acids Res 2022, 50, 5401–5423.

[37] Miroshnichenko SK, Patutina OA, Burakova EA, Chelobanov BP, Fokina AA, Vlassov VV, Altman S, Zenkova MA, Stetsenko DA. Mesyl phosphoramidate antisense oligonucleotides as an alternative to phosphorothioates with improved biochemical and biological properties. Proc Natl Acad Sci USA 2019, 116, 1229–1234.

[38] Rait V, Sergueev D, Summers J, He K, Huang F, Krzyzanowska B, Ramsey Shaw B. Boranophosphate nucleic acids—A versatile DNA backbone. Nucleosides Nucleotides 1999, 18, 1379–1380.

[39] Chen J-K, Schultz RG, Lloyd DH, Gryaznov SM. Synthesis of oligodeoxyribonucleotide N3'-P5' phosphoramidates. Nucleic Acids Res 1995, 23, 2661–2668.

[40] Kumar P, Caruthers MH. DNA analogues modified at the nonlinking positions of phosphorus. Acc Chem Res 2020, 53, 2152–2166.

[41] Summerton J, Weller D. Morpholino antisense oligomers: Design, preparation, and properties. Antisense Nucleic Acid Drug Dev 1997, 7, 187–195.

[42] Nan Y, Zhang Y-J. Antisense phosphordiamidate morpholino oligomers as novel antiviral compounds. Front Microbiol 2018, 9, 750 (15 pages).

[43] Egholm M, Buchardt O, Christensen L, Behrens C, Freier SM, Driver DA, Berg RH, Kim SK, Norden B, Nielsen PE. PNA hybridizes to complementary oligonucleotides obeying the Watson-Crick hydrogen-bonding rules. Nature 1993, 365, 566–568.

[44] Nielsen PE. Peptide nucleic acids (PNA) in chemical biology and drug discovery. Chem Biodiv 2010, 7, 786–804.

[45] Zheng H, Botos I, Clausse V, Nikolayevskiy H, Rastede EE, Fouz MF, Mazur SJ, Appella DH. Conformational constraints of cyclopentane peptide nucleic acids facilitate tunable binding to DNA. Nucleic Acids Res 2021, 49, 713–725.

[46] Damidov VV, Potaman VN, Frank-Kamenetskii MD, Egholm M, Buchard O, Sönnichsen SH, Nielsen PE. Stability of peptide nucleic acids in human serum and cellular extracts. Biochem Pharmacol 1994, 48, 1310–1313.

[47] Liczner C, Duke K, Juneau G, Egli M, Wilds CJ. Beyond ribose and phosphate: Selected nucleic acid modifications for structure-function investigations and therapeutic applcations. Beilstein J Org Chem 2021, 17, 908–931.

[48] Manoharan M. 2-Carbohydrate modifications in antisense oligonucleotide therapy: Importance of conformation, configuration and conjugation. Biochim Biophys Acta 1999, 1489, 117–130.

[49] Obika S, Nanbu D, Hari Y, Morio K-I, In Y, Ishida T, Imanishi T. Synthesis of 2'-O,4'-C-methyleneuridine and -cytidine. Novel bicyclic nucleosides having a fixed C3'-endo sugar puckering. Tetrahedron Lett 1997, 38, 873–878.

[50] Koshkin AA, Singh SK, Nielsen P, Rajwanshia VK, Kumara R, Meldgaard M, Olsen CE, Wengel J. LNA (Locked nucleic acids): Synthesis of the adenine, cytosine, guanine, 5-methylcytosine, thymine and uracil bicyclonucleoside monomers, oligomerisation, and unprecedented nucleic acid recognition. Tetrahedron 1998, 54, 3607–3630.

[51] Swayze EE, Siwkowski AM, Wancewicz EV, Migawa MT, Wyrzykiewicz TK, Hung G, Monia BP, Bennett CF. Antisense oligonucleotides containing locked nucleic acid improve potency but cause significant hepatotoxicity in animals. Nucleic Acids Res 2007, 35, 687–700.

[52] Rajwanshi VK, Håkansson AE, Dahl BM, Wengel J. LNA stereoisomers: Xylo-LNA (β-D-xylo configured locked nucleic acid) and α-L-LNA (α-L-ribo configured locked nucleic acid). Chem Commun 1999, 0, 1395–1396.

[53] Fluiter K, Frieden M, Vreijling J, Rosenbohm C, De Wissel MB, Christensen SM, Koch T, Orum H, Baas F. On the in vitro and in vivo properties of four locked nucleic acid nucleotides incorporated into an anti-H-Ras antisense oligonucleotide. ChemBioChem 2005, 6, 1104–1109.

[54] El-Khoury R, Damha MJ. 2'-Fluoro-arabinonucleic Acid (FANA): A Versatile Tool for Probing Biomolecular Interactions. Acc Chem Res 2021, 54, 2287–2297.

[55] Verbeure B, Lescrinier E, Wang J, Herdewijn P. RNase H mediated cleavage of RNA by cyclohexene nucleic acid (CeNA). Nucleic Acids Res 2001, 29, 4941–4947.

[56] Jana SK, Harikrishna S, Sudhakar S, El-Khoury R, Pradeepkumar, Damha MJ. Nucleoside analogues with a seven-membered sugar ring: Synthesis and structural compatibility in DNA−RNA Hybrids. J Org Chem 2022, 87, 2367–2379.

[57] Kajino R, Ueno Y. (S)-5′-C-Aminopropyl-2′-O-methyl nucleosides enhance antisense activity in cultured cells and binding affinity to complementary single-stranded RNA. Bioorg Med Chem 2021, 30, 115925 (7 pages.

[58] Frei S, Katolik AK, Leumann CJ. Synthesis, biophysical properties, and RNase H activity of 6′-difluoro [4.3.0]bicyclo-DNA. Beilstein J Org Chem 2019, 15, 79–88.

[59] Leydier C, Bellon L, Barascut JL, Morvan F, Rayner B, Imbach JL. 4′-Thio-RNA: Synthesis of mixed base 4′-thio-oligoribonucleotides, nuclease resistance, and base pairing properties with complementary single and double strand. Antisense Res Dev 1995, 5, 167–174.

[60] Opalinska JB, Kalota A, Gifford LK, Lu P, Jen K-Y, Pradeepkumar PI, Barman J, Kim TK, Swider CR, Chattopadhyaya J, Gewirtz AM. Oxetane modified, conformationally constrained, antisense oligodeoxyribonucleotides function efficiently as gene silencing molecules. Nucleic Acids Res 2004, 32, 195791–195799.

[61] Srivastava P, Barman J, Pathmasiri W, Plashkevych O, Wenska M, Chattopadhyaya J. Five- and six-membered conformationally locked 2′,4′-carbocyclic ribo-thymidines: Synthesis, structure, and biochemical studies. J Am Chem Soc 2007, 129, 8362–8379.

[62] Seth PP, Vasquez G, Allerson CA, Berdeja A, Gaus H, Kinberger GA, Prakash TP, Migawa MT, Bhat B, Swayze EE. Synthesis and biophysical evaluation of 2′,4′-constrained 2′O-methoxyethyl and 2′,4′-constrained 2′O-ethyl nucleic acid analogues. J Org Chem 2010, 75, 1569–1581.

[63] Koizumi M. 2′-O,4′-C-Ethylene-bridged nucleic acids (ENA) as next-generation antisense and antigene agents. Biol Pharm Bull 2004, 27, 453–456.

[64] Prakash TP, Siwkowski A, Allerson CR, Migawa MT, Lee S, Gaus HJ, Black C, Seth PP, Swayze EE, Bhat B. Antisense oligonucleotides containing conformationally constrained 2′,4′-(N-methoxy) aminomethylene and 2′,4′-aminooxymethylene and 2′-O,4′-C-aminomethylene bridged nucleoside analogues show improved potency in animal models. J Med Chem 2010, 53, 1636–1650.

[65] Rahman SMA, Seki S, Obika S, Yoshikawa H, Miyashita K, Imanishi T. Design, synthesis, and properties of 2′,4′-BNANC: A bridged nucleic acid analogue. J Am Chem Soc 2008, 130, 4886–4896.

[66] Osawa T, Obika S, Hari Y. Synthesis and properties of novel 2′-C,4′C-ethyleneoxy-bridged 2′-deoxyribonucleic acids with exocyclic methylene groups. Org Biomol Chem 2016, 14, 9481–9484.

[67] Ito Y, Nishida K, Tsutsui N, Fuchi Y, Hari Y. Synthesis and properties of oligonucleotides containing 2′-O,4′-C-ethylene-bridged 5-methyluridine with exocyclic methylene and methyl groups in the bridge. Eur J Org Chem 2021, 0, 4993–5002.

[68] Habuchi T, Yamaguchi T, Obika S. Thioamide-bridged nucleic acid (thioAmNA) containing thymine or 2-thiothymine: Duplex-forming ability, base discrimination, and enzymatic stability. ChemBioChem 2019, 20, 1060–1067.

[69] Pandey SK, Wheeler TM, Justice SL, Kim A, Younis HS, Gattis D, Jauvin D, Puymirat J, Swayze EE, Freire SM, Bennett CF, Thornton CA, MacLeod AR. Identification and characterization of modified antisense oligonucleotides targeting DMPK in mice and nonhuman promates for the treatment of myotonic dystrophy type 1. J Pharmacol Exp Ther 2015, 355, 329–340.

[70] Prakash TP, Johnston JF, Graham MJ, Condon TP, Manoharan M. 2′-O-[2-[(N,N-Dimethylamino)oxy] ethyl]-modified oligonucleotides inhibit expression of mRNA in vitro and in vivo. Nucleic Acids Res 2004, 32, 828–833.

[71] Prakash TP, Kawasaki AM, Lesnik EA, Owens SR, Manoharan M. 2′-O-[2-(Amino)-2-oxoethyl] oligonucleotides. Org Lett 2003, 5, 403–406.

[72] Herdewijn P. Heterocyclic modifications of oligonucleotides and antisense technology. Antisense Nucleic Acid Drug Dev 2000, 10, 297–310.

[73] Shen L, Sikowski A, Wancewicz EV, Lesnik E, Butler M, Witchell D, Vasuez G, Ross B, Acevedo O, Inamati G, Sasmor H, Manoharan M, Monia BP. Evaluation of C-5 propynyl pyrimidine-containing oligonucleotides in vitro and in vivo. Antisense Nucleic Acid Drug Dev 2003, 13, 129–142.

[74] Gutierrez AJ, Terhorst TJ, Matteucci MD, Froehler BC. 5-Heteroaryl 2′-deoxyuridine analogs. Synthesis and incorporation into high-affinity oligonucleotides. J Am Chem Soc 1994, 116, 5540–5544.

[75] Lin K-Y, Jones RJ, Matteucci MD. Synthesis and incorporation into oligodeoxynucleotides which have enhanced binding to complementary RNA. J Am Chem Soc 1995, 117, 3873–3874.

[76] Flanagan WM, Wolf JJ, Olson P, Grant D, Lin K-Y, Wagner RW, Matteucci MD. A cytosine analog that confers enhanced potency to antisense oligonucleotides. Proc Natl Acad Sci USA 1999, 96, 3513–3518.

[77] Geary RS, Henry SP, Grillone LR. Fomivirsen: Clinical pharmacology and potential drug interactions. Clin Pharmacokinet 2002, 41, 255–260.

[78] Duell PB, Santos RD, Kirwan B-A, Witzum JL, Tsimikas S, Kastelein JJP. Long term mipomersen treatment is associated with a reduction in cardiovascular events in patients with familial hypercholesterolemia. J Clin Lipidol 2016, 10, 1011–1021.

[79] Raal FJ, Santos RD, Blom DJ, Marais AD, Charng M-J, Cromwell WC, Lachmann RH, Gaudet D, Tan JL, Chasan-Taber S, Tribble DL, Flaim JD, Crooke ST. Mipomersen, an apolipoprotein B synthesis inhibitor, for lowering of LDL cholesterol concentrations in patients with homozygous familial hypercholesterolaemia: A randomised, double-blind, placebo-controlled trial. Lancet 2010, 375, 998–1006.

[80] Benson MD, Dasgupta NR, Monia BP. Inotersen (transthyretin-specific antisense oligonucleotide) for treatment of transthyretin amyloidosis. Neurodegener Dis Manag 2019, 9, 25–30.

[81] Witztum JL, Gaudet D, Freedman SD, Alexander VJ, Digenio A, Williams KR, Yang Q, Hughes SG, Geary RS, Arca M, Stroes ESG, Bergeron J, Soran H,MD, Civeira F, Hemphill L, Tsimikas S, Blom DJ, O'Dea L, Bruckert E. Volanesorsen and triglyceride levels in familial chylomicronemia syndrome. New Engl J Med 2019, 381, 541–542.

[82] Ouellette MM, Wright WE, Shay JW. Targeting telomerase-expressing cancer cells. J Cell Mol Med 2011, 15, 1433–1442.

[83] Herbert B, Pitts AE, Baker SI, Hamilton SE, Wright WE, Shay JW, Corey DR. Inhibition of human telomerase in immortal human cells leads to progressive telomere shortening and cell death. Proc Natl Acad Sci USA 1999, 96, 14276–14281.

[84] Ozsolak F, Milos PM. RNA sequencing: Advances, challenges and opportunities. Nat Rev Genet 2011, 12, 87–98.

[85] Kole R, Krainer AR, Altman S. RNA therapeutics: Beyond RNA interference and antisense oligonucleotides. Nat Rev Drug Discov 2012, 11, 125–140.

[86] Havens MA, Hastings ML. Splice-switching antisense oligonucleotides as therapeutic drugs. Nucleic Acids Res 2016, 44, 6549–6563.

[87] Raguraman P, Wang T, Ma L, Jørgensen PT, Wengel J, Veedu RN. Alpha-L-locked nucleic acid-modified antisense oligonucleotides induce efficient splice modulation In vitro. Int J Mol Sci 2020, 21, 2434 (12 pages).

[88] Evequoz D, Verhaart IEC, van de Vijver D, Renner W, Aartsma-Rus A, Leumann CJ. 7′5′-Alpha-bicyclo-DNA: New chemistry for oligonucleotide exon splicing modulation therapy. Nucleic Acids Res 2021, 49, 12089–12105.

[89] Kole R, Krieg AM. Exon skipping therapy for Duchenne muscular dystrophy. Adv Drug Deliv Rev 2015, 87, 104–107.

[90] Scoto M, Finkel R, Mercuri E, Muntoni F. Genetic therapies for inherited neuromuscular disorder. The Lancet. Child & Adolescent Health 2018, 2, 600–609.

[91] Frank DE, Schnell FJ, Akana C, El-Husayni SH, Desjardins CA, Morgan J, Charleston JS, Sardone V, Domingos J, Dickson G, Straub V, Guglieri M, Mercuri E, Servais L, Muntoni F. Increased dystrophin production with golodirsen in patients with Duchenne muscular dystrophy. Neurology 2020, 94, e2270–e2282.

[92] Clemens PR, Rao VK, Connolly AM, Harper AD, Mah JK, Smith EC, McDonald CM, Zaidman CM, Morgenroth LP, Osaki H, Satou Y, Yamashita T, Hoffman EP. Safety, tolerability, and efficacy of Viltolarsen in boys With Duchenne muscular dystrophy amenable to exon 53 skipping. JAMA Neurol 2020, 77, 982–991.

[93] Shirley M, Casimersen. First approval. Drugs 2021, 81, 875–879.

[94] Faravelli I, Nizzardo M, Comi GP, Corti S. Spinal muscular atrophy – Recent therapeutic advances for an old challenge. Nat Rev Neurosci 2015, 11, 351–359.

[95] Lorson CL, Hahnen E, Androphy EJ, Wirth BA. Single nucleotide in the SMN gene regulates splicing and is responsible for spinal muscular atrophy. Proc Natl Acad Sci USA 1999, 96, 6307–6311.

[96] Hua Y, Vickers TA, Baker BF, Bennett CF, Krainer AR. Enhancement of SMN2 exon 7 inclusion by antisense oligonucleotides targeting the exon. PLoS Biol 2007, 5, e73, 729–744.

[97] Dominski Z, Kole R. Restoration of correct splicing in thalassemic pre-mRNA by antisense oligonucleotides. Proc Natl Acad Sci USA 1993, 90, 8673–8677.

[98] Bennett CF, Baker BF, Pham N, Swayze E, Geary RS. Pharmacology of antisense drugs. Annu Rev Pharmacol Toxicol 2017, 57, 81–105.

[99] Xodo LE, Manzini G, Quadrifoglio F, van der Marel GA, van Boom JH. Effect of 5-methylcytosine on the stability of triple-stranded DNA-a thermodynamic study. Nucleic Acids Res 1991, 19, 5625–5631.

[100] Hildbrand S, Blaser A, Parel SP, Leumann CJ. 5-Substituted 2-aminopyridine C-nucleosides as protonated cytidine equivalents: Increasing efficiency and selectivity in DNA triple-helix formation. J Am Chem Soc 1997, 119, 5499–5511.

[101] Thoung NT, Helene C. Sequence-specific recognition and modification of double-helical DNA by oligonucleotides. Angew Chem Int Ed Engl 1993, 32, 666–690.

[102] Milligan JF, Krawczyk SH, Wadwani S, Matteucci MD. An anti-parallel triple helix motif with oligodeoxynucleotides containing 2′-deoxyguanosine and 7-deaza-2′-deoxyxanthosine. Nucleic Acids Res 1993, 21, 327–333.

[103] Olivas WM, Maher LJ, III. Overcoming potassium-mediated triplex inhibition. Nucleic Acids Res 1995, 23, 1936–1941.

[104] Aubert Y, Perrouault L, Helene C, Giovannangeli C, Asseline U. Synthesis and proreties of triple helix-forming oligodeoxyribonucleotides containing 7-chloro-7-deaza-2′-deoxyguanosine. Bioorg Med Chem 2001, 9, 1617–1624.

[105] Hari Y, Obika S, Imanishi T. Towards the sequence-selective recognition of double-stranded DNA containing pyrimidine-purine Interruptions by triplex-forming oligonucleotides. Eur J Org Chem 2012, 0, 2875–2887.

[106] Huang C-Y, Bi G, Miller PS. Triplex formation by oligonucleotides containing novel deoxycytidine derivatives. Nucleic Acids Res 1996, 24, 2606–2613.

[107] Wang Y, Rusling DA, Powers VEC, Lack O, Osborne SD, Fox KR, Brown T. Stable recognition of TA interruptions by triplex forming oligonucleotides containing a novel nucleoside. Biochemistry 2005, 44, 5884–5892.

[108] Taniguchi Y, Uchida Y, Takaki T, Aoki E, Sasaki S. Recognition of CG interrupting site by W-shaped nucleoside analogs (WNA) having the pyrazole ring in an anti-parallel triplex DNA. Bioorg Med Chem 2009, 17, 6803–6810.

[109] Taniguchi Y, Nakamura A, Senko F, Nagatsugi F, Sasaki S. Effects of halogenated WNA derivatives on sequence dependency for expansion of recognition sequences in non-natural-type triplexes. J Org Chem 2006, 71, 2115–2122.

[110] Rao TS, Hogan ME, Revankar GR. Synthesis of triple helix forming oligonucleotides containing 2′-deoxyformycin. Nucleosides Nucleotides 1994, 13, 95–107.

[111] Garde S, Peters MS, Serrano A, Schrader T. A synthetic methodology toward pyrrolo[2,3-b]pyridones for GC base pair recognition. Org. Lett 2018, 20, 6961–6964.

[112] Alavijeh NS, Serrano A, Peters MS, Wölper C, Schrader T. Design and synthesis of artificial nucleobases for sequence-selective DNA recognition within the major groove. Chem Asian J 2023, 18, e202300637 (9 pages).

[113] Cuenoud B, Casset F, Hüsken D, Natt F, Wolf RM, Altmann KH, Martin P, Moser HE. Dual recognition of double-stranded DNA by 2-aminoethoxy-modified oligonucleotides. Angew Chem Intl Ed Eng 1998, 37, 1288–1291.

[114] Sun BW, Babu BR, Sorensen MD, Zakrzewska K, Wengel J, Sun JS. Sequence and pH effects of LNA-containing triple helix-forming oligonucleotides: Physical chemistry, biochemistry, and modeling studies. Biochemistry 2004, 43, 4160–4169.

[115] Asensio JL, Carr R, Brown T, Lane AN. Conformational and thermodynamic properties of parallel intramolecular triple helices containing a DNA, RNA, or 2-OMe DNA third strand. J Am Chem Soc 1999, 121, 11063–11070.

[116] Gryaznov SM, Lloyd DH, Chen J-K, Schultz RG, De-Dionisio LA, Ratmeyer L, Wilson WD. Oligonucleotide N3′->P5′ phosphoramidates. Proc Natl Acad Sci USA 1995, 92, 5798–5802.

[117] Escude C, Giovannangeli C, Sun JS, Lloyd DH, Chen JK, Gryaznov SM, Garestier T, Helene C. Stable triple helices formed by oligonucleotide N3′→P5′ phosphoramidates inhibit transcription elongation. Proc Natl Acad Sci USA 1996, 93, 4365–4369.

[118] Danielsen MB, Christensen NJ, Jørgensen PT, Jensen KJ, Wengel J, Lou C. Polyamine-functionalized 2′-amino-LNA in oligonucleotides: Facile synthesis of new monomers and high-affinity bibnding towards ssDNA and dsDNA. Chem Eur J 2021, 27, 1416–1422.

[119] Osborne SD, Powers VEC, Rusling DA, Lack O, Fox KR, Brown T. Selectivity and affinity of triplex-forming oligonucleotides containing 2′-aminoethoxy-5-(3-aminoprop-1-ynyl)uridine for recognizing AT base pairs in duplex DNA. Nucleic Acids Res 2004, 32, 4439–4447.

[120] Bentin T, Hansen GI, Nielsen PE. Structural diversity of target-specific homopyrimidine peptide nucleic acid-dsDNA complexes. Nucleic Acids Res 2006, 34, 5790–5799.

[121] Nielsen PE, Egholm M, Buchardt O. Evidence for (PNA)2/DNA triplex structure upon binding of PNA to dsDNA by strand displacement. J Mol Recogn 1994, 7, 165–170.

[122] Talbott JM, Tessier BR, Harding EE, Walby GD, Hess KJ, Baskevics V, Katkevics M, Rozners E, MacKay JA. Improved Triplex-Forming Isoorotamide PNA Nucleobases for A-U Recognition of RNA Duplexes. Chem Eur J 2023, 29, e202302390 (1 of 11).

[123] Economos NG, Oyaghire S, Quijano E, Ricciardi AS, Saltzman EM, Glazer PM. Peptide nucleic acids and gene editing: Perspectives on structure and repair. Molecules 2020, 25, 735 (21 pages.

[124] Rusling DA, Broughton-Head VJ, Brown T, Fox KR. Towards the targeted modulation of gene expression by modified triplex-forming oligonucleotides. Curr Chem Biol 2008, 2, 1–10.

[125] Kolevzon N, Yavin E. Site-specific DNA photocleavage and photomodulation by oligonucleotide conjugates. Oligonucleotides 2010, 20, 263–275.

[126] Nakao J, Mikame Y, Eshima H, Yamamoto T, Dohno C, Wada T, Yamayoshi A. Unique crosslinking properties of psoralen-conjugated oligonucleotides developed by novel psoralen N-hydroxysuccinimide esters. ChemBioChem 2023, 24, e202200789 (1 of 10).

[127] Panattoni A, El-Sagheer AH, Brown T, Kellett A, Hocek M. Oxidative DNA cleavage with clip-phenanthroline triplex-forming oligonucleotide hybrids. ChemBioChem 2020, 21, 991–1000.

[128] Fantoni NZ, McGorman B, Molphy Z, Singleton D, Walsh S, El-Sagheer AH, McKee V, Brown T, Kellett A. Development of gene-targeted polypyridyl triplex-forming oligonucleotide hybrids. ChemBioChem 2020, 21, 3563–3574.

[129] Li Y, Syed J, Sugiyama H. RNA-DNA triplex formation by long noncoding RNAs. Cell Chem Biol 2016, 23, 1325–1333.

[130] Akira S, Takeda K, Kaisho T. Toll-like receptors: Critical proteins linking innate and acquired immunity. Nat Immunol 2001, 2, 675–680.

[131] Krieg AM, Yi AK, Matson S, Waldschmidt TJ, Bishop GA, Teasdale R, Koretzky GA, Klinman DM. CpG motifs in bacterial DNA trigger direct B-cell activation. Nature 1995, 374, 546–549.

[132] Trinchieri G, Sher A. Cooperation of Toll-like receptor signals in innate immune defence. Nat Rev Immunol 2007, 7, 179–190.

[133] Uhlmann E. Immune stimulatory oligonucleotides. In Kurreck J, eds. RSC Biomolecular Sciences. Therapeutic Oligonucleotides. RSC (Royal Society of Chemistry) 2008, Ch 6, 142–162.

[134] Hornung V, Ellegast J, Kim S, Brzozka K, Jung A, Kato H, Poeck H, Akira S, Conzelmann KK, Scheele M, Endres S, Hartmann G. 5′-Triphosphate RNA is the ligand for RIG-I. Science 2006, 314, 994–997.

[135] Hartmann G, Weeratna RD, Ballas ZK, Payette P, Blackwell S, Suparto I, Rasmyssen WL, Waldschmidt M, Sjuthi D, Purcell RH, Davis HL, Krieg AM. Delineation of a CpG phosphorothioate oligodeoxynucleotide for activating primate immune responses in vitro and in vivo. J Immunol 2000, 164, 1617–1624.

[136] Yu D, Zhao Q, Kandimalla ER, Agrawal S. Accessible 5′-end of CpG-containing phosphorothioate oligodeoxynucleotides is essential for immunostimulatory activity. Bioorg Med Chem Lett 2000, 10, 2585–2588.

[137] Elias F, Flo J, Lopez RA, Zorzopulos J, Montaner AD, Rodriguez J. Strong cytosine-guanosine independent immunostimulation in humans and other primates by synthetic oligodeoxynucleotides with PyNTTTTGT motifs. J Immunol 2003, 171, 3697–3704.

[138] Pollak AJ, Zhao L, Crooke ST. Systematic analysis of chemical modifications of phosphorothioate antisense oligonucleotides that modulate their innate immune response. Nucleic Acid Therapeutics 2023, 33, 95–107.

[139] Vollmer J, Weeratna R, Payette P, Jurk M, Schetter C, Laucht M, Wader T, Tluk S, Liu M, Davis HL, Krieg AM. Characterization of three CpG oligodeoxynucleotide classes with distinct immunostimulatory activities. Eur J Immunol 2004, 34, 251–262.

[140] Ballas ZK, Rasmussen WL, Krieg AM. Induction of NK activity in murine and human cells by CpG motifs in oligodeoxynucleotides and bacterial DNA. J Immunol 1996, 157, 1840–1845.

[141] Krieg AM, Guga P, Stec W. P-Chirality-dependent immune activation by phosphorothioate CpG oligodeoxynucleotides. Oligonucleotides 2003, 13, 491–499.

[142] Zhao Q, Temsamani J, Iadarola PL, Jiang Z, Agrawal S. Effect of different chemically modified oligodeoxynucleotides on immune stimulation. Biochem Pharmacol 1996, 51, 173–182.

[143] Yu D, Kandimalla ER, Zhao Q, Cong Y, Agrawal S. Immunostimulatory activity of CpG oligonucleotides containing non-ionic methylphosphonate linkages. Bioorg Med Chem 2001, 9, 2803–2808.

[144] Krieg AM. Mechanisms and applications of immune stimulatory CpG oligodeoxynucleotides. Biochim Biophys Acta 1999, 1489, 107–116.

[145] Kandimalla ER, Yu D, Zhao Q, Agrawal S. Effect of chemical modifications of cytosine and guanine in a cpg-motif of oligonucleotides: Structure–immunostimulatory activity relationships. Bioorg Med Chem 2001, 9, 807–813.

[146] Jurk M, Kritzler A, Debelak H, Vollmer J, Krieg AM, Uhlmann E. Structure-activity relationship studies on the immune stimulatory effects of base-modified CpG toll-like receptor 9 agonists. ChemMedChem 2006, 1, 1007–1014.

[147] Vollmer J, Jepsen JS, Uhlmann E, Schetter C, Jurk M, Wader T, Wullner M, Krieg AM. Modulation of CpG oligodeoxynucleotide-mediated immune stimulation by locked nucleic acid (LNA). Oligonucleotides 2004, 14, 23–31.

[148] Henry S, Stecker K, Brooks D, Monteith D, Conklin B, Bennett CF. Chemically modified oligonucleotides exhibit decreased immune stimulation in mice. J Pharmacol Exp Ther 2000, 292, 468–479.

[149] Krieg AM. Therapeutic potential of Toll-like receptor 9 activation. Nat Rev Drug Discov 2006, 5, 471–484.

[150] Vollmer J. Progress in drug development of immunostimulatory CpG oligodeoxynucleotide ligands for TLR9. Expert Opin Biol Ther 2005, 5, 673–682.

[151] Bodera P. Immunostimulatory oligonucleotides. Recent Pat Inflamm Allergy Drug Discov 2011, 5, 87–93.

[152] Ruden M, Puri N. Novel anticancer therapeutics targeting telomerase. Cancer Treat Rev 2013, 39, 444–456.

[153] O'Hagan MP, Morales JC, Galan MC. Binding and beyond: What else can G-quadruplex ligands do?. Eur J Org Chem 2019, 0, 4995–5017.

[154] Rhodes D, Lipps HJ. Survey and summary G-quadruplexes and their regulatory roles in biology. Nucleic Acids Res 2015, 43, 8627–8637.

[155] Cadoni E, De Paepe L, Manicardi A, Madder A. Beyond small molecules: Targeting G-quadruplex structures with oligonucleotides and their analogues. Nucleic Acids Res 2021, 49, 6638–6659.

[156] Bisoi A, Sarkar S, Singh PC. Hydrophobic interaction-induced topology-independent destabilization of G-quadruplex. Biochemistry 2023, 62, 23, 3430–3439.

[157] Yazdian-Robati R, Bayat P, Oroojalian F, Zargari M, Ramezani M, Taghdisi SM, Abnous K. Therapeutic applications of AS1411 aptamer, an update review. Int J Biol Macromol 2020, 155, 1420–1431.

[158] Santos T, Pereira P, Campello MPC, Paulo A, Queiroz JA, Cabrita E, Cruz C. RNA G-quadruplex as supramolecular carrier for cancer-selective delivery. Eur J Pharm Biopharm 2019, 142, 473–479.

# 12 Nucleic acids as drugs and drug targets 2: RNA in chemotherapy

## 12.1 Small interfering RNAs (siRNAs)

Eukaryotic cells have a conserved machinery with which to respond to penetration of foreign double-stranded RNA into the cell. This machinery protects them against exogenous pathogenic nucleic acids [1]. Exposure to only a few molecules of dsRNA per cell triggers systemic silencing, that is, silencing of gene expression throughout the whole animal. The dsRNA is first cleaved by an enzyme called Dicer to around 21-nucleotide-long fragments that typically consist of a 19-nucleotide duplexed region, phosphorylated 5'-end, free hydroxyl group at the 3'-end and 2 nucleotide overhangs at both termini (Figure 12.1) [2, 3]. These fragments are known as small interfering RNAs (siRNAs). They mediate gene silencing by formation of a ribonucleoprotein complex, RNA-induced silencing complex (RISC) [4]. During the course of this process, one of the strands in siRNA (sense strand or passenger strand) becomes cleaved, while the other strand (antisense or guide strand) remains bound to the protein complex and serves as a template for recognition of mRNA [5]. An essential component of this complex is Argonaut 2 (Ago2) RNase that cleaves mRNA at the middle of the sequence complementary to the antisense strand [6]. Human Ago2 is an 859-amino acid-residue-long protein that contains binding domains for the 5'-terminus and 3'- overhang of the antisense strand and a catalytic domain for the cleavage of the sense strand or mRNA [7]. Which one of the strands in siRNA serves as an antisense strand is determined by the thermodynamic stability of the duplex close to the terminus [8]. More precisely, siRNA becomes bound to the RISC machinery in such a way that the 5'-end of the antisense strand pairs less tightly with the sense strand than does the 3'-end. The reason is that dicer acts in concert with another protein, in mammal cells with HIV-transactivating response RNA-binding protein (TRBP). This protein binds to the thermodynamically more stable terminus and, hence, determines the orientation of loading to the RISC complex.

Dicer-catalyzed cleavage of a long dsRNA yields a large number of various siRNAs, resulting in silencing of multiple genes and most likely cell death, apoptosis. The same machinery may, however, be used for medical purposes as well. If instead of a long dsRNA only a single siRNA is introduced, a selective gene silencing takes place. The sequence complementarity between the 5'-end of the antisense strand and mRNA determines the selectivity of gene silencing [9]. The critical nucleotides are those in positions 2–8, counting from the 5'-terminus of the antisense strand. They form the so-called seed region that largely determines mRNA specificity [10] (Figure 12.2A). Owing to the fact that the seed region consists of only seven nucleotides, the risk for off-target effects, that is, silencing of genes not aimed at being silenced, is obvious. Fortunately, the risk appears to be reduced by introduction of a modified nucleotide within the seed region [11–13]. According to a recent report [14], replacement of one phosphodiester linkage in

https://doi.org/10.1515/9783111325637-012

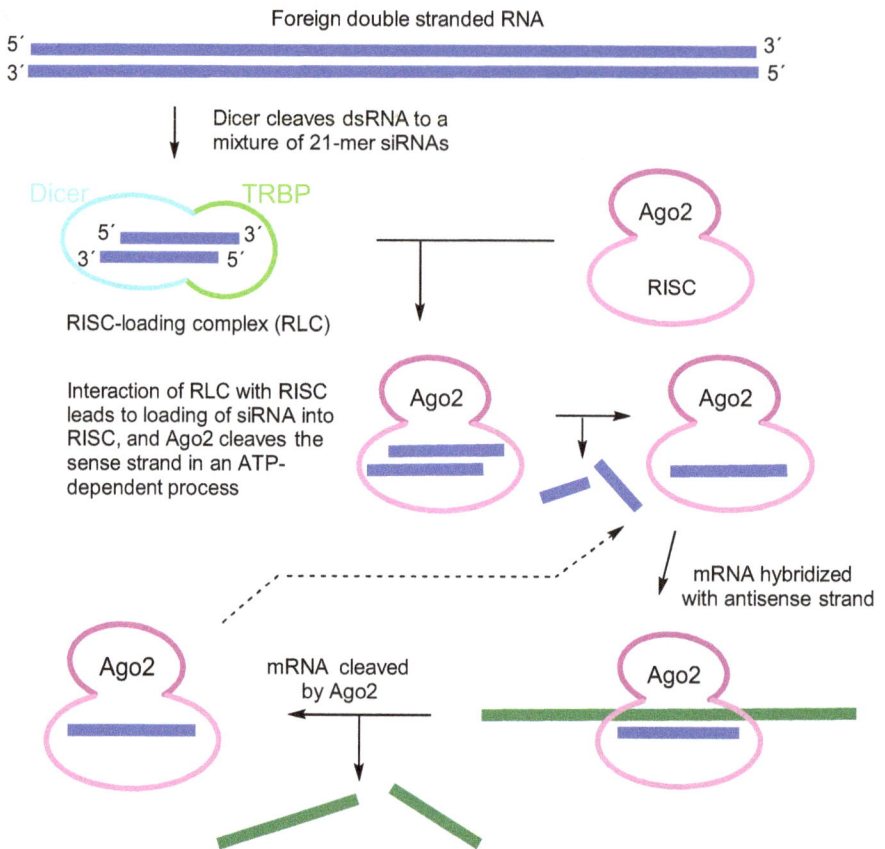

**Figure 12.1:** Mechanism of gene silencing triggered by foreign dsRNA.

the seed region with an amide linkage suppresses the off-target activity, leaving the on-target activity unchanged. Besides the seed region, complementarity outside this region still plays a role. In particular, nucleotides 10 and 16 in the antisense strand have been argued to be of special importance [15]. The preferred cleavage site is U paired with nucleotide 10 counted from the 5′-terminus of the antisense strand.

Like antisense oligonucleotides, siRNAs are potential drugs with which gene expression could be inhibited in a selective manner. In fact, siRNA induced gene silencing appears to be even more efficient than that obtained with antisense oligonucleotides. Selectivity still is a partly open question; how severe problem off-target effect really is. The problems that should be overcome on the way to medical applications are largely the same as in antisense strategy. The half-life of unmodified siRNA is only minutes in plasma [16] and, hence, stabilization toward nucleases obviously is of primary importance. Another important aspect is the influence of modifications on the interaction of both the sense and antisense strands with the enzymatic machinery, above all with

A

Higher duplex stability     Cleavage site     Lower duplex stability

Sense    5′ $^{2-}O_3PO$ - N N N N N N N N N N N N N N N N N N N N - OH   3′

Antisense   3′ HO - N N N N N N N N N N N N N N N N N N N N - $OPO_3^{2-}$   5′

**N** marked influence on dupled stability      **N** seed region

B

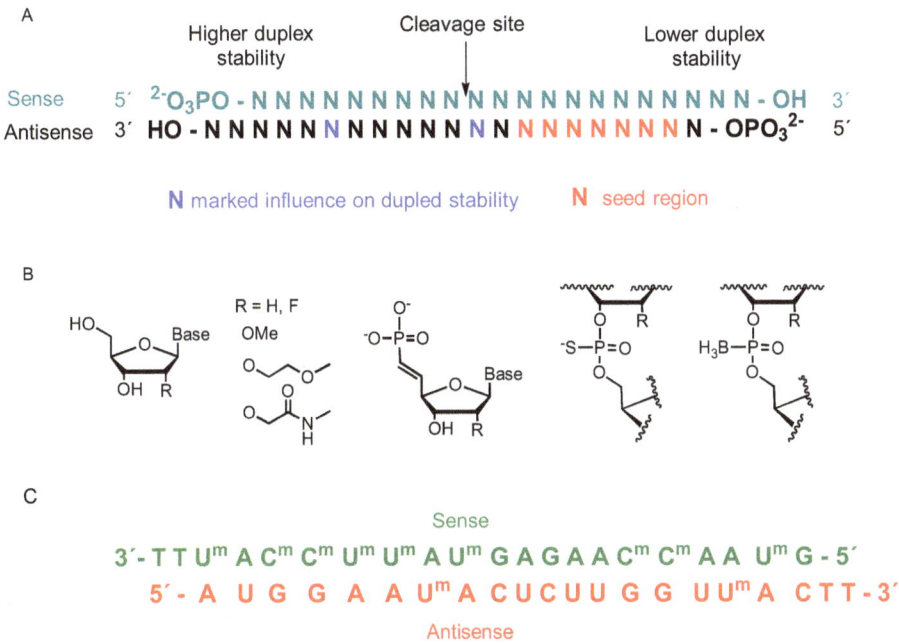

C

Sense

3′- T T U$^m$ A C$^m$ C$^m$ U$^m$ U$^m$ A U$^m$ G A G A A C$^m$ C$^m$ A A U$^m$ G - 5′

5′- A U G G A A U$^m$ A C U C U U G G U U$^m$ A C T T - 3′

Antisense

**Figure 12.2:** (A) Structure of siRNA, (B) modifications frequently used in candidates of therapeutic siRNAs and (C) structure of patisiran, the first siRNA approved for clinical use. U$^m$ and C$^m$ refer to 5-methyl-uridine and -cytidine, respectively.

Ago2 [17]. 2′-*O*-Me, 2′-*O*-MOE, 2′-deoxy-2′-fluoro, and 5-methyluridine and -cytidine are frequently used to warrant sufficient stability (Figure 12.2B) [18, 19]. LNA is also stabiliz-ing but the stabilization is not sufficient. The sense strand can often be modified more extensively than the antisense (guide) strand. The 2′-deoxy-2′-fluoro RNA modification is, however, tolerated rather well even in the antisense-strand. The sugar modifications also retard immune stimulation, except the sometimes used 4′-thio substitution [20].

    The 5′-phosphate group is essential for recognition and binding of the antisense strand to Ago2 [5]. The group is, however, susceptible to enzymatic hydrolysis. To avoid dephosphorylation, 5′-(*E*)-vinylphosphonate group has been introduced as a metaboli-cally stable surrogate of the 5′-phosphate group [21]. Another important observation is that 2′-deoxyribo- or arabino-nucleotides having S-type ring puckering are well toler-ated as the 5′-terminal nucleotides of the antisense strand [22]. Binding of the 5′-phosphorylated terminus of the antisense strand to Ago2 evidently forces the first nu-cleoside to adopt an S-type conformation instead of the normal N-conformation of ribo-nucleosides. H-bonding of the 2′-*O*-substituent to Ago2 may also facilitate the required conformational change from N to S. This likely is the reason for positive effect of 2′-*O*-(2-metylamino-2-oxoethyl) substituent on gene-silencing in vivo [23].

Phosphorothioate linkages are used in overhang positions to protect against exonucleases and to strengthen the interaction with blood components, which is important for pharmacokinetics [24]. The stereochemistry of these phosphorothioate linkages appears to play a role. In mouse model, the efficiency of the antisense effect was enhanced by $R_P$ stereochemistry at the 5'- and $S_P$ stereochemistry at the 3'-terminus of the antisense strand [25]. Boranophosphates have also been used at overhangs and additionally even in double helical region, with the exception of the central part of the antisense-strand [26, 27]. Cell-specific targeting and cellular uptake are enhanced by covalent conjugation, as discussed in more detail in Chapter 13. All termini, except the 5'-end of the antisense strand, may be used for conjugation. Circularization of siRNA stabilized the sense strand, but weakened binding to the guide strand [28]. In vivo studies on the effect of circularization still are limited.

So far five siRNA drugs, patisiran (brand name ONPATTRO), givosiran (GIV-LAARI), lumasiran (OXLUMO), inclisiran (LEQVIO) and vuttrisiran (AMVUTTRA) have been approved for clinical use [19]. Patisiran is a duplex of two 21-mer oligonucleotides that consist of a 19-mer duplex ORN sequence and a 3'-terminal TpT overhang [29] (Figure 12.2C). All internucleosidic linkages are unmodified phosphodiester bonds. All pyrimidine ribonucleosides in the sense strand and two uridines in the antisense strand are 2'-O-methylated. The purine nucleosides are unmodified. The drug is used for treatment of hereditary transthyretin amyloidosis, that is, for the same purpose as inotersen AON (cf. Section 11.2). In both cases the drug target is the transthyretin mRNA that is cleaved either by RISC (patisiran) or by RNase H (inotersen). Patisiran is delivered as lipid nanoparticles (cf. Section 13.2).

Inclisiran [30] has been developed for treatment of hypercholesterolemia by inhibition of the synthesis of PCSK9 protein (proprotein convertase subtilisin–kexin type 9), resulting in lowering in the level of low-density lipoprotein (LDL) cholesterol. Givosiran [31], in turn, is targeted to hepatic 5-aminolevulinic acid synthase 1 (ALAS1) to treat a rare type of acute hepatic porphyrias, diseases in which porphyrins build up negatively affects nervous system. Both are 21-mer ORN duplexes having a two nucleotide overhang either at the 3'-terminus of the sense strand (inclisiran) or at the 3'-terminus of the antisense strand (givosiran). The 3'-terminus of the sense (inclisiran) or antisense strand (givosiran) additionally bears a trivalent N-acetylgalactosamine conjugate group to target the drug to asialoglycoprotein receptor in liver (cf. Section 13.3). All the nucleosides are either 2'-O-methyl or 2'-deoxy-2'-fluoro derivatives, making the siRNA highly stable toward nuclease degradation, but increasing the toxicity risk. Inclisiran contains one 2'-deoxynucleoside in the middle of the antisense strand. The internucleosidic linkages are largely unmodified phosphodiester bonds. The two terminal linkages are, however, phosphorothioates, except at the 3'-terminus that bears the glyco conjugate.

Lumasiran is aimed at treatment of primary hyperoxaluria, i.e., increased excretion of oxalate, a rare disease resulting in formation of oxalate stones [32]. Vutrisiran, the fifth siRNA drug, is used for treatment of the polyneuropathy of hereditary transthyretin-mediated (hATTR) amyloidosis. It interferes with the expression of the trans-thyretin (TTR) gene that encodes transthyretin, a serum protein in liver whose major function is transport of vitamin A and thyroxine [33]. Mutations in this gene, that are rather rare, result in accumulation of amyloid protein aggregates. Lumasiran and Vutrisiran also bear a triantennary *N*-acetyl-galactosamine group at the 3'-terminus of the sense strand.

## 12.2 MicroRNAs

While siRNAs are potential drugs, the RNAi machinery is at the same time an attractive drug target. Numerous genes in human genome encode synthesis of micro RNAs (miRNA), that is, short hairpin RNAs having usually an internal loop within the double helical stem (Figure 12.3) [34]. Strictly speaking, the miRNA genes encode precursors of miRNA, 1–3 kDa size primary miRNA (pri-miRNA) [35]. This is a single stranded RNA chain incorporating stem-loops that eventually become processed to miRNA. An RNase enzyme Drosha, complexed with a double-stranded RNA-binding protein Pasha, first cleaves, still in nucleus, pri-miRNAs into hairpin structures consisting of 70–100 nucleotides [36]. The pre-miRNAs thus formed are transported into cytoplasm through an Exportin-5 dependent pathway [37]. In cytoplasm, they are further processed by Dicer to 18- to 24-nucleotide-long miRNA duplexes [38]. Like siRNAs, these miRNA duplexes have 2 nucleotide overhangs but, unlike siRNAs, the double helix is not fully complementary but contains an internal loop. One of the strands is then incorporated into a miRNA-induced silencing complex (miRISC) and the other strand is rejected, often by degradation [39]. Thermodynamic properties of the duplex again determine which one of the strands becomes accepted. The miRISC complex then binds to the 3'-untranslated region (3'-UTR) of mRNA. In case mRNA and miRNA are fully complementary, mRNA is cleaved. Otherwise, mRNA becomes passively arrested, not cleaved [40]. Passive arresting also results in partial translational inhibition and is more common among these two mechanisms. In addition, several non-canonical Drosha- or Dicer-independent pathways for formation of miRNAs have been identified [41]. For example, spliced introns may serve as pre-miRNA avoiding Drosha-catalyzed tailoring.

Human genome encodes 600 miRNAs [42], although even higher figures have also been reported [43]. According to bioinformatics estimations, they influence on even 60% of protein encoding genes in the human genome [44]. The reason is that a single miRNA might interact with multiple genes, since imperfect match may be sufficient to trigger the biological response. miRNAs play a role in numerous biological processes including, for example, cellular differentiation [45], proliferation [46] and apoptosis [47]. It is, hence, only natural that abnormal expression levels of miRNAs serve as indicators of various diseases, and miRNAs are potential drug targets. Increases and decreases in levels of miRNA can, for example, either cause cancer or suppress it [48]. Intensive efforts have been made to restore normal miRNA spectrum [49]. Single

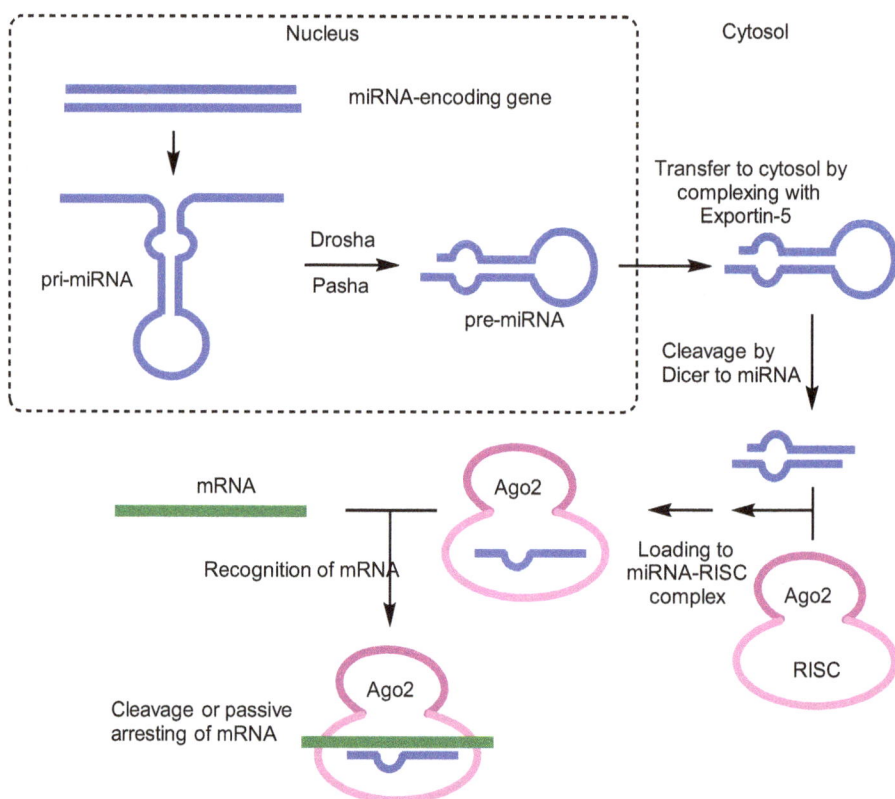

**Figure 12.3:** Biosynthesis and function of micro RNAs.

stranded oligonucleotides can be used to bind to miRISC instead of mRNA and, hence, to inhibit the function of miRNA. Alternatively, miRNA mimics can be used to increase the miRNA levels.

Oligonucleotides that are aimed at hybridizing with miRNA engaged in the RISC complex are called antagomirs [50]. They do not activate RNase H but prevent the action of miRISC complex by passive arresting. The modifications commonly used include 2′-deoxy-2′-fluoro-RNA, 2′-O-Me-RNA, 2′-O-MOE-RNA, LNA and DNA nucleosides, and phosphorothioate linkages. With antagomirs, the risk of off-target effects is at least as obvious as with siRNA [51]. Since antagomirs have to recognize miRNA already engaged in a protein complex, the role of the seed region, that is, nucleotides 2–8 from the 5′-terminus, dominates, leading to a reduced specificity [52]. Only few antagomirs have so far entered phase II clinical trials [53]. Miravirsen (RG-101), is a 15-mer phosphorotioate ON containing LNA and DNA and 5-methylcytosine modifications. It is an antagomir for miR-122, aimed at treatment of Hepatitis C infection [54]. The clinical trials have, however, been halted. The other phase II studies include Antagomir RG-102 targeted to miR-21 against Alport syndrome, a genetic kidney disease

[55], MRG-106 targeted to miR-155 against cutaneous T cell lymphoma [56] and MRG-201 targeted to miR-29 against cutaneous fibrosis [57].

MiRNAs are also promising biomarkers of cancers [58, 59]. They can be oncogenic (oncomiRs) or tumor suppressors (TS-miRs). Most cancers display oncomiRs (e.g., miR-17–22, miR-125b, and miR-125) that block the translation of TS-miRs (e.g., miR-133a, miR-145, and miR-143), which leads to development of cancer. Accordingly, antisense-miRNAs that repress oncomiRs are potential cancer drugs, but still largely in pre-clinical phase [60].

## 12.3 2′,5′-Oligoadenylates (2-5A-ORNs)

Short 2′,5′-linked oligoribonucleotides (2–5A-ORNs) comprise still one class of oligonu-cleotides expected to have potential in chemotherapy. Such oligomers play role in one of the principle pathways of interferon (INF) response to viral infections [61]. As de-picted in Figure 12.4, INF activates two enzymes, 2′,5′-oligoadenylate synthetase (OAS) [62] and ribonuclease L (RNase L) [63]. Double stranded replicates of viral RNA acti-vate OAS that catalyze polymerization of ATP to short 2–5As-ORNs, above all 5′-triphosphate of trimer (A2′p5′)$_2$A [64]. The latter then activates normally latent RNase L to a dimeric endoribonuclease [65] that cleaves single stranded RNA, both viral and native RNA, within U-rich sequences [66]. 2′-Phosphodiester phosphatase, in turn, hy-drolyzes 2–5A-ORNs, resulting in termination of single stranded RNA degradation upon extinguishment of INF activation.

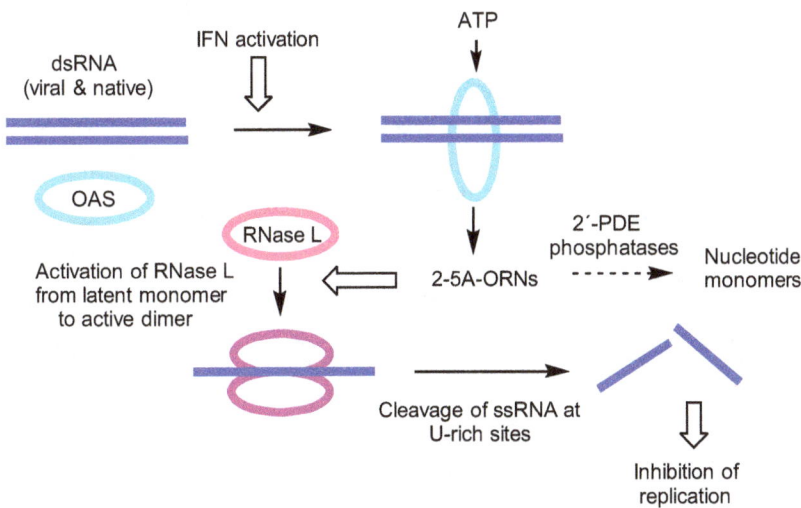

**Figure 12.4:** Inhibition of RNA replication via an interferon induced pathway by activation of 2′,5′-oligoadenylate synthetase and ribonuclease L.

2–5A-ORNs have been shown to inhibit infections of many RNA viruses [61, 67–69]. Their biological life-time, however, is short due to cleavage by 2'-phosphodiesterases and 5'-phosphatases [70]. Masking of the phosphate groups is required to enhance cellular uptake. The therapeutic potency depends on success of structural modifications. The structural requirements for binding and activation of RNase L are rather well known (Figure 12.5). Trimeric (or longer) structure is essential [71]. 5'-Phosphate and internucleosidic 2',5'-phosphodiester linkages are also required [72], but may be replaced with phosphoromonothioate groups [73]. Among the hydroxyl functions, the 3'-OH of the intervening unit is the only important one [74]. Additionally, the N1 and $N^6$ sites of the 5'- and 2'-terminal adenosines are required for binding and activation of RNase L, respectively [72, 74]. Accordingly, the number of allowed structural modifications is rather limited. In fact, only the 2'-terminal sugar moiety can be extensively modified. Figure 12.5 shows examples of 2–5A-ORN analogs with enhanced biological stability without compromise with RNase L activation. To minimize nonspecific RNA

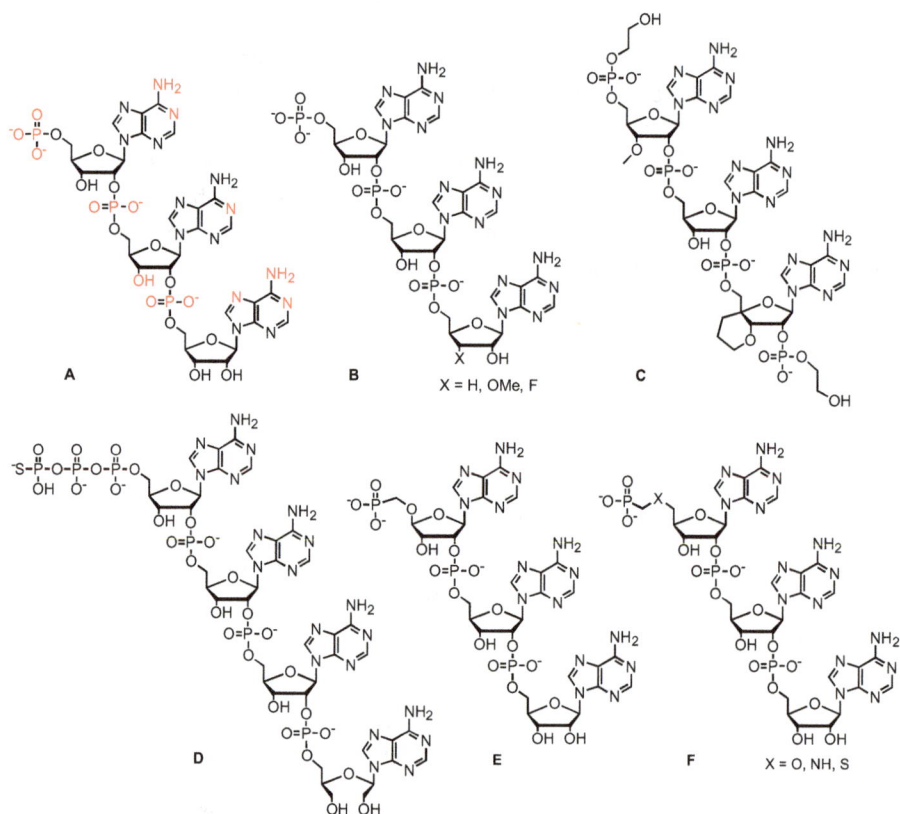

**Figure 12.5:** (A) Structure of trimeric 2–5A-ORN 5'-monophosphate. Atoms essential for binding and/or activation for RNase L are shaded in red. (B-F) Structural analogs shown to activate RNase L: B, X = H [76], B, X = OMe [77], B, X = F [78], C [79], D [80], E [81] and F[81].

degradation, 2–5A-ORNs have been conjugated to antisense oligonucleotides. Since activation of RNase L occurs through binding to 2–5A trimer, the target RNA of the conjugated ASO becomes predominantly cleaved [75].

2–5A-ORNs have been used to combat against numerous viral infections. Several mechanisms have been suggested to operate in parallel [61]. In addition to cleavage of viral single stranded RNA genome, viral encoded mRNAs as well as cellular mRNAs and rRNA required for viral replication are degraded. Infected cells undergo apoptosis and the small RNAs generated by RNase L enhance INF production. The therapeutic influence has been demonstrated in case of respiratory syncytial virus with primate animal models. Applications against cancer have additionally been described [75]. None of the drug candidates have so far passed to major clinical trials.

## 12.4 Site-directed RNA editing

More than half of genetic disorders are consequences of point mutations [82]. Site-directed editing of RNA offers an approach for correction of some of them, not permanently at the genomic level but transiently at mRNA level. Some of common modifications, such as 2′-O-methylation and oxidation of adenosine to inosine, can be obtained by RNA-guided pathways, which means that they potentially are of interest for therapeutic applications.

Enzymatic deamination of adenosine (A) to inosine (I, becomes read as G) or cytidine (C) to uridine (U) has been of special interest [83]. As regards drug development, the crucial question is how to do this site specifically, which evidently is a prerequisite for drug development. In mammals, two enzymes, adenosine deaminases acting on double stranded RNA (ADAR1 and ADAR 2), catalyze A to I conversion [84]. Among these, ADAR2 has been mainly used for development of therapeutic applications. It is a modular enzyme that consists of the catalytic deaminase domain at C-terminus and two dsRNA-recognizing domains at N-terminus. The deamination is unspecific [85], but has been converted site-specific by replacing the dsRNA-recognizing domain with an antisense oligonucleotide serving as a guideRNA (gRNA) [86]. To achieve this, the deaminase domain has been fused with λ-phage N protein that is an only 22-amino acid-long polypeptide known to bind tightly to short RNA stem loops. When gRNA bears a terminal stem loop, a stable deaminase-(λ-N)-gRNA construct is obtained. Unfortunately, some off-target deamination has still occurred [87].

Several other techniques have also been used for linking ADAR deaminase to gRNA [83]. For example, ADAR deaminase has been fused to a SNAP-tag instead of λ-N. SNAP-tag is a 182-amino acid-long polypeptide that binds to $O^6$-benzylguanine bearing targets by displacing guanine and forming a covalent sulfide linkage to benzylated target [88]. Accordingly, this approach enables covalent binding of ADAR-SNAP conjugate to $O^6$-benzylguanine-gRNA. ADAR deaminase has also been fused to bacteriophage MS2 coat protein (MCP) that specifically binds to a special RNA stem loop called MS2 [89]. Accordingly, the gRNA when derivatized with this MS2 stem loop becomes non-covalently anchored to ADAR-MS2 conjugate. A slightly different approach consists of utilization of a gRNA that not only finds the correct sequence in target RNA, but is also able to recruit the endogenous ADAR. The latter aim is realized by attaching the antisense moiety of gRNA to a 28 nucleotides stem loop known to be recognized by ADAR2 [90]. As regards site-directed deamination of cytidine to uridine, a cytidine deaminase (APOBEC3A) has been fused to a deactivated dCas13 RNase that does not catalyze RNA cleavage but still binds to single stranded RNA directed by a CRISPR RNA guide (see Section 13.5) [91].

In addition to deaminase-based RNA editing, 2′-O-methylation and pseudouridylation have received interest as modifications that can be site-specifically generated by an RNA-guided mechanism [92]. Investigations toward their therapeutic applications are still at a very early stage.

## 12.5  In vitro transcribed mRNA in chemotherapy

The potential of in vitro transcribed mRNA as a therapeutic agent is obvious, but several hurdles related to preparation, stability, immune activation and, above all, delivery into cells still have to be overcome on the way to mRNA drugs. One important step has already been taken. The success of mRNA vaccines in combat against SARS2 virus has greatly strengthened confidence in mRNA, not only as a vaccine, but also as a drug.

Like eukaryotic mRNA, mRNA vaccines consist of a 7-methyl guanosine ($m^7G$) cap at the 5′-terminus, a poly(A) tail at the 3′-terminus, and a single stranded RNA sequence that contains a translated open reading frame (ORF) region flanked by 3′- and 5′-untranslated regions (3′- and 5′-UTRs) [93]. The $m^7G$-cap is linked to the 5′-terminal nucleoside via a 5′,5′-triphosphate bridge. The cap may occur in two different forms: the 5′-terminal nucleoside (the one linked to $m^7G$ triphosphate) is either in 2′-OH (called cap 0) or in 2′-OMe (called cap 1) form. Additionally, the next nucleoside may also be 2′-O-methylated (called cap 2). In case the first nucleoside is adenosine, which often is the case, it may additionally be $N^6$-methylated (called $m^6Am^{O2'}$) [94]. The RNA sequence is obtained by transcription on a chemically synthesized DNA template. RNA polymerases T7 and T3, or SP6 bacteriophages are commonly used. $m^7$-Cap may be introduced enzymatically by Vaccinia Capping Enzyme (VCE) after transcription, but more often during transcription using a so-called anti reverse 3′-O-methylated ARCA-cap ($m_2^{7,O3'}$-GpppG) [95] or a trinucleotide cap ($m^7GpppAm^{O2'}pG$) [96] as a building block (Figure 12.6). In the commonly used Comirnaty Covid-19 vaccine, the cap is an even more heavily modified trinucleotide, $m_2^{7,O3'}$-GpppAm$^{O2'}$pG. 3′-Poly(A) tail may also be introduced by transcription or enzymatically. The length of 3′-tail in mammalian varies from a few dozen be even 250 nucleotides. In mRNA vaccines, a 100-nucleotide tail is normal [97]. The RNA sequence is GC-rich and contains 5-methylcytidine and pseudouridine or $N^1$-pseudouridine modifications that reduce immunogenicity and enhance translation [98–100]. Pseudouridine modifications can be introduced during transcription by using their 5'-triphosphates in place of UTP [101].

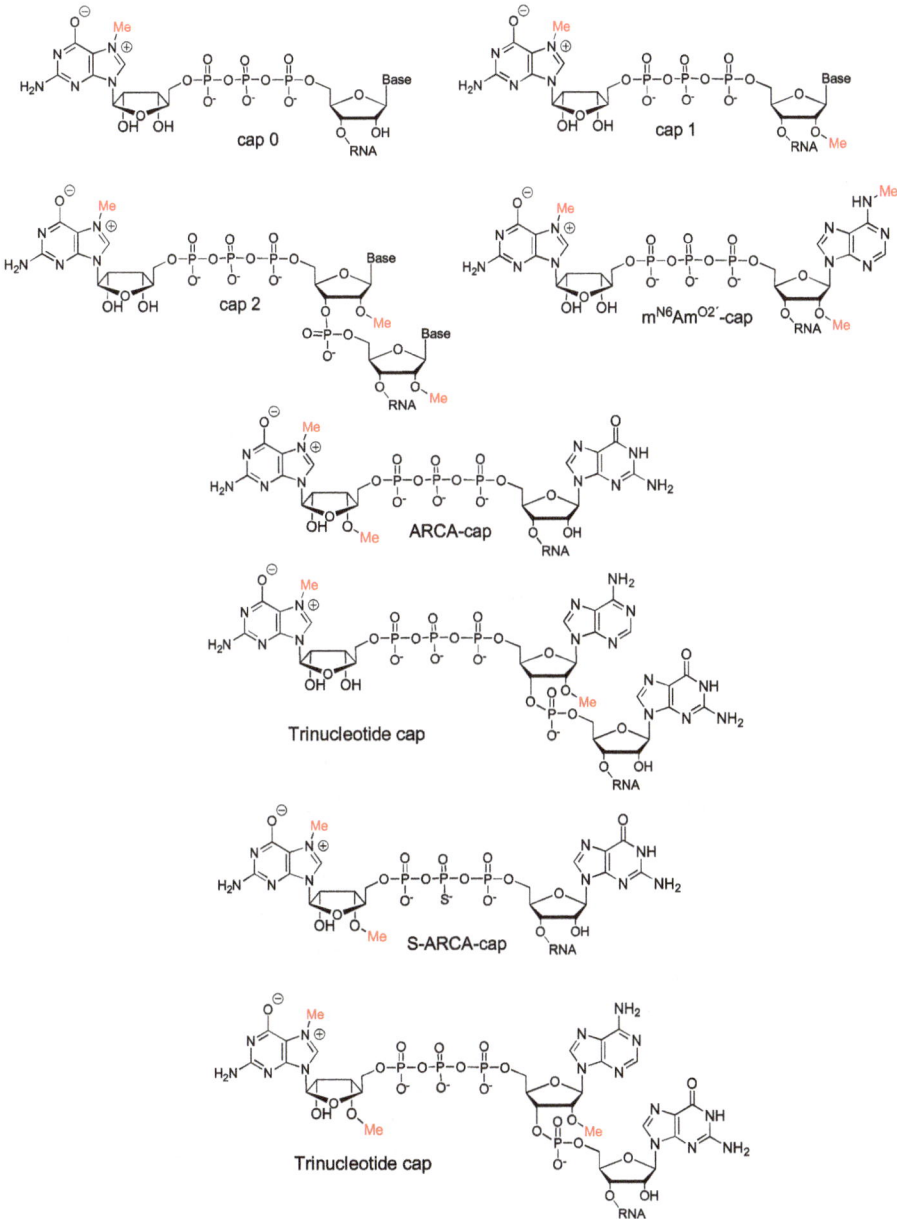

**Figure 12.6:** Alternative cap structures of natural and transcribed mRNA.

Besides vaccination, in vitro transcribed mRNA is believed to find applications in several fields of chemotherapy, such as infectious diseases, metabolic genetic diseases, cancer, cardiovascular disease and cerebrovascular diseases [102, 103]. In these fields, individual drug development projects have often proceeded to phase 2, sometimes even to phase 3 clinical studies. Advances in mRNA chemotherapy undoubtedly largely depend on success in combining various known structural modifications in such a way that biological half-life and translational activity are prolonged, penetration through membranes is improved and innate immune responses are avoided [104]. At the same time, base-pairing within the ORF sequence must remain unchanged. That is why modifications are restricted to the cap-structure, secondary structure, sequence elements, length of 5'- and 3'-UTRs, and the length of 3'-poly(U). All these have received interest as means to stabilize mRNA and increase its translation efficiency [105]. A combinatorial approach was recently utilized to find sequence-based rules for stabilization [106]. According to this study "structured superfolder mRNAs can be designed to improve both stability and expression with further enhancement through pseudouridine nucleoside modification." As regards site-specific mutations, phosphorothioate substitution at the intervening phosphorus of ARCA-cap has been argued to increase stability and translational efficiency of RNA vaccines in immature dendritic cells, and to induce superior immune responses in vivo [107–109]. Within the trinucleotide cap, O2'–C4 bridging of $m^7G$ with a methylene group increases translational activity compared to the commonly used caps discussed above [110]

A highly important issue common to all oligonucleotide drugs is delivery, how to get the drug internalized into cell. Recently developed techniques for preparation of RNA–lipid nanoparticles that penetrate through cell membranes were crucial for the success of Covid-19 vaccination by mRNA vaccines [111]. This subject, together with alternative delivery approaches, is discussed in more detail in Chapter 13.

## 12.6 Targeting of RNA with small molecules

In spite of the success of oligonucleotide drugs, the possibility of targeting a certain RNA by a small molecule has also received attention. The tertiary structure of RNA to some extent resembles that of proteins, and proteins are commonly targeted by small molecules. However, the interactions essential for formation of small molecule adducts with RNA and proteins are different. With RNA, stacking and hydrogen bonding interactions are most important, whereas hydrophobic interactions, which are crucial for binding to proteins, play a less dominant role [112].

So far only one small molecule that targets RNA has been approved as a drug. This is Risdiplam (Figure 12.7A), sold under name Evrysdi. It is used for treatment of spinal muscular atrophy (SMA) [113]. This is a disease caused by mutations in a gene (SMN1) that encodes SMN (survival of motor neuron) protein. Risdiplam does not restore activity of SMN1 gene but works by influencing on splicing of the transcript of a related gene, SMN2. SMN2 is normally inactive because it contains a base substitution at the 5'-splice site at exon 7. Risdiplam stabilizes the mutual interaction between the 5'-splice site of exon 7 and U1 small nuclear ribonucleoprotein of the spliceosome (cf. Section 10.3). Consequently, the ability of SMN2 to produce SMN protein is restored, at least partially.

However, the number of structural motifs of RNA that obviously could be recognized by a small molecule is rather limited. One such motif is a triple helical segment in RNA. The most extensively studied example is offered by MALAT1 (metastasis-associated lung adenocarcinoma transcript 1). This is a long noncoding RNA overexpressed in several cancers. Its 3'-terminus forms by back-folding a triple-helical segment that consists of UAU triplets inserted by a $C^+GC$ triplet and a CG doublet [114]. This triple helical structure evidently enhances the overexpression MALAT1 in cancer. Two compounds have been identified that effectively influence on triple helix dynamics of MALAT1. SM5 (Figure 12.7B) reduces MALAT1 transcript abundance [115], whereas DPFp8 has been identified as a triple helix stabilizer that in vitro prevents its exonucleolytic degradation [116].

The other structural motifs that have received interest include G-quadruplexes [117], repeated three letter codes [118], bacterial riboswitches [119] and internal ribosome entry sites (IRESs) of viruses [120], that is, sequences within the RNA molecules that allow initiation of translation by a cap-independent mechanism. RNA G-quadruplexes are obvious candidates for drug development. They are associated with several important cellular events, and a number of small molecules have already been developed to target them, mainly for analytical purposes [117]. Studies on therapeutic applications, however, still are at early stage. With three letter code expansions, the situation is more advanced. Myotonic muscular dystrophy 1 is a muscular disease caused by RNA repeat expansions. Increased number of internal 5'CUG/3'GUC loops occurs at 3'-UTR of mRNA that codes dystrophia myotonica protein kinase (DMPK). A construct that binds to two neighboring internal loops and is additionally equipped with Bleomycin 5 has been shown to cleave these CUG repeat expansions [118]. As regards binding to viral IRESs, compound D in Figure 12.7 has been shown to bind to a stemloop of Enterovirus 71 IRES and to suppress interaction with human RNA binding proteins [120].

Riboswitches also are noteworthy targets of drug discovery. They occur predominantly in bacteria mRNA that code vital metabolic pathways. In principle, synthesis of the product that a given riboswitch codes can be blocked by a structural analog of this product. This offers a novel approach for development of antibiotics. A riboflavin analog Ribocil-B (12.7E), for example, binds to riboflavin riboswitch at nanomolar concentrations inhibiting riboflavin synthesis, which leads to bacterial death [119].

Micro RNAs, discussed above in Section 12.2, still offers one potential target for small molecular drugs. Naphthyridine dimer (F in Figure 11.7) has been shown to bind to hairpin loop of two hairpin RNAs resulting in their dimerization. This effectively impeded the cleavage of hairpins to siRNA by dicer and, hence, the subsequent binding to target mRNA [121]. No therapeutic applications have been yet reported.

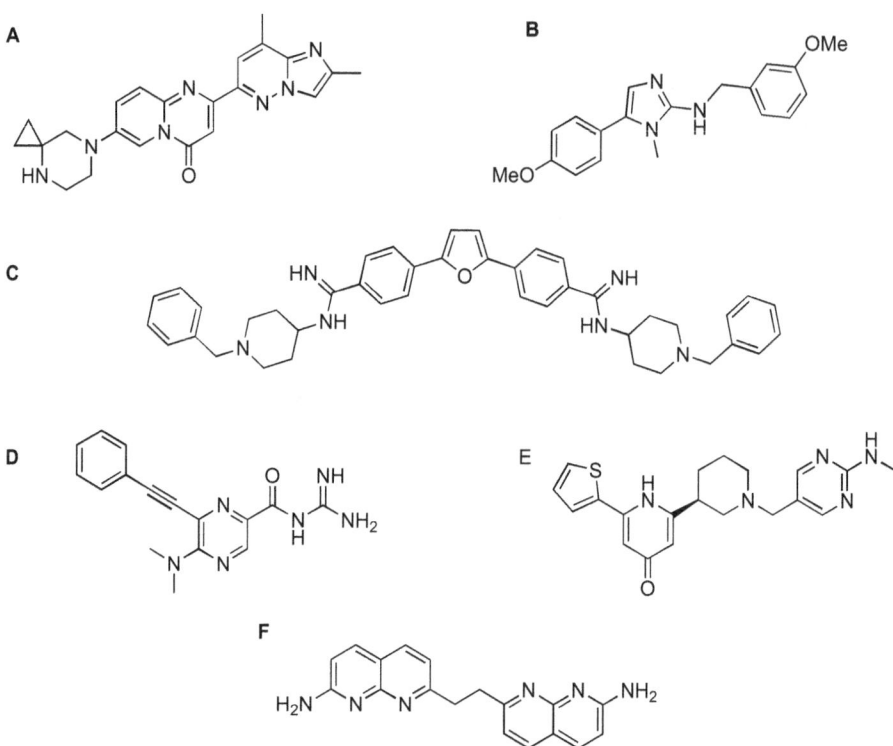

**Figure 12.7:** Small molecules used for RNA targeting: (A) Risdiplam correcting RNA splicing related to spinal muscular atrophy [113], (B) SM5, inhibitor of formation of MALAT1 [115], (C) DPFp8, stabilizer of trihelical structural motif of MALAT1 [116], (D) Binder of internal ribosome entry sites (IRESs) of viruses [120], (E) Ribocil-B, inhibitor of riboflavin riboswitch biosynthesis [119] and (F) naphthyridine dimer targeting miRNAs [121].

## 12.7 DNAzymes as therapeutic cleaving agents of RNA

Ribozymes and DNAzymes that in principle could cleave the target mRNA without participation of any intracellular enzymes, have also received interest as potential antisense drugs. Hammerhead ribozymes stabilized by structural modifications against biodegradation were developed already in 1990s [122], but no ribozyme drug candidates ended up to major clinical trials [123]. More recently, DNAzymes that are somewhat more stable and easier to prepare than ribozymes have been studied for the same purpose [124]. However, the challenges still are the same: insufficient biostability and cleaving activity at low metal ion concentrations. Some encouraging steps have anyway been taken. DNAzyme 8–17 (cf. Section 10.4) containing a 2′-O-Me modification at C10 of the catalytic core, and additionally LNA and 2′-O-Me modifications in the phosphorothioate wings, has been shown to cleave MALAT1 RNA in vitro cultured cells [125]. The cleaving activity was about 10% of that obtained with an AON. DNA-zyme 10–23, in turn, cleaved of mRNA transcripts in cultured mammalian cells when stabilized with 2′-fluoroarabino and α-L-threofuranosyl modifications in optimal positions within binding arms [126].

Another promising approach is in vitro chemical evolution of RNA endonucleases composed entirely of 2′-deoxy-2′-fluoro-β-D-arabino (FANA) nucleosides [127]. Phosphorothioate modifications have been used to still enhance the biostability. These artificial endonucleases knocked down KRAS mRNA at physiological $Mg^{2+}$ concentrations. Similar artificial endonucleases have additionally been engineered to form nanostructures that could cleave genomic SARS-CoV-2 RNA under physiological conditions [128].

So far only half a dozen variants of DNAzyme 10–23 have passed to phase II clinical trials [129]. In addition, DNAzymes have received interest as intracellular biosensors. Owing to metal ion–specific cleavage, they have been used for determination of biodistribution of various metal ions [130, 131].

# Further reading

Adachi H, Hengesbach M, Yu Y-T, Morais P. From antisense RNA to RNA modification: Therapeutic potential of RNA-based technologies. Biomedicines 2021, 9, 550 (26 pages).

Childs-Disney JL, Yang X, Gibaut QMR, Tong Y, Batey RT, Disney MD. Targeting RNA structures with small molecules. Nat Drug Discov 2022, 21, 736–762.

Deigan KE, Ferre-D'Amare AR. Riboswitches: Discovery of drugs that target bacterial gene-regulatory RNAs. Acc Chem Res 2011, 44, 1329–1338.

Deleavey GF, Damha MJ. Designing modified oligonucleotides for targeted gebe silencing. Chem Biol 2012, 19, 937–954.

Dominska M, Dykxhoorn DM. Breaking down the barriers: SiRNA delivery and endosome escape. J Cell Sci 2010, 123, 1183–1189.

Dong H, Lei J, Ding L, Wen Y, Ju H, Zhang X. MicroRNA: Function, detection, and bioanalysis. Chem Rev 2013, 113, 6207–6233.

Egli M, Manoharan M. Critical reviews and perspectives, chemistry, structure and function of approved oligonucleotide therapeutics. Nucleic Acids Res, 2023, 51, 6, 2529–2573.

Egli M, Manoharan M. Re-engineering RNA molecules into therapeutic agents. Acc Chem Res 2019, 52, 1036–1047.

Falese JP, Donlic, Hargrove AE. Targeting RNA with small molecules: From fundamental principles towards the clinic. Chem Soc Rev 2021, 50, 2224–2243.

Garst AD, Edwards AL, Batey RT. Riboswitches: Structures and mechanisms. Cold Spring Harb Perspect Biol 2011, 3, a003533 (13 pages).

Glazier DA, Liao J, Roberts BL, Li X, Yang K, Stevens CM, Tang W. Chemical synthesis and biological application of modified oligonucleotides. Bioconjugate Chem 2020, 31, 1213–1233.

Hallberg ZH, Su Y, Kitto RZ, Hammond MC. Engineering and in vivo applications of riboswitches. Annu Rev Biochem 2017, 86, 515–539.

Li B, Qu L, Yang J. RNA-Guided RNA Modifications: Biogenesis, Functions, and Applications. Acc Chem Res 2023, 56, 3198–3210.

Mo J, Weng X, Zhou X. Detection, clinical application, and manipulation of RNA modifications. Acc Chem Res 2023, 56, 2788–2800.

Pfeiffer LS, Stafforst T. Precision RNA base editing with engineered and endogenous effectors. Nat Biotechnol 2023, 41, 1526–1542.

Serganov A. Nudler EA Decade of riboswitches. Cell 2013, 152, 17–24.

Silverman RH. Viral Encounters with 2′,5′-Oligoadenylate Synthetase and RNase L during the Interferon Antiviral Response. J Virol 2007, 81, 12720–12729.

Sletten RL, Rossi JJ, Han S-P. The current state and future directions of RNAi-based therapeutics. Nat Rev Drug Discov 2019, 18, 421–446.

Warminski M, Mamot A, Depaix A, Kowalska J, Jemielity J. Chemical modifications of mRNA ends for therapeutic applications. Acc Chem Res 2023, 56, 2814–2826.

# References

[1] Fire A, Xu S, Montgomery MK, Kostas SA, Driver SE, Mello CC. Potent and specific genetic interference by double-stranded RNA in Caenorhabditis elegans. Nature 1998, 39, 806–811.

[2] Bernstein E, Caudy AA, Hammond SM, Hannon GJ. Role for a bidentate ribonuclease in the initiation step of RNA interference. Nature 2001, 409, 363–366.

[3] Okamura K, Chung WJ, Ruby JG, Guo H, Bartel DP, Lai EC. The Drosophila hairpin RNA pathway generates endogenous short interfering RNAs. Nature 2008, 453, 803–806.

[4] Nakanishi K. Anatomy of RISC: How do small RNAs and chaperones activate Argonaute proteins. Wiley Interdiscip Rev RNA 2016, 7, 637–660.

[5] Gregory RI, Chendrimada TP, Cooch N, Shiekhattar R. Human RISC couples microRNA biogenesis and posttranscriptional gene silencing. Cell 2005, 123, 631–640.

[6] Rand TA, Ginalski K, Grishin NV, Wang X. Biochemical identification of Argonaute 2 as the sole protein required for RNA-induced silencing complex activity. Proc Natl Acad Sci USA 2004, 101, 14385–14389.

[7] Sheu-Gruttadauria J, MacRae IJ. Structural foundations of RNA silencing by Argonaute. J Mol Biol 2017, 429, 2619–2639.

[8] Schwarz DS, Hutvagner G, Du T, Xu Z, Aronin N, Zamore PD. Asymmetry in the assembly of the RNAi enzyme complex. Cell 2003, 115, 199–208.

[9] Jackson AL, Burchard J, Schelter J, Chau BN, Cleary M, Lim L, Linsley PS. Widespread siRNA "off-target" transcript silencing mediated by seed region sequence complementarity. RNA 2006, 12, 1179–1187.

[10] Lewis BP, Burge CB, Bartel DP. Conserved seed pairing, often flanked by adenosines, indicates that thousands of human genes are microRNA targets. Cell 2005, 120, 15–20.

[11] Jackson AL, Burchard J, Leake D, Reynolds A, Schelter J, Guo J, Johnson JM, Lim L, Karpilow J, Nichols K, Marshall W, Khvorova A, Linsley PS. Position-specific chemical modification of siRNAs reduces "off-target" transcript silencing. RNA 2006, 12, 1197–1205.

[12] Rydzik AM, Gottschling D, Simon E, Skronska-Wasek W, Rippmann JF, Riether D. Epigenetic modification 6-methyladenosine can impact the potency and specificity of siRNA. ChemBioChem 2021, 22, 491–495.

[13] Hofmeister A, Jahn-Hofmann K, Brunner B, Helms M, Metz-Weidmann C, Krack A, Kurz M, Heubel C, Scheidler S. Small interfering RNAs containing dioxane- and morpholino-derived nucleotide analogues show improved off-target profiles and chirality-dependent in vivo knock-down. J Med Chem 2022, 65, 13736–13752.

[14] Richter M, Viel JA, Kotikam V, Gajula PK, Coyle L, Pal C, Rozners E. Amide modifications in the seed region of the guide strand improve the on-target specificity of short interfering RNA. ACS Chem Biol 2023, 18, 1, 7–11.

[15]  Schwarz DS, Ding H, Kennington L, Moore JT, Schelter J, Burchard J, Linsley PS, Aronin N, Xu Z, Zamore PD. Designing siRNA that distinguish between genes that differ by a single nucleotide. PLoS Genetics 2006, 2, e140.

[16]  Layzer JM, McCaffrey AP, Tanner AK, Huang Z, Kay MA, Sullenger BA. In vivo activity of nuclease-resistant siRNAs. RNA 2004, 10, 766–771.

[17]  Egli M, Manoharan M. Re-engineering RNA molecules into therapeutic agents. Acc Chem Res 2019, 52, 1036–1047.

[18]  Bumcrot D, Manoharan M, Koteliansky V, Sah DWY. RNAi therapeutics: A potential new class of pharmaceutical drugs. Nat Chem Biol 2006, 2, 711–719.

[19]  Egli M, Manoharan M. Critical reviews and perspectives chemistry, structure and function of approved oligonucleotide therapeutics. Nucleic Acids Res 2023, 51, 6, 2529–2573.

[20]  Judge AD, Sood V, Shaw JR, Fang D, McClintock K, MacLachlan I. Sequence-dependent stimulation of the mammalian innate immune response by synthetic siRNA. Nat Biotechnol 2005, 23, 457–462.

[21]  Elkayam E, Parmar R, Brown CR, Willoughby JLS, Theile CS, Manoharan M, Joshua-Toe L. siRNA carrying an (E)-vinylphosphonate moiety at the 5′-end of the guide strand augments gene-silencing by enhanced binding to human argonaute-2. Nucleic Acids Res 2017, 45, 3528–3536.

[22]  Deleavey GF, Frank F, Hassler M, Wisnovsky S, Nagar B, Damha MJ. The 5′-binding MID domain of human Argonaut-2 tolerates chemically modified nucleotide analogues. Nucleic Acid Ther 2013, 23, 81–87.

[23]  Parmer R, Brown C, Matsuda S, Willoughby J, Theile C, Charisse K, Foster DJ, Zlatev I, Jadhav V, Maier M, Egli M, Manoharan M, Rajeev KG. Facile synthesis, deometry and 2′-substituent-dependent in vivo activity of 5′-(E)- and 5′-(Z)-vinylphosphonate –modified siRNA conjugates. J Med Chem 2018, 61, 734–744.

[24]  Choung S, Kim YJ, Kim S, Park HO, Choi YC. Chemical modification of siRNAs to improve serum stability without loss of efficacy. Biochem Biophys Res Commun 2006, 342, 919–927.

[25]  Jahns H, Taneja N, Willoughby JLS, Akabane-Nakata M, Brown CR, Nguyen T, Bisbe A, Matsuda S, Hettinger M, Manoharan RM, Rajeev KG, Maier MA, Zlatev I, Charisse K, Egli M, Manoharan M. Chirality matters: Stereo-defined phosphorothioate linkages at the termini of small interfering RNAs improve pharmacology in vivo. Nucleic Acids Res 2022, 50, 1221–1240.

[26]  Hall AH, Wan J, Shaughnessy EE, Ramsay Shaw B, Alexander KA. RNA interference using boranophosphate siRNAs: Structure-activity relationships. Nucleic Acids Res 2004, 32, 5991–6000.

[27]  Hall AH, Wan J, Spesock A, Sergueeva Z, Ramsay Shaw B, Alexander KA. High potency silencing by single-stranded boranophosphate siRNA. Nucleic Acids Res 2006, 34, 2773–2781.

[28]  Jahns H, Degaonkar R, Podbevsek P, Gupta S, Bisbe A, Aluri K, Szeto J, Kumar P, LeBlanc S, Racie T, Brown CR, Castoreno A, Guenther DC, Jadhav V, Maier MA, Plavec J, Egli M, Manoharan M, Zlatev I. Small circular interfering RNAs (sciRNAs) as a potent therapeutic platform for gene-silencing. Nucleic Acids Res 2021, 49, 10250–10264.

[29]  Sheridan C. With Alnylam's amyloidosis success, RNAi approval hopes soar. Nat Biotechnol 2017, 35, 995–997.

[30]  Kosmas CE, Estrella AM, Sourlas A, Silverio D, Hilario E, Montan PD, Guzman E. Inclisiran: A new promising agent in the of hypercholesterolemia. Diseases 2018, 6, 63(6 pages).

[31]  Chan A, Liebow A, Yasuda M, Gan L, Racie T, Maier M, Kuchimanchi S, Foster D, Milstein S, Charisse K, Sehgal A, Manoharan M, Meyers R, Fitzgerald K, Simon A, Desnick RJ, Querbes W. Preclinical development of a subcutaneous ALAS1 RNAi therapeutic for treatment of hepatic porphyrias using circulating RNA quantification. Mol Ther Nucleic Acids 2015, 4, e263 (9 pages).

[32]  Hulton S-A. Lumasiran: Expanding the treatment options for patients with primary hyperoxaluria type 1. Expert Opin Orphan Drugs 2021, 9, 189–198.

[33] Aimo A, Castiglione V, Rapezzi C, Franzini M, Panichella G, Vergaro G, Gillmore J, Fontana M, Passino C, Emdin M. RNA-targeting and gene editing therapies for transthyretin amyloidosis. Nat Rev Cardiology 2022, 19, 655–667.

[34] Dong H, Lei J, Ding L, Wen Y, Ju H, Zhang X. MicroRNA: Function, detection, and bioanalysis. Chem Rev 2013, 113, 6207–6233.

[35] Lee Y, Kim M, Han J, Yeom KH, Lee S, Baek SH. MicroRNA genes are transcribed by RNA polymerase II. EMBO J 2004, 23, 4051–4060.

[36] Lee Y, Ahn C, Han J, Choi H, Kim J, Yim J. The nuclear RNase III Drosha initiates microRNA processing. Nature 2003, 425, 415–419.

[37] Bohnsack MT, Czaplinski K, Gorlich D. Exportin 5 is a RanGTP-dependent dsRNA-binding protein that mediates nuclear export of pre-miRNAs. RNA 2004, 10, 185–191.

[38] Hammond SM, Bernstein E, Beach D, Hannon GJ. An RNA-directed nuclease mediates post-transcriptional gene silencing in Drosophila cells. Nature 2000, 404, 293–296.

[39] Bartel DP. MicroRNAs: Target recognition and regulatory functions. Cell 2009, 136, 215–233.

[40] Liu J, Valencia-Sanchez MA, Hannon GJ, Parker R. MicroRNA-dependent localization of targeted mRNAs to mammalian P-bodies. Nat Cell Biol 2005, 7, 719–723.

[41] Treiber T, Treiber N, Meister G. Regulation of microRNA biogenesis and its crosstalk with other cellular pathways. Nat Rev Molec Cell Biology 2019, 20, 7–20.

[42] Fromm B, Billipp T, Peck LE, Johansen M, Tarver JE, King BL, Newcomb JM, Sempere LF, Flatmark K, Hovig E, Peterson KJ. A uniform system for the annotation of human microRNA genes and the evolution of the human microRNAome. Annu Rev Genet 2015, 49, 213–242.

[43] Homo sapiens miRNAs in the miRBase at Manchester University. http://www.mirbase.org/cgi-bin/mirna_summary.pl?org=hsa.

[44] Friedman RC, Farh KK-H, Burge CB, Bartel DP. Most mammalian mRNAs are conserved targets of microRNAs. Genome Res 2009, 19, 92–105.

[45] Dostie J, Mourelatos Z, Yang M, Sharma A, Dreyfuss G. Numerous microRNPs in neuronal cells containing novel microRNAs. RNA 2003, 9, 180–186.

[46] Wang YL, Keys DN, Au-Young JK, Chen CF. MicroRNAs in embryonic stem cells. J Cell Physiol 2009, 218, 251–255.

[47] Xu P, Vernooy SY, Guo M, Hay BA. The Drosophila microRNA Mir-14 suppresses cell death and is required for normal fat metabolism. Curr Biol 2003, 13, 790–795.

[48] Ryan BM, Robles AI, Harris CC. Genetic variation in microRNA networks: The implications for cancer research. Nat Rev Cancer 2010, 10, 389–402.

[49] Farazi TA, Spitzer JI, Morozov P, Tuschl T. miRNAs in human cancer. J Pathol 2011, 223, 102–115.

[50] Li Z, Rana TM. Therapeutic targeting of microRNAs: Current status and future challenges. Nat Rev Drug Discov 2014, 13, 622–638.

[51] Li Z, Yang CS, Nakashima K, Rana TM. Small RNA-mediated regulation of iPS cell generation. EMBO J 2011, 30, 823–834.

[52] Obad S, dos Santos CO, Petri A, Heidenblad M, Broom O, Ruse C, Fu C, Lindow M, Stenvang J, Straarup EM, Hansen HF, Koch T, Pappin D, Hannon GJ, Kauppinen S. Silencing of microRNA families by seed-targeting tiny LNAs. Nat Genet 2011, 43, 371–378.

[53] Diener C, Keller A, Meese E. Emerging concepts of miRNA therapeutics: From cells to clinic. Trends Genetics 2022, 38, 613–625.

[54] Ree MH, Meer AJ, Nuenen AC, Bruijne J, Ottosen S, Janssen HL, Kootstra NA, Reesink HW. Miravirsen dosing in chronic hepatitis C patients results in decreased microRNA-122 levels without affecting other microRNAs in plasma. Aliment Pharmacol Ther 2016, 43, 102–113.

[55] Gomez IG, MacKenna DA, Johnson BG, Kaimal V, Roach AM, Ren S, Nakagawa N, Xin C, Newitt R, Pandya S, Xia TH, Liu X, Borza DB, Grafals M, Shankland SJ, Himmelfarb J, Portilla D, Liu S, Chau BN,

Duffield JS. Anti–microRNA-21 oligonucleotides prevent Alport nephropathy progression by stimulating metabolic pathways. J Clin Invest 2015, 125, 141–156.

[56] Seto AG, Beatty X, Lynch JM, Hermreck M, Tetzlaff M, Duvic M, Jackson AL. Cobomarsen, an oligonucleotide inhibitor of miR-155, co-ordinately regulates multiple survival pathways to reduce cellular proliferation and survival in cutaneous T-cell lymphoma. Br J Haematol 2018, 183, 428–444.

[57] Gallant-Behm CL, Piper J, Lynch JM, Seto AG, Hong SJ, Mustoe TA, Maari C, Pestano LA, Dalby CM, Jackson AL, Rubin P, Marshall WS. A microRNA-29 mimic (Remlarsen) represses extracellular matrix expression and fibroplasia in the skin. J Invest Dermatol 2019, 139, 1073–1081.

[58] Forterre A, Komuro H, Aminova S, Harada M. A comprehensive review of cancer microRNA therapeutic delivery strategies. Cancers 2020, 12, 1852, (21 pages).

[59] Otmani K, Lewalle P. Tumor suppressor miRNA in cancer cells and the tumor microenvironment: Mechanism of deregulation and clinical implications. Front Oncol 2021, 11, 708765, (15 pages).

[60] Kong YW, Ferland-McCollough D, Jackson TJ, Bushell M. microRNAs in cancer management. Lancet Oncol 2012, 13, e249–e258.

[61] Silverman RH. Viral encounters with 2′,5′-oligoadenylate synthetase and RNase L during the interferon antiviral response. J Virol 2007, 81, 12720–12729.

[62] Kristiansen H, Gad HH, Eskildsen-Larsen S, Despres P, Hartmann RJ. The Oligoadenylate synthetase family: An ancient protein family with multiple antiviral activities. Interf Cytok Res 2011, 31, 41–47.

[63] Liang S-L, Quirk D, Zhou A. RNase L: Its biological roles and regulation. IUBMB Life 2006, 58, 508–514.

[64] Anderson BR, Muramatsu H, Jha BK, Silverman RH, Weissman D, Kariko K. Nucleoside modifications in RNA limit activation of 2′-5′-oligoadenylate synthetase and increase resistance to cleavage by RNase L. Nucleic Acids Res 2011, 39, 9329–9338.

[65] Dong B, Silverman RH. 2-5A-Dependent RNase molecules dimerize during activation by 2-5A. J Biol Chem 1995, 279, 4133–4137.

[66] Castelli J, Wood KA, Youle RJ. The 2-5A system in viral infection and apoptosis. Biomed Pharmacother 1998, 52, 386–390.

[67] Zhou A, Paranjape J, Brown TL, Nie H, Naik S, Dong B, Chang A, Trapp B, Fairchild R, Colmenares C, Silverman RH. Interferon action and apoptosis are defective in mice devoid of 2′,5′-oligoadenylate-dependent RNase L. EMBO J 1997, 16, 6355–6363.

[68] Han J-Q, Barton DJ. Activation and evasion of the antiviral 29–59 oligoadenylate synthetase/ribonuclease L pathway by hepatitis C virus mRNA. RNA 2002, 8, 512–525.

[69] Defilippi P, Huez G, Verhaegen-Levalle M, De Clercq E, Imai J, Torrence P, Content J. Antiviral activity of a chemically stabilized 2-5A analog upon microinjection into H8eLa cells. FEBS Lett 1986, 198, 326–332.

[70] Kubota K, Nakahara T, Ohtsuka T, Yoshida S, Kawaguchi J, Fujita Y, Ozeki Y, Hara A, Yoshimura C, Furukawa H, Haruyama H, Ichikawa K, Yamashita M, Matsuoka T. Identification of 2′-phosphodiesterase, which plays a role in the 2–5A system regulated by interferon. J Biol Chem 2004, 279, 7832–7841.

[71] Dong B, Xu L, Zhou A, Hassel BA, Lee X, Torrence PF, Silverman RH. Intrinsic molecular activities of the interferon-induced 2-5A-dependent RNase. J Biol Chem 1994, 269, 14153–14158.

[72] Tanaka N, Nakanishi M, Kusakabe Y, Goto Y, Kitade Y, Nakamura KT. Structural basis for recognition of 2′,5′-linked oligoadenylates by human ribonuclease L. EMBO J 2004, 23, 3929–3938.

[73] Xiang Y, Wang Z, Murakami J, Plummer S, Klein EA, Carpten JD, Trent JM, Isaacs WB, Casey G, Silverman RH. Effects of RNase L mutations associated with prostate cancer on apoptosis Induced by 2,5-Oligoadenylates. Cancer Res 2003, 63, 6795–7801.

[74] Player MR, Torrence PF. The 2-5A system: Modulation of viral and cellular processes through acceleration of RNA degradation. Pharmacol Ther 1998, 78, 55–113.

[75]   Adah SA, Bayly SF, Cramer H, Silverman RH, Torrence PF. Chemistry and biochemistry of 2′, 5′-oligoadenylate-based antisense strategy. Curr Med Chem 2001, 8, 1189–1212.

[76]   Torrence PF, Brozda D, Alster D, Charubala R, Pfleiderer W. Only one 3′-hydroxyl group of ppp5′A2′p5′A2′p5′A (2-5A) is required for activation of the 2-5A-dependent endonuclease. J Biol Chem 1988, 263, 1131–1139.

[77]   Hartog JAJ, Wijnands RA, van Boom JH. Chemical synthesis of pppA2′p5′A2′p5′A, an interferon-induced inhibitor of protein synthesis, and some functional analogs. J Org Chem 1981, 46, 2242–2251.

[78]   Kalinichenko EN, Podkopaeva TL, Kelve M, Saarma M, Mikhailopulo IA. 3′-Fluoro-3′-deoxy analogs of 2–5A 5′-monophosphate: Binding to 2–5A-dependent endoribonuclease and susceptibility to (2′-5′) phosphodiesterase degradation. Biochem Biophys Res Commun 1990, 167, 20–26.

[79]   Morita K, Kaneko M, Obika S, Imanishi T, Kitade Y, Koizumi M. Biologically stable 2-5A analogues containing 3′-O,4′-C-bridged adenosine as potent RNase L agonists. ChemMedChem 2007, 2, 1703–1707.

[80]   Bisbal C, Silhol M, Lemaître M, Bayard B, Salehzada T, Lebleu B, Perrée TD, Blackburn MG. 5′-Modified agonist and antagonist (2′-5′)(A)n analogues: Synthesis and biological activity. Biochemistry 1987, 26, 5172–5178.

[81]   Lasek T, Petrova M, Kosiova I, Simak O, Budesínsky M, Kozak J, Snasel J, Vavrina Z, Birkus G, Rosenberg I, Pav ZO. 5′-Phosphonate modified oligoadenylates as potent activators of human RNase L. Bioorg Med Chem 2022, 56, 116632 (9 pages).

[82]   Rees HA, Liu DR. Base editing: Precision chemistry on the genome and transcriptome of living cells. Nat Rev Genet 2018, 19, 770–788.

[83]   Khosravi HM, Jantsch MF. Site-directed RNA editing: Recent advances and open challenges. RNA Biology 2021, 18, 41–50.

[84]   Nishikura K. Functions and regulation of RNA editing by ADAR deaminases. Annu Rev Biochem 2010, 79, 321–349.

[85]   Bazak L, Haviv A, Barak M, Jacob-Hirsch J, Deng P, Zhang R, Isaacs FJ, Rechavi G, Li JB, Eisenberg E, Levanon EY. A-to-I RNA editing occurs at over a hundred million genomic sites, located in a majority of human genes. Genome Res 2014, 24, 365–376.

[86]   Montiel-Gonzalez MF, Vallecillo-Viejo I, Yudowski DA, Rosenthal JJC. Correction of mutations within the cystic fibrosis transmembrane conductance regulator by site-directed RNA editing. PNAS 2013, 110, 18285–18290.

[87]   Vallecillo-Viejo IC, Liscovitch-Brauer N, Montiel-Gonzalez MF, Eisenberg E, Rosenthal JJC. Abundant off-target edits from site-directed RNA editing can be reduced by nuclear localization of the editing enzyme. RNA Biol 2018, 15, 104–114.

[88]   Vogel P, Schneider MF, Wettengel J, Stafforst T. Improving site-directed RNA editing in vitro and in cell culture by chemical modification of the guideRNA. Angew Chem Int Ed 2014, 53, 6267–6271.

[89]   Azad MTA, Bhakta S, Tsukahara T. Site-directed RNA editing by adenosine deaminase acting on RNA for correction of the genetic code in gene therapy. Gene Ther 2017, 24, 779–786.

[90]   Fukuda M, Umeno H, Nose K, Nishitarumizu A, Noguchi R, Nakagawa H. Construction of a guide-RNA for site-directed RNA mutagenesis utilising intracellular A-to-I RNA editing. Sci Rep 2017, 7, 41478, (13 pages).

[91]   Huang X, Lv J, Li Y, Mao S, Li Z, Jing Z, Sun Y, Zhang X, Shen S, Wang X, Di M, Ge J, Huang X, Zuo E, Chi T. Programmable C-to-U RNA editing using the human APOBEC3A deaminase. The EMBO Journal 2020, 39, e104741 (12 pages).

[92]   Adachi H, Hengesbach M, Yu Y-T, Morais P. From antisense RNA to RNA modification: Therapeutic potential of RNA-based technologies. Biomedicines 2021, 9, 550 (26 pages).

[93]   Rosa SS, Prazeres DMF, Azevedo AM, Marques MPC. mRNA vaccines manufacturing: Challenges and bottlenecks. Vaccine 2021, 39, 2190–2200.

[94] Wang Y-S, Kumari M, Chen G-H, Hong M-H, Yuan JP-Y, Tsai J-L, H-c W. mRNA-based vaccines and therapeutics: An in-depth survey of current and up-coming clinical applications. J Biomed Sci 2023, 30, 84, (35 pages).

[95] Jemielity J, Fowler T, Zuberek J, Stepinski J, Lewdorowicz M, Niedzwiecka A, Stolarski R, Darzynkiewicz E, Rhoads RE. Novel "anti-reverse" cap analogs with superior translational properties. RNA 2003, 9, 1108–1122.

[96] Henderson JM, Ujita A, Hill E, Yousif-Rosales S, Smith C, Ko N, McReynolds T, Cabral CR, Escamilla-Powers JR, Houston ME. Cap 1 messenger RNA synthesis with Cotranscriptional CleanCap® analog by In vitro transcription. Current Protocols 2021, 1, e39 (17 pages).

[97] Schlake T, Thess A, Fotin-Mleczek M, Kallen KJ. Develping mRNA-vaccine technologies. RNA Biol 2012, 9, 1319–1330.

[98] Weng Y, Li C, Yang T, Hu B, Zhang M, Guo S, Xiao H, Liang XJ, Huang Y. The challende and prospect of mRNA therapeutics landscape. Biotechnol Adv 2020, 40, 107534, (23 pages).

[99] Parr CJC, Wada S, Kotake K, Kameda S, Matsuura S, Sakashita S, Park S, Sugiyama H, Kuang Y, Saito H. N1-Methylpseudouridine substitution enhances the performance of synthetic mRNA switches in cells. Nucleic Acids Res 2020, 48, e35 (9 pages).

[100] Nance KD, Meier JL. Modifications in an emergency: The role of N1-methylpseudouridine in COVID-19 vaccines. ACS Cent Sci 2021, 7, 748–756.

[101] Morais P, Adachi H, Yu Y-T. The critical contribution of pseudouridine to mRNA COVID-19 vaccines. Front Cell Develop Biology 2021, 9, 789427, (9 pages).

[102] Kariko K. In vitro-transcribed mRNA therapeutics: Out of the shadows and into the spotlight. Mol Ther 2019, 27, 691–692.

[103] Qin S, Tang X, Chen Y, Chen K, Fan N, Xiao W, Zheng Q, Li G, Teng Y, Wu M, Song X. mRNA-based therapeutics: Powerful and versatile tools to combat diseases. Signal Transduct Target Ther 2022, 7, 166 (35 pages).

[104] Warminski M, Mamot A, Depaix A, Kowalska J, Jemielity J. Chemical modifications of mRNA ends for therapeutic applications. Acc Chem Res 2023, 56, 2814–2826.

[105] Jia L, Qian S-B. Therapeutic mRNA engineering from head to tail. Acc Chem Res 2021, 54, 4272–4282.

[106] Leppek K, Byeon GW, Kladwang W, Wayment-Steele HK, Kerr CH, Xu AF, Kim DS, Topkar VV, Choe C, Rothschild D, Tiu GC, Wellington-Oguri R, Fujii K, Sharma E, Watkins AM, Nicol JJ, Romano J, Tunguz B, Diaz F, Cai H, Guo P, Wu J, Meng F, Shi S, Participants E, Dormitzer PR, Solorzano A, Barna M, Das R. Combinatorial optimization of mRNA structure, stability, and translation for RNA-based therapeutics. Nat Commun 2022, 13, 1536, (22 pages).

[107] Kuhn AN, Diken M, Kreiter S, Selmi A, Kowalska J, Jemielity J, Darzynkiewicz E, Huber C, Tureci O, Sahin U. Phosphorothioate cap analogs increase stability and translational efficiency of RNA vaccines in immature dendritic cells and induce superior immune responses in vivo. Gene Ther 2010, 17, 961–971.

[108] Strenkowska M, Grzela R, Majewski M, Wnek K, Kowalska J, Lukaszewicz M, Zuberek J, Darzynkiewicz E, Kuhn AN, Sahin U, Jemielity J. Cap analogs modified with 1,2-dithiodiphosphate moiety protect mRNA from decapping and enhance its translational potential. Nucleic Acids Res 2016, 44, 9578–9590.

[109] Warminski M, Kowalska J, Nowak E, Kubacka D, Tibble R, Kasprzyk R, Sikorski PJ, Gross JD, Nowotny M, Jemielity J. Structural insights into the interaction of clinically relevant phosphorothioate mRNA cap analogs with initiation Factor 4E reveal stabilization via Eeectrostatic thio-effect. ACS Chem Biol 2021, 16, 334–343.

[110] Senthilvelan A, Vonderfecht T, Shanmugasundaram M, Pal I, Potter J, Kore AR. Trinucleotide cap analogue bearing a locked nucleic acid moiety: Synthesis, mRNA modification, and translation for therapeutic applications. Org Lett 2021, 23, 4133–4136.

[111] Buschmann MD, Carrasco MJ, Alishetty S, Paige M, Alameh MG, Weissman D. Nanomaterial delivery systems for mRNA vaccines. Vaccines 2021, 9, 65, (30 pages).

[112] Falese JP, Donlic, Hargrove AE. Targeting RNA with small molecules: From fundamental principles towards the clinic. Chem Soc Rev 2021, 50, 2224–2243.

[113] FDA.gov, 2020, FDA Approves Oral Treatment for Spinal Muscular Atrophy, Retrieved from https://www.fda.gov/newsevents/press-announcements/fda-approves-oral-treatment-spinalmuscular-atrophy.

[114] Brown JA, Bulkley D, Wang J, Valenstein ML, Yario TA, Steitz TA, Steitz JA. Structural insights into the stabilization of MALAT1 noncoding RNA by a bipartite triple helix. Nat Struct Mol Biol 2014, 21, 633–640.

[115] Abulwerdi FA, Xu W, Ageeli AA, Yonkunas MJ, Arun GH, Nam H, Schneekloth JS, Jr, Dayie TK, Spector D, Baird N, Le Grice SFJ. Selective small-molecule targeting of a triple helix encoded by the long noncoding RNA, MALAT1. ACS Chem Biol 2019, 14, 223–235.

[116] Donlic A, Zafferani M, Padroni G, Puri M, Hargrove AE. Regulation of MALAT1 triple helix stability and in vitro degradation by diphenylfurans. Nucleic Acids Res 2020, 48, 7653–7664.

[117] Tao Y, Zheng Y, Zhai Q, Wei D. Recent advances in the development of small molecules targeting RNA G-quadruplexes for drug discovery. Bioorganic Chemistry 2021, 110, 104804, (10 pages).

[118] Benhamou RI, Abe M, Choudhary S, Meyer SM, Angelbello AJ, Disney MD. Optimization of the linker domain in a dimeric compound that degrades an r(CUG) repeat expansion in Cells. J Med Chem 2020, 63, 7827–7839.

[119] Howe JA, Wang H, Fischmann TO, Balibar CJ, Xiao L, Galgoci AM, Malinverni JC, Mayhood T, Villafania A, Nahvi A, Murgolo N, Barbieri CM, Mann PA, Carr D, Xia E, Zuck P, Riley D, Painter RE, Walker SS, Sherborne B, de Jesus R, Pan W, Plotkin MA, Wu J, Rindgen D, Cummings J, Garlisi CG, Zhang R, Sheth PR, Gill CJ, Tang H, Roemer T. Selective small-molecule inhibition of an RNA structural element. Nature 2015, 526, 672–677.

[120] Davila-Calderon J, Patwardhan N, Chiu L, Sugarman A, Cai Z, Penutmutchu SR, Li M, Brewer G, Hargrove AE, Tolbert BS. IRES-targeting small molecule inhibits enterovirus 71 replication via allosteric stabilization of a ternary complex. Nat Commun 2020, 11, 4775, 13.

[121] Murata A, Mori Y, Di Y, Sugai A, Das B, Takashima Y, Nakatani K. Small molecule-induced dimerization of hairpin RNA interfered with the dicer cleavage reaction. Biochemistry 2021, 60, 245–249.

[122] Usman N, Blatt LM. Nuclease-resistant synthetic ribozymes: Developing a new class of therapeutics. J Clin Invest 2000, 106, 1197–1202.

[123] Grassi G, Dawson P, Guarnieri G, Kandolf R, Grassi M. Therapeutic potential of hammerhead ribozymes in the treatment of hyper-proliferative diseases. Curr Pharmaceut Biotechnol 2004, 5, 369–386.

[124] Zhou W, Ding J, Liu J. Theranostic DNAzymes. Theranostics 2017, 7, 1010–1025.

[125] Chiba K, Yamaguchi T, Obika S. Development of 8–17 XNAzymes that are functional in cells. Chem Sci 2023, 14, 7620–7629.

[126] Wang Y, Nguyen K, Spitale RC, Chaput JC. A biologically stable DNAzyme that efficiently silences gene expression in cells. Nat Chem 2021, 13, 319–326.

[127] Taylor AI, Wan JK, Donde MJ, Peak-Chew S-Y, Holliger PA. Modular XNAzyme that cleaves long, structured RNAs under physiological conditions enables allele-specific gene silencing in cells. Nat Chem 2022, 14, 1295–1305.

[128] Gerber PP, Donde MJ, Matheson NJ, Taylor AI. XNAzymes targeting the SARS-CoV-2 genome inhibit viral infection. Nat Commun 2022, 13, 6716 (12 pages).

[129] Xiao L, Zhao Y, Yang M, Luan G, Du T, Deng S, Jia X. A promising nucleic acid therapy drug: DNAzymes and its delivery system. Front Mol Biosci 2023, 10, 1270101 (16 pages).

[130] Huang P-J-J, Liu J. In vitro selection of chemically modified DNAzymes. Chemistry Open 2020, 9, 1046–1059.

[131] Gao M, Wei D, Chen S, Qin B, Wang Y, Li Z, Yu H. Selection of RNA-cleaving TNA enzymes for cellular Mg2+ imaging. ChemBioChem 2023, 24, e202200651 (7 pages).

# 13 Nucleic acids as drugs and drug targets 3: target recognition and delivery

## 13.1 Aptamers

Aptamers are single-stranded oligonucleotides that exhibit high and selective affinity toward a given target, and the size of which may vary from small molecules to macromolecules or even specific regions on the cell surface [1]. They are isolated from a random pool of synthetic oligonucleotides by in vitro selection, a method known as SELEX (systematic evolution of ligands by exponential enrichment) [2, 3]. Aptamers can be either oligodeoxyribonucleotides (ODNs) or oligoribonucleotides (ORNs), the latter ones being more popular. The intrastrand interactions are stronger with ORNs than ODNs. This together with more diverse three-dimensional structure of ORN expectedly results in higher affinity and selectivity [4].

The selection of ODN and ORN aptamers is somewhat different, although the underlying principle is still the same. While ODNs may be amplified enzymatically by PCR (polymerase chain reaction) [5], ORNs must be first copied to ODNs by reverse transcriptase and after amplification, transcribed back to ORNs. The generation of ORN aptamers by conventional SELEX methodology [2] is depicted in Figure 13.1A. A pool of single-stranded ODNs is first obtained by chemical solid-supported ODN synthesis, or enzymatically by primer extension reaction. These ODNs contain a fixed primer region of known sequence in both termini and an intervening 20- to 50-nucleotide-long random sequence. The random sequence is obtained by using a mixture of all canonical nucleoside phosphoramidites in coupling steps. The pool is then amplified by PCR and transcribed enzymatically to a corresponding ORN library. The ORN pool obtained is equilibrated with the target, and the sequences bound to the target are separated from the unbound ones. Depending on the identity of the target, various techniques based on the partition between liquid and solid phases are used [6, 7]. With protein targets, nitrocellulose membrane filtration is often applied, since nitrocellulose captures protein/ORN adducts much more efficiently than free ORNs. Affinity and resin chromatography, capillary electrophoresis and magnetic beads coated with the target are other frequently used techniques. The ORNs bound to the target are then released, converted to ODN sequences by reverse transcriptase and subjected to reconversion to ORN library by PCR amplification and subsequent transcription. When this cycle is repeated several times, the most tightly binding ORN is gradually enriched. The identity of the final aptamer is clarified by sequencing.

Aptamers can additionally be generated to recognize structures on cell surface, such as glycans or proteins, expressed as an indication of some abnormal state of the cell. In other words, whole cells are used as targets even when the target modification on the cell surface is unknown [1, 7]. On applying whole-cell targeting, the ORN library has to be subjected to both positive and negative selections, as depicted in Figure 13.1B.

https://doi.org/10.1515/9783111325637-013

ORNs that bind to control cells not expressing the target modification are first removed. In the next step, cells expressing the target modification are treated with the remaining ORNs. The bound ORNs are released and after amplification taken to the next round.

Interestingly enough, the whole-cell targeting approach has been applied even in vivo to generation of aptamers for cancer metastases [8]. The ORN library is administered in a mouse model of the cancer under interest. After circulation, the organ of interest is collected, and the bound ORNs are liberated, amplified and reinjected for the next round.

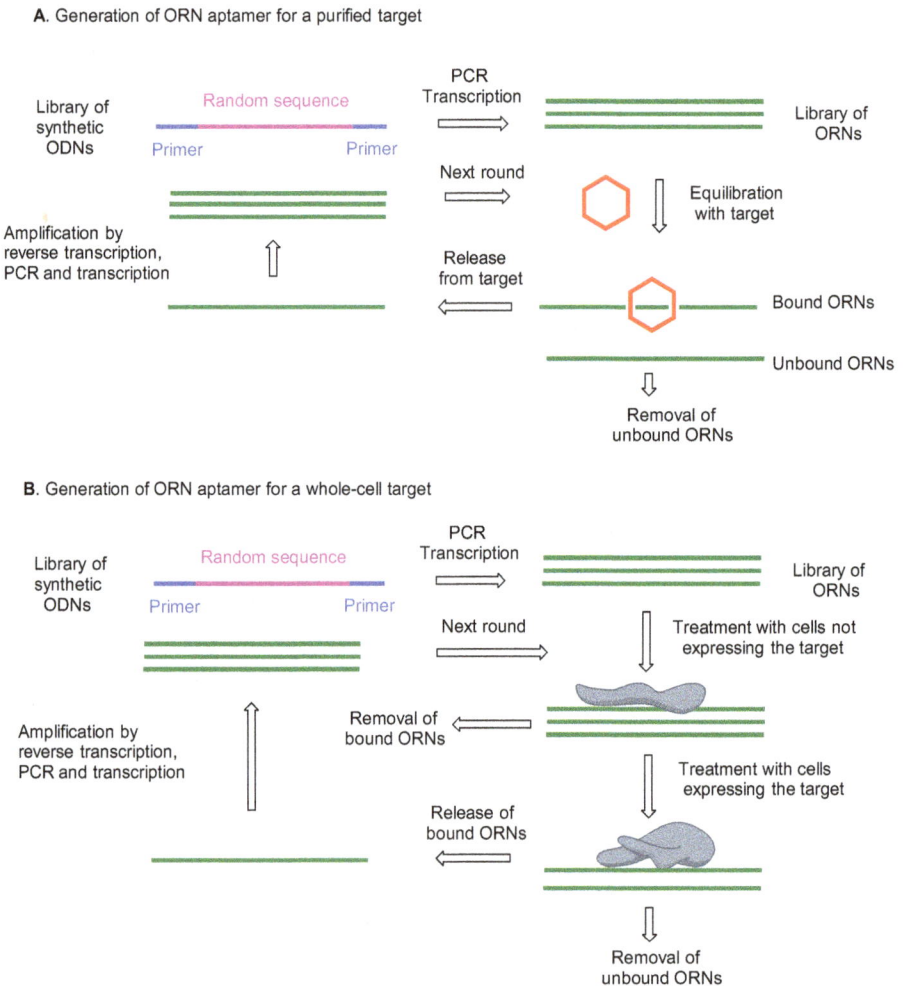

**A.** Generation of ORN aptamer for a purified target

**B.** Generation of ORN aptamer for a whole-cell target

**Figure 13.1:** Generation of ORN aptamers for purified (**A**) and whole-cell-based (**B**) targets.

Aptamers are potential drug candidates, very much like antibodies. They, in fact, have even some clear advantages over antibodies [9]. Their production by cell-free chemical synthesis is rather cost-effective and the variability between batches is small. They are relatively small, penetrate rapidly in tissues and, what is important, they seldom are immune stimulatory. But aptamers suffer from two major shortcomings. Although they are generally regarded as high-affinity binding agents, the affinity still may be too low for efficient chemotherapy. One possible way to increase affinity is introduction of base moiety modifications such as amino acids resembling side chains that expectedly exert stabilizing nonbonding interactions between aptamers and the target protein [10]. Unfortunately, attempts to achieve very high affinity often lead to sacrifices in selectivity.

Another challenge is biological half-life. Aptamers like oligonucleotide drugs, in general, suffer from rapid excretion and susceptibility to nuclease degradation [11]. The biological half-life of chemically unmodified aptamers is as short as 10 min. The most common sugar modifications include LNA and replacement of the 2'-OH with a fluoro, amino or methoxy substituent [12]. These modifications stabilize toward nucleases and may additionally enhance binding. Among them, 2'-O-methylribo (Figure 13.2A) [13] and LNA nucleotides (Figure 13.2B) [14] can be incorporated during the SELEX process, whereas 2'-fluoro [15] and 2'-amino [16] modifications (Figure 13.2C) have been successfully introduced only when present in pyrimidine nucleotides. In other words, a canonical nucleoside 5'-triphosphate is replaced with one of these structural analogs during the transcription step [17]. Base moiety modifications are rather well tolerated by DNA polymerases, and numerous DNA-based aptamers containing C5-derivatized 2'-deoxyuridines or C8-drivatized 2'-deoxyadenosines have been selected [18]. Sugar modifications, except those discussed above, are not accepted by polymerases conventionally used in SELEX. Engineered polymerases have, however, enabled the selection process for some heavily sugar-modified nucleoside 5'-triphosphates. 2'-Choro and 2'-azido (Figure 13.2D) [19], 2'-fluoro-arabino (Figure 13.2E) [20], hexitol nucleic acid (Figure 13.2F) [21], α-L-threose (Figure 13.2G) [22] and 4'-thio (Figure 13.2H) [23] monomers have been successfully introduced by the SELEX protocol.

Modifications may also be incorporated after the SELEX process by replacing the constituent of a preselected aptamer with its structural analog during the final chemical synthesis of the aptamer. Introduction of unlocked nucleic acid monomer (Figure 13.2I) [24] offers an example. A phosphodiester linkage can be replaced with a phosphorothioate or dithiophosphate linkage to increase the biological stability, or with a neutral methylphosphonate, alkylphosphate or triazole linkage to improve the delivery [25]. 3'- or 5'-terminus may be truncated and subsequently elongated. For example, a handle containing a functionality, such as phenyl azide or diazirine group, for covalent attachment to the target may be added [26]. Post-SELEX 3',5'-circularization of a linear aptamer may also enhance target-binding [27]. However, there is always a risk for lowering the affinity by a post-SELEX modification. Another way to increase the affinity to a specific target is to generate an additional aptamer toward a site in close vicinity on the cell surface and tether the two aptamers to each other with an appropriate linker [28].

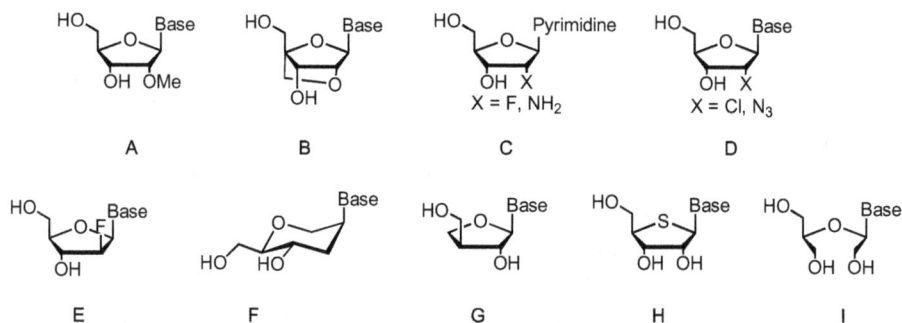

**Figure 13.2:** Modified nucleosides introduced into aptamers by conventional polymerases (A–C), by engineered polymerases (D–H) in SELEX process, or by replacing a nucleoside in a preselected aptamer with its structural analog in the final synthesis (I).

Spiegelmers constitute a special class of biologically stable aptamers. They are ORNs that contain L-sugars instead of the normal D-sugars. For this reason, spiegelmers are poor substrates of nucleases [29]. Their generation, however, is rather demanding, since enantiomer of the ultimate target analog is required for selection. This markedly limits the scope of spiegelmer applications. Generation of spiegelmer to a protein target, for instance, requires its enantiomer consisting of D-amino acids only.

Even if stabilized toward nuclease degradation, aptamers are rapidly excreted by renal filtration. To prolong circulation in the blood stream, they are often conjugated to high-molecular-weight polyethylene glycol (PEG) [30]. Formulation with cholesterol [31], proteins [32], liposomes [33] or nanoparticles [34] has also been utilized for the same purpose. Covalent clustering of aptamers also prolongs their circulation and improves pharmacokinetics [35].

As drugs, aptamers could in principle serve either as antagonists by preventing protein–protein or receptor–ligand interactions, or as agonists by activating the target. However, all aptamers that so far have entered in major clinical trials are antagonists. Only two aptamers have received approval for medicinal use. The first one, Macugen (brand name "Pegaptanib sodium" approved in 2004), is a PEG conjugate of a 27-mer RNA aptamer consisting of 2′-F- and 2′-O-Me-modified nucleosides, with the exception of two unmodified ones, and an inverted thymidine at the 3′-terminus. It is an antiangiogenic drug that stops the growth of new blood vessels. Macugen is used for the treatment of neovascular (wet) age-related macular degeneration (AMD) [36]. Its molecular target is vascular endothelial growth factor (VEGF) isoform 165. The drug is administered by intravitreal injection into the eyes. The second one, Avacincaptad pegol (brand name Izervay), is also a medication for the treatment of AMD. It is a 39-mer pegylated

RNA aptamer with similar modifications as Macugen. The target is a protein (human complement C5) that switches on VEGF [37]. As regards applications of spiegelmers outside drug development, an L-RNA aptamer has been developed for the recognition of RNA G-quadruplex structure [38]. An appropriately selected L-RNA aptamer is even selective for G4 conformers, binding parallel G4 conformer considerably more tightly than antiparallel and hybrid ones [39].

Coagulation is another indication area under active development of aptamer therapeutics. Four aptamers are in clinical trials. First one, ARC1779, a 49-mer DNA/RNA aptamer PEG conjugate, binds to a glycoprotein (von Willebrand factor) that plays a role in platelet adhesion and thrombosis [40]. Second one, REG1, is a two-component system that consists of aptamers RB006 and RB007. RB006 is a pegylated 31-mer RNA aptamer with 2′-fluoro modifications and 3′-thymidine cap. It binds to coagulation factor IXa and, hence, prevents thrombin formation [41]. The other component, RB007, is a complementary 15-mer sequence that binds to RB006 reversing its action. Third one, BAX499, a 32-mer pegylated RNA aptamer, is aimed at improving the coagulation of hemophilia patients. It binds to the tissue factor pathway inhibitor [42]. Fourth one, NU172, is an unmodified 26-mer DNA aptamer that has been developed to bind thrombin [43].

Two aptamers are in clinical trials as anticancer drug candidates. AS1411 is a 26-mer unmodified guanine-rich DNA aptamer having a G-quadruplex structure, which stabilizes it against enzymatic degradation. The aptamer binds to nucleolin protein expressed on the surface of cancer cells [44]. This leads to internalization, resulting in apoptosis by inhibition of the synthesis of BCL2-mRNA-binding protein involved in cell survival and proliferation. AS1411 in combination with cytarabine (*arabino* cytosine) has passed to phase II clinical trials against acute myeloid leukemia. The other anticancer drug candidate, NOX-A12, is a pegylated 45-nucleotide RNA spiegelmer [45]. It has been generated to bind to the chemokine C-X-C motif ligand 12 (CXCL12) that plays an important role in cancer cell proliferation and angiogenesis. NOX-A12, in combination with small molecular drugs, has passed to phase II clinical trials against relapsed chronic lymphocytic leukemia or multiple myeloma [1]. Two other pegylated RNA spiegelmers, 40-mer NOX-E36 and 44-mer NOX-H94, have passed to clinical trials for the treatment of diabetic nephropathy and anemia of chronic disease, respectively. NOX-E36 binds to human chemokine CCL2 [46] and NOX-H94 to hepcidin [47], a peptide playing a key role in iron hemostasis. A 2′-fluoro-RNA aptamer has recently been selected against the receptor-binding domain of SARS-CoV-2 spike protein [48, 49].

Since aptamers can be generated for the recognition of proteins or glycans expressed on the cell surface, it is quite natural that they have received interest as cell-specific drug carriers [7, 50–52]. The results of massive work on cell-type recognition have been collected in a single database [53]. The targets for which aptamers are available are numerous [51]: prostate-specific membrane antigen (PSMA), CD4 receptor, HIV-1 envelope glycoprotein gp120, the epidermal growth factor receptor (EGFR), tenasin-C glycoprotein overexpressed during tumor growth and angiogenesis, just to mention a few examples. Usually the drug is conjugated covalently to the aptamer, but noncovalent binding of aromatic planar drug molecules by intercalation is also possible. The conjugated drug is often another oligonucleotide. These aptamer-targeted oligonucleotide conjugates are discussed later in Section 13.3. Besides their role as drug carriers, aptamers have found numerous applications in clinical diagnostics. In particular, pathogen recognition, detection of various cancer-related biomarkers and molecular recognition of cancers are under active research [37, 54–56].

In diagnostics, aptamers dissected into two parts have received interest as biosensors [57]. Simultaneous binding of both fragments to a specific target results in formation of entire aptamer that triggers a detectable signal. A fluorescent sensor of thrombin serves as an illustrative example. The system consists of three components: a water-soluble polycationic highly fluorescent polymer, known as PEEP, and two fragments of thrombin-binding DNA aptamer, one of them conjugated to fluorescein. These two fragments light up high fluorescence emission upon binding to PEEP by electrostatic interactions. In the presence of thrombin, the fluorescence intensity is marked reduced, since thrombin triggers the formation of a G-quadruplex structure that retards interaction between PEEP and the fluorescein conjugate group [58]. Binding of the two aptamer components to their target can be enhanced by cooperative formation of a covalent linkage between the two splits. Three reactions have been reported for this purpose: strain-promoted alkyne–azide cycloaddition [59], pH-responsive $N$-methoxyoxazolidine formation between 3′-terminal 2′-deoxy-2′-$N$-(methoxyamino) and 5′-terminal 5′-deoxy-5′-(2-oxoethyl) nucleosides [60], and boronic ester formation between 3′-terminal ribonucleoside and 5′-terminal boronic acid [61].

## 13.2 Transfection of naked therapeutic oligonucleotides

Delivery of therapeutic oligonucleotides is one of the major problems on the way to oligonucleotide drugs. Even if injected into a certain organ, the oligonucleotide has to pass the cell membrane. This cellular uptake takes place via an endocytic mechanism [62]. In other words, oligonucleotide internalized into the cell remains trapped in an endocytic vesicle formed from the cell membrane [63]. The process is reversible; endosome may again become fused with the cell membrane, releasing its content outside the cell. However, only a few percent of the vesicles that stay in cytoplasm release their cargo there. The rest undergo slow acidification; pH inside the endosome drops from 6.3 to 5.5. Eventually the endosome is fused with lysosome, and its content becomes degraded by hydrolytic enzymes.

Phosphorothioate oligonucleotides show marked tendency of binding to the cell surface proteins, which enhances endocytosis [64] compared to their phosphodiester counterparts, which are taken up through entrapment of a minor amount of extracellular liquid phase into tiny vesicles, a process known as pinocytosis. A possible way to enhance the cellular uptake of oligonucleotides is their delivery as nanoscale complex with cationic complexing agents (polyplex complexes) [65] or entrapped in liposomes [66]. Polyplex complexes are taken up by another endocytic mechanism, phagocytosis, and liposomes either by endocytosis, or possibly by fusion with the cell membrane which releases the oligonucleotide directly into the cytoplasm. Evidently, the efficiency of release from the endosomes/liposomes plays a major role in all the uptake mechanisms discussed above.

Poly(ethylenimine) (PEI) serves as an example of a rather extensively studied transfection agent that has been assumed to enhance endosomal release [67]. It has been argued that PEI could possibly work as a proton sponge in the pH range around 6. In other words, during acidification of endosome, an increasing amount of the nitrogen atoms of PEI becomes protonated, which increases the osmotic pressure inside endosome and might lead to rupture of endosome and release of cargo into the cytoplasm [68]. However, the relevance of this proton sponge concept has been recently challenged [62, 69]. Another method expected to enhance the endosomal escape is derivatization of polyimines with fusogenic peptides that mimic the fusion domain of influenza viruses [70], or with endosomolytic peptide from HIV gp41 protein [71]. The major limitation for the use of PEI or its congeners in vivo is that the polyamine component is required in excess to achieve efficient transfection, and this tends to lead to liver toxicity.

Increasing the lipophilicity of oligonucleotide by lipid formulation has turned out to be a successful approach [72, 73]. The therapeutic oligonucleotide, often siRNA, is complexed with cationic fusogenic lipids that facilitate the release from endosomes by interaction with the anionic membrane. The most commonly used formulations include lipid nanoparticles (LNPs), lipidoid nanoparticles and lipoplexes. Synthesis, characterization and formulation methods of various lipid-based carriers used for delivery of oligonucleotide drugs have recently been reviewed [74].

LNPs are composed of ionizable cationic lipids, usually ammonium ions bearing long alkoxy chains, cholesterol, phospholipids and PEG lipids [73, 75]. The ionizable cationic lipid entraps the negatively charged oligonucleotide into the LNP, yielding about 80 nm particles and the surface of which is electrically close to neutral. Under the slightly acidic conditions inside endosomes, the lipids remain positively charged and by interaction with endosome membrane facilitate release the oligonucleotide cargo into the cytoplasm. The efficiency of this process largely depends on the overall shape of the lipid. Figure 13.3 shows the structures of some lipids optimized for the purpose. Cholesterol and phospholipids stabilize the LNP structure and shape. PEG lipids prevent particle aggregation. Approval of patisiran as the first siRNA drug was a breakthrough for LNP formulation. Patisiran is administered as LNPs, the key component of which is an ionizable cationic lipid, MC3 (Figure 13.3). The particles additionally contain phospholipid, cholesterol and PEG lipids, and the optimal composition on mol% scale being 50% MC3, 10% phospholipid, 38.5% cholesterol and 1.5% PEG lipids [76]. This formulation has also formed the basis for the mRNA-LNPs used successfully for the delivery of SARS-CoV-2 mRNA vaccines of various companies, although the lipids used in place of MC3 vary [77]. Examples are given in Figure 13.3. In case accumulation in the liver is desired, as with patisiran, the surface of siRNA-LNP is coated with apolipoprotein E (ApoE), since ApoE receptors are overexpressed in hepatocytes [78, 79].

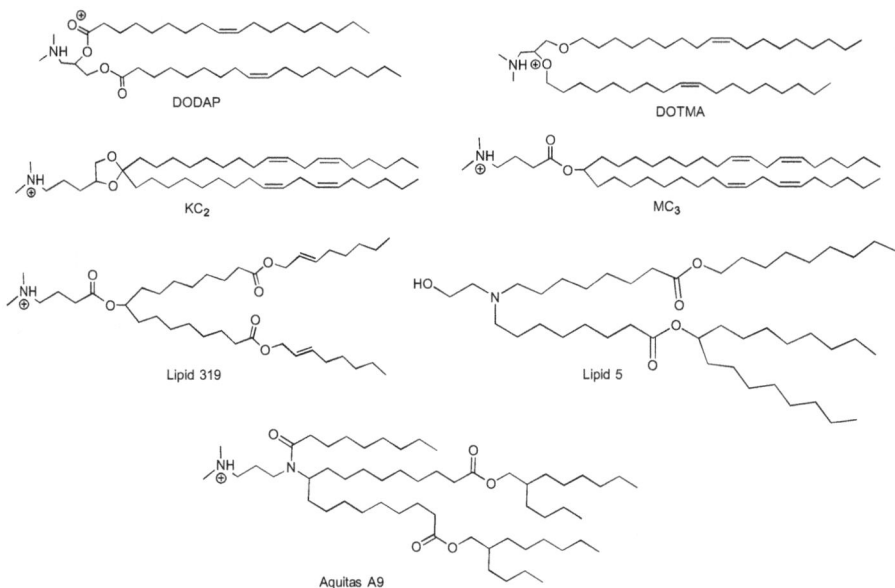

**Figure 13.3:** Ionizable cationic lipids used as constituents of RNA-lipid nanoparticles. Lipid 319, Lipid 5 and Aquitas A9 are used in mRNA-LNPs of SARS-CoV2 vaccines.

Lipidoid nanoparticles are somewhat smaller than LNPs [80]. They differ from LNPs by the identity of lipid constituents. In addition to cholesterol or some other neutral lipid, they contain PEG lipids and various lipidoids, that is, lipid-like materials, such as amines bearing long-chain alkoxycarbonyl or alkylaminocarbonyl side chains.

Lipoplexes differ from LNPs and lipidoid nanoparticles in the sense that siRNA is not encapsulated inside the LNP but bound electrostatically on the surface of positively charged LNP [81]. The size of these nanoscale complexes is about 100 nm.

Besides LNPs, exosomes (liposomes) have received interest as carriers of nucleic acid drugs [82, 83]. They are around 100-nm-sized spherical particles encapsulated with one or several lipid bilayers. Cells continuously release exosomes into the extracellular space, but they can also be prepared artificially in aqueous solution from phospholipids, for instance, from phosphatidylcholine and cholesterol. A single lipid bilayer usually encapsulates exosomes used as drug carries. When this kind of a carrier reaches the target cell, the exosome membrane fuses with the cell membrane releasing the cargo into the cytoplasm. As a proof of concept, exosomes conjugated with a PSMA deliver siRNA into tumor cells and induce tumor regression [84], and siRNA-loaded exosomes bearing cholesterol-anchored EGFR RNA aptamers resulted in the regression of nonsmall-cell lung cancer in animal trials [85].

Different types of nanoparticles have been used for the systemic delivery of an antisense oligonucleotide into the brain through the blood–brain barrier (BBB) [86]. The particles were obtained by self-assembly of the oligonucleotide with the glucose conjugate of a PEG–(poly-lysine) copolymer, and the amino functions of which were mostly derivatized with 3-mercapto-propylamidine and 2-thiolaneimine groups. Internalization takes place by binding of glucose ligands on the surface of nanoparticles to the glucose transporter 1, known to carry glucose through the BBB. This is one of the few examples of extrahepatic carriers of oligonucleotide drugs.

Nanoparticles made up entirely of nucleic acid components are gaining increasing attention as another delivery method of therapeutic oligonucleotides [87–89]. The size and shape of the particles can be controlled by the so-called DNA origami technique, that is, by making use of hybridization of long DNA sequences with each other and with short "staple" sequences. DNA nanoparticles do not markedly accumulate in the liver; hence, they may serve as extrahepatic carriers. Tetrahedral DNA structures, for example, are of interest since they cross the plasma membrane without assistance of any transfection agent by a so-called caveolin-mediated endocytosis, that is, by initial binding into small invaginations of plasma membrane [90]. Some of the chains may additionally display targeting ligands, such as aptamers. Successful examples include targeting of toll-like receptor (TLR9) receptors with CpG oligonucleotides [91], nucleolin receptor with aptamer [92] and folate receptor with folic acid conjugation [93].

In addition to nucleic acid nanoparticles, inorganic nanoparticle carriers have received interest [94]. Oligonucleotides functionalized with a terminal thiol group form a dense layer around the particle, owing to high affinity of sulfur to gold [95]. Such particles, called spherical nucleic acids, are internalized without transfection agents. When the surface density is sufficiently high, the oligonucleotide chains are rather well protected against nucleases and they do not undergo exchange reaction with glutathione, while the high density does not prevent their hybridization with complementary targets. Even higher stability against enzymatic degradation is achieved by using phosphorothioate oligonucleotides instead of their unmodified counterparts [96]. SiRNA in the form of spherical nucleic acid has silenced a luciferase reporter gene twice as efficiently as achieved by transfection agents [97]. The gene silencing efficiency is enhanced when both siRNA strands are immobilized to the core, giving a hairpin-like structure that prevents dissociation of the strands [98]. Spherical nucleic acids that consist of a polymeric core structure and mixed ASO/tyrosinase coating have been shown to exhibit antimelanogenic effect in a mouse model [99].

## 13.3 Receptor-mediated delivery of oligonucleotide conjugates

A viable approach for targeting oligonucleotide drugs to a specific organ or cell type is utilization of covalent conjugation [100]. A conjugate group attached covalently to the therapeutic oligonucleotide is aimed at binding to a receptor on the cell surface, resulting in enrichment of oligonucleotide outside the cell. The cellular uptake then takes place through receptor-mediated endocytosis. Accordingly, efficiency of endosomal escape still plays an important role. Lipid, carbohydrate, peptide, antibody, aptamer and small molecular conjugates have been used for targeting.

Lipid conjugates enhance delivery above all to the liver. They exhibit prolonged circulation time in plasma, enhanced cellular uptake and facile escape from endosomes [100, 101]. The most commonly used conjugate group is cholesterol that, in particular, has been used for targeting of siRNAs and antagomirs. Conjugation of cholesterol to the 3′-terminus of siRNA sense strand (Figure 13.4A) enhances interaction with lipoproteins and, hence, cellular uptake via lipoprotein–receptor-mediated pathways [102]. A spacer between cholesterol and oligonucleotide seems to be essential. With 5′-terminal conjugates of the sense strand, a 6–10 atom linker was observed to be optimal [103]. The 5′-terminus of the antisense strand is not generally used for conjugation since the 5′-phosphate is essential for recognition by Ago2. However, this site has been used for construction of a photoresponsive cholesterol conjugate (Figure 13.4B) [104]. The underlying idea is that the conjugate is internalized in the absence of light, and the actual inhibitory action is triggered by photochemical cleavage and o-nitrophenylethylene linker between the 5′-phosphate and the cholesterol conjugate. As regards antagomir conjugates, a 3′-cholesterol conjugate of a 2′-O-Me antagomir silenced the liver-specific miR-12 in mice [105]. Another conjugate that additionally contained several phosphorothioate linkages was shown to suppress breast cancer metastasis [106].

Interestingly, cholesterol conjugation has additionally shown promise for gene silencing in the central nervous system (CNS) when injected directly into the brain. SiRNA bearing cholesterol at the 3′-terminus of the sense strand silenced the Huntington's disease gene in the striatum of mouse brain, when administered directly into neurons, but not when administered systemically [107]. Expression of 2′,3′-cyclic nucleotide 3′-phosphodiesterase, an enzyme specific for a different CNS cell type, namely oligodendrocytes, was silenced with a cholesterol-conjugated siRNA administered to the rat brain (corpus callosum) [108]. A 3′-cholesterol conjugate of a 2′-O-Me antagomir that contained several thioate modifications efficiently targeted miRNAs when injected locally into the mouse cortex [109].

α-Tocopherol (vitamin E), which has its own physiological transport pathway, is another lipid successfully used for liver targeting with mouse models [110, 111]. The conjugate group was attached to the 5′-terminus of the antisense strand of a somewhat elongated (27/29) siRNA (Figure 13.4C). The conjugate became cleaved by Dicer to an unconjugated 21/21 siRNA in cytoplasm. The level of target mRNA (ApoB mRNA) was reduced even more markedly than with a cholesterol conjugate. On using a gapmer-type antisense oligonucleotide for the same purpose, α- tocopherol at the 5′-terminus enhanced the antisense effect, but only when attached via a sequence of 4–7 unlocked nucleic acid monomers (cf. Figure 13.2I) that evidently were cleaved inside the cell [112]. Both cholesterol and α-tocopherol conjugates with a DNA/RNA heteroduplex, bearing the conjugate group at the 5′-terminus of the RNA strand, could even cross the BBB after intravenous administration in mice and rats [113].

Among other lipids, palmitic acid, anandamide and squalene deserve attention. Palmitoyl group attached via a 2-hydroxy-3-aminopropyl linker to the 5′-phosphorothioate function of a thiophosphoramidate antisense oligonucleotide (AON) enforced telomerase inhibition in several cancers (Figure 13.4D) [114]. Anandamide that binds to cannabinoid receptors in neuronal and immune cells caused efficient RNA interference when conjugated to a dendritic siRNA nanostructure depicted in Figure 13.4E [115]. 3′-Squalene conjugation of the siRNA sense strand, in turn, results in spontaneous nanoparticle formation (Figure 13.4F). The particles have been shown to stabilize the siRNA, enhance uptake in cancer cells and markedly inhibit tumor growth [116, 117].

Besides lipids, sugar clusters have been utilized for targeting oligonucleotides to the liver since late 1990s. These early studies showed that a tetravalent galactose [118] and a trivalent N-acetylgalactosamine (NAG) conjugate groups [119] markedly enhanced the accumulation of oligonucleotides into parenchymal liver cells by binding to the asialoglycoprotein receptor. More recently, a trivalent NAG conjugate of a 2′-O-MOE RNA gapmer (Figure 13.5A) was observed to exhibit 10-fold increase in the antisense effect compared to the unconjugated oligonucleotide [120]. According to the positron emission studies with rats, approximately one third of 2′-O-Me RNA bearing two trivalent galactose clusters (Figure 13.5B) accumulates in the liver, the rest being mainly excreted through the kidneys to urine [121]. In striking contrast, the accumulation of the same oligonucleotide bearing only one galactose is negligible. As discussed in Section 12.1, siRNAs bearing a trivalent NAG conjugate group at the 3′-terminus of either the sense (inclisiran) or antisense (givosiran) strand have been accepted for clinical use. Replacement of the glycosidase-sensitive β-glycosidic linkages with more resistant S- and C-glycosidic bonds or their α-anomers did not enhance binding to asialoglycoprotein receptors [122]. The same triantennary structure also internalizes PNA when conjugated to the C-terminus. Interestingly, the internalization is more efficient with a linear triantennary structure, that is, when three consecutive C-terminal N-(2-aminoethyl)glycine units each bears a single NAG ligand [123].

Peptide conjugates are used for dual purposes: targeting to a specific receptor or facilitation of unspecific cellular uptake by conjugation to a cell-penetrating peptide (CPP). CPPs are short peptides that are usually rich in basic amino acids, Arg and Lys [124]. When present at a high concentration, they are able to penetrate directly through the membrane of eukaryotic cells. At low concentrations, and especially when used as carriers of other biomolecules, they are taken up by a less efficient endocytic mechanism [125]. Presence of hydrophobic amino acids, Trp and Phe, seems to enhance endocytosis [126]. CPP conjugation particularly improves the uptake of neutral oligonucleotides, not

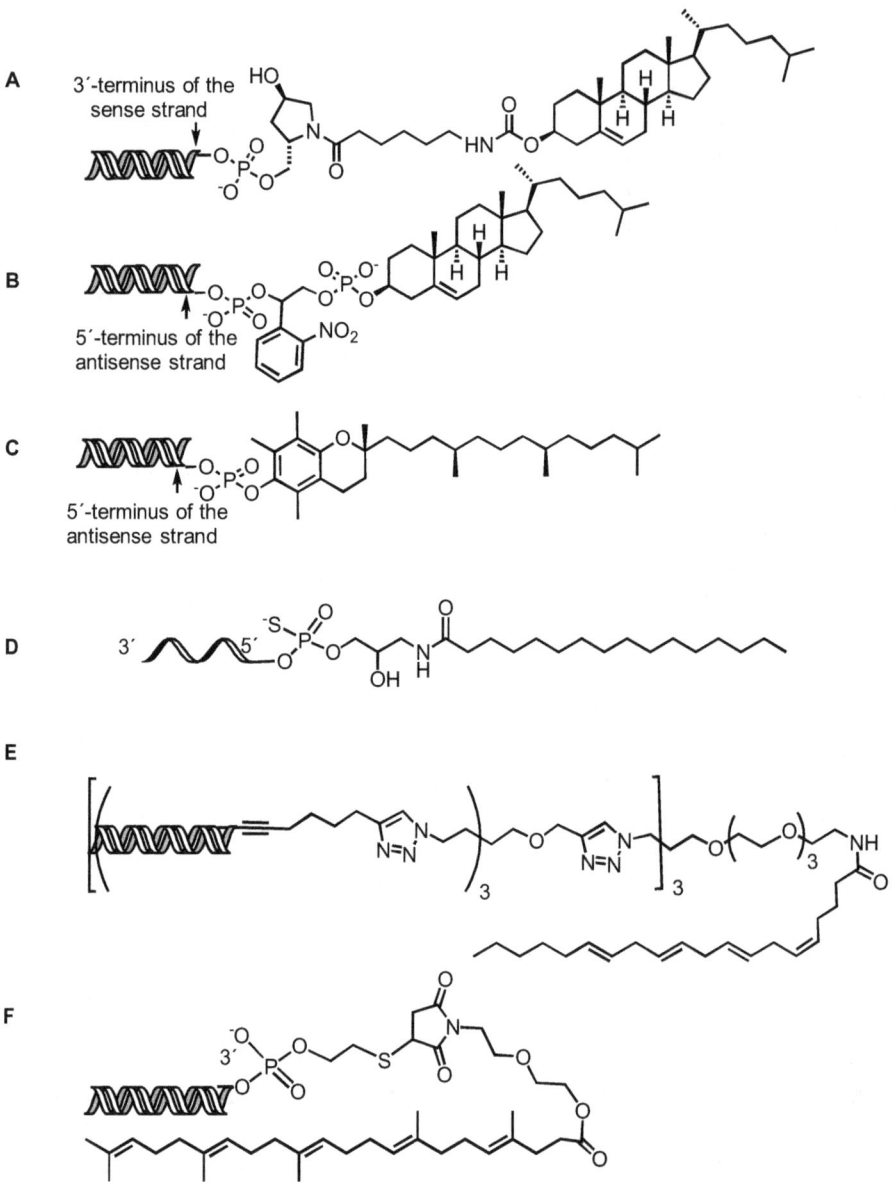

**Figure 13.4:** Lipid conjugates of therapeutic oligonucleotides: (A) cholesterol conjugate of siRNA [102], (B) photoresponsive cholesterol conjugate of siRNA [104], (C) α-tocopherol conjugate of siRNA [110, 111], (D) palmitoyl conjugate of thiophosphoramidate AON [114], (E) anandamide conjugate of a dendritic siRNA [115] and (F) squalene conjugate of siRNA [116, 117].

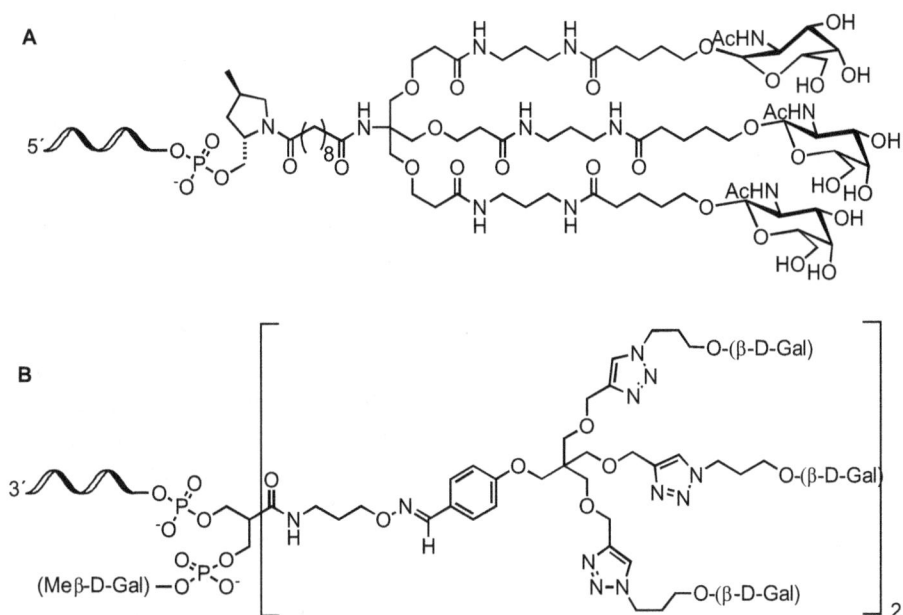

**Figure 13.5:** Sugar cluster conjugates for receptor-mediated targeting of therapeutic oligonucleotides: A [120] and B [121].

that of anionic ones. An early example is offered by the transportan conjugate of PNA. Transportan is a 27-amino acid-long CPP that consists of a 12-mer amino terminal sequence of neuropeptide galanin, linked via lysine residue to a 14-mer carboxy terminal sequence of a peptide toxin from wasp venom. When PNA–transportan conjugate was administered into the spinal canal of a rat, a galanin-type receptor in the spinal cord became downregulated [127]. A real breakthrough for medical use of CPP conjugates has, however, been achieved with neuromuscular diseases [128]. Both PNA and morpholino oligomers have been successfully used as splice-switching oligonucleotides that eliminate the influences of out-of-frame mutations within an exon. A peptide conjugate of a splice-switching morpholino oligonucleotide restored the dystrophin protein expression in a mice model [129]. The conjugate group consisted of a muscle-targeting pentapeptide core, Ile-Leu-Phe-GLN-Tyr, and flanking CPP wings containing Arg, β-Ala and 6-aminohexanoic acid (Figure 13.6A). Related conjugates are under continuous effort for exon skipping base medication [128, 130].

Integrin receptors have been targeted by trivalent cyclic Arg-Gly-Asp (cRGD) conjugates. An siRNA conjugate bearing the trivalent cRGD conjugate group at the 3′-terminus of the sense strand (Figure 13.6B) was bound selectively to αvβ3 integrin receptor in the cell line. This receptor is overexpressed in angiogenic vasculature and in certain tumors [131]. The potential of cRGD–siRNA molecules in antitumor therapy has been demonstrated with tumor-bearing mice [132]. A monovalent conjugate (Figure 12.6C) targeted to vascular EGFR downregulated the corresponding mRNA (45–44%) and reduced the tumor volume.

Another peptide shown to have potential for receptor-mediated cell targeting is bombesin, Asn-Gln-Trp-Ala-Val-Gly-His-Leu-Met, that binds to gastrin-releasing peptide receptor, a member of the G-protein-coupled receptor superfamily. Receptor-mediated uptake of a splice-switching 2'-O-Me RNA phosphorothioate oligonucleotide (Figure 13.6D) [133] and its triantennary bombesin-His$_6$ analog (Figure 12.6E) [134] has been demonstrated in cell lines.

Antibodies are extensively used for targeting of small-molecule cancer drugs. This attractive alternative has not, however, gained similar success with therapeutic oligonucleotides. The real drug development of such conjugates is still at an early phase in spite of some promising observations reported. Treatment of leukemia cells with a disulfide-linked conjugate of a phosphorothioate AON and a monoclonal antibody specific for CD19 receptor retarded the cell growth both in vitro and on using a mouse model [135]. In another study [136], a monoclonal antibody conjugate (αCD22 Ab) of an AON knockdown protein MXD3, a transcriptional repressor expressed during the S phase of cell cycle, resulted in leukemia cell apoptosis in vitro. siRNA conjugated with Fab' fragment of transferrin receptor CD71 antibody showed long-lasting gene-silencing in the heart and skeletal muscles in mice [137]. Conjugates of AONs with antibodies of antigens expressed in glioblastoma stem cells (CD44, EphA2) downregulated a key gene (DRR/FAM107A) in patient-derived cancer stem cells [138]. Release of antibody conjugates from endosomes is possibly the factor that most severely limits their therapeutic efficiency [139].

Aptamers are attractive candidates for targeting of oligonucleotides since they can be selected for recognition of special structures on cell membrane. Their utilization for the purpose, however, is still somewhat limited. Most extensive studies deal with PSMA. This is a protein that is expressed on the surface of some prostate cancer cells. The aptamer moiety enhances internalization into PSMA-expressing cells and the siRNA moiety, after processing with Dicer, downregulates gene(s) vital for the cell. Inhibition of tumor growth has been demonstrated in prostate cancer cell lines [140]. A previously selected aptamer [141] was linked to siRNAs that targeted two essential genes, PLK1 and BCL2, overexpressed in human tumors. The structurally optimized conjugates resulted in pronounced regression of PSMA-expressing tumors in a mouse model after systemic administration, and the antitumor activity was further enhanced by PEG conjugation [142]. Another example is an siRNA conjugated to an aptamer that had been selected for HIV-1 envelope (gp120) protein. The antiviral activity was tested in a humanized mouse model where HIV-1 replication mimicked the situation in human HIV-infected patients [143, 144]. Treatment with the aptamer alone suppressed HIV-1 replication, but the aptamer–siRNA combination prolonged the inhibition, resulting in an antiviral effect that extended several weeks beyond the last injected dose.

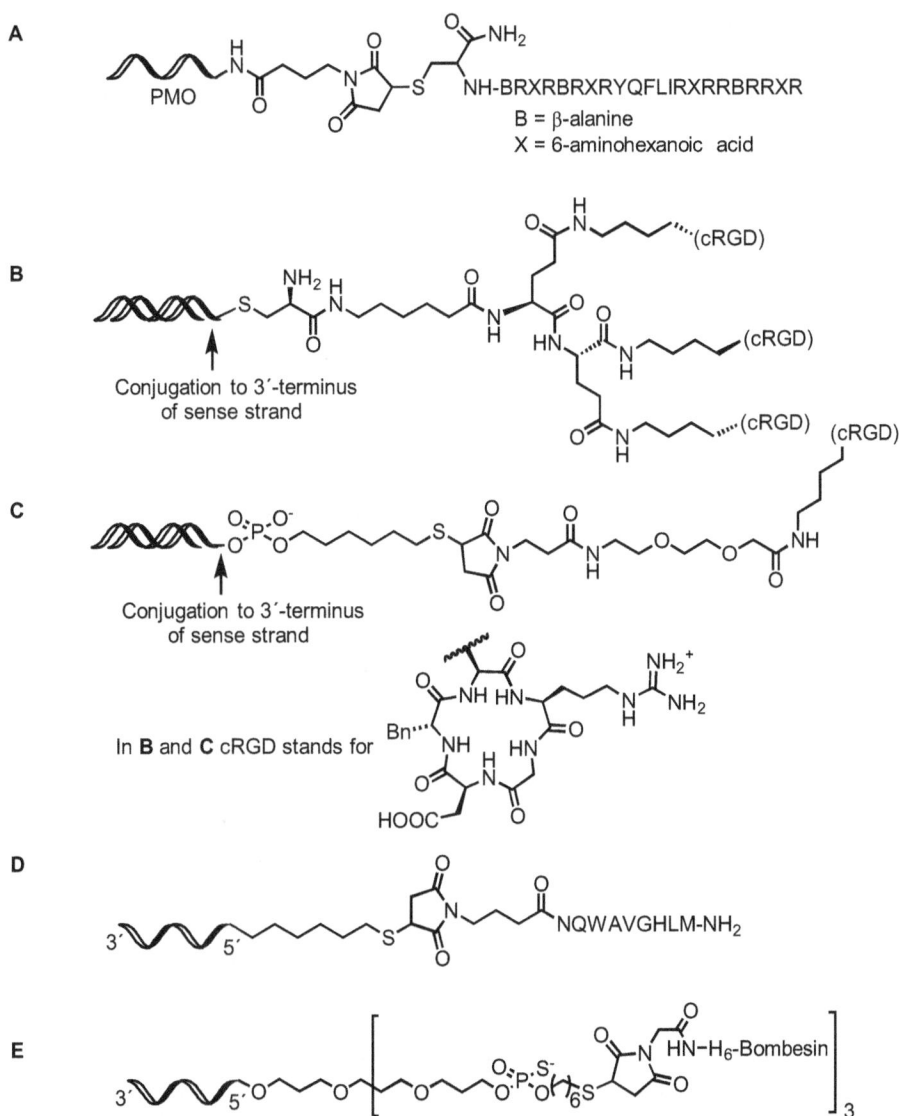

**Figure 13.6:** Peptide conjugates for receptor-mediated targeting of therapeutic oligonucleotides: A [129], B [131], C [132], D [133] and E [134].

Immune stimulatory CpG–oligonucleotides that serve as agonists of TLR9 still constitute a family of oligonucleotides useful for targeting. CpG–siRNA conjugates have been targeted to the immune suppressor gene Stat3 in order to stimulate anticancer immune response [145, 146]. Silencing of Stat3 in mice leads to activation of tumor-associated immune cells and ultimately to potent antitumor immune responses.

Among small molecules utilized for targeting, anisamide and folic acid deserve attention. Anisamide is a high-affinity ligand of sigma receptors that play a role in regulation of ion channels. A trivalent anisamide conjugate of a splice-switching antisense oligonucleotide (Figure 13.7A) has displayed enhanced cellular uptake in cell lines [147]. Folic acid, in turn, binds to folic receptor α that is overexpressed in many human cancers. A folic acid–siRNA conjugate (Figure 13.7B) was specifically internalized into folate receptor expressing cells in cell lines, but silencing of a reporter gene took place only when delivered as a complex with a polycationic transfection agent [148]. Folic acid has been conjugated both to siRNA and to various carriers used to enhance its delivery [149, 150]. Internalization to HeLa cells and knockdown of the reporter gene in 30–80% efficiency has been documented, but no real breakthrough in chemotherapy has been achieved so far.

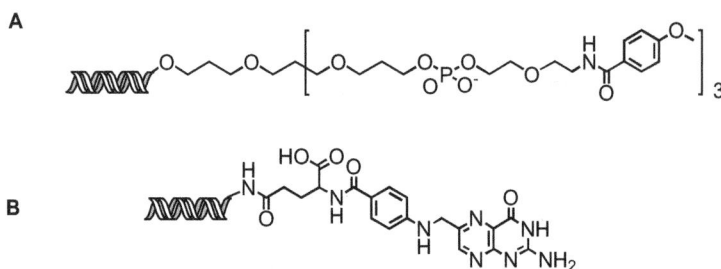

A

B

**Figure 13.7:** Small-molecule conjugates of siRNA used for receptor-mediated targeting: (A) anisamide conjugate [147] and (B) folic acid conjugate [148].

Besides receptor-selective targeting, delivery and pharmacokinetics of therapeutic oligonucleotides can be improved by unspecific carriers attached covalently. A common method is covalent PEGylation. This prolongs the circulation time of oligonucleotides, owing to reduced renal filtration. A more complex carrier polymer is poly(butylaminovinylether) (PBAVE) depicted in Figure 13.8. A part of the primary amino groups is PEGylated and another part bears a receptor-targeting ligand NAG in Figure 13.8B. The rest of the amino groups are acylated with maleic anhydride, which prevents protonation of the amino groups at physiological pH. The therapeutic oligonucleotide is attached to this carrier polymer via a biodegradable linker, such as a disulfide linker [151]. When internalized into endosomes, pH drops, resulting in cleavage of the maleoyl groups and concomitant protonation of the exposed amino groups. This, in turn, is assumed to increase the osmotic pressure inside the endosome and, hence, enhance release of the cargo into the cytoplasm. An alternative approach is coinjection of oligonucleotide with this polymer [152]. Oligonucleotides bearing a linear tail that contained several disulfide linkages (Figure 12.8B) have been reported to be internalized into the cytoplasm in 10 min after addition into a cell culture, that is, twice as rapidly as with an efficient transfection agent [153]. Very recently, perfluorocarbon-conjugated oligonucleotides (Figure 13.8C) have been reported to penetrate HeLa cells and several types of human cells without damaging the cells [154].

Instead of PBAVE-like synthetic polymers, proteins can be used as carriers of therapeutic oligonucleotides. Conjugates bearing a targeting ligand, for example, have been tethered to albumin via a biodegradable disulfide linkage [155], or the targeting ligand has been anchored to albumin via a PEG chain [156]. A fusion protein of peptide transduction domain and dsRNA-binding domain (DRBD) has been shown to deliver siRNA to tumors in a mouse model [157]. Binding of siRNA to DRBD masks its negative charges facilitating the cellular uptake. Protamine is a polycationic protein that binds siRNA by electrostatic interactions. A targeted system is obtained by fusion with an antibody [158]. Delivery to tumor cells has been verified in vivo. Polycationic protamine possibly enhances the escape from endosomes.

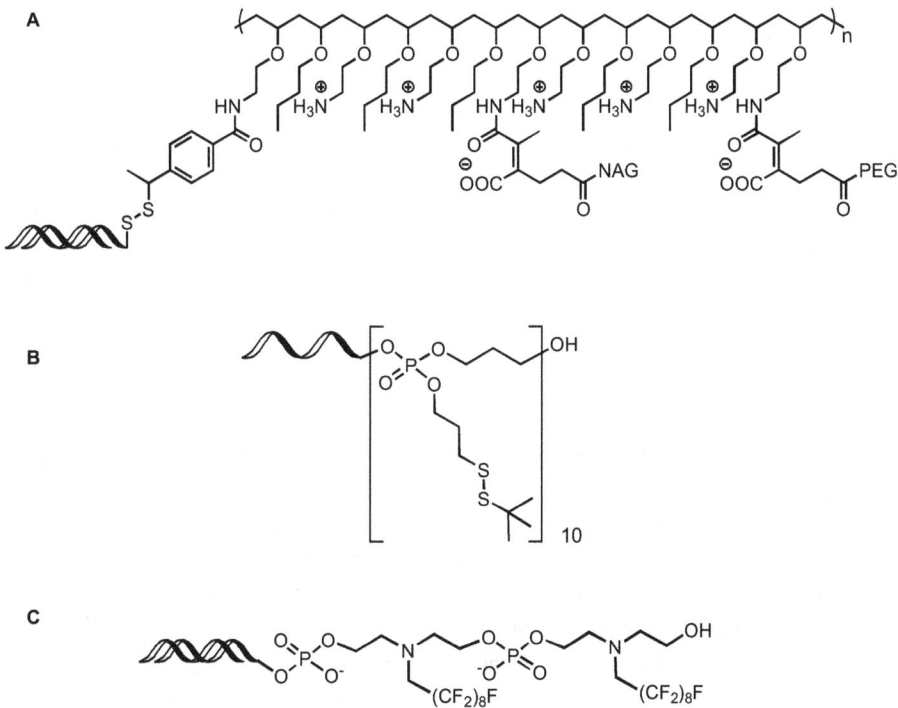

**Figure 13.8:** (A) Poly(butylaminovinylether) carrier conjugated to siRNA via a cleavable disulfide linker [151], (B) linear oligo(disulfide) carrier conjugated to an antisense oligonucleotide [153] and (C) perfluorocarbon-conjugated siRNA [154].

## 13.4 Prodrugs of therapeutic oligonucleotides

A possible way to enhance the cellular uptake of therapeutic oligonucleotides and to protect them against nuclease degradation is to administer them in a protected form, as prodrugs. An obvious approach is protection of phosphodiester or phosphorothioate linkages, but in case of siRNA, protection of the 2'-OH is also an option. The protecting groups are aimed at being removable by intracellular enzymes, reductive intracellular conditions, photochemically or simply by heat without participation of any external agent. All these approaches are still at rather early stage of development, but they may well have considerable potential.

A prerequisite for utilization of an enzyme-labile protecting group is that the intracellular activity of the enzyme in question is much higher in cytoplasm than in plasma. Carboxyesterases, for example, are these kinds of enzymes, and hence, esterase-labile protecting groups have played a pioneering role in development of prodrug strategies for oligonucleotides [159]. The most extensively studied approach is protection of internucleosidic phosphodiester linkages by various S-acyl-2-thioethyl (SATE) groups [160]. Intracellular esterases catalyze hydrolysis of the thioester linkage and the exposed thiol group attacks on C1, resulting in departure of the protecting group remnants as episulfide (Figure 13.9A). The shortcoming of this approach is that the episulfide released is a highly efficient alkylating agent. In addition, negative charge is accumulated on oligonucleotide upon removal of increasing number of protecting groups, and this retards the subsequent enzymatic deacylation steps [161]. In addition, aqueous solubility of a fully protected oligomer is poor.

A modification of the SATE strategy has, however, given interesting results regarding delivery of siRNA [162]. One fourth of phosphodiester linkages were protected with different combinations of S-pivaloyl-, S-(2-hydroxy-1,1-dimethylethylcarbonyl)- and S-(4-formylbenzoyl)-2-thioethyl groups. The 2'-hydroxy functions neighboring these protected phosphodiester linkages were either methylated or replaced with a fluoro substituent. The partly protected siRNAs were quite well soluble and stable toward nucleases, but they were not internalized. For this reason, the formyl functionalities of four S-(4-formylbenzoyl)-2-thioethyl groups in the sense strand were derivatized with a TAT-peptide delivery domain. Inside the cell, the phosphate-protecting groups were removed and RNA interference was induced. Systemic delivery leading to dose-dependent RNAi response in a mouse model was demonstrated by conjugating a hepatocyte-specific trivalent NAG unit to a S-(4-formylbenzoyl)-2-thioethyl group at the 5'-terminus of the sense strand (Figure 13.9B).

Oligonucleotides sensitive to nitroreductases constitute another potential family of prodrugs. Nitroreductases become activated under oxygen-deficient (hypoxic) conditions that are typical for intracellular environment in solid tumors. Hypoxia-activated antisense oligonucleotides are, hence, attractive candidates for cancer chemotherapy. ODNs bearing 5-nitro-2-furylmethyl [163], 5-nitro-2-(thiophen-2-yl)methyl [163] or 3-(2-nitrophenyl)propyl groups [164] on phosphodiester linkages (Figure 13.10A and B) ex-

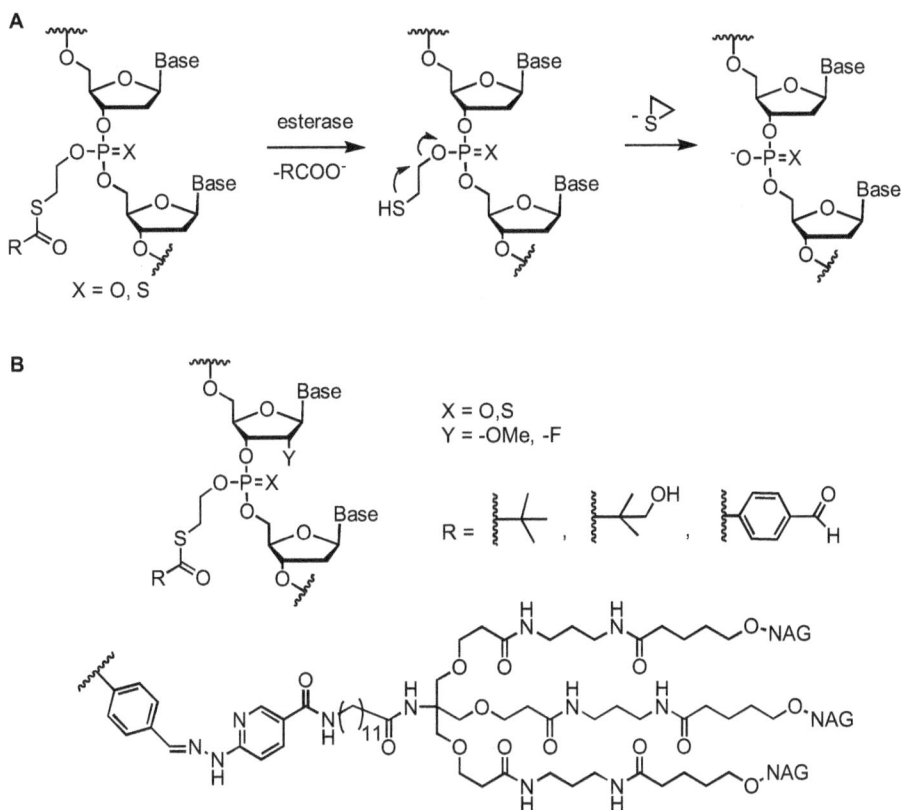

**Figure 13.9:** (A) Mechanism of esterase-dependent removal of S-acyl-2-thioethyl protecting groups from internucleosidic phosphodiester linkages [160]. (B) S-Acyl-2-thioethyl groups used for partial protection of siRNA and targeting of siRNA to asialoprotein receptor in mouse liver. NAG stands for N-acetylgalactosamine [162].

hibit good nuclease resistance and cellular uptake, and they undergo deprotection by nitroreductases in tumor cell extracts.

Oligomers of $O^4$-(4-nitrobenzyl)floxouridine (2'-deoxy-5-fluorouridine) or its [4-(4-nitrophenylazido)benzyl analog (Figure 13.10C) offer an alternative approach for utilization of hypoxia activation [165]. The protected oligomer is nontoxic, but intracellular reductases reduce under hypoxic conditions nucleobases to 5-fluoro-4-aminobenzyl)uracils and catabolic breakdown of the oligomer produces toxic 5-fluorouracil [166]. As a proof of concept, floxuridine oligomer suppressed the growth of solid tumors in mice.

Intracellular concentration of glutathione, a peptide-like SH antioxidant (GSH), falls in the range of 1–10 mM. Accordingly, one might expect that disulfides are reductively cleaved to thiols in the intracellular environment. Based on this expectation, internucleosidic phosphodiesters have been protected with a *trans*-5-benzyloxy-1,2-dithian-4-yl group

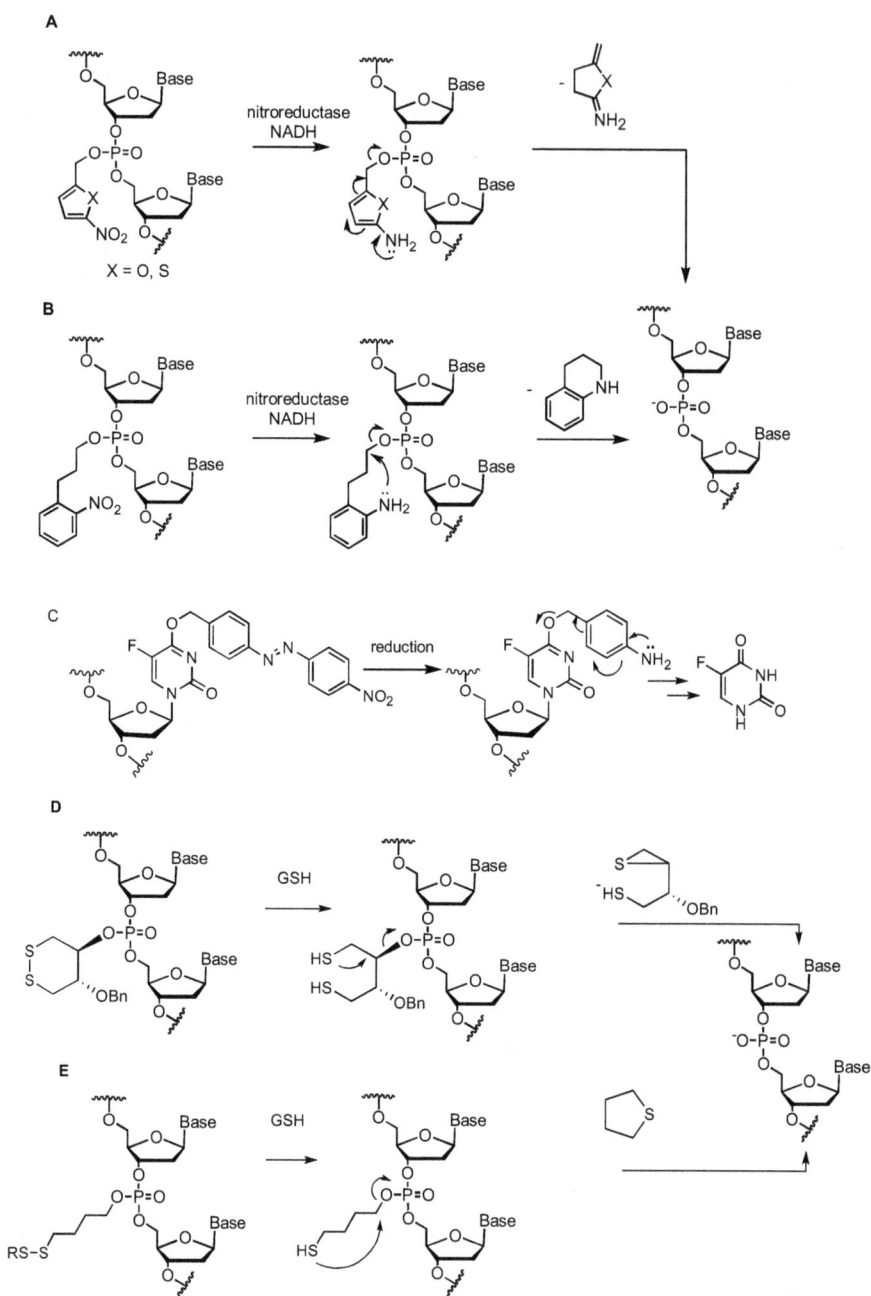

**Figure 13.10:** Removal of 5-nitro-2-furylmethyl (A) and 3-(2-nitrophenyl)propyl (B) groups by nitroreductase under hypoxic conditions [164]. Reductase induced the release of toxic 5-fluorouracil from protected floxouridine oligomers under hypoxic conditions (C) [166]. (D and E) Glutatione induced deprotection of disulfide-protected oligomers [167, 168].

(Figure 13.10D) [167]. Protected ODNs were stable against nucleases and penetrated through the cell membrane. The protecting groups could be removed from a 10-mer in 10 mM GSH at pH 7.0, although not very rapidly. In fact, the reaction took 75 h at 37 °C. Evidently, for this reason, the antisense effect was weaker than with unprotected phosphorothioate AON. A slightly higher effect has been reported for the open-chain dithio analogue that is deprotected four times faster (Figure 13.10E) [168].

Thermolabile groups have been used for the protection of internucleosidic phosphodiester linkages of immune stimulatory CpG oligonucleotides [169]. N-Formyl-N-methyl-2-aminoethyl-protected phosphorothioate ODNs (Figure 13.11A) that underwent thermolytic deprotection with a half-life of 73 h at 37 °C markedly increased immunoprotection against various pathogen-caused diseases in mouse models. Numerous thermolabile groups exhibiting faster or slower removal kinetics have been developed [170, 171] since this pioneering study. Illustrative examples are depicted in Figure 13.11B–D.

4-Acetylthio-2,2-dimethyl-3-oxobutyl group has been introduced as a thermolabile protecting group that is additionally removable by esterases (Figure 13.11E) [172]. The underlying idea is that the group initially serves as an esterase-labile group. The esterase-catalyzed reaction, however, is decelerated with accumulation of negatively charged phosphodiester linkages. Finally, the thermolytic ester hydrolysis takes over. Unfortunately, the desired thermolytic departure of the protecting group is accompanied with cleavage of the O–C5′ bond as a minor side reaction, which limits the application of this approach to relatively short ODNs.

Esterase- and reduction-responsive groups have been used for the protection of 2′-OH of siRNA. Blocking of the 2′-OH function stabilizes siRNA toward nucleases by preventing the attack of 2′-OH on phosphorus. In addition, cellular uptake is expectedly enhanced due to increased lipophilicity. On using acyloxymethyl groups, the ester linkage is hydrolyzed by intracellular esterases, and the remaining 2′-O-hydroxymethyl group is removed spontaneously as formaldehyde. siRNAs bearing several 2′-O-pivaloyloxymethyl groups on the sense strand (Figure 13.12A) have been shown to enhance cellular uptake and control gene expression in human cell lines in the presence of a transfection agent [173]. Somewhat unexpectedly, the bulky group did not markedly destabilize the RNA duplex, and the A-type conformation was still maintained [174]. Cellular uptake without a transfection agent was achieved by more hydrophobic 2′-O-phenylisobutyloxymethyl protections [175]. The shortcoming is that the esterase-catalyzed deprotection is quite slow.

2′-O-Methyldithiomethyl protection is another sugar moiety modification that has been shown to exhibit enhanced RNA interference in cell line [176]. The disulfide bond is reductively cleaved by intracellular glutathione, and the 2′-OH is exposed by spontaneous departure of the mercaptomethyl group as thioformaldehyde (Figure 13.12B).

Numerous pro-ODN strategies based on photolabile protecting groups have been introduced and validated in cell lines by silencing of a reporter gene [177]. One should, however, bear in mind that on going to animal models the situation is more complicated. The tissue diffusion of light is very limited, and lengthy UV irradiation, even with a long UV wavelength, causes side reactions lowering the efficiency of deprotec-

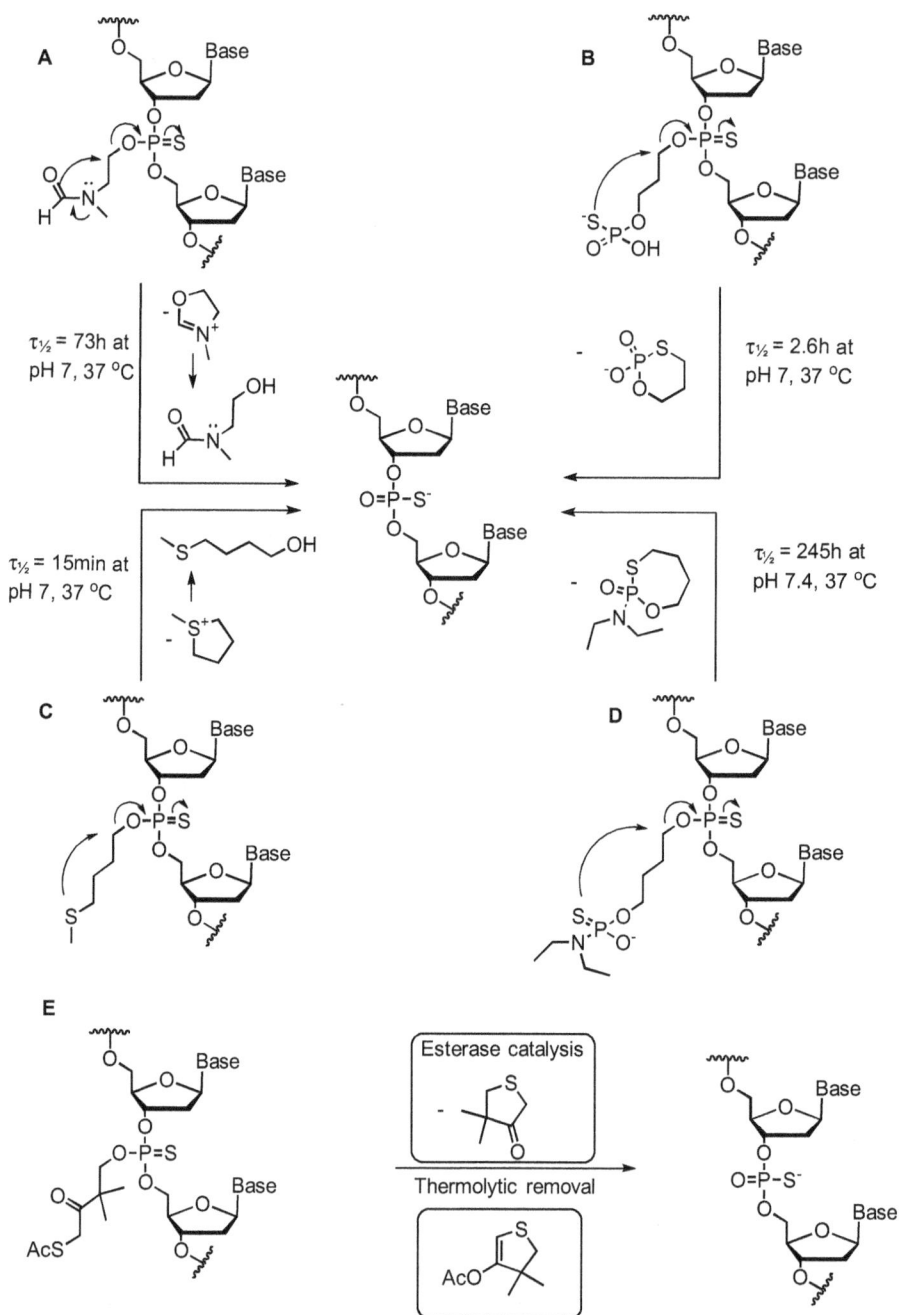

**Figure 13.11:** Thermolabile protecting groups of internucleosidic phosphodiester linkages (A–D) [169–171] and an esterase- and thermo-labile group (E) [172].

**A**

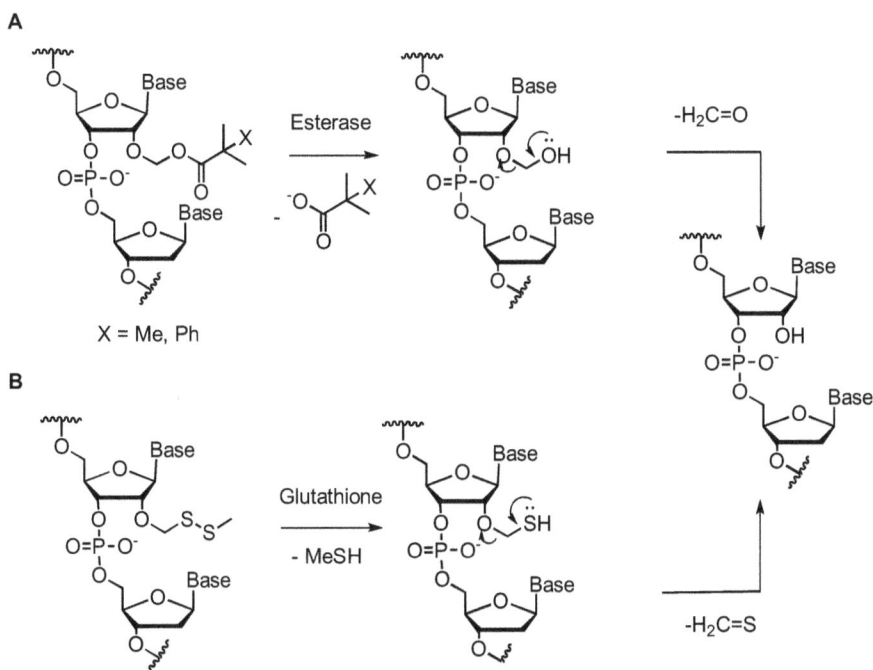

Esterase

-H₂C=O → (shown as -H$_2$C=O)

X = Me, Ph

**B**

Glutathione

- MeSH

-H₂C=S

**Figure 13.12:** Deprotection of siRNAs bearing esterase-labile (A) [173] or reductively cleavable (B) [176] 2′-O-protecting groups.

tion and producing potentially toxic products. Accordingly, the future of real therapeutic applications is difficult to predict.

Photocatalytic triggering of RNA cleavage has been demonstrated in cell lines by both RNA interference and antisense mechanism. As regards siRNA technology, introduction of 2-(2-nitrophenyl)propyl-protected thymine and guanine bases (Figure 13.13B and C) close to the Ago2 cleavage site converts siRNA totally inactive, and the original catalytic activity is fully recovered by irradiation at 366 nm [178]. Alternatively, 6-nitropiperonyloxymethyl (NPOM) protections (Figure 13.13D and E) in the seed region turned the siRNA inactive and photochemical deprotection returned the activity [179]. The RNase H-based antisense effect has been shown to be abolished by three NPOM protections along the phosphorothioate ODN and recovered by irradiation at 365 nm [180].

## 13.5 Genome targeting by CRISPR/Cas9

CRISPR/Cas9 is a system of "the clustered regularly interspaced short palindromic repeats" (CRISPR) and "the CRISPR-associated protein 9" (Cas9). It is a defense mechanism of *Streptococcus pyogenes* bacteria [181] against foreign dsDNA. Related CRISPR-based mechanisms are typical adaptive immunity systems in bacteria and archaea. In a modi-

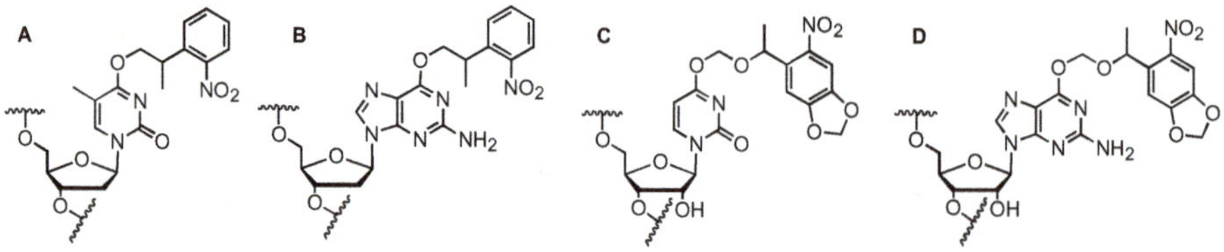

**Figure 13.13:** Photolabile protections in the vicinity of Ago2 cleaving site (A and B) [178] and in the seed region (C and D) of siRNA [179].

fied form, this system is extensively used for site-selective targeting of genome [182]. *S. pyogenes* uses a three-component nucleoprotein complex to cleave foreign DNS: Cas9 apoenzyme, CRISPR-RNA (crRNA) and trans-activating CRISPR-RNA (tracrRNA). 5′-Terminal nucleotides (1–20) of crRNA form the guide sequence that recognizes the DNA strand, while the 3′-terminal sequence (21–40) hybridizes with the 5′-terminal sequence of tracrRNA and activates the Cas9 machinery. Both crRNA and tracrRNA are produced by the cellular machinery during the immunization process triggered by foreign DNA. When the CRISPR/Cas9 system is used for diagnostic or therapeutic purposes, the two short RNAs, crRNA and tracrRNA, are merged to a longer single-guide RNA (sgRNA), which complexes with apo-Cas9. The complex is schematically depicted in Figure 13.14.

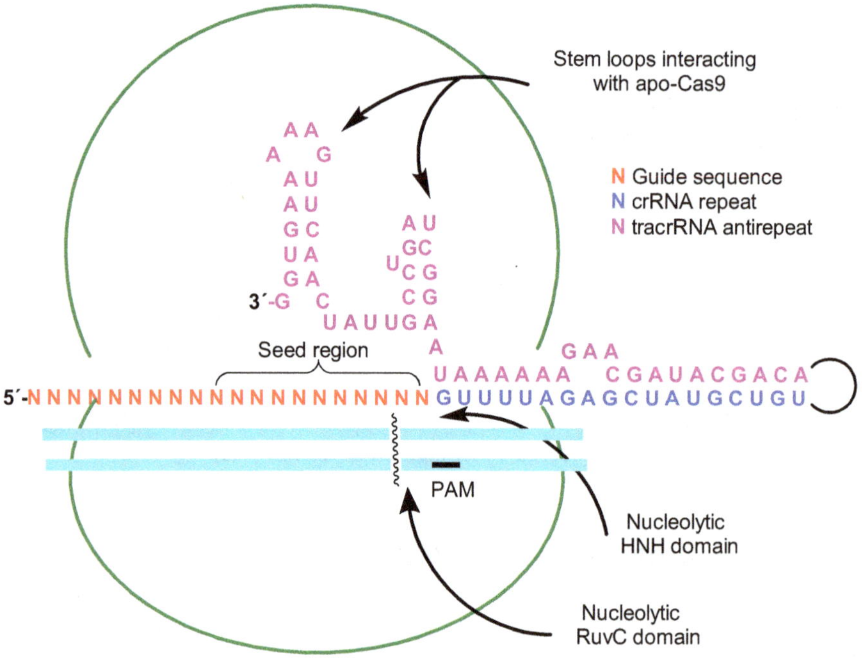

**Figure 13.14:** Schematic presentation of recognition and cleavage of dsDNA by a complex of Cas9 enzyme and single-guide RNA.

For anchoring of sgRNA to Cas9, the two stem loops within the tracrRNA sequence and the linker sequence between them play an important role [183, 184]. Additionally, several direct contacts, most likely through 2′-OH groups, occur in the borderline region of the guide sequence and crRNA repeat. Cas9 is a large DNA endonuclease that contains two catalytic domains [185]. The HNH domain cleaves the DNA strand that is complementary to the guide sequence of sgRNA and the RuvC domain cuts the complementary strand. Both strands are cleaved at the same base pair, giving a blunt ended product.

By the Cas/sgRNA complex, virtually any sequence in DNA may be targeted. About 10–12 nucleotides in the 3′-terminus of the guide sequence form a seed region that is of primary importance for target recognition. Mismatches in seed region are exceptionally destabilizing. A crucial structural requirement for cleavage is the presence of specific 3–5 nucleotide base sequence (protospacer adjacent motif, PAM) in DNA close to the desired cleavage site. The structure of PAM varies from one Cas to another. For the most frequently used SpyCas9, it is 5′-NGG-3′ in the nontarget strand.

CRISPR/Cas technology clearly shows phenomenal promise as a technique of predesigned manipulation of genome, even human genome [186, 187]. Many thresholds have, however, to be overcome on the way to real therapeutic applications [188]. Owing to the relatively short guide sequence and the even shorter seed region of crRNA, elimination of off-target effects undoubtedly is the major problem to be solved, as far as human genome is concerned [189]. In this respect, antiviral therapy by manipulation of viral genome appears less problematic [190]. Structural modifications have enabled therapeutic applications of antisense oligonucleotides, siRNA and aptamers. Most likely, structural studies will also solve some of the problems related to biological stability [191, 192], delivery [193–195] and possibly also selectivity of targeting [196, 197]. Recently, 2′-O-methyl-3′-phosphonoacetate modification at 3′-end of sgRNA has been reported to enhance the efficiency of gRBA/Cas9 editing by an order of magnitude compared to more common 2′-O-methyl-3′-phosphorothioate modification [198].

# Further reading

Benizri S, Gissot A, Martin A, Vialet B, Grinstaff MW, Barthelemy P. Bioconjugated oligonucleotides: Recent developments and therapeutic applications. Bioconjugate Chem 2019, 30, 366–383.

Debart F, Dupouy C, Vasseur -J-J. Stimuli-responsive oligonucleotides in prodrug-based approaches for gene silencing. Beilstein J Org Chem 2018, 14, 436–469.

Gallas A, Alexander C, Davies MC, Purib S, Allen S. Chemistry and formulations for siRNA therapeutics. Chem Soc Rev 2013, 42, 7983–7997.

Hu Q, Li H, Wang L, Gu H, Fan C. DNA nanotechnology-enabled drug delivery systems. Chem Rev 2019, 119, 6459–6506.

Jiang F, Doudna JA. CRISPR-Cas9 structures and mechanisms. Annu Rev Biophys 2017, 46, 505–529.

Kaur H, Bruno JG, Kumar A, Sharma TK. Aptamers in the therapeutic and diagnostic pipelines.
Theranostics 2018, 8, 4016–4032.

Kruspe S, Giangrande PH. Aptamer-siRNA chimeras: Discovery, progress, and future prospects.
Biomedicines 2017, 5, e45 (20 pages).

Kulkarni JA, Witzigmann D, Chen S, Cullis PR, van der Meel R. Lipid nanoparticle technology for clinical
translation of siRNA therapeutics. Acc Chem Res 2019, 52, 2435–2444.

Lachelt U, Wagner E. Nucleic acid therapeutics using polyplexes: a journey of 50 years (and beyond).
Chem Rev 2015, 115, 11043–11078.

Mallick AM, Tripathi A, Mishra S, Mukherjee A, Dutta C, Chatterjee A, Roy RS. Emerging approaches for
enabling RNAi therapeutics. Chem Asian J 2022, 17, e202200451 (1 of 35).

Ming X, Laing B. Bioconjugates for targeted delivery of therapeutic oligonucleotides. Adv Drug Deliv Rev
2015, 87, 81–89.

Pei D, Buyanova M. Overcoming endosomal entrapment in drug delivery. Bioconjugate Chem 2019, 30,
273–283.

Röthlisberger P, Hollenstein M. Aptamer chemistry. Adv Drug Deliv Rev 2018, 134, 3–21.

Setten RL, Rossi JJ, Han Sp. The current state and future directions of RNAi-based therapeutics. Nat Rev
Drug Discov 2019, 18, 421–446.

Wang C, Zhang Y, Dong Y. Lipid nanoparticle–mRNA formulations for therapeutic applications. Acc Chem
Res 2021, 54, 4283–4293.

Xiao Y, Tang Z, Huang X, Chen W, Zhou J, Liu H, Liu C, Kong N, Tao W. Emerging mRNA technologies:
delivery strategies and biomedical applications. Chem Soc Rev 2022, 51, 3828–3845.

Zhang Y, Lai BS, Juhas M. Recent advances in aptamer discovery and applications. Molecules 2019, 24, 941
(22 pages.

Zhou J, Rossi J. Aptamers as targeted therapeutics: current potential and challenges. Nat Rev Drug Discov
2017, 16, 181–202.

## References

[1]    Zhou J, Rossi J. Aptamers as targeted therapeutics: current potential and challenges. Nat Rev Drug
       Discov 2017, 16, 181–202.

[2]    Tuerk C, Gold L. Systematic evolution of ligands by exponential enrichment: RNA ligands to
       bacteriophage T4 DNA polymerase. Science 1990, 249, 505–510.

[3]    Ellington AD, Szostak JW. In vitro selection of RNA molecules that bind specific ligands. Nature 1990,
       346, 818–822.

[4]    Shu Y, Pi F, Sharma A, Rajabi M, Haque F, Shu D, Leggas M, Evers BM, Guo P. Stable RNA
       nanoparticles as potential new generation drugs for cancer therapy. Adv Drug Deliv Rev 2014, 66,
       74–89.

[5]    Bartlett JMS, Stirling D. A short history of the polymerase chain reaction. In Bartlett JMS, Stirling D
       eds. PCR Protocols. Methods in Molecular Biology™. Humana Press, vol. 226, 3–6.

[6]    Ozer A, Pagano JM, Lis JT. New technologies provide quantum changes in the scale, speed, and
       success of SELEX methods and aptamer characterization. Mol Ther Nucleic Acids 2014, 3, e183 (18
       pages).

[7]    Darmostuk M, Rimpelova S, Gbelcova H, Ruml T. Current approaches in SELEX: An update to
       aptamer selection technology. Biotechnol Adv 2015, 33, 1141–1161.

[8]    Mi J, Liu Y, Rabbani ZN, Yang Z, Urban JH, Sullenger BA, Clary BM. In vivo selection of tumor-
       targeting RNA motifs. Nat Chem Biol 2010, 6, 22–24.

[9]    Keefe AD, Pai S, Ellington A. Aptamers as therapeutics. Nat Rev Drug Discov 2010, 9, 537–550.

[10] Chen Z, Luo H, Gubu A, Yu S, Zhang H, Dai H, Zhang Y, Zhang B, Ma Y, Lu A, Zhang G. Chemically modified aptamers for binding affinity to the target proteins via enhanced non-covalent bonding. Front Cell Dev Biol 2023, 11, 1091809 (12 pages).

[11] Healy JM, Lewis SD, Kurz M, Boomer RM, Thompson KM, Wilson C, McCauley TG. Pharmacokinetics and biodistribution of novel aptamer compositions. Pharm Res 2004, 21, 2234–2246.

[12] Kratschmer C, Levy M. Effect of chemical modifications on aptamer stability in serum. Nucleic Acid Ther 2017, 27, 335–344.

[13] Burmeister PE, Lewis SD, Silva RF, Preiss JR, Horwitz LR, Pendergrast PS, McCauley TG, Kurz JC, Epstein DM, Wilson C, Keefe AD. Direct in vitro selection of a 2'-O-methyl aptamer to VEGF. Chem Biol 2005, 12, 25–33.

[14] Veedu RN, Wengel J. Locked nucleic acid nucleoside triphosphates and polymerases: On the way towards evolution of LNA aptamers. Mol Biosyst 2009, 5, 787–792.

[15] Ruckman J, Green LS, Beeson J, Waugh S, Gillette WL, Henninger DD, Claesson-Welsh L, Janjic N. 2'-Fluoropyrimidine RNA-based aptamers to the 165-amino acid form of vascular endothelial growth factor (VEGF165). J Biol Chem 1998, 273, 32, 20556–20567.

[16] Lin Y, Qiu Q, Gill SC, Jayasena SD. Modified RNA sequence pools for in vitro selection. Nucleic Acids Res 1994, 22, 5229–5234.

[17] Keefe AD, Cload ST. SELEX with modified nucleotides. Curr Opin Chem Biol 2008, 12, 448–456.

[18] Röthlisberger P, Hollenstein M. Aptamer chemistry. Adv Drug Deliv Rev 2018, 134, 3–21.

[19] Chen TJ, Romesberg FE. Enzymatic synthesis, amplification, and application of DNA with a functionalized backbone. Angew Chem Int Ed 2017, 56, 14046–14051.

[20] Lietard J, Abou Assi H, Gomez-Pinto I, Gonzalez C, Somoza MM, Damha MJ. Mapping the affinity landscape of thrombin-binding aptamers on 2' F-ANA/DNA chimeric G-Quadruplex microarrays. Nucleic Acids Res 2017, 45, 1619–1632.

[21] Eremeeva E, Fikatas A, Margamuljana L, Abramov M, Schols D, Groaz E, Herdewijn P. Highly stable hexitol based XNA aptamers targeting the vascular endothelial growth factor. Nucleic Acids Res 2019, 47, 4927–4939.

[22] Yu HY, Zhang S, Chaput JC. Darwinian evolution of an alternative genetic system provides support for TNA as an RNA progenitor. Nat Chem 2012, 4, 183–187.

[23] Minakawa N, Sanji M, Kato Y, Matsuda A. Investigations toward the selection of fully-modified 4'-thioRNA aptamers: optimization of in vitro transcription steps in the presence of 4'-thioNTPs. Bioorg Med Chem 2008, 16, 9450–9456.

[24] Pasternak A, Hernandez FJ, Rasmussen LM, Vester B, Wengel J. Improved thrombin binding aptamer by incorporation of a single unlocked nucleic acid monomer. Nucleic Acid Res 2011, 39, 1155–1164.

[25] Li L, Xu S, Yan H, Li X, Yazd HS, Li X, Huang T, Cui C, Jiang J, Tan W. Nucleic acid aptamers for molecular diagnostics and therapeutics: Advances and perspectives. Angew Chem Int Ed 2021, 60, 2221–2231.

[26] Elskens JP, Elskens JM, Madder A. Chemical modification of aptamers for increased binding affinity in diagnostic applications: Current status and future prospects. Int J Mol Sci 2020, 21, 4522 (31 pages.

[27] Ji D, Lyu K, Zhao H, Kwok CK. Circular L-RNA aptamer promotes target recognition and controls gene activity. Nucleic Acids Res 2021, 49, 7280–7291.

[28] Vorobyeva M, Vorobjev P, Venyaminova A. Multivalent aptamers: Versatile tools for diagnostic and therapeutic applications. Molecules 2016, 21, 1613 (21 pages).

[29] Vater A, Klussmann S. Toward third-generation aptamers: Spiegelmers and their therapeutic prospects. Curr Opin Drug Discov Devel 2003, 6, 253–261.

[30] Dougan H, Lyster DM, Vo CV, Stafford A, Weitz JI, Hobbs JB. Extending the lifetime of anticoagulant oligodeoxynucleotide aptamers in blood. Nucl Med Biol 2000, 27, 289–297.

[31]    Lee CH, Lee SH, Kim JH, Noh YH, Noh GJ, Lee SW. Pharmacokinetics of a cholesterol conjugated aptamer against the hepatitis C virus (HCV) NS5B protein. Mol Ther Nucleic Acids 2015, 4, e254 (11 pages).

[32]    Heo K, Min S-W, Sung HJ, Kima HG, Kima HJ, Kim YH, Choi BK, Han S, Chung S, Lee ES, Chung J, Kima I-H. An aptamer-antibody complex (oligobody) as a novel delivery platform for targeted cancer therapies. J Control Release 2016, 229, 1–9.

[33]    Willis MC, Collins B, Zhang T, Green LS, Sebesta DP, Bell C, Kellogg E, Gill SC, Magallanez A, Knauer S, Bendele RA, Gill PS, Janjic N. Liposome-anchored vascular endothelial growth factor aptamers. Bioconjug Chem 1998, 9, 573–582.

[34]    Zhou J, Soontornworajit B, Martin J, Sullenger BA, Gilboa E, Wang Y. A hybrid DNA aptamer-dendrimer nanomaterial for targeted cell labeling. Macromol Biosci 2009, 9, 831–835.

[35]    Borbas KE, Ferreira CS, Perkins A, Bruce JI, Missailidis S. Design and synthesis of mono- and multimeric targeted radiopharmaceuticals based on novel cyclen ligands coupled to antiMUC1 aptamers for the diagnostic imaging and targeted radiotherapy of cancer. Bioconjug Chem 2007, 18, 1205–1212.

[36]    Macugen (pegaptanib) (PDF). European Medicines Agency: 1–3. 2010. Retrieved 2013-12–08.

[37]    Kaur H, Bruno JG, Kumar A, Sharma TK. Aptamers in the therapeutic and diagnostic pipelines. Theranostics 2018, 8, 4016–4032.

[38]    Chan C-Y, Kwok CK. Specific binding of a D-RNA G-quadruplex structure with an L-RNA aptamer. Angew Chem Int Ed 2020, 59, 5293–5297.

[39]    Ji D, Yuan J-H, Chen S-B, Tan J-H, Kwok CK. Selective targeting of parallel G-quadruplex structure using L-RNA aptamer. Nucleic Acids Res 2023, 51, 11439–11452.

[40]    Jilma B, Paulinska P, Jilma-Stohlawetz P, Gilbert JC, Hutabarat R, Knobl P. A randomized pilot trial of anti-von Willebrand factor aptamer ARC1779 in patients with type 2b von Willebrand disease. Thromb Haemost 2010, 104, 563–570.

[41]    Rusconi CP, Scardino E, Layzer J, Pitoc GA, Ortel TL, Monroe D, Sullenger BA. RNA aptamers as reversible antagonists of coagulation factor IXa. Nature 2002, 419, 90–94.

[42]    Waters EK, Genga RM, Thomson HA, Kurz C, Schaub RG, Scheiflinger F, McKinness KE. Aptamer BAX 499 mediates inhibition of tissue factor pathway inhibitor via interaction with multiple domains of the protein. J Thromb Haemost 2013, 11, 1137–1145.

[43]    Waters E, Richardson J, Schaub R, Kurz J. Effect of NU172 and bivalirudin on ecarin clotting time in human plasma and whole blood. J Thromb Haemost Supplement 2, Abstract PPWE-1682009, 7.

[44]    Soundararajan S, Chen W, Spicer EK, Courtenay-Luck N, Fernandes DJ. The nucleolin targeting aptamer AS1411 destabilizes Bcl-2 messenger RNA in human breast cancer cells. Cancer Res 2008, 68, 2358–2365.

[45]    Hoellenriegel J, Zboralski D, Maasch C, Rosin NY, Wierda WG, Keating MJ, Kruschinski A, Burger JA. The Spiegelmer NOX-A12, a novel CXCL12 inhibitor, interferes with chronic lymphocytic leukemia cell motility and causes chemosensitization. Blood 2014, 123, 1032–1039.

[46]    Oberthür D, Achenbach J, Gabdulkhakov A, Buchner K, Maasch C, Falke S, Rehders D, Klussmann S, Betzel C. Crystal structure of a mirror-image L-RNA aptamer (Spiegelmer) in complex with the natural L-protein target CCL2. Nat Commun 2015, 6, 6923 (11 pages.

[47]    Schwoebel F, van Eijk LT, Zboralski D, Sell S, Buchner K, Maasch C, Purschke W, Humphrey M, Zöllner S, Eulberg D, Morich F, Pickkers P, Klussmann S. The effects of the anti-hepcidin spiegelmer NOX-H94 on inflammation-induced anemia in cynomolgus monkeys. Blood 2013, 121, 2311–2315.

[48]    Schmitz A, Weber A, Bayin M, Breuers S, Fieberg V, Famulok M, Mayer G. A SARS-CoV-2 spike binding DNA aptamer that Inhibits pseudovirus infection by an RBD-independent mechanism. Angew Chem Int Ed 2021, 60, 10279–10285.

[49]  Valero J, Civit L, Dupont DM, Selnihhin D, Reinert LS, Idorn M, Israels BA, Bednarz AM, Bus C, Asbach B, Peterhoff D, Pedersen FS, Birkedal V, Wagner R, Paludan SR, Kjems J. A serum-stable RNA aptamer specific for SARS-CoV-2 neutralizes viral entry. PNAS 2021, 118, e2112942118 (10 pages).

[50]  Nimjee SM, Rusconi CP, Sullenger BA. Aptamers: an emerging class of therapeutics. Annu Rev Med 2005, 56, 555–583.

[51]  Zhou J, Rossi JJ. Cell-specific aptamer-mediated targeted drug delivery. Oligonucleotides 2011, 21, 1–10.

[52]  Zhou J, Rossi JJ. Cell-type-specific, aptamer functionalized agents for targeted disease therapy. Mol Ther Nucleic Acids 2014, 3, e169 (17 pages).

[53]  http://aptamer.icmb.utexas.edu/

[54]  Zhang Y, Lai BS, Juhas M. Recent advances in aptamer discovery and applications. Molecules 2019, 24, 941 (22 pages.

[55]  Sun H, Tan W, Zu Y. Aptamers: Versatile molecular recognition probes for cancer detection. Analyst 2016, 141, 403–415.

[56]  Wu L, Zhang Y, Wang Z, Zhang Y, Zou J, Qiu L. Aptamer-based cancer cell analysis and treatment. Chem Open 2022, 11, e202200141 (11 pages).

[57]  Debiais M, Lelievre A, Smietana M, Muller S. Splitting aptamers and nucleic acid enzymes for the development of advanced biosensors. Nucleic Acids Res 2020, 48, 3400–3422.

[58]  Liu X, Shi L, Hua X, Huang Y, Su S, Fan Q, Wang L, Huang W. Target-induced conjunction of split aptamer fragments and assembly with a water-soluble conjugated polymer for improved protein detection. ACS Appl Mater Interfaces 2014, 6, 3406–3412.

[59]  Sharma AK, Heemstra JM. Small-molecule-dependent split aptamer ligation. J Am Chem Soc 2011, 133, 12426–12429.

[60]  Aho A, Virta P. Assembly of split aptamers by a dynamic pH-responsive covalent ligation. Chem Commun 2023, 59, 5689–5692.

[61]  Lelievre-Büttner A, Schnarr T, Debiais M, Smietana M, Müller S. Boronic acid assisted self-assembly of functional RNAs. Chem Eur J 2023, 29, e202300196 (9 pages).

[62]  Pei D, Buyanova M. Overcoming endosomal entrapment in drug delivery. Bioconjugate Chem 2019, 30, 273–283.

[63]  Ming X, Laing B. Bioconjugates for targeted delivery of therapeutic oligonucleotides. Adv Drug Deliv Rev 2015, 87, 81–89.

[64]  Ming X. Cellular delivery of siRNA and antisense oligonucleotides via receptor mediated endocytosis. Expert Opin Drug Deliv 2011, 8, 435–449.

[65]  Lachelt U, Wagner E. Nucleic acid therapeutics using polyplexes: a journey of 50 years (and beyond). Chem Rev 2015, 115, 11043–11078.

[66]  Pattini BS, Chupin VV, Torchilin VP. New developments in liposomal drug delivery. Chem Rev 2015, 115, 10938–10966.

[67]  Rettig GR, Behlke MA. Progress toward in vivo use of siRNAs. Mol Ther 2012, 40, 483–512.

[68]  Akinc A, Thomas M, Klibanov AM, Langer R. Exploring polyethylenimine-mediated DNA transfection and the proton sponge hypothesis. J Gen Med 2005, 7, 657–663.

[69]  Benjaminsen RV, Mattebjerg MA, Henriksen JR, Moghimi SM, Andresen TL. The possible "proton sponge" effect of polyethyleneimine (PEI) does not include change in lysosomal pH. Mol Ther 2013, 21, 149–157.

[70]  Oliveira S, van Rooy I, Kranenburg O, Stor G, Schiffelers RM. Fusogenic peptides enhance endosomal escape improving siRNA-induced silencing of oncogenes. Int J Pharm 2007, 331, 211–214.

[71]  Kwon EJ, Bergen JM, Pun SH. Application of an HIV gp41-derived peptide for enhanced intracellular trafficking of synthetic gene and siRNA delivery vehicles. Bioconjugate Chem 2008, 19, 920–927.

[72]  Gallas A, Alexander C, Davies MC, Purib S, Allen S. Chemistry and formulations for siRNA therapeutics. Chem Soc Rev 2013, 42, 7983–7997.

[73] Kulkarni JA, Witzigmann D, Chen S, Cullis PR, van der Meel R. Lipid nanoparticle technology for clinical translation of siRNA therapeutics. Acc Chem Res 2019, 52, 2435–2444.

[74] Zhang Y, Sun C, Wang C, Jankovic KE, Dong Y. Lipids and lipid derivatives for RNA delivery. Chem Rev 2021, 121, 12181–12277.

[75] Zimmermann TS, Lee ACH, Akinc A, Bramlage B, Bumcrot D, Fedoruk MN, Harborth J, Heyes JA, Jeffs LB, John M, Judge AD, Lam K, McClintock K, Nechev LV, Palmer LR, Racie T, Rohl I, Seiffert S, Shanmugam S, Sood V, Soutschek J, Toudjarska I, Wheat AJ, Yaworski E, Zedalis W, Koteliansky V, Manoharan M, Vornlocher HP, MacLachlan I. RNAi-mediated gene silencing in non-human primates. Nature 2006, 441, 111–114.

[76] Jayaraman M, Ansell SM, Mui BL, Tam YK, Chen J, Du X, Butler D, Eltepu L, Matsuda S, Narayanannair JK, Rajeev KG, Hafez IM, Akinc A, Maier MA, Tracy MA, Cullis PR, Madden TD, Manoharan M, Hope MJ. Maximizing the potency of siRNA lipid nanoparticles for hepatic gene silencing in vivo. Angew Chem Int Ed 2012, 51, 8529–8533.

[77] Buschmann MD, Carrasco MJ, Alishetty S, Paige M, Alameh MG, Weissman D. Nanomaterial delivery systems for mRNA vaccines. Vaccines 2021, 9, 65 (30 pages).

[78] Akinc A, Querbes W, De S, Qin J, Frank-Kamenetsky M, Jayaprakash KN, Jayaraman M, Rajeev KG, Cantley WL, Dorkin JR, Butler JS, Qin L, Racie T, Sprague A, Fava E, Zeigerer A, Hope MJ, Zerial M, Sah DW, Fitzgerald K, Tracy MA, Manoharan M, Koteliansky V, Fougerolles A, Maier MA. Targeted delivery of RNAi therapeutics with endogenous and exogenous ligand-based mechanisms. Mol Ther 2010, 18, 1357–1364.

[79] Johannes L, Lucchino M. Current challenges in delivery and cytosolic translocation of therapeutic RNAs. Nucleic Acid Ther 2018, 3, 178–193.

[80] Akinc A, Zumbuehl A, Goldberg M, Leshchiner ES, Busini V, Hossain N, Bacallado SA, Nguyen J, Fuller R, Alvarez A, Borodovsky T, Borland R, Constien A, de Fougerolles JR, Dorkin DN, Jayaprakash K, Jayaraman M, John M, Koteliansky V, Manoharan M, Nechev L, Qin J, Racie T, Raitcheva D, Rajeev KG, Sah DWY, Soutschek J, Toudjarska I, Vornlocher HP, Zimmermann TS, Langer R, Anderson DG. A combinatorial library of lipid-like materials for delivery of RNAi therapeutics. Nat Biotechnol 2008, 26, 561–569.

[81] Strumberg D, Schultheis B, Traugott U, Vank C, Santel A, Keil O, Giese K, Kaufmann J, Drevs. Phase I clinical development of Atu027, a siRNA formulation targeting PKN3 in patients with advanced solid tumors. J Int J Clin Pharmacol Ther 2012, 50, 76–78.

[82] Mendonça MCP, Kont A, Rodriguburto MR, Cryan JF, O'Driscoll CM. Advances in the design of (nano) formulations for delivery of antisense oligonucleotides and small Interfering RNA: Focus on the central nervous system. Mol Pharmaceutics 2021, 18, 1491–1506.

[83] Huang L, Wu E, Liao J, Wei Z, Wang J, Chen Z. Research advances of engineered exosomes as drug delivery carrier. ACS Omega 2023, 8, 43374–43387.

[84] Pi F, Binzel DW, Lee TJ, Li Z, Sun M, Rychahou P, Li H, Haque F, Wang S, Croce CM, Guo B, Evers BM, Guo P. Nanoparticle orientation to control RNA loading and ligand display on extracellular vesicles for cancer regression. Nat Nanotechnol 2018, 13, 82–89.

[85] Li Z, Yang L, Wang H, Binzel DW, Williams TM, Guo P. Non-small-cell lung cancer regression by siRNA delivered through exosomes that display EGFR RNA aptamer. Nucleic Acid Ther 2021, 31, 364–374.

[86] Min HS, Kim HJ, Naito M, Ogura S, Toh K, Hayashi K, Kim BS, Fukushima S, Anraku Y, Miyata K, Kataoka K. Systemic brain delivery of antisense oligonucleotides across the blood–brain barrier with a glucose-coated polymeric nanocarrier. Angew Chem Int Ed 2020, 59, 8173–8180.

[87] de Vries JW, Zhang F, Herrmann A. Drug delivery systems based on nucleic acid nanostructures. J Control Release 2013, 172, 467–483.

[88] Hu Q, Li H, Wang L, Gu H, Fan C. DNA nanotechnology-enabled drug delivery systems. Chem Rev 2019, 119, 6459–6506.

[89]  Roberts TC, Langer R, Wood MJA. Advances in oligonucleotide drug delivery. Nat Rev Drug Discov 2020, 19, 673–694.

[90]  Copp W, Pontarelli A, Wilds CJ. Recent advances of DNA tetrahedra for therapeutic delivery and biosensing. ChemBioChem 2021, 22, 2237–2246.

[91]  Li J, Pei H, Zhu B, Liang L, Wei M, He Y, Chen N, Li D, Huang Q, Fan C. Self-assembled multivalent DNA nanostructures for noninvasive intracellular delivery of immunostimulatory CpG oligonucleotides. ACS Nano 2011, 5, 8783–8789.

[92]  Charoenphol P, Bermudez H. Aptamer-targeted DNA nanostructures for therapeutic delivery. Mol Pharm 2014, 11, 1721–1725.

[93]  Haque F, Shu D, Shu Y, Shlyakhtenko LS, Rychahou PG, Evers BM, Guo P. Ultrastable synergistic tetravalent RNA nanoparticles for targeting to cancers. Nano Today 2012, 7, 245–257.

[94]  Vigderman L, Zubarev ER. Therapeutic platforms based on gold nanoparticles and their covalent conjugates with drug molecules. Adv Drug Deliv Rev 2013, 65, 663–676.

[95]  Rosi NL, Giljohann DA, Thaxton CX, Lytton-Jean AKR, Han MS, Mirkin CA. Oligonucleotide-modified gold nanoparticles for intracellular gene regulation. Science 2006, 312, 1027–1030.

[96]  Kyriazi M-E, El-Sagheer AH, Medintz IL, Brown T, Kanaras AG. An investigation into the resistance of spherical nucleic acids against DNA enzymatic degradation. Bioconjugate Chem 2022, 33, 219–225.

[97]  Giljohann DA, Seferos DS, Prigodich AG, Patel PC, Mirkin CA. Gene regulation with polyvalent siRNA–nanoparticle conjugates. J Am Chem Soc 2009, 131, 2072–2073.

[98]  Vasher MK, Yamankurt G, Mirkin CA. Hairpin-like siRNA-based spherical nucleic acids. J Am Chem Soc 2022, 144, 3174–3181.

[99]  Fang Y, Lu X, Wang D, Cai J, Wang Y, Chen P, Ren M, Lu H, Union J, Zhang L, Sun Y, Jia F, Kang X, Tan X, Zhang K. Spherical nucleic acids for topical treatment of hyperpigmentation. J Am Chem Soc 2021, 143, 1296–1300.

[100] Benizri S, Gissot A, Martin A, Vialet B, Grinstaff MW, Barthelemy P. Bioconjugated oligonucleotides: recent developments and therapeutic applications. Bioconjugate Chem 2019, 30, 366–383.

[101] Wang S, Allen N, Prakash TP, Liang XH, Crooke ST. Lipid conjugates enhance endosomal release of antisense oligonucleotides into cells. Nucleic Acid Ther 2019, 29, 245–255.

[102] Wolfrum C, Shi S, Jayaprakash KN, Jayaraman M, Wang G, Pandey RK, Rajeev KG, Nakayama T, Charrise K, Ndungo EM, Zimmermann T, Koteliansky V, Manoharan M, Stoffel M. Mechanisms and optimization of in vivo delivery of lipophilic siRNAs. Nat Biotechnol 2007, 25, 1149–1157.

[103] Petrova NS, Chernikov IV, Meschaninova MI, Dovydenko IiS, Venyaminova AG, Zenkova MA, Vlassov VV, Chernolovskaya EL. Carrier-free cellular uptake and the gene-silencing activity of the lipophilic siRNAs is strongly affected by the length of the linker between siRNA and lipophilic group. Nucleic Acids Res 2012, 40, 2330–2344.

[104] Yang J, Chen C, Tang X. Cholesterol-modified caged siRNAs for photoregulating exogenous and endogenous gene expression. Bioconjugate Chem 2018, 29, 1010–1015.

[105] Krützfeldt J, Rajewsky N, Braich R, Rajeev KG, Tuschl T, Manoharan M, Stoffel M. Silencing of microRNAs in vivo with 'antagomirs'. Nature 2005, 438, 685–689.

[106] Ma L, Reinhardt F, Pan E, Soutschek J, Bhat B, Marcusson EG, Teruya-Feldstein J, Bell GW, Weinberg RA. Therapeutic silencing of miR-10b inhibits metastasis in a mouse mammary tumor model. Nat Biotechnol 2010, 28, 341–347.

[107] DiFiglia M, Sena-Esteves M, Chase K, Sapp E, Pfister E, Sass M, Yoder J, Reeves P, Pandey RK, Rajeev KG, Manoharan M, Sah DWY, Zamore PD, Aronin N. Therapeutic silencing of mutant huntingtin with siRNA attenuates striatal and cortical neuropathology and behavioral deficits. Proc Natl Acad Sci 2007, 104, 17204–17209.

[108] Chen Q, Butler D, Querbes W, Pandey RK, Ge P, Maier MA, Zhang L, Rajeev KG, Nechev L, Kotelianski V, Manoharan M, Sah DWY. Lipophilic siRNAs mediate efficient gene silencing in oligodendrocytes with direct CNS delivery. J Control Release 2010, 144, 227–232.

[109] Krützfeldt J, Kuwajima S, Braich R, Rajeev KG, Pena J, Tuschl T, Manoharan M, Stoffel M. Specificity, duplex degradation and subcellular localization of antagomirs. Nucleic Acids Res 2007, 35, 2885–2892.

[110] Nishina K, Unno T, Uno Y, Kubodera T, Kanouchi T, Mizusawa H, Yokota T. Efficient in vivo delivery of siRNA to the liver by conjugation of alpha-tocopherol. Mol Ther 2008, 16, 734–740.

[111] Asami Y, Yoshioka K, Nishina K, Nagata T, Yokota T. Drug delivery system of therapeutic oligonucleotides. Drug Discov Ther 2016, 10, 256–262.

[112] Nishina T, Numata J, Nishina K, Yoshida-Tanaka K, Nitta K, Piao W, Iwata R, Ito S, Kuwahara H, Wada T, Mizusawa H, Yokota T. Chimeric antisense oligonucleotide conjugated to α-tocopherol. Mol Ther Nucleic Acids 2015, 4, e220 (10 pages).

[113] Nagata T, Dwyer CA, Yoshida-Tanaka K, Ihara K, Ohyagi M, Kaburagi H, Miyata H, Ebihara S, Yoshioka K, Ishii T, Miyata K, Miyata K, Powers B, Igari T, Yamamoto S, Arimura N, Hirabayashi H, Uchihara T, RI, Wada T, Bennett CF, Seth PP, Rigo F, Yokota T. Cholesterol-functionalized DNA/RNA heteroduplexes cross the blood–brain barrier and knock down genes in the rodent CNS. Nat Biotechnol 2021, 39, 1529–1536.

[114] Herbert B-S, Gellert GC, Hochreiter A, Pongracz K, Wright WE, Zielinska D, Chin AC, Harley CB, Shay JW, Gryaznov SM. Lipid modification of GRN163, an N3′→ P5′ thio-phosphoramidate oligonucleotide, enhances the potency of telomerase inhibition. Oncogene 2005, 24, 5262–5268.

[115] Brunner K, Harder J, Halbach T, Willibald J, Spada F, Gnerlich F, Sparrer K, Beil A, Meckl L, Bruchle C, Conzelmann K, Carell T. Cell-penetrating and neurotargeting dendritic siRNA nanostructures. Angew Chem Int Ed Engl 2015, 54, 1946–1949.

[116] Raouane M, Desmaele D, Gilbert-Sirieix M, Gueutin C, Zouhiri F, Bourgaux C, Lepeltier E, Gref R, Ben Salah R, Clayman G, Massaad-Massade L, Couvreur P. Synthesis, characterization, and in vivo delivery of siRNA-squalene nanoparticles targeting fusion oncogene in papillary thyroid carcinoma. J Med Chem 2011, 54, 4067–4076.

[117] Raouane M, Desmaële D, Urbinati G, Massaad-Massade L, Couvreur P. Lipid conjugated oligonucleotides: A useful strategy for delivery. Bioconjugate Chem 2012, 23, 1091–1104.

[118] Biessen EAL, Vietsch H, Rump ET, Fluiter K, Kuiper J, Bijsterbosch MK, Van Berkel TJC. Targeted delivery of oligodeoxynucleotides to parenchymal liver cells in vivo. Biochem J 1999, 340, 783–792.

[119] Hamzavi R, Dolle F, Tavitian B, Dahl O, Nielsen PE. Modulation of the pharmacokinetic properties of PNA: preparation of galactosyl, mannosyl, fucosyl, N-acetylgalactosaminyl, and N-acetylglucosaminyl derivatives of aminoethylglycine peptide nucleic acid monomers and their incorporation into PNA oligomers. Bioconjugate Chem 2003, 14, 941–954.

[120] Prakash TP, Graham MJ, Yu J, Carty R, Low A, Chappell A, Schmidt K, Zhao C, Aghajan M, Murray HF, Riney S, Booten SL, Murray SF, Gaus H, Crosby J, Lima WF, Guo S, Monia BP, Swayze EE, Seth PP. Targeted delivery of antisense oligonucleotides to hepatocytes using triantennary N-acetyl galactosamine improves potency 10-fold in mice. Nucleic Acids Res 2014, 42, 8796–8807.

[121] Mäkilä J, Jadhav S, Kiviniemi A, Käkelä M, Liljenbäck H, Poijärvi-Virta P, Laitala-Leinonen T, Lönnberg H, Roivainen A, Virta P. Synthesis of multi-galactose-conjugated 2′-O-methyl oligoribonucleotides and their in vivo imaging with positron emission tomography. Bioorg Med Chem 2014, 22, 6806–6813.

[122] Kandasamy P, Mori S, Matsuda S, Erande N, Datta D, Willoughby JLS, Taneja N, O'Shea J, Bisbe A, Manoharan RM, Yucius K, Nguyen T, Indrakanti R, Gupta S, Gilbert JA, Racie T, Chan A, Liu J, Hutabarat R, Nair JK, Charisse K, Maier MA, Rajeev KG, Egli M, Manoharan M. Metabolically stable anomeric linkages containing GalNAc–siRNA conjugates: An interplay among ASGPR, glycosidase, and RISC pathways. J Med Chem 2023, 66, 2506–2523.

[123] Bhingardeve P, Madhanagopal BR, Naick H, Jain P, Manoharan M, Ganesh K. Receptor-specific delivery of peptide nucleic acids conjugated to three sequentially linked N-acetyl galactosamine moieties into hepatocytes. J Org Chem 2020, 85, 8812–8824.

[124] Kurrikoff K, Gestin M, Langel U. Recent in vivo advances in cell-penetrating peptide-assisted drug delivery. Expert Opin Drug Deliv 2016, 13, 373–387.

[125] Palm-Apergi C, Lonn P, Dowdy SF. Do cell-penetrating peptides actually "penetrate" cellular membranes?. Mol Ther 2012, 20, 695–697.

[126] Ye J, Liu E, Gong J, Wang J, Huang Y, He H, Yang VC. High-yield synthesis of monomeric LMWP(CPP)-siRNA covalent conjugate for effective cytosolic delivery of siRNA. Theranostics 2017, 7, 2495–2508.

[127] Pooga M, Soomets U, Hällbrink M, Valkna A, Saar K, Rezaei K, Kahl U, Hao JX, Xu XJ, Wiesenfeld-Hallin Z, Hökfelt T, Bartfai T, Langel U. Cell penetrating PNA constructs regulate galanin receptor levels and modify pain transmission in vivo. Nat Biotechnol 1998, 16, 857–861.

[128] Gait MJ, Arzumanov AA, McClorey G, Godfrey C, Betts C, Hammond S, Wood MJA. Cell-penetrating peptide conjugates of steric blocking oligonucleotides as therapeutics for neuromuscular diseases from a historical perspective to current prospects of treatment. Nucleic Acid Ther 2019, 29, 1 (12 pages).

[129] Yin HF, Saleh AF, Betts C, Camelliti P, Seow Y, Ashraf S, Arzumanow A, Hammond S, Merritt T, Gait MJ, Wood MJA. Pip5 transduction peptides direct high efficiency oligonucleotide-mediated dystrophin exon skipping in heart and phenotypic correction in mdx mice. Mol Ther 2011, 19, 1295–1303.

[130] Betts CA, McClorey G, Healicon R, Hammond SM, Manzano R, Muses S, Ball V, Godfrey C, Merritt TM, Westering T. Cmah-dystrophin deficient Mdx mice display an ccelerated cardiac phenotype that is improved following peptide-PMO exon skipping treatment. Hum Mol Genet 2019, 28, 396–406.

[131] Alam MR, Ming X, Fisher M, Lackey JG, Rajeev KG, Manoharan M, Juliano RL. Multivalent cyclic RGD conjugates for targeted delivery of small interfering RNA. Bioconjugate Chem 2011, 22, 1673–1681.

[132] Liu X, Wang W, Samarsky D, Liu L, Xu Q, Zhang W, Zhu G, Wu P, Zuo X, Deng H, Zhang J, Wu Z, Chen X, Zhao L, Qiu Z, Zhang Z, Zeng Q, Yang W, Zhang B, Ji A. Tumor-targeted in vivo gene silencing via systemic delivery of cRGD-conjugated siRNA. Nucleic Acids Res 2014, 42, 11805–11817.

[133] Ming X, Alam MR, Fisher M, Yan Y, Chen X, Juliano RL. Intracellular delivery of an antisense oligonucleotide via endocytosis of a G protein-coupled receptor. Nucleic Acids Res 2010, 38, 6567–6576.

[134] Nakagawa O, Ming X, Carver K, Juliano R. Conjugation with receptor-targeted histidine-rich peptides enhances the pharmacological effectiveness of antisense oligonucleotides. Bioconjugate Chem 2014, 25, 165–170.

[135] Uckun FM, Qazi S, Dibirdik I, Myers DE. Rational design of an immunoconjugate for selective knock-down of leukemia-specific E2A-PBX1 fusion gene expression in human Pre-B leukemia. Integr Biol 2013, 5, 122–132.

[136] Satake N, Duong C, Yoshida S, Oestergaard M, Chen C, Peralta R, Guo S, Seth PP, Li Y, Beckett L, Chung J, Nolta J, Nitin N, Tuscano JM. Novel targeted therapy for precursor B-cell acute lymphoblastic leukemia: anti-CD22 antibody-MXD3 antisense oligonucleotide conjugate. Mol Med 2016, 22, 632–642.

[137] Sugo T, Terada M, Oikawa T, Miyata K, Nishimura S, Kenjo E, Ogasawara-Shimizu M, Makita Y, Imaichi S, Murata S, Otake K, Kikuchi K, Teratani M, Masuda Y, Kamei T, Takagahara S, Ikeda S, Ohtaki T, Matsumoto H. Development of antibody-siRNA conjugate targeted to cardiac and skeletal muscles. J Control Release 2016, 237, 1–13.

[138] Arnold AE, Malek-Adamian E, Le PU, Meng A, Martínez-Montero S, Petrecca K, Damha MJ, Shoichet MS. Antibody-antisense oligonucleotide conjugate downregulates a key enge in glioblastoma stem cells. Mol Ther Nucleic Acids 2018, 11, 518–527.

[139] Cuellar TL, Barnes D, Nelson C, Tanguay J, Yu S-F, Wen X, Scales SJ, Gesch J, Davis D, van Brabant Smith A, Leake D, Vandlen R, Siebel CW. Systematic evaluation of antibody-mediated siRNA delivery using an industrial platform of THIOMAB–siRNA Conjugates. Nucleic Acids Res 2015, 43, 1189–1203.

[140] McNamara JO 2nd, Andrechek ER, Wang Y, Viles KD, Rempel RE, Gilboa E, Sullenger BA, Giangrande PH. Cell type–specific delivery of siRNAs with aptamer siRNA chimeras. Nat Biotechnol 2006, 24, 1005–1015.

[141] Lupold SE, Hicke BJ, Lin Y, Coffey DS. Identification and characterization of nuclease-stabilized RNA molecules that bind human prostate cancer cells via the prostate-specific membrane antigen. Cancer Res 2002, 62, 4029–4033.

[142] Dassie JP, Liu XY, Thomas GS, Whitaker RM, Thiel KW, Stockdale KR, Meyerholz DK, McCaffrey AP, McNamara JO2, Giangrande PH. Systemic administration of optimized aptamer-siRNA chimeras promotes regression of PSMA-expressing tumors. Nat Biotechnol 2009, 27, 839–849.

[143] Neff CP, Zhou J, Remling L, Kuruvilla J, Zhang J, Li H, Smith DD, Swidersk P, Rossi JJ, Akkina R. An aptamer-siRNA chimera suppresses HIV-1 viral loads and protects from helper CD4(+) T cell decline in humanized mice. Sci Transl Med 2011, 3, 66ra6 (10 pages).

[144] Zhou J, Rossi J. Cell-type specific aptamer and aptamer-siRNA conjugates for targeted HIV-1 therapy. J Invest Med 2014, 62, 914–919.

[145] Kortylewski M, Swiderski P, Herrmann A, Wang L, Kowolik C, Kujawski M, Lee H, Scuto A, Liu Y, Yang C, Deng J, Soifer HS, Raubitschek A, Forman S, Rossi JJ, Pardoll DM, Jove R, Yu H. In vivo delivery of siRNA to immune cells by conjugation to a TLR9 agonist enhances antitumor immune responses. Nat Biotechnol 2009, 27, 925–932.

[146] Zhang Q, Hossain DMS, Nechaev S, Kozlowska A, Zhang W, Liu Y, Kowolik CM, Swiderski P, Rossi JJ, Forman S, Pal S, Bhatia R, Raubitschek A, Yu H, Kortylewski M. TLR9-mediated siRNA delivery for targeting of normal and malignant human hematopoietic cells in vivo. Blood 2013, 121, 1304–1315.

[147] Nakagawa O, Ming X, Huang L, Juliano RL. Targeted intracellular delivery of antisense oligonucleotides via conjugation with small-molecule ligands. J Am Chem Soc 2010, 132, 8848–8849.

[148] Dohmen C, Fröhlich T, Lächelt U, Röhl I, Vornloche H-P, Hadwiger P, Wagner E. Defined folate-PEG-siRNA conjugates for receptor-specific gene silencing. Mol Ther Nucleic Acids 2012, 1, e7 (6 pages).

[149] Salim L, Desaulniers J-P. To conjugate or to package? A Look at targeted siRNA delivery through folate receptors. Nucleic Acid Ther 2021, 31, 21–38.

[150] Gangopadhyay S, Nikam RR, Gore KR. Folate receptor-mediated siRNA delivery: Recent edvelopments and future directions for RNAi Therapeutics. Nucleic Acid Ther 2021, 31, 245–270.

[151] Rozema BD, Lewis DL, Wakefield DH, Wong SC, Klein JJ, Roesch PL, Bertin SL, Reppen TW, Chu Q, Blokhin AV, Hagstrom JE, Wolff JA. Dynamic polyconjugates for targeted in vivo delivery of siRNA to hepatocytes. Proc Natl Acad Sci USA 2007, 104, 12982–12987.

[152] Wooddell CI, Rozema DB, Hossbach M, John M, Hamilton HL, Chu Q, Hegge JO, Klein JJ, Wakefield DH, Oropeza CE, Deckert J, Roehl I, Jahn-Hofmann K, Hadwiger P, Vornlocher HP, McLachlan A, Lewis DL. Hepatocyte-targeted RNAi therapeutics for the treatment of chronic hepatitis B virus infection. Mol Ther 2013, 21, 973–985.

[153] Shu Z, Tanaka I, Ota A, Fushihara D, Abe N, Kawaguchi S, Nakamoto K, Tomoike F, Tada S, Ito Y, Kimura Y, Abe H. Disulfide-unit conjugation enables ultrafast cytosolic internalization of antisense DNA and siRNA. Angew Chem Int Ed 2019, 58, 6611–6615.

[154] Takatsu M, Morihiro K, Watanabe H, Yuki M, Hattori T, Noi K, Aikawa K, Noguchi K, Yohda M, Okazoe T, Okamoto A. Cellular penetration and intracellular dynamics of perfluorocarbon-conjugated DNA/RNA as a potential means of conditional nucleic acid delivery. ACS Chem Biol 2023, 18, 2590–2598.

[155] Ming X, Carver K, Wu L. Albumin-based nanoconjugates for targeted delivery of therapeutic oligonucleotides. Biomaterials 2013, 34, 7939–7949.

[156] Kang H, Alam MR, Dixit V, Fisher M, Juliano RL. Cellular delivery and biological activity of antisense oligonucleotides conjugated to a targeted protein carrier. Bioconjugate Chem 2008, 19, 2182–2188.

[157] Eguchi A, Meade BR, Chang YC, Fredrickson CT, Willert K, Puri N, Dowdy SF. Efficient siRNA delivery into primary cells by a peptide transduction domain dsRNA binding domain fusion protein. Nat Biotechnol 2009, 27, 567–571.

[158] Song E, Zhu P, Lee SK, Chowdhury D, Kussman S, Dykxhoorn DM, Feng Y, Palliser D, Weiner DB, Shankar P, Marasco WA, Lieberman J. Antibody mediated in vivo delivery of small interfering RNAs via cell-surface receptors. Nat Biotechnol 2005, 23, 709–717.

[159] Debart F, Dupouy C, Vasseur -J-J. Stimuli-responsive oligonucleotides in prodrug-based approaches for gene silencing. Beilstein J Org Chem 2018, 14, 436–469.

[160] Tosquellas G, Alvarez K, Dell'Aquila C, Morvan F, Vasseur J-J, Imbach J-L, Rayner B. The pro-oligonucleotide approach: Solid phase synthesis and preliminary evaluation of model pro-dodecathymidylates. Nucleic Acids Res 1998, 26, 2069–2074.

[161] Ora M, Taherpour S, Linna R, Leisvuori A, Hietamäki E, Poijärvi-Virta P, Beigelman L, Lönnberg H. Biodegradable protections for nucleoside 5′-monophosphates: comparative study on the removal of O-acetyl and O-acetyloxymethyl protected 3-hydroxy-2,2-bis(ethoxycarbonyl)propyl groups. J Org Chem 2009, 74, 4992–5001.

[162] Meade BR, Gogoi K, Hamil AS, Palm-Apergi C, van den Berg A, Hagopian JC, Springer AD, Eguchi A, Kacsinta AD, Dowdy CF, Presente A, Lönn P, Kaulich M, Yoshioka N, Gros E, Cui X-S, Dowdy SF. Efficient delivery of RNAi prodrugs containing reversible charge-neutralizing phosphotriester backbone modifications. Nat Biotechnol 2014, 32, 1256–1261.

[163] Zhang N, Tan C, Cai P, Zhang P, Zhao Y, Jiang Y. The design, synthesis and evaluation of hypoxia-activated pro-oligonucleotides. Chem Commun 2009, 0, 3216–3218.

[164] Saneyoshi H, Iketani K, Kondo K, Saneyoshi T, Okamoto I, Ono A. Synthesis and characterization of cell-permeable oligonucleotides bearing reduction-activated protecting groups on the internucleotide linkages. Bioconjugate Chem 2016, 27, 2149–2156.

[165] Jiho Y, Kurihara R, Kawai K, Yamada H, Uto Y, Tanabe K. Enzymatic activation of indolequinone-substituted 5-fluorodeoxyuridine prodrugs in hypoxic cells. Bioorg Med Chem Lett 2019, 29, 1304–1307.

[166] Morihiro K, Ishinabe T, Takatsu M, Osumi H, Osawa T, Okamoto A. Floxuridine oligomers activated under hypoxic environment. J Am Chem Soc 2021, 143, 3340–3347.

[167] Hayashi J, Samezawa Y, Ochi Y, Wada S-I, Urata H. Syntheses of prodrug-type phosphotriester oligonucleotides responsive to intracellular reducing environment for improvement of cell membrane permeability and nuclease resistance. Bioorg Med Chem Lett 2017, 27, 3135–3138.

[168] Sugimoto N, Hayashi J, Funaki R, Wada S-I, Wada F, Harada-Shiba M, Urata H. Prodrug-type phosphotriester oligonucleotides with linear disulfide pro-moieties responsive to reducing environment. ChemBioChem 2023, 24, e202300526 (1 of 10).

[169] Grajkowski A, Pedras-Vasconcelos J, Wang V, Ausín C, Hess S, Verthelyi D, Beaucage SL. Thermolytic CpG-containing DNA oligonucleotides as potential immunotherapeutic prodrugs. Nucleic Acids Res 2005, 33, 3550–3560.

[170] Grajkowski A, Ausín C, Kauffman JS, Snyder J, Hess S, Lloyd JR, Beaucage SL. Solid-phase synthesis of thermolytic DNA oligonucleotides functionalized with a single 4-hydroxy-1-butyl or 4-phosphato-/thiophosphato-1-butyl thiophosphate protecting group. J Org Chem 2007, 72, 805–815.

[171] Ausín C, Kauffman JS, Duff RJ, Shivaprasad S, Beaucage SL. Assessment of heat-sensitive thiophosphate protecting groups in the development of thermolytic DNA oligonucleotide prodrugs. Tetrahedron 2010, 66, 68–79.

[172] Leisvuori A, Lönnberg H, Ora M. 4-Acetylthio-2,2-dimethyl-3-oxobutyl group as an esterase- and thermo-labile protecting group for oligomeric phosphodiester. Eur J Org Chem 2014, 0, 5816–5826.

[173] Biscans A, Bos M, Martin AR, Ader N, Sczakiel G, Vasseur -J-J, Dupouy C, Debart F. Direct synthesis of partially modified 2′-O-pivaloyloxymethyl RNAs by a base-labile protecting group strategy and their potential for prodrug-based gene-silencing applications. ChemBioChem 2014, 15, 2674–2679.

[174] Baraguey C, Lescrinier E, Lavergne T, Debart F, Herdewijn P, Vasseur J-J. The biolabile 2′-O-pivaloyloxymethyl modification in an RNA helix: an NMR solution structure. Org Biomol Chem 2013, 11, 2638–2647.

[175] Biscans A, Bertrand J-R, Dubois J, Rüger J, Vasseur -J-J, Sczakiel G, Dupouy C, Debart F. Lipophilic 2′-O-Acetal ester RNAs: synthesis, thermal duplex stability, nuclease resistance, cellular uptake, and siRNA activity after spontaneous naked delivery. ChemBioChem 2016, 17, 2054–2062.

[176] Ochi Y, Nakagawa O, Sakaguchi K, Wada S-I, Urata H. A post-synthetic approach for the synthesis of 2′-O-methyldithiomethyl-modified oligonucleotides responsive to a reducing environment. Chem Commun 2013, 49, 7620–7622.

[177] Liu Q, Deiters A. Optochemical control of deoxyoligonucleotide function via a nucleobase-caging approach. Acc Chem Res 2014, 47, 45–55.

[178] Mikat V, Heckel A. Light-dependent RNA interference with nucleobase-caged siRNAs. RNA 2007, 13, 2341–2347.

[179] Govan JM, Young D, Lusic H, Liu Q, Lively MO, Deiters A. Optochemical control of RNA interference in mammalian cells. Nucleic Acids Res 2013, 41, 10518–10528.

[180] Young DD, Lusic H, Lively MO, Yoder JA, Deiters A. Gene silencing in mammalian cells with light-activated antisense agents. ChemBioChem 2008, 9, 2937–2940.

[181] Jinek M, Chylinski K, Fonfara I, Hauer M, Doudna JA, Charpentier E. A programmable dual-RNA-guided DNA endonuclease in adaptive bacterial immunity. Science 2012, 337, 816–821.

[182] Jiang F, Doudna JA. CRISPR-Cas9 structures and mechanisms. Annu Rev Biophys 2017, 46, 505–529.

[183] Jiang F, Taylor DW, Chen JS, Kornfeld JE, Zhou K, Thompson AJ, Nogales E, Doudna JA. Structures of a CRISPR-Cas9 R-loop complex primed for DNA cleavage. Science 2016, 351, 867–871.

[184] Nishimasu H, Ran FA, Hsu PD, Konermann S, Shehata SI, Dohmae N, Ishitani R, Zhang F, Nureki O. Crystal structure of Cas9 in complex with guide RNA and target DNA. Cell 2014, 156, 935–949.

[185] Chen H, Choi J, Bailey S. Cut site selection by the two nuclease domains of the Cas9 RNA-guided endonuclease. J Biol Chem 2014, 289, 13284–13294.

[186] Wright AV, Nunez JK, Doudna JA. Biology and applications of CRISPR systems: Harnessing nature's toolbox for genome engineering. Cell 2016, 164, 29–44.

[187] Mali P, Yang L, Esvelt KM, Aach J, Guell M, DiCarlo JE, Norville JE, Church GM. RNA-guided human genome engineering via Cas9. Science 2013, 339, 823–826.

[188] Fellmann C, Gowen BG, Lin PC, Doudna JA, Corn JE. Cornerstones of CRISPR-Cas in drug discovery and therapy. Nat Rev Drug Discov 2017, 16, 89–100.

[189] Barkau CL, O'Reilly D, Rohilla KJ, Damha MJ, Gagnon KT. Rationally designed anti-CRISPR nucleic acid inhibitors of CRISPR-Cas9. Nucleic Acid Ther 2019, 29, 136–147.

[190] Lee C. CRISPR/Cas9-based antiviral strategy: current status and the potential challenge. Molecules 2019, 24, 1349 (28 pages).

[191] Brown W, Zhou W, Deiters A. Regulating CRISPR/Cas9 function through conditional guide RNA control. ChemBioChem 2021, 22, 63–72.

[192] O'Reilly D, Kartje ZJ, Ageely EA, Malek-Adamian E, Habibian M, Schofield A, Barkau CL, Rohilla KJ, DeRossett LB, Weigle AT, Damha MJ, Gagnon KT. Extensive CRISPR RNA modification reveals chemical compatibility and structure-activity relationships for Cas9 biochemical activity. Nucleic Acids Res 2019, 47, 546–558.

[193] Duan L, Ouyang K, Wang J, Xu L, Xu X, Wen C, Xie Y, Liang Y, Xia J. Exosomes as targeted delivery platform of CRISPR/Cas9 for therapeutic genome editing. ChemBioChem 2021, 22, 3360–3368.

[194] Madigan V, Zhang F, Dahlman JE. Drug delivery systems for CRISPR-based genome editors. Nat Rev Drug Discov 2023, 22, 875–894.

[195] Qiu M, Glass Z, Chen J, Haas M, Jin X, Zhao X, Rui X, Ye Z, Li Y, Zhang F, Xu Q. Lipid nanoparticle-mediated codelivery of Cas9 mRNA and single-guide RNA achieves liver-specific in vivo genome editing of *Angptl3*. PNAS 2021, 118, e2020401118 (10 pages).

[196] Barkau CL, O'Reilly D, Eddington SB, Damha MJ, Gagnon KT. Small nucleic acids and the path to the clinic for anti-CRISPR. Biochem Pharmacol 2021, 189, 114492 (10 pages).

[197] Ryan DE, Taussig D, Steinfeld I, Phadnis SM, Lunstad BD, Singh M, Vuong X, Okochi KD, McCaffrey R, Olesiak M, Roy S, Yung CW, Curry B, Sampson JR, Bruhn L, Dellinger DJ. Improving CRISPR-Cas specificity with chemical modifications in single-guide RNAs. Nucleic Acids Res 2018, 46, 792–803.

[198] Ryan DE, Diamant-Levi T, Steinfeld I, Taussig D, Visal-Shah S, Thakker S, Lunstad BD, Kaiser RJ, McCaffrey R, Ortiz M, Townsend J, Welch WRW, Singh M, Curry B, Dellinger DJ, Bruhn L. Phosphonoacetate modifications enhance the stability and editing yields of guide RNAs for Cas9 editors. Biochemistry 2023, 62, 3512–3520.

# Abbreviations

| | |
|---|---|
| 2′,3′-cAMP | Adenosine 2′,3′-cyclic monophosphate |
| 2′-F-ANA | 2′-Deoxy-2′-fluoroarabino nucleic acid |
| 2′-NMP | Nucleoside 2′-monophosphate |
| 3′-AMP | Adenosine 3′-monophosphate |
| 5′-ATP | Adenosine 5′-triphosphate |
| 5′-CTP | Cytidine 5′-triphosphate |
| 5′-GMP | Guanosine 5′-monophosphate |
| 5′-IMP | Inosine 5′-monophosphate |
| 5′-UMP | Uridine 5′-monophosphate |
| 5′-UTP | Uridine 5′-triphosphate |
| AdCl | Adamantane-1-carbonyl chloride |
| Ade | Adenine |
| Ado | Adenosine |
| Ago2 RNase | Argonaut 2 ribonuclease |
| AIDS | Acquired immune deficiency syndrome |
| ALAS | 5-Aminolevulinic acid synthase |
| AMD | Age-related macular degeneration |
| AON | Antisense oligonucleotide |
| ApoE | Apolipoprotein E |
| AZT | 3′-Azido-3′-deoxythymidine |
| BN | Bicyclic nucleoside |
| BNA | Bridged nucleic acid |
| Boc | *tert*-Butyloxycarbonyl |
| Bpoc | 2-(Biphenyl-4-yl)propan-2-yloxycarbonyl |
| bpy | Bipyridyl |
| BVDU | 5-(2-Bromovinyl)-2′-deoxyuridine |
| Cas9 | CRISPR-associated protein 9 |
| CD | Circular dichroism |
| CeNA | Cyclohexene nucleic acid |
| cEt-BNA | 2′,4′-Constrained-2′-*O*-ethyl nucleic acid |
| chrysi | 5,6-Chrysenediimine |
| CNS | Central nervous system |
| CPG | Controlled pore glass |
| CPP | Cell-penetrating peptide |
| CRISPR | Clustered regularly interspaced short palindromic repeats |
| cRGD | Cyclic Arg-Gly-Asp tripeptide |
| crRNA | CRISPR-RNA |
| Ctd | Cytidine |
| CycloSal | Cyclosaligenyl |
| Cyt | Cytosine |
| dabcyl | 4-(Dimethylaminoazo)benzene-4-carboxylic acid |
| dAdo | 2′-Deoxyadenosine |
| dba | Dibenzylideneacetone |
| DBU | 1,8-Diazabicyclo[5.4.0]undec-7-ene |
| DCC | *N,N*′-Dicyclohexylcarbodiimide |
| DCM | Dichloromethyl |
| dCtd | 2′-Deoxycytidine |

https://doi.org/10.1515/9783111325637-014

| | |
|---|---|
| DDD | Diethyldithiocarbonate disulfide |
| DDTT | *N,N*-Dimethyl-*N'*-(3-thioxo-3*H*-1,2,4-dithiazol-5-yl)methanimidamide |
| DFT | Density functional theory |
| dGuo | 2'-Deoxyguanosine |
| DIC | Diisopropylcarbodiimide |
| DIPEA | Diisopropylethylamine |
| DMAA | Dimethylacetamide |
| DMAOE | Dimethylaminooxyethyl |
| DMD | Duchenne muscular dystrophy |
| DME | Dimethyl ether |
| DMF | *N,N*-Dimethylformamide |
| Dmoc | 1-(1,3-Dithian-2-yl)-1-methylethoxycarbonyl |
| DMOCP | 2-Chloro-5,5-dimethyl-1,3,2-dioxaphosphinane 2-oxide |
| DMSO | Dimethylsulfoxide |
| DMTr | 4,4'-Dimethoxytrityl |
| DNA | Deoxyribonucleic acid |
| dppf | 1,1'-Bis(diphenylphosphino)ferrocene |
| dppz | Dipyridophenazine |
| DRBD | Double-stranded RNA-binding domain |
| DTD | Tetraethylthiuram disulfide |
| EC | European Commission |
| EDITH | 3-Ethoxy-1,2,4-dithiazoline-5-one |
| EGFR | Epidermal growth factor receptor |
| ENA | 2'-*O*,4'-*C*-Ethylene bridged nucleic acid |
| ESE | Exonic splicing enhancer site |
| ESS | Exonic splicing silencer site |
| FDA | U.S. Food and Drug Administration |
| FMN | Flavin mononucleotide |
| Fmoc | 9-Fluorenylmethoxycarbonyl |
| glmS | Glucosamine-6-phosphate riboswitch (ribozyme) |
| GRPR | Gastrin-releasing peptide receptor |
| Gua | Guanine |
| Guo | Guanosine |
| HBV | Hepatitis B virus |
| HCV | Hepatitis C virus |
| HIV | Human immunodeficiency virus |
| HCMV | Human cytomegalovirus |
| HDV | Hepatitis delta virus (ribozyme) |
| HNE | *trans*-4-Hydroxy-2-nonenal |
| HNH | A nuclease domain in Cas9 |
| HpNP | 2-Hydroxypropyl *p*-nitrophenyl phosphate |
| HSV | Herpes simplex virus |
| IGS | Internal guiding sequence |
| INF | Interferon |
| LCAA-CPG | Long-chain aminoalkyl controlled pore glass |
| LDA | Lithium diisopropylamide |
| LDL | Low-density lipoprotein |
| LNA | Locked nucleic acid |
| LNP | Lipid nanoparticle |

| MCE | 2-*N*-Methylcarbamoylethyl |
| miRISC | miRNA-induced silencing complex |
| miRNA | Micro-RNA |
| MOE | 2-Methoxyethyl |
| MPPS | Macroporous polystyrene |
| mRNA | Messenger ribonucleic acid |
| MSNT | 1-(2-Mesitylenesulfonyl)-3-nitro-1*H*-1,2,4-triazolide |
| NAG | *N*-Acetylgalactosamine |
| NBS | *N*-Bromosuccinimide |
| NDP | Nucleoside diphosphate |
| NEAR | Nicking enzyme amplification reaction |
| NIS | *N*-Iodosuccinimide |
| NMA | 2-Amino-2-oxoethyl |
| NMR | Nuclear magnetic resonance |
| nOe | Nuclear Overhauser effect |
| NOESY | Nuclear Overhauser effect spectroscopy |
| Np | Nucleoside 3′-monophosphate |
| NpN | Dinucleoside(3′,5′)monophosphate |
| NPOM | 6-Nitropiperonyloxymethyl |
| NPP | 2-(2-Nitrophenyl)propyl |
| N2S2 | Disodium-2-carbamoyl-2-cyanoethylene-1,1-dithiolate |
| NTP | Nucleoside 5′-triphosphate |
| OAS | 2′,5′-Oligoadenylate synthetase |
| ODN | Oligodeoxyribonucleotide |
| ORN | Oligoribonucleotide |
| *o*-tolyl | 2-Methylphenyl |
| OXE | Oxepane nucleic acid |
| OXP | Bis(2-oxooxazolidin-3-yl)phosphinic chloride |
| PADS | Phenylacetyl disulfide |
| PAM | Protospacer adjacent motif |
| PBAVE | Poly(butylaminovinylether) |
| pby | Bipyridyl |
| PCR | Polymerase chain reaction |
| PCSK9 | Proprotein convertase subtilisin-kexin type 9 |
| $Pd_2(dba)_3$ | Tris(dibenzylideneacetone)dipalladium(0) |
| PEG | Polyethylene glycol |
| PEI | Poly(ethylenimine) |
| PDGF | Platelet-derived growth factor |
| PMF | Potential of mean force |
| PMO | Phosphorodiamidate morpholino oligomer |
| PMPA | 9-[2-(phosphonomethoxy)propyl]adenine |
| pN | Nucleoside 5′-monophosphate |
| PNA | Peptide nucleic acid |
| POC | Isopropoxycarbonyloxymethyl |
| POM | Pivaloyloxymethyl |
| pre-miRNA | Precursor miRNA |
| PreQ1 | 7-Aminomethyl-7-deazaguanine |
| pri-miRNA | Primary miRNA |
| PSMA | Prostate-specific membrane antigen |

| | |
|---|---|
| PTD | Peptide transduction domain |
| Py | Pyridine |
| PyBOP | (Benzotriazol-1-yloxy)tripyrrolidinophosphonium hexafluorophosphate |
| QM/MM | Quantum mechanics/molecular mechanics |
| RISC | RNA-induced silencing complex |
| RNA | Ribonucleic acid |
| rRNA | Ribosomal ribonucleic acid |
| RSV | Respiratory syncytial virus |
| RuvC | A nuclease domain in Cas9 |
| SAM | *S*-Adenosylmethionine |
| SATE | *S*-Acyl-2-thioethyl |
| SELEX | Systematic evolution of ligands by exponential enrichment |
| sgRNA | Single-guide RNA |
| siRNA | Small interfering RNA |
| SMA | Spinal muscular atrophy |
| SMN | Survival motor neuron |
| snRNA | Small nuclear ribonucleic acid |
| snRNP | Small nuclear ribonucleoprotein particle |
| TBAF | Tetrabutylammonium fluoride |
| TBDMS | *tert*-Butyldimethylsilyl |
| TCBOC | 2,2,2-Trichloro-*tert*-butoxycarbonyl |
| TdT | Terminal deoxynucleotidyl transferase |
| TFO | Triple helix-forming oligonucleotide |
| TFPI | Tissue factor pathway inhibitor |
| Thd | Thymidine |
| THF | Tetrahydrofuran |
| thioAmNA | Thioamide-bridged nucleic acid |
| Thy | Thymine |
| TLR | Toll-like receptor |
| TOCSY | Total coherence transfer spectroscopy |
| TOM | Triisopropylsilyloxymethyl |
| TPP | Thiamine pyrophosphate |
| tracrRNA | *trans*-Activating CRISPR-RNA |
| TRBP | HIV-*trans*-activating response RNA-binding protein |
| tRNA | Transfer ribonucleic acid |
| trpn | 3,3′,3″-Triaminotripropylamine |
| TS | Twister-sister (ribozyme) |
| Ts | *p*-Toluenesulfonyl (tosyl) |
| Ura | Uracil |
| Urd | Uridine |
| UTR | Untranslated region |
| VEGF | Vascular endothelial growth factor |
| VEGFR | Vascular endothelial growth factor receptor |
| VS | Varkud satellite (ribozyme) |
| VZV | Varicella zoster virus |

# Index

https://doi.org/10.1515/9783111325637-015